SOME 19th CENTURY ENGLISH WOODWORKING TOOLS

Edge and Joiner Tools and Bit Braces

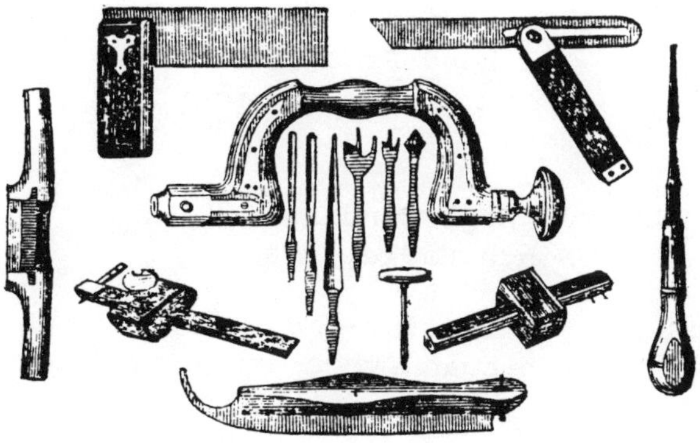

by

KENNETH D. ROBERTS

Fitzwilliam, N.H.

Published by
Ken Roberts Publishing Co.
Fitzwilliam, N.H. 03447

First Published 1980

ISBN 0-913602-40-X

Printed by
The Bond Press, Inc.
Hartford, Connecticut

Bound by
General Book Binding Co. Inc.
South Hadley, Mass.

Cover Material *Skivertex-Sanigal*, Serie 3 - 2064
Furnished by Whitman Products, Ltd., Johnston, RI

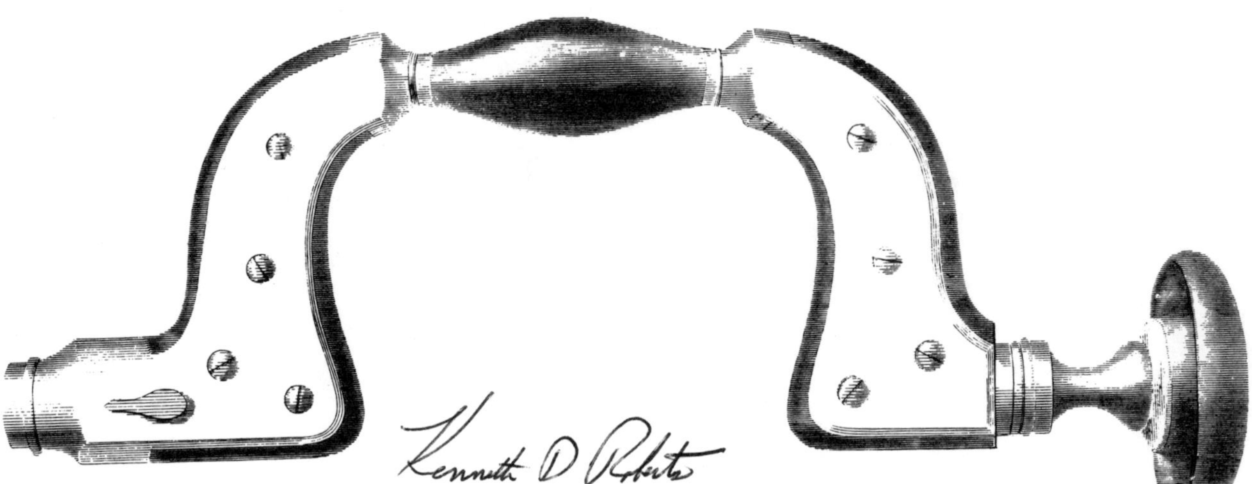

Patron IAH & MERR LIBRARY Book Number #50 of 750 Copies Date Nov. 26, 1980

Preface

During the course of my earlier writings concerning wooden planes, I became aware of the significant influences of English planemaking on early 19th Century American practices. During my first visit to the Sheffield Reference Library in 1967, I obtained a photocopy of their Joseph Smith's *1816 Key to Sheffield Manufactories* and also first became aware of the later *Sheffield Illustrated Lists*. These works kindled a desire to discover additional history of this subject and to determine possible influences on other tools in America. Only limited published material concerning the historical development of these tools was available. Twelve subsequent visits to Sheffield followed, during which time I have spent over a year of my life in study at Great Britain.

Tools for the Trades and Crafts related the industrial development of these tools at Birmingham. This parallel study of Sheffield tools is by no means intended as a final word on this subject. This summarizes my observations concerning some English tools. The principals for the most part appear to have been too busy "grubbing out" a living to record their performances in any great detail. Consequently it has been necessary to infer certain trends based on primary records from directory data, public records, trade catalogues and the tools themselves.

Leadership in edge tool production at Sheffield was achieved through development of crucible steel making. As background materials a historical survey and consideration of some of the problems in steel making are presented in the introductary chapter. Saws and files, edge tools, joiner tools, planes, and finally a study in depth of the 19th Century English brace follow. Check lists of manufacturers or dealers appear as appendices to each of these chapters.

Those desiring further details regarding the functional uses or construction of English tools are referred to R.A.Salaman's detailed and descriptive work *DICTIONARY of TOOLS used in the woodworking and allied trades,c1700-1970*. Similarly W.L.Goodman's excellent writings: *The History of Woodworking Tools* and *British Planemakers from 1700* [2nd edition,1978] provide broad historical background to this subject. I am indeed grateful for the encouragement and assistance extended to me by both of these gentlemen over the last thirteen years.

Hopefully this publication will enable collectors, students of history and technology, museums and historical societies to gain a better appreciation of these artifacts, their makers, and provide an indication of dates of their manufacture.

Grateful acknowledgement is made to the following for use of primary data and their answering numerous requests through correspondence:

Sheffield Central Library: Mr. Robert F. Atkins , Director, Mr. Martin Olive, Local History Librarian and Staff
Birmingham Reference Library, Miss Dorothy McCulla, Local Studies Librarian and Staff
Manchester Central Library; Archives
Science Reference Library, The British Library,Chancery Lane, London, Miss B.M.Rimmer, Industrial Property Section
The Patent Office,London, Dr. R.T.Smith
Patent Sales Office, Orpington, Kent
Guild Hall Library, London
Public Records Offices, Kew, Richmond and London

K.W.Hawley, Sheffield, Trade Catalogues
K.C.Barraclough, Sheffield, Steel Manufacturing Data
R.A.Salaman, Harpenden, Trade Catalogues
John Sainsbury, Sheffield, Trade Catalouges
Donald Wing, Marion, Massachusetts, Trade Catalouges and 1870 & 1876 Sheffield Lists

Additionally I acknowledge with thanks information received from the following persons,most of whom are tool collectors:

United States		England
Frederick R. Bair	David L. Parke	Edward Acland
Kendall H. Bassett	Edward Parish	Roy Arnold
Carl E. Bopp	Russell G. Parish	Tony Barwick
Winthrop L. Carter	Eric M. Peterson	Reg Eaton
Jack Clouser	Walter Phelps	Michael Falkner
James H. Cooley	Roger B. Phillips	K.W. Hawley
Richard DeAvila	Edward Smith	Michael Holmes
William A. Downes	Roger K. Smith	Jennifer Hughes
Joseph Dziadul	Richard J. Souza	Richard M. Maude
Kenneth Gleason	Joseph Torras	June Murdock
Arnold Gordon	George D. Tuttle	W.B. Oxley
Jerry Glaser	Donald B. Wing	Dermot Roche
Robert D. Graham	Henry Wing	Terry Smith
William Holden	Raymond E. Wilson	Michael Swede
Paul B. Kebabian	Thomas A. Witte	Philip Walker
John Meloney		Jim Wareham
J. Lee Murray		D.W. Woodhouse

This publication is dedicated to all above who furnished assistance and encouraged me while working on this project.

The Index is divided into two sections: I - Names of Individuals and Firms; and II - Subject Matter. Names listed in the Check Lists of Sawmakers, Edge Tool and Joiner Tool Firms and Brace and Bit Manufacturers, appearing in the Appendices, are not included in this Index. Names appearing in various Tables, unless considered important to the subject matter, are not indexed, nor is the material in the footnotes.

It was necessary to reproduce some of the primary documentary data from photocopies. Unfortunately this has resulted in some loss of original detail.

Unless otherwise identified all photographs were the work of John Abbott of Keene, New Hampshire to whom I am most grateful for his artistic accomplishments. Thanks are also expressed to my wife, Jane W. Roberts for constructive criticism organizing the text and for typesetting the chapter titles.

Kenneth D. Roberts, August 1980

Table of Contents

LIST OF ILLUSTRATIONS

LIST OF PLATES

LIST OF TABLES

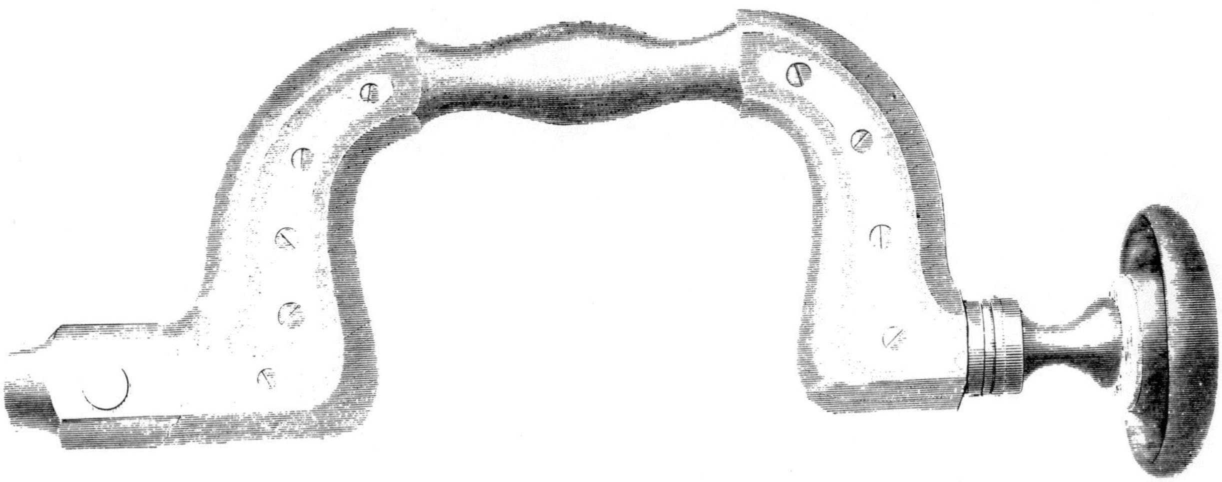

FRONTISPIECE - Some 19th Century English Joiner Tools and a Skate
[Various joiner tools and skate shown in 19th Century Wood Cut (See Title Page); also Plates CLXXIII & CCXI]

Chapter I

Introduction
Sheffield History and Nature of Steel

At the end of the 18th Century Sheffield was the foremost center in cutlery in Great Britain. Supremacy there in the manufacture of edge tools soon followed. During the first quarter of the 19th Century leadership in the production of joiner tools was also achieved there. What were the principal four factors contributing to this development? These were: a favorable location, abundant natural resources, the established cutlery trade and abilities to produce iron and steel. The former lead in production of these articles had been at Birmingham.[1]

Sheffield is located in the southwestern part of Yorkshire, just east of the Pennines. This mountain range is known as the "back-bone" of England. It is approximately 140 miles NNW of London; 70 miles NE of Birmingham and 50 miles SW of Hull, a shipping port on the River Humber. The River Don flows SE into Sheffield, joining the River Sheaf, at the center and then turns NE through Doncaster into the Humber at Goole, about 50 miles distance.

In the early part of the 17th Century, when the iron industry was first being developed around Sheffield, transportation was principally by pack horse. From Doncaster, approximately 20 miles NE of Sheffield, the River Don was seasonably navigable to Hull on the River Humber. Bawtry, approxiamtely 25 miles east of Sheffield, was similarly navigable to Hull by the Rivers Idle and Trent. A revolution in transport accompanied the industrial growth during the 18th Century.[2] This brought about a system of canals, supplemented by turnpikes. Rotherham, five miles from Sheffield, was connected by canal to Doncaster by 1743. This was extended to Tinsley, three miles from Sheffield, and by 1819 to the center of Sheffield. In 1838 the first railway opened in South Yorkshire, the Sheffield and Rotherham Railway.[3] In 1845 railway connections were completed to Manchester.

In addition to the River Don, Sheffield water power was augmented from four tributary streams: the Sheaf; the Porter; the Loxely and the Rivelin. This abundant source of energy provided power for many mills.

Important raw materials mined from the South Yorkshire hills included: ironstone; coal; pot clay; refractory sandstone and ganister. Wood provided for production of charcoal. Fortunately the coal was of coking quality. Incidentally, the demands for coal in other parts of England was probably the most important factor in developing the transportation system adjacent to Sheffield. Denudation of forest for the production of charcoal had been a national problem as early as 1600. While Abraham Darby is generally credited as the first to successfully smelt iron using coke in 1709, this practice did not become general until after 1750, and was not introduced into the Sheffield area until 1765.[4] After James Watts' 1781 patent for rotary motion, an increased demand resulted in the industrial applications for steam engines.

This necessitated substantial increased requirements for coal production. The first steam engine was introduced at Sheffield in 1772 to provide power for blowing air to furnaces. The first steam powered rolling mill was installed in 1789.[5] Water power was subject to seasonal climate.

The local iron ore unfortunately contained a high amount of phosphorus, which made it unsuitable for steel making, until later refining methods were developed. However, wrought iron, suitable for production of nails, wagon tires, bar and other constructional use could be smelted by proper refining. The process of making wrought iron first involved melting ironstone in a charcoal fired blast furnace and casting the impure iron into "pigs". These were broken, reheated in an oxidizing atmosphere and hammered, thus forming a "bloom". Production of wrought iron markedly increased after Henry Cort introduced his "puddling" process. Wrought iron was made at the Top Forge, Wortley, about ten miles north of Sheffield from 1640 - 1912.

The cutlery industry is believed to have been established in the Sheffield area during the 13th and 14th Centuries.[7] A familiar 1340 reference to Chaucer's works noted *"Sheffield thwitel"*, an all purpose knife. The manufacturing processes accompanying the cutlery trades are divided into three separate occupations: forgers, grinders and cutlers. Sheared blanks cut from a steel rod, heated in a charcoal fire, were hand forged on an anvil into rough shaped blades. Small knives of the pocket type had blades made from a single piece of steel. Large knives, sickels and scythes were made of composites of iron and steel, effecting economy by conserving the more expensive steel. In this event the forged steel blade was reheated, laid on a rod of iron and hammer forged together at a welding heat. The tang was then shaped by reheating and forging. Hardening was effected by reheating to the proper temperature, followed by quenching in oil or water. This treatment rendered steel hard, but brittle. In order to restore toughness, it was necessary to undergo a tempering treatment. Hardened steel is dull gray, but reheating through low temperatures brings about color changes. Its surface oxides proressively change in color from yellow straw, brown, purple, dark blue to light blue. Slight loss in hardness results with an increase in tempering temperature, which can be judged by this color change.

Grinding wheels were cut from locally quarried sandstone and ganister. These were mounted on shafts and driven from waterwheels. In many instances such grinding shops, known as"wheels",were located apart from the forgers. These were divided into rooms called "hulls", which might be rented out to different establishments. It was the abundance of available water power that contributed largely to the early prominance of the cutlery industry at Sheffield. Hand power was used at London. However, water power was somewhat seasonal at Sheffield. In 1786 Messrs. Proctor introduced the first steam grinding mill on the east bank of the Sheaf.[8] By 1840 the number of grinding establishments powered by steam had increased to 50, and 40 remained by water power.[9]

The finishing stage of cutlery manufacture involved the assembly, handling (hafting) and completing the finished article. This required machining methods for making springs, linings and applying the materials for covering. Forging work and heat treatment were necessary to make the springs. Considerable skill was demanded. It was to this trade that the term "cutler" was specified.[10]

The cutlery trade at Sheffield was essentially a "cottage" industry. Small workshops were attached to individual worker's living residences, with the exception of the "grinding wheels" (shops). An enterprising workman could become a small entrepreneur and thus advance to the status of a "little master". Thus he assumed the responsibilities of procurement of materials, production, management of manufacturing, payments for these services and the sales of the final products. Such an arrangement was only possible with such small scale industries where services could be hired temporarily to complete a given order or contract. Payments for such services could be made in cash, goods or in services by the "little master" or credit to a factor who sold the finished goods. A "little master" might have been apprenticed to any of the three trades of forging, grinding or cutler, and served his trade as a journeyman. The existence of numerous "little masters" had the effect of holding down the prices of cutlery as offered by the larger manufacturers. At the same time their existence with insignificant fixed charges held down the price of labor.

An Act of Parliment in 1624 incorporated the Cutler's Company in Hallamshire.[11] This Guild provided for the election of officers, supervision and maintenence of high quality of goods produced by its Members, regulation of number and services of apprentices and jurisdiction of marks or stamps. The 1st Article of their Constitution concerned the period of holidays in which no work was performed. Articles 2 - 15 are abstracted in Table I. (See Page 4) Subsequent acts modified these Rules and generally lowered the standards. During the period 1780 - 1814 control of the Company shifted to the larger manufacturers, resulting in considerable dissatisfaction among Members. By Act of 1814 any person who took out a mark was eligible to become a Member and restrictions as to the number of apprentices working under a Master were abolished. After this date the principal activity of the Company was the continuance of supervising the marks issued to freemen. The motto of the Hallamshire Corporation of Cutlers was *" Pour Parvenier A Bonne Foy"*, which literally translated reads - "to succeed in business take care to keep up your credit". Thus from a legal regulating body, formerly protecting freemen, their professional status and welfare, the Cutler's Company became more influencial as a manufacturer's association. Many separate local trade societies were formed from 1770 to 1814 after artisans broke away from the Company.[12] These local trade societies developed into unions. Highly skilled labor required by most of the cutlery trades was not generally available outside of Sheffield. With this realization these unions were in a good bargaining position to make demands for increases in wages, especially in times of prosperity. At such times after wage demands were made, rather than to combat these unions and undergo a strike, the manufacturers generally agreed to raise prices, thus passing the burden on to the customers.[13]

The growth in population at Sheffield is shown in Table II. (See page 4). The industrial growth at Sheffield was largely due to the expansion in the cutlery trade. However, another contribution factor was successful manufacture of 'Sheffield silver plate', introduced by Thomas Boulsover in 1742.[14] This involved rolling together plates of copper and silver, thus producing a bi-metal. The surfaces of the two metals were fusion welded together. By further rolling this composite, a relatively thin coating of silver on copper could be effected, thus resulting in an economy. Such material was first used for buttons. Soon after products such as urns, tea-pots, candlesticks and dishes were frabricated from rolled sheet.[15] By 1765 there were six firms manufacturing such goods.[16]

TABLE I

Articles 2 - 15 of Cutler's Company Constitution

2. No person to excercise the said trade who has not served an apprenticeship of seven years, or been instructed by his father for that term; penalty 40s.
3. No person to have more than one apprentice, nor to engage another until the former be in his last year, nor take any for less than seven years; penalty 40s
4. No person occupying any wheel for the grinding of knives to allow any work to be done there during the holiday months; penalty as before.
5. No occupier of a wheel to suffer any person to grind or glaze any knives there who does not reside within the Lordship and liberties; on the same penalty
7. No person to strike any mark upon his wares, but that which is assigned to him in the Lord's Court; penalty 10s.
8. No hafter shall haft any knives for any chapman, hardwareman or dagger-maker, or other person, not dwelling within the liberties; penalty 20s.
9. Nor shall any knife blades be sold to any person not dwelling in the liberties; penalty 8d.
10. No journeyman to be employed under the age of 20, except that such that shall be allowed by the jury, or who have been apprenticed and taught by their fathers; penalty 40s.
11. No person who has not served an apprenticeship or been instructed by his father, to set up in trade, except he first pay to the Jury of Cutlers, £ 5, the one half for the Earl's use, the other for the poor of said Corporation; penalty 40s.
12. Every apprentice to be presented to the Jury within one year and the indentures to be sealed before them: At the expiration of the term each apprentice to bring his indenture to the Jury, and to subscribe here established; penalty 10s.
13. All persons summoned to appear on the Jury, to appear or forfeit 6s 8d.
14. Each Juryman to appear when summoned by the foreman to settle questions touching the ordinances; in the same penalty.
15. The Jury with concurrence of the Lord or his "learned steward" to make such regulations as may be found necessary.

[source of above: Cutlers,under Earls of Shrewsbury, *Sheffield Directory,1837*]

TABLE II

Population Growth at Sheffield, 1700 - 1901

[Source: G.I.H. Lloyd: *The Cutlery Trades*, [London, 1913], pp.151-152]

Year	Population
1700	6,000
1725	10,000
1750	20,000
1775	32,000
1801	46,000
1831	92,000
1851	135,000
1871	240,000
1901	409,000

[Note; Figures rounded off to nearest 1000]

However, the most significant factor affecting the growth of Sheffield was the development of crucible steel making by Benjamin Huntsman, 1740-1750. Huntsman, a Doncaster clockmaker, had come to the Sheffield area in 1742 in a effort to make better quality steel springs than could be produced from either available shear or German steel. His plan was to directly melt steel in a crucible, cast this into an ingot, and subsequently work this into wrought material. After ten years of experimentation he achieved this invention. The importance of this development was noted by the famous English metallurgist R.A. Hatfield in 1895; *"The invention of cast steel is second to no previous invention in the world's history, unless it be the invention of printing."*[18] Several accounts of this development have been previously published.[19] Before discussing this subject further, the nature of iron and steel and their previous production methods will be briefly considered.

Much of the iron recovered from the earth's surface is combined as an oxide compound, generally mixed with various clay or silicious materials. In order to remove the oxygen, the ore is heated in a blast furnace in the precense of carbon. During combustion of carbon in the presence of oxygen (21% of air), carbon dioxide and heat are produced. At high temperatures the carbon dioxide is reduced by carbon to form carbon monoxide. Since at high temperatures carbon has more affinity than iron for oxygen, iron oxide is reduced to iron. Carbon in turn is dissolved and the melting point of the mixture is lowered. A flux of limestone, which floats on top of the molten mixture, dissolves much of the silicon and other impurities present in the ore. After sufficient molten material has accumulated, the furnace is tapped and the resulting product cast into moulds. This is known as "pig iron" and contains about 5% carbon and an equal amount of impurities with the balance of 90% iron. As such this material is very brittle and has no engineering use. By further refining during a second melting, reducing the carbon content to about 4% and fluxing off more of the silicon, the molten material can be cast into useful structural shapes. This is cast iron. The proper composition can be readily machined to final shape, but is too brittle to be plastically deformed by mechanically working, thus has limited applications.

Subjecting pig iron to further oxidization refining, the carbon content can be reduced to a very low level. This was accomplished by reheating and burning off the carbon by exposure to air. As the carbon is depleted, the melting temperature increases and the impurities combine in a pasty slag. After the proper time and temperature had developed, this pasty material with the iron was worked up into a ball and mechanically forged into strips or rods. The resulting material was very low in carbon, containing stringers of slag. It could be readily forged at elevated temperatures and had some plasticity for limited cold working at normal temperatures. It could be readily welded by hammering pieces together at elevated temperatures (fire welding). However, since it had insufficient carbon, it could not be hardened by heat treatment and therefore was not suitable material for a cutting edge, required by tools. In this manner wrought iron was made.

Steel, an alloy of iron and carbon (up to 1.4%C), has properties intermediate between wrought and cast iron.[20] Unlike cast iron it can be mechanically worked, similar to wrought iron; but it can be melted and cast, unlike wrought iron. The properties of steel may be greatly modified by heat treatment; from glass hard above cast iron to dead soft, comparable with wrought iron. While the understanding of hardening of steel is technical and rather complicated, it is in part due to the fact that iron exists in different crystalline structures at different temperatures. (allotropic) Heating pure iron to 890° C. changes the crystalline structure from body-centered to face-centered cubic. In the former lower temperature form carbon has very limited solubility, but in the latter higher temerature form is soluble up to about 1.7% by weight. As the amount of dissolved carbon increases to 0.8%, this allotropic change is linearly proportionally reduced to 690° Cent. At temperatures between these two limits with alloys of up to 0.8% carbon, both crystalline forms of iron can co-exist in fixed proportions. In slowly cooled alloys below 690° Cent. only the body-centered cubic iron can exist, the carbon having been thrown out of solution as cementite, (Fe_3C). With carbon contents of between 0.6 and 1.4% whereby the carbon is dissolved in the face-centered cubic iron at a high temperature and then followed by rapidly cooling by water quenching below room temperature, a transitional structure is developed that is extremely hard and of high strength.[21] This consists of a supersaturated solution of submicroscopic particle of carbon in iron. The optimum amount of carbon to produce maximum hardness by such quenching treatment is of the order of 0.6 - 0.8%C. Excess undissolved carbides in the form of cementite contributes to wear resistance of hardened steel. Tool steel for knives and wood working cutting edges are made from steel containing 0.9 - 1.1%Carbon. This assures by proper heat treatment maximun hardness can be developed with added wear resistance from excess carbides. File steel has even higher carbon content. After heat treating to maximum hardness, it is necessary to temper to restore proper toughness by reheating to some temperature between 150 - 300° Cent. This results in some loss or sacrifice to hardness, but increases ductility. Fully annealed (dead soft) steel has been spheroidized so that the particles of iron carbide (cementite) are aglomerated and distributed through the matrix of body centered cubic iron. In this condition steel can be plastically deformed by cold working a limited amount, similar to wrought iron.

Before the end of the 15th Century it had been discovered in Germany that prolonged heating of wrought iron at elevated temperature packed in charcoal in clay containers brought about considerable penetration of carbon.[22] The exact relationship between the amount of carbon and its response to heat treatment was not known. Therefore, such steel making was an emperical or trial by error process. Today it is an exact science. At that time in order to make steel, it was first necessary to make wrought iron and then carburize it. It was soon learned that the presence of the two deliterious elements of sulfur and phosphorus must be avoided in wrought iron to produce quality steel. Sulfur causes iron or steel to be hot short by incipient melting of a compound formed with iron at normal forging temperatures. Phosphorus results in cold shortness (brittleness or low impact strength) at normal temperatures.

Early in the 17th Century this steel making process was introduced into England. Heating was accomplished by burning coal. Absorption of harmful sulfur was prevented, since the iron was protected within a sealed container. This steel making method became known as "cementation". The product of heating wrought iron bars for about a week packed in clay containers at 1000°Cent. was known as "blister steel", due to the resulting character of the surface. In this condition the steel was brittle and readily fractured into small pieces. From the appearance of the fracture and by sparking on a high speed rotating grinding wheel, it was possible to estimate the carbon content. By reheating such steel, it could be forged or rolled into desirable size. However, considerable gradient in carbon existed from the surface to the center. An improvement was achieved by forging together several pieces of blister steel, thus producing layers of high and low carbon material through the cross section. This material became known as "shear" steel. Being more uniform throughout its section in carbon content, such material had an improved cutting edge in comparison to "blister" steel directly forged. By repeating this process a second time "double shear" steel was produced with further improvements in uniformity.

German steel differed in that it was produced directly from molten pig iron, having low impurity content and smelted from carbonate ore peculiar to that country. Quenched pellets of such material were drawn out on a forge and welded into "faggots", much the same as later English shear steel was combined.[23] Considerable amount of German steel was imported to Sheffield for use in cutlery and edge tools before crucible melting was developed.

The cutlery industry at Sheffield was dependent on importation of Sweedish or Russian wrought iron for subsequent production into blister steel. Sweedish iron was remarkedly pure and entirely refined by charcoal, coal not being available in Sweeden.[34] As previously noted the wrought iron produced in Great Britain was largely phosphorus bearing and not suitable for making "blister" steel. The alternative was to import German steel, which was more expensive and not readily available. Sweedish iron and German steel were brought by boat to Hull, taken on barges on rivers and finally about 25 miles by pack horse into Sheffield. Some steel came down from the Newcastle area, but this was also made from Sweedish wrought iron by cementation.

Benjamin Huntsman contemplated that by melting "blister" steel, casting an ingot and forging this into rod would achieve a more uniform and therefore superior product for clock springs. What were the problems? First the development of a crucible that would withstand a temperature of 1600° Cent. necessary to hold the liquid metal was needed. This was accomplished after experimenting with local and Stourbridge clays.[24] Next a furnace had to be constructed to attain this temperature. A natural draft high chimney, having a flue 30 - 40 feet height, achieved this result. Fractured pieces of "blister" steel were melted in a crucible covered with a glass flux.[25] This was available from local glass manufactories. After melting and proper refining, the crucible was pulled from the furnace and teemed into an ingot by pouring the contents into a cast iron mould.[26]

Huntsman did not patent his process. Others copied his technique and were soon also producing cast steel. At first the cutlery trade did not accept such crucible steel, claiming that it was difficult to forge im comparison to shear steel. Huntsman's material found ready market in France for cutlery uses. Agitation by Sheffield manufacturers then attempted in vain to prohibit export of cast steel.[27] Had Huntsman accepted an invitation from Mathew Boulton to set up his works at Birmingham, Sheffield cast steel production might have been stifled. However, Huntsman's cast steel found demands in England for files, needles, dies and edge tools. Samuel Walker with John Booth built a Huntsman refining furnace in 1750 and became the first competitors.[28] Huntsman died in 1786. The second directory of Sheffield, published in 1787 listed seven cast steel refiners and 13 converters. (blister steel)

An individual melt, called a 'heat', within a crucible was limited to approximately 60 pounds. If heavier ingots were to be cast, it was necessary to combine the contents of other crucibles. To enable continuous flow of metal while pouring (teeming) required synchronous timing during melting. Fig. 1 shows an ingot being teemed into a cast iron mould. A melter is seen in the background with his pot ready to supplement the contents being poured in the foreground. This method of making tool steel survived well into the 20th Century, until ultimately being replaced by carbon arc and high frequency electric melting.

Fig. 1 Pouring or Teeming a Cast Steel Ingot
[Collection of K.W. Hawley, Sheffield]

The Bessemer proces for making steel was introduced at Sheffield during the late 1850's, followed by the Open Hearth (Siemens-Martin) method. Individual heats of the former amounted to 25 tons melted in half and hour and over 100 tons made in 10 - 12 hours by the latter. However these methods were not applicable for tool steel which required high purity, very low phosphorus and sulfur contents, as well as freedom from non-metallic inclusions. Therefore such steel making methods are not discussed further in this study.[29]

Various recipes were empirically concocted for making different heats of varying carbon content for different applications of cast steel. Among those noted on a Price List of about 1850 by Marsh Brothers & Co. were cast steels for: Calico Engraver's tools; Cold Chisels; Turning Tools; Taps & Dies; Spring Steel for Railway and Other Carriage; Drift Bolt Steel; Pin or Broach; Razor; Penknife; and Spindle.[30] An 1854 Catalogue of S.S. Brittian & Co. offered advice to their customers for selecting the proper steel for a particular use. (See Plate I, below)

REMARKS TO USERS OF STEEL.

PRACTICAL EXPERIENCE has taught us the folly of trying to produce *One Steel* of such a temper as to be suitable for *all purposes.* We have, therefore, to request our friends, when ordering Steel, always to state the purpose for which it is intended.

The Quality of Steel means the absence of Phosphorous, Sulphur, and other impurities. This can only be secured by the selection of Pure Iron Ores, which are costly in proportion to their purity. The grain of Steel is no indication of quality; Hard Steel shows a close fine grain, and a Milder Steel of the same quality shows a coarse grain.

Temper of Steel means percentage of Carbon combined with the Iron, and has no relation to price. High prices will not guarantee users getting suitable Steel, unless the correct Temper is supplied.

Much Steel is rejected as bad in quality, because it has been used for a purpose for which the Temper was unsuitable. Much also is spoiled by over-heating in working it, and much is condemned because sufficient care has not been exercised in Hardening and Tempering.

In order to prevent the foregoing results, we would recommend that *no kind of Steel* should be *heated more than is absolutely necessary to forge* it to the required shape, and that it be *well hammered* before it is thrown down. In Hardening, no more heat should be given than is *just necessary to throw off the scale,* and the tool should be *thoroughly yet slowly soaked through* to one uniform heat before it is immersed in the cold water after which the Temper should be reduced according to the purpose for which it is required. One or two trials will serve to show the necessary Temper.

Plate I - Remarks to Users of Steel *S.S. Brittain & Co. Catalogue*
[c.1854, Local History Archives, Central Library, Sheffield]

In order to shape the cast steel ingot into a size and form more convenient for further processing into a tool, it was necessary to hot work by forging or rolling. Upon solidification the cast ingot was stripped from the mould and heated to a suitable elevated temperature so that the material could be readily plastically deformed to reduced cross section. In such forging three types of trip hammers were used: belly helve, head helve and tilt.[31] In each type a hammer weighing of the order of 200 pounds or more was suspended from one end of a beam. With the helve arrangement the other end of the beam was hinged to a fixed platform. A rotating drum with cogs spaced around its perifery lifted either the middle (belly) or the hammer end (head helve) allowing the hammer to be alternately raised and fall, striking an anvil as the cogs passed and released the beam. Figs. 2 & 3 illustrate two belly helves that were formerly in operation at Wortley Iron Works Top Forge. With the trip arrangement the beam was suspended in the middle like a seawaw. Cogs on a rotating drum lifted one end of the beam, allowing it to fall as they passed. A hammer attached to the other end struck on an anvil. Depending upon the speed of rotation and the number of cogs, 50 to 200 blows per minute might be struck. Thus an ingot was forged into a rod. Several heatings might be required.

Fig. 2 Large Belly Helve Hammer, Top Forge, Wortley
Note Added Spring Beam

Fig. 3 Small Belly Helve Hammer, Top Forge, Wortley
[Figs. 2 & 3 Courtesy of the Sheffield Trades Historical Society]

Henry Cort in 1784, concurrently with developing his puddling process for converting pig iron into wrought iron in an air furnace, first introduced grooved rolls for shaping round sections from iron bars.[32] The 1837 Sheffield Directory noted: -

> ". . . . In 1769 Mr. Thomas Boulsover, the founder of the silver-plated trade, having turned his attention to the manufacture of edge tools &c., built the Whiteley Wood Works, where he established the first machinery for rolling iron and steel into plates, suitable for saws, fenders, spades shovels, &c., which had been previously formed soley by the laborius and more expensive operation of the hammer, . . "

The same 1837 Directory also noted in the Introduction: -

> ". . . The first Steam Grinding Wheel was errected by Messrs. Proctor in 1786 on the bank of the Sheaf; there are now upwards to twenty similar establishments in the town and suburbs, besides many smaller ones attached to manufactories. . . "

No doubt helve and trip hammers were soon after powered from steam engines. James Nasmyth in June 1840, then at Manchester, patented his steam hammer that was soon to revolutionize the art of forging and such installations were widely practices at Sheffield. (see Plate II)[33]

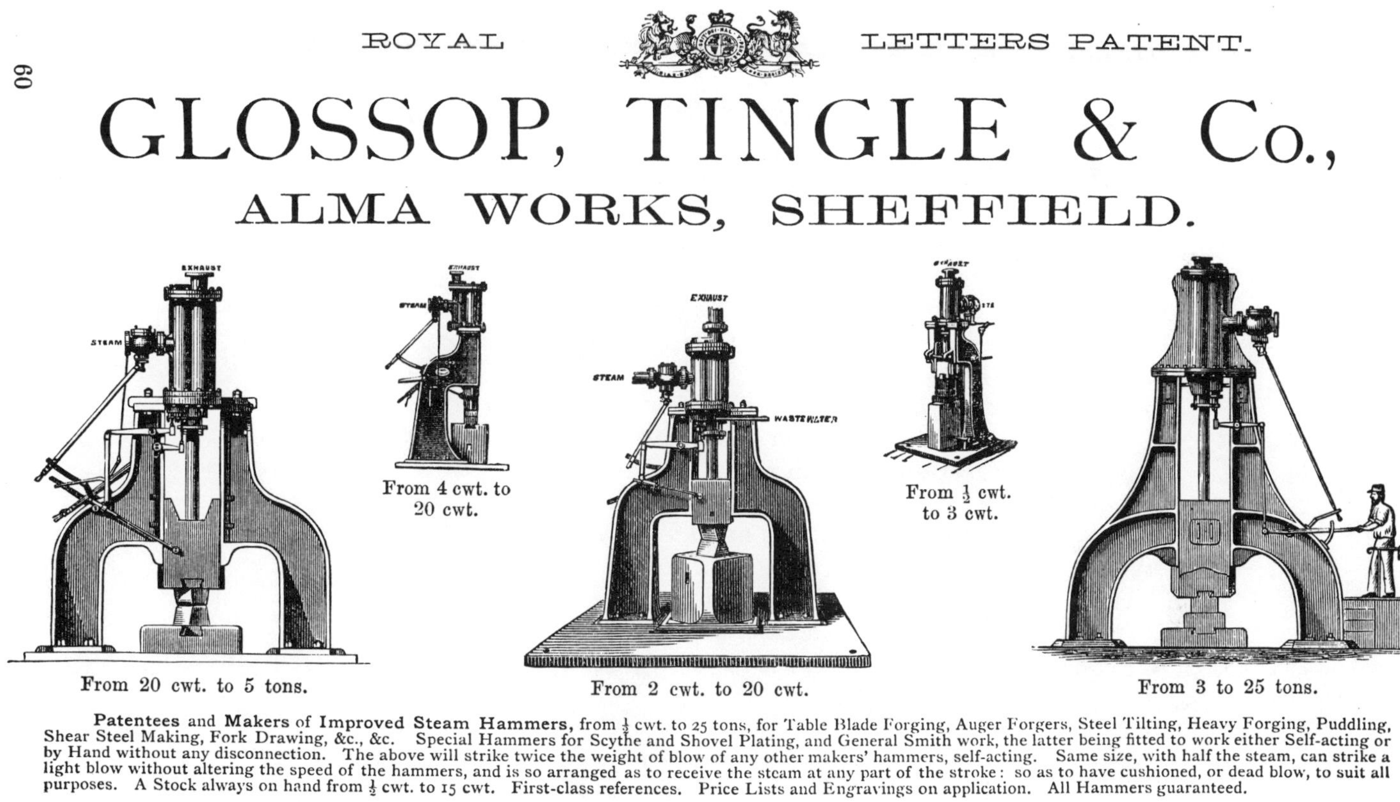

Plate 2 - Glossop, Tingle & Co. Advertisement in *1879 Illustrated Guide to Sheffield* [page 60, Advertising Section]

The process of forging cast steel ingots and bars into rods and plates was called tilting. This was a very important step in enhancing the quality of cast steel by refining the grain structure from coarse columnar to fine equi-axial. Such working also homogenized the internal composition from segregation and non-uniform gradients concommitant with solidification. Care had to be exercised to prevent folding the surface that might otherwise result in seams or laps. Excessive temperature was to be avoided to prevent coarse grain scaling from oxidization and decarburization of the surface. It was essential to uniformly work the section and continue deforming to complete deformation so that residual heat would not result in grain growth upon finishing to size. Indeed proper technique was a real art requiring considerable skill, experience and patience.

1. See text and Table II, *Tools for the Trades & Crafts,* Kenneth D. Roberts [Fitzwilliam, NH, 1976. pp. 18 - 19.
2. P.H. Abell, *Transport and Industry in South Yorkshire,*[Barnsley,1977], pp. 19-27.
3. W.White, *1864 Sheffield Directory,* Introduction, p.8
4. P.H. Abell, *op. cit.,* p.30
5. *ibid,* p.35
6. C.R. Andrews, *The Story of Wortley Ironworks* [Nottingham, 3rd Edition, 1975]
7. R.A. Mott, *Journal of the Iron and Steel Institute,* "The Sheffield Crucible Steel Industry and its Founder Benjamin Huntsman", [England, March 1965], p.231
8. *1837 Sheffield Directory,* Introduction, p.37
9. G.I.H. Lloyd, *The Cutlery Trades,* [London, 2nd Edition, 1968], p.179
10. *ibid,* p.53. Pp37 - 56 give an excellent presentation of knife making.
11. *ibid, p.114*
12. R.E. Leader, *History of the Cutler's Company,* Vol. I, p.79-p.89
13. S. Pollard, *A History of Sheffield Labour,* [Liverpool, 1959], p.68 See also: V.L. Melton, *Sheffield City Museums Information Sheet #6,* [Sheffield, 1973] Trade Unionism and the Sheffield Outrages, four pages.
14. *Sheffield Directory, 1837,* p.35 Introduction
15. N.H. Moore, *Old Pewter,Brass,Copper and Sheffield Plate,*[New York,1905], Fig.99
16. M. Walton, *Sheffield - Its Story and its Achievements,*[Sheffield, 1968], p.118
17. Lloyd, *op. cit.,* p.151-152
18. R.A.Hatfield, *Transactions of American Institute of Mining Engineers,* [New York, 1895] "Benjamin Huntsman of Sheffield,The Inventor of Crucible Steel",pp.170-180
19. *ibid;*also, Mott, *op. cit.,*[see reference No. 7 above; also, K.C.Barraclough, *Benjamin Huntsman, 1704-1776,* Sheffield City Libraries, Local Studies Leaflet,1976
20. W.K.V. Gale, *The British Iron & Steel Industry,* [Devon,1967], p.3
21. This micro-constituent is known as *Martensite.*
22. F.Overman, *The Manufacture of Iron,* [Philadelphia, 1850], p.466
23. K.C.Barraclough, *op. cit.,[reference No. 19],* p.7
24. R.A. Mott, *op. cit,* p.236. 25. *ibid, p.236*
26. There is an excellent description of crucible steel melting in Charles Tomlinson *Cyclopedia of Useful Arts,* [London,1866], p.924-927.
27. K.C.Barraclough, *Journal Historical Metallurgical Society,* [January 1977], p. 27
28. R.A. Mott, *op. cit.,* p.235
29. K.C. Barraclough, *Sheffield Steel,* [Buxton, 1976], p.14
30. Archives Local History, Central Library, Sheffield, *Marsh 127*
31. J.W.H. Silvester,*Sheffield City Museums Information Sheet,* #3, p.3
32. W.K.V. Gale, *op. cit.,* p.47
33. S.Smiles, *Industrial Biography: Iron Workes and Tool-makers,*[Boston,1864],p.348
34. Typical analysis of Sweedish Bar reported by Wm. Jessop & Sons showed: 0.10% Carbon; .028% Silicon; trace Manganese; .005% Sulfur; 0.008% Phosphorus. *A Visit to a Sheffiel Steel Works* [Sheffield, 1913], p.5

Two extraneous fallacious statements about steel.

1. Often examining a broken piece of steel an observer will state the steel"crystallized" and therefore fractured. Actually the steel was crystalline all the time, and the fracture revealed the character of its structure.

2. Often people will state that the steel in tools is not as good as made made during the 19th Century. Actually the steel made today is much freer from impurities and inclusions than in any time in history. It should be pointed out that only the tools made from good steel,properly forged and hardened have survived. But thousands have broken and have been remelted. What is lacking in steel today is knowledge of the art of forging to develop the proper grain and fibre. The problem is the lack of time in today's competitive market for such hand work and finishing.

Chapter II

Sheffield Steel Making - Saws and Files

The format of this Chapter will be to republish significant material from the Pawson and Brailsford 1879 *Illustrated Guide to Sheffield*. This work was originally published in 1862, which edition was reprinted in 1971. The second enlarged 1879 edition contains a well written section concerning the manufacturing of steel and tools. Since this edition is now difficult to obtain and offers such excellent material relative to this study, these sections are reprinted with footnotes commenting, correcting or enlarging some statements of the original text. There are a few minor deletions. Supplementary data will include some important firms not mentioned in the original text. Several advertisements from this 1879 edition and others from Sheffield directories follow. These attest to the important relationship of Sheffield production of steel and subsequent tool manufacturing. Later editions of this were also published in 1889 and 1899, but these do not offer the excellent detail and illustrations that appear in the 1879 edition.

While an introductory background concerning the making of English iron and steel has been presented in the previous Chapter, those desiring more details are referred to *Sheffield Steel* by K.C. Barraclough (Moorland Publishing Co., Buxton, Derbys., 1976) and *The British Iron & Steel Industry* by W.K.V. Gale (David & Charles Publisher, Newton Abbot, Devons., 1967).

Continuing with material from the 1879 *Illustrated Guide to Sheffield* pp. 201 - 229 with some deletions: -

MANUFACTURES OF SHEFFIELD

" We now come to the manufacturing processes of the town, which are, after all, the most interesting objects in a busy hive of industry like Sheffield. The number of trades carried on in Sheffield has increased very greatly during the present century. They now include the manufacture of iron, steel, Bessemer and Siemens-Martin steel; armour plates, heavy ordnance, rifle barrels, shot and shell; railway engines, rails, buffers, springs, tyres, axles, carriages and wagons; large steel castings for marine engines and heavy machinery generally; steel bells; iron and steel foundry work of all kinds; iron and steel wire for ropes, cables, needles, &c.; sheet steel for pens and other such purposes; ship plates, bolts; &c.; boilers, engines, lathes; planing, drilling, boring, slotting and other machines, and engineers' tools of all kinds; scythes, sickles, reaping machine and other kinds of knives; rakes, forks, and agricultural implements generally; garden tools; saws, files, axes, hatchets, hammers; joiners', engravers' and other edge tools; spades, shovels, mattocks, and other excavators' tools; table and pocket cutlery, scissors and cutlery of all kinds; sheep shears, garden shears, and lawn mowers; silver, silver-plated, and Britannia metal wares; stove grates, fenders; cooking, hot-air and other stoves; brass and other chandeliers; optical instruments, combs; powder flasks; wire and iron fencing; bicycles; surgical instruments of all kinds; portmanteaus, cabinet cases, and other such articles; tape and other measures; skates, &c., &c.

IRON MAKING

The smelting of iron is one of the oldest Sheffield industries. The Romans are believed to have begun here the manufacture of iron, found in many places near the surface, their example being followed by their successors the Danes and Saxons. It is said that Lord Waltheof, who made so gallant a stand against the Norman invaders, armed his followers with weapons procured from the iron foundries and forges on his Sheffield estates. That iron was made in Hallamshire in the middle ages is proved by a charter granted by Richard de Busli, one of its earliest Norman lords, to the Monks of Kirkstead, to erect forges at Kimberworth for the smelting and fabricating of iron. Foreign iron began to be used in Sheffield more than 300 years ago, entries appearing in the accounts of the Church Burgesses of the purchase of Danish and Spanish iron in 1557, the latter being at that time the dearer. As foreign irons came more into use for purposes of manufacturing, the smelting of iron ceased to be carried on in the town, though it flourished in the neighborhood. Messrs. John Brown and Co. were the first to re-introduce the manufacture, adding it to their other industries about twenty years ago.[1] The process is so well known that no detailed description is necessary. Iron is chiefly made now in huge round furnaces constructed of iron plates, and lined with ganister, or other heat-resisting substances. A furnace is filled with iron ore and coke in alternate layers, lime being added as a flux; and the fire having been lighted, a strong current or blast is applied by means of steam power to get up the requisite heat; hence the name "blast" furnace. On a level with the bottom of the furnace are beds of sand, in which indentations("moulds"they are called) are made with pieces of wood, three or four feet long and several inches thick. When sufficiently melted, the ore is run from the furnace into the moulds, where it is left to cool and harden. The blocks of rough iron made in the sand moulds are called "pigs"--a name which is said to have originated from the resemblance, in old times, of these iron castings in their sand gutters to a sow and her sucking pigs.

The process of manufacturing pig into bar iron is simple. The pigs are put into a reverberating furnace and melted to a soft, spongy substance, then stirred about to burn out the carbon, silica, and other impurities. The process is called puddling; and is done by machinery in the iron districts. The iron is taken from the puddling furnaces in soft, spongy balls, and beaten into bars under the steam hammer. The iron made in Sheffield is of the medium quality, known as forge iron. Spiegeleisen is also made in considerable quantities. The latter iron contains from 10 to 30 per cent, of manganese, according to the purposes for which it is required, and is largely used in the manufacture of crucible and Bessemer steel.[2] In making spiegeleisen, manganiferous ores imported from Spain are chiefly used.

Messrs. John Brown and Co. commenced the making of spiegeleisen some four years ago. Up to that time spiegeleisen was imported in large quantities from Germany, but the home product is much superior, and is now generally preferred.

Among leading houses in the iron trade are Charles Cammell and Co., William Cooke and Co., Thomas Andrew and Co. (Wortley), Burrows and Co., and other considerable firms.

THE PROCESSES OF STEEL MAKING

The use of steel dates back to a very remote period. It was made in smelting iron with wood, but was of much inferior quality to the best steel now made. The following brief outline of the present process of steel making, taken from a paper read before the members of the Literary and Philosophical Society some years ago by Mr. Henry Seebohm, will serve as an introduction to the more detailed descriptions which follow:--"The iron is received by the Sheffield steel manufacturer in the form of bars about three inches wide, five-eighths of an inch thick, and from six to twelve feet long. The iron ore has been melted in a blast furnace and run into pigs of cast metal, an average sample of which may roughly be stated to consist of 95 per cent. of iron and 5 per cent. of carbon. The pig metal has been melted and deprived of nearly the whole of its carbon in a furnace constructed either on the Walloon or Lancashire system, which in Sweden takes the place of the puddling furnace in this country, and the spongy mass of ball of pure iron left has been hammered into the bars of wrought iron of the dimensions described. This process is performed in the converting furnace, from which it issues in the same shape as it went in; it is then sent into the melting furnace, is there broken into small pieces, and melted into an ingot of cast steel, which is afterwards sent to the forge or rolling mill to be hammered or rolled into a bar of the shape and dimensions required."

This outline description it will be observed assumes that Swedish iron is used in making cast steel; and, in fact, this is largely so. The best iron for steel purposes is produced from the Dannemora mines of Sweden. Swedish iron generally, and Russian iron, have long been in high repute for the best steel; but iron made from native hematite and other ores, or from imported Spanish ores, is now mixed with the foreign iron for some qualities of steel.

The series of processes described above by which carbon is taken out of the cast iron and a portion of it returned again, is (Mr. Seebohm observes) sufficiently complicated and expensive, and attended with serious loss by diminution in weight, but until the invention of a better and cheaper mode of arriving at the desired result, it must be accepted as the only known way of making best cast steel.

STEEL CONVERTING

The process is thus described in Mr. Seebhom's very interesting paper:--

"The process of converting, or as it is generally called in the Encyclopaedias, cementing iron into steel, is carried on in a converting furnace.

"The converting furnace consists of two stone troughs, technically called converting pots, about four feet wide, four feet deep, and twelve feet long each, placed side by side with a fire underneath them, the flues of which conduct the heat all round each pot. Over the pot is a vault of brick, and the whole is enclosed in a brick dome to prevent the heat from escaping.

"A layer of charcoal broken up into pieces from a quarter to half-inch square, is placed at the bottom of a pot, a row of bars of iron is then laid over it, this is covered with charcoal, another row of bars of iron follow, and so on until the pot is filled with alternate layers of charcoal and iron; it is then carefully closed with a thick cover of 'wheel swarf,' a species of mud or clay which accumulates at the trough of the grinding wheel, and is of course the material of the grindstone worn away in the process of grinding, a substance which will resist long exposure to great heat, and keep the top of the pot as near as possible air-tight.

"These pots, full of iron and charcoal, are raised to nearly a white heat, and the fire is kept burning for about a week or rather more after the furnace is 'fired,' according to the amount of carbonization required. After the fire is allowed to go out, the furnace takes a week or rather more to cool, when the cover is removed, and the bars of iron which were placed in the furnace are taken out of it bars of 'blister steel,' so called from the bubbles or blisters which have arisen on the surface during the process of conversion.

"A chemical change in the composition of the bars has taken place. They were originally pure iron, or nearly so, were fibrous in their structure, and would bend double without breaking. After the process of conversion, they are carburet of iron or steel, containing from half to one-and-a-half per cent, of carbon, according to the length of time they have been in the furnace, and the degree of heat to which they have been subjected. They are now more or less crystallized in their structure, and can be broken by a slight blow of a hammer.

"The process by which iron becomes steel is known to chemists as an example of the occlusion of gases. As soon as the interior of the converting pot is hot enough to ignite the charcoal, a fractional part is consumed, and what little air there is in the pot becomes carbonic oxide. As the iron becomes hot it absorbs this carbonic oxide. The carbon having a greater affinity for iron than oxygen, unites with the iron and liberates the oxygen, which is immediately seized upon by the carbon of the charcoal and becomes again carbonic oxide. The oxygen thus becomes a sort of messenger, continually carrying the carbon or the charcoal into the iron.

"The outside of the bars becomes carbonized first, and finally, if the iron remain long enough in the furnace, the centre of bars as thick as 1½ inch will become completely crystallized."[3]

The converting furnaces in use in Sheffield vary in size, some holding only 15 tons, and others as much as 30 tons. Three bars, called "tap" bars, are allowed to project through a hole in the end of one of the pots, called the "tap hole," which is kept air-tight by ramming sand round the tap bars, and these bars are drawn out of the furnace at intervals so as to guide the converter in ascertaining the exact point of time at which his fire may be let out.

SHEAR STEEL

In the early days when Sheffield first acquired a reputation for the quality of its cutlery, "shear steel" was universally used for cutlery and edge tools. Shear steel is made from blister steel. Bars of medium temper are selected, and having been cut into short lengths are heated to a high temperature, and then beaten together under a tilt or forge hammer until they are as completely welded as if they had originally formed one block. When a specially good quality is required, the welded bars are cut in short lengths, heated, and thoroughly welded together a second time. After one welding the steel is called "single shear," after a second welding it is called "double shear." The chief purpose for which shear steel is now used is for table cutlery and tools of several kinds.

CRUCIBLE OR REFINED STEEL

The manufacture of cast steel was discovered in 1740 by Mr. Benjamin Huntsman, of Handsworth, near Sheffield. The process is known as steel refining. The secret was kept for some years, Mr. Huntsman making steel by his new process at works near Attercliffe, still used by his descendants. The premises being rigidly closed against all persons but regular workmen, a trade rival disguised himself in the commonest wayfaring attire, and presented himself during a stormy night, craving shelter and rest.[4] The commiseration of one of the workman overcame his prudence. The stranger was admitted, in the belief, from his appearance, that a trade process would be the last thing he would be likely to understand. He was allowed to lie on the cinder heap, and, pretending to sleep, watched the operations with a vigilance which put him in possession of the invaluable secret. This ruse is attributed to one of the Walkers of Grenoside, who were afterwards large iron and steel makers at Masbrough, and who seem to have erected steel furnaces as early as 1749. The process is thus described by Mr. Seebohm:-

"In the manufacture of cast steel, the process of melting is of the utmost importance. The melting furnace consists of a flat stack. containing a flue in each three feet, or rather less. To each flue there is a melting 'hole' wide enough to contain two melting pots, and deep enough to allow of sufficient cokes to cover the lids. The top of the melting holes is on a level with the floor of the furnace, the grate bars below being accessible from the cellar. The pots are usually made of a mixture of Burton or Derby clay, and sometimes Stourbridge clay, Stannington clay, and Devonshire or China clay, with a small addition of ground cokes and old pots ground. Each pot lasts one day, and is used three times, containing severally, about 50, 44 and 38 lbs. of steel each 'round.'

"The bar steel is first carefully selected of the exact temper required, broken up into small pieces, and conveyed to the pot (which has already been placed in the melting hole) through a kind of iron funnel, called a 'charger.' The degree of heat to which the furnace is allowed to go is carefully watched by the 'puller out,' who is technically said to 'work' the holes, and the exact period at which the steel is ready to be 'teemed' or poured into the mould is noted by the melter. The pots are lifted out of the holes by means of a pair of iron 'pulling-out' tongs. As soon as the lid is removed with the lid tongs, the scum or flux is removed from the surface of the molten steel, which is then poured into a cast-iron ingot mould, formed of two halves, tightly ringed and wedged together. The interior of the mould has been previously 'reeked' or covered with a coat of coal tar soot to prevent the ingot from adhering to the mould.

"The melting of steel is a process requiring the greatest skill. If the molten steel does not remain long enough in the fire, it will 'teem' or pour 'fiery,' or like boiling water full of vapour or gas, and the ingot will, when cold, be more or less full of bubbles or holes, technically called 'honeycombs.' When the ingot is hammered down into a bar, these honeycombs will appear as cracks or seams, technically called 'roaks.' If the steel remains too long in the fire it will teem 'dead,' as if all the gas had escaped, and when used afterwards in the bar will be of inferior quality. Should a piece of coke accidentally find its way into the pot, the ingot will show a bright sparkling fracture: technically speaking, it will be said to 'stare.' This ingot will be 'hot short,' and will crumble to pieces when red hot under the hammer. If the molten steel becomes chilled before it is poured into the mould, which may be detected by the stream 'skimming over' as it is teemed, the fracture of the ingot will appear dull in colour and full of small holes or honeycombs.'

The melting of the steel occupies about three hours. The lid of the furnace, often lifted during that time to see how the melting goes on, is at length removed. A stranger's dazzled eye will fall with a wavering glance on the white glowing mass within; but the workman--his legs protected by wet sacking--goes to the hole and looks down with unshrinking gaze as he seizes the crucible with his tongs to carry it to the moulds. When the crucible is turned, the steel runs out in a thin white stream and throws out brilliant coruscations, very beautiful to look upon when the operation is performed at night. Plate III shows the melter in the act of pouring the steel into the mould, which, to a stranger, is the most interesting part of the process.

STEEL MELTING.

Plate III - STEEL MELTING [*Illustrated Guide to Sheffield, 1879*, p.207]

In the floor on his left are seen the tops of the holes or furnaces in which the melting takes place, and on the shelves round the room are crucibles ready for use. Manganese, wolfram,[5] titanium, and other ingredients are put into the crucible in the belief that they improve the steel. Nearly every manufacturer has a different specific. A good deal of secrecy is often maintained as to the exact nature and quantity of the "physic" (so it is called) put into the melting pot, but its value is doubted by many engaged in the trade, workmen telling stories of the physic having been thrown aside without any perceptible difference in the quality or temper of the steel. Of the value of spiegeleisen, which has come into general use of late years, no doubt is entertained. Its tendency is to harden the temper of the steel and make the ingot sound by preventing the formation of honey-combs.[6]

Great variety is necessary in the quality and temper of steel, according to the purpose for which it is required.[7] These depend mainly on the quality and temper of the bar steel used in its manufacture, but both depend very greatly also upon the skill of the melter. Half the secret of making good cast steel lies, practical men assert, in knowing the exact moment for taking it out of the fire. The same skill is required in hardening and tempering, which is declared to be one of the fine arts, so much nicety is necessary in the degree of heat applied and in the manipulation of the steel. The difficulties of manufacture successfully overcome, the difficulties of selection follow. The practised eye judges of the quality and temper of steel according as the grain of the newly fractured end is coarse or fine, dull or bright; and the varieties of quality and temper being very great indeed, the shades of difference in the appearance of the grain are proportionately minute and difficult to distinguish. Only the practised eye can tell the full meaning of the different textures either of cast or blister steel.

EXTENSION OF THE STEEL TRADE

The extension of the steel trade already alluded to claims a more than passing notice. Thirty years ago steel was used chiefly for cutlery, files, saws, and for the edges of joiners' and other cutting tools made principally of iron. The great development which has since taken place in the manufacture of machinery and tools of all kinds has increased a hundred-fold the purposes for which steel of the old description is required. Iron and steel, in forgings and castings of all sizes and shapes--from the hundred-ton gun and the enormous armour plate, down to the smallest bolt and screw--have to be cut, planed, shaved, slotted, bored, or sawn; and the cutting edge of the tools used for all such purposes, however different in appearance from the fine edge of a penknife, must be of the same high-class steel to stand the work. That, however, is not all. The greater strength, hardness, and tenacity of steel, has led to its being largely used as a substitute for iron in general manufactures as well as in cutting tools, a development which has worked great changes already, and threatens to revolutionize the iron trade in a few years. In the steel trade, as now carried on, there are four pretty clearly defined, though not quite distinct divisions. (1) The old trade in shear and crucible steel, for cutlery, edge tools, &c. (2) Crucible steel in large castings. (3) Bessemer and Siemens-Martin steel.[8] (4) Steel founding.

The multifarious purposes for which steel of the first class is used have already been indicated. The products of the second division are cranks, cylinders and other large castings for marine and other engines, tyres, steel bells, and various other articles. To some of these purposes, and to rails, axles, rollers, and many other purposes, Bessemer steel is also applied. The fourth division indicates the most recent development of the steel trade--the application of steel to the thousand small castings hitherto made by the ironfounder--a branch of growing importance. It will be obvious that the purpose of the last three divisions of the steel trade is, for the most part, one and the same,--the substitution of steel for iron. The one impediment in the way of this change is the greater first-cost of steel. In regard to Bessemer steel this impediment can scarcely be said to exist now, so little do the prices exceed those of good iron. In regard to crucible steel, the difference in price is more serious, but is counterbalanced by the vastly increased wear and strength attained. A great future awaits the steel trade in the new as in the old departments.

For steel, Sheffield is still the world's great workshop. It is true that of late years France, Germany, Belgium, and the United States have entered the lists as rival manufacturers, and we hear much from time to time of the success of foreign steel makers; but Sheffield steel, especially the best cast steel, has still an unrivalled reputation for quality throughout the world.[9] This is sufficiently evidenced in the fact that for their best tools foreign manufacturers use Sheffield steel, and parade the fact before consumers as indisputable proof of the excellence of their goods. The quantity produced in Sheffield is, moreover, vastly in excess of that made in any foreign country. Important improvements have been made in the manufacture of steel here of late years, and it is becoming much more of a science than formerly. The conditions of success in producing specific qualities and tempers are better known, and the best Sheffield makers are now able to make a given temper and quality of steel with an accuracy and certainty to which the chief foreign manufacturers are far from having attained.[10]

STEEL FORGING, TILTING AND ROLLING

After leaving the converting furnace or the mould of the caster, steel is forged, tilted, or rolled before being made into cutlery, tools, and the various other forms in which it reaches the general public. Tilts and rolling mills are considered necessary adjuncts of the larger steel works, and give manufacturers more complete control of the operations of their trade, but tilting and rolling are also carried on as separate businesses.

Forging or tilting is the older process, and, where both processes are applied to the same material, takes the precedence. Before the discovery of steam power, forge and tilt hammers, like grinding wheels, were built on the rivers that run through the town, and worked by means of water-wheels. Many such forges and tilts still exist, and in them the process is both interesting and picturesque. The hammers, as will be seen from the illustration, (Plate IV), are very peculiar objects. The staff or "heave" is a huge length of rough timber, with a block of hard composite metal bound to one end for a head, and works on a pivot fixed near the lower end. The hammers, arranged in a row, are raised by the pressure of an irregular revolving drum on the pivot end, and fall in succession with a great thud on the hot, yielding steel. To the visitor, at first sight, they look like the heads of

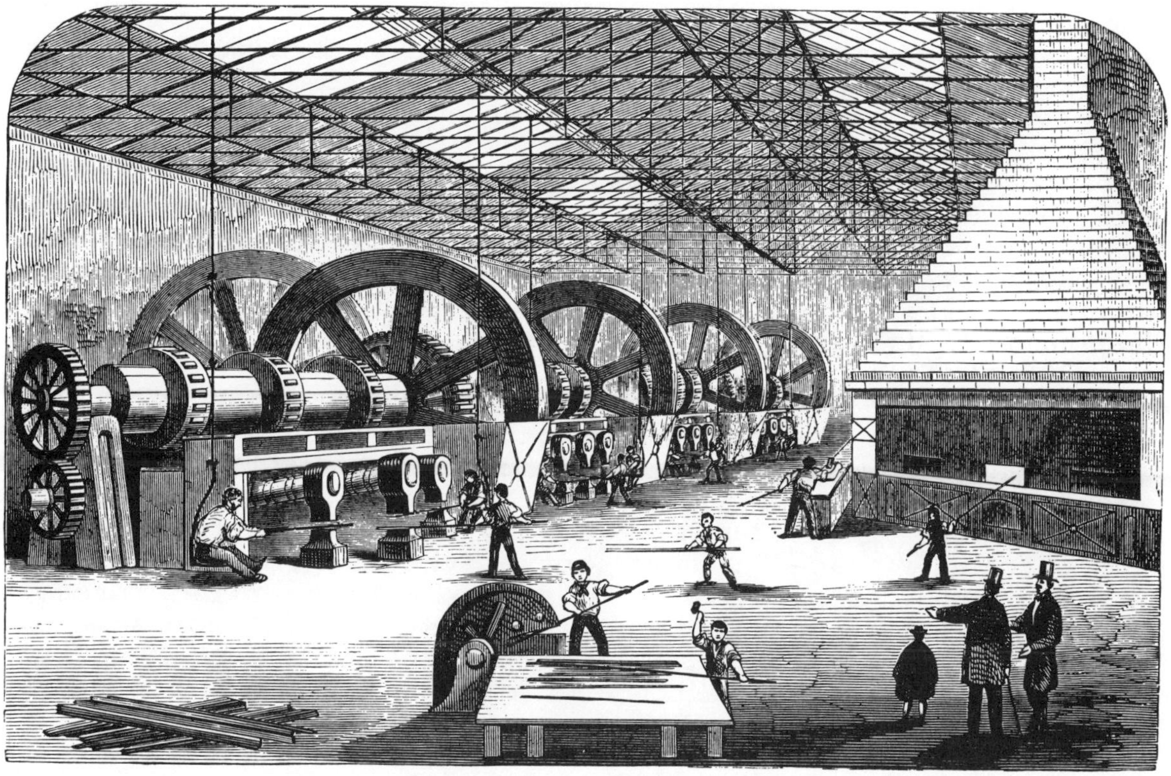

STEEL TILTING.

Plate IV - Steel Tilting [*Illustrated Guide to Sheffield, 1879*, p.211]

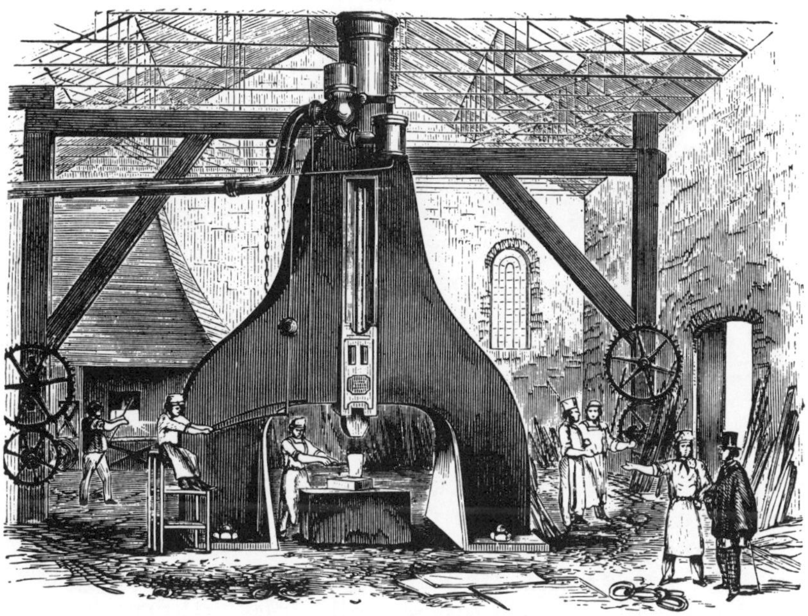

STEAM HAMMER.

Plate V - STEAM HAMMER [*Illustrated Guide to Sheffield, 1879*, p.213]

antediluvian monsters, nodding, some deliverately, others with frantic energy. Tilt hammers work the more rapidly, and with an incessant, deafening noise, quite irritating to unaccustomed ears. They are used for light work. The forge hammer, which is longer and heavier, is worked at a slower rate, but delivers a more crushing blow. It is used for the manipulation of larger masses of steel. The tilter sits in front on a suspended seat, which swings to and fro to accommodate his action, as he moves and turns the steel dexterously on the anvil, so that the blows are distributed evenly, until the desired size is obtained. Tilt hammers vary in weight of blow from 2½ to 5 cwt. A forge hammer, driven by water power, delivers about 150; a tilt hammer about 300 strokes a minute; driven by steam, they deliver nearly twice this number of strokes. Forge and tilt hammers are being rapidly superseded by the more tractable steam hammer, invented some 25 years ago by Nasmyth, and now made in great variety and of all sizes. The steam hammer, as will be seen from the illustration, (Plate V), moves up and down in a heavy iron frame, requiring neither water-wheel nor revolving drum; it can be placed in any part of the shop, and can be so regulated, by the motion of a handle, as to strike with terrific force, or gently crack a nut. For the welding of shear steel, and for some other purposes, some practical men still prefer the more ponderous and less rebounding stroke of the tilt and forge hammer, but for the great majority of purposes the steam hammer is now universally used.

Rolling on a large scale is a business requiring a considerable outlay of capital, owing to the great variety of sizes and shapes which steel has to be reduced for the multifarious purposes of the general manufacturer. It is important to minimize the work of the forger, by supplying him with steel of the most convenient size and shape for the implements he has to make. There is obviously a wide difference between the most convenient size and shape of steel for an axe head and a lancet blade. Each ingot, moreover, must pass through a graduated succession of grooves before it can enter the groove which has to fix its final shape and size. It is generally passed through two trains of rolls - the "clogging" train, by which it is drawn out and rougly shaped, and the finishing train, in the successive grooves of which it is reduced to the exact size and shape required.

A train consists of two or three rolls, fixed one above another, in strong iron frames. For short lengths a two high train is generally used; the ingot being put through the rolls and returned over them. For medium lenghts a three-high train is preferred, it being easier and safer to return the ingot between the two upper rolls.

But the shaping of steel is only one of the objects of hammering and rolling. A main and essential object is to close the pores, and give the steel a closer and more perfect grain, and thereby greater strength and tenacity. For these purposes, hammering is the more effective process, and, though for all ordinary purposes rolling is sufficient, much of the steel used for the highest quality of cutlery, and for the steeling of edge-tools and some other purposes, is forged to size and shape.

The skill and rapidity with which the forger draws out a short, thick, steel casting, into a long, thin, straight rod, perfectly smooth and even, for the making of such small articles as pen knives, by dexterously moving and turning it under the hammer, is remarkable, and is the result of long and careful training. With steam hammers, a further point is gained. By the insertion of dies in the face of the hammer or the anvil, or both, the steel is shaped and fashioned as in the more elaborate grooves of the rolling mill; but this is done chiefly in the subsequent forging of implements, not in the original tilting and forging of the steel itself.

It will be understood that steel is heated to a high temperature for hammering and rolling, as for subsequent manipulation by the tool and cutlery forger. For this purpose, furnaces are built along the sides of the mills and forges, as near as practicable to the rolls or hammers as the case may be.

MANUFACTURERS OF STEEL

Before proceeding to notice some of the leading works where the interesting and important processes we have described are carried on, a few words are necessary with reference to the general features of the vast establishments which have grown up during the last quarter of a century.

The advantage of having works for the heavier descriptions of manufacture in close proximity to the railways was early recognized by Sheffield manufacturers. It was seen that the railway siding was almost as much a necessity to the steel manufacturer of the present day as the dam and the goit were to the grinder and tilter before the days of steam power. The main extensions of late years have consequently followed the lines of railway, and the once beautiful valley of the Don, from Owlerton above the town to Tinsley and Brightside below it--a distance of four or five miles--has been turned by the prosperity of the town into a pandemonium, the glare and sulphurous smoke of which horrify strangers, while rejoicing the hearts of those to whom they are the sure sign of good wages and growing wealth. Factories are beginning to invade the still charming Abbeydale now that the Midland Company have carried their main line through it. Many of the largest works are intersected from end to end by lines of rails, working on which the locomotive engine and the steam crane do all the internal carrying of fuel, material, and manufactured goods, effecting thereby a great saving of cost in labour. For the most part the larger works are so arranged that the raw material is received at one extremity, and is delivered in the shape of finished merchandise at the other extremity--an arrangement by which labour is saved and work facilitated. Another feature of Sheffield factories is their completeness. The larger steel manufacturers--many of whom also make files, tools, and other goods--have not only tilts and rolling mills in their works; they also consider engineering shops an indispensable adjunct, and make much of the machinery they use, doing all the repairs; attaining thus the most complete control over all the departments and operations of their trade.

Messrs. WILLIAM JESSOP AND SONS Limited.--The largest works engaged exclusively in the manufacture of steel are those of Messrs. William Jessop and Sons Limited, of Brightside. Their extent may be judged from the fact that they cover an area of over thirty acres; they are intersected by lines of rails exceeding three miles in length, and in addition to six water wheels, no less than twenty-eight steam boilers are required to drive the numerous engines, hammers and machines. It will be seen from our illustration, (Plate VI), that the River Don flows

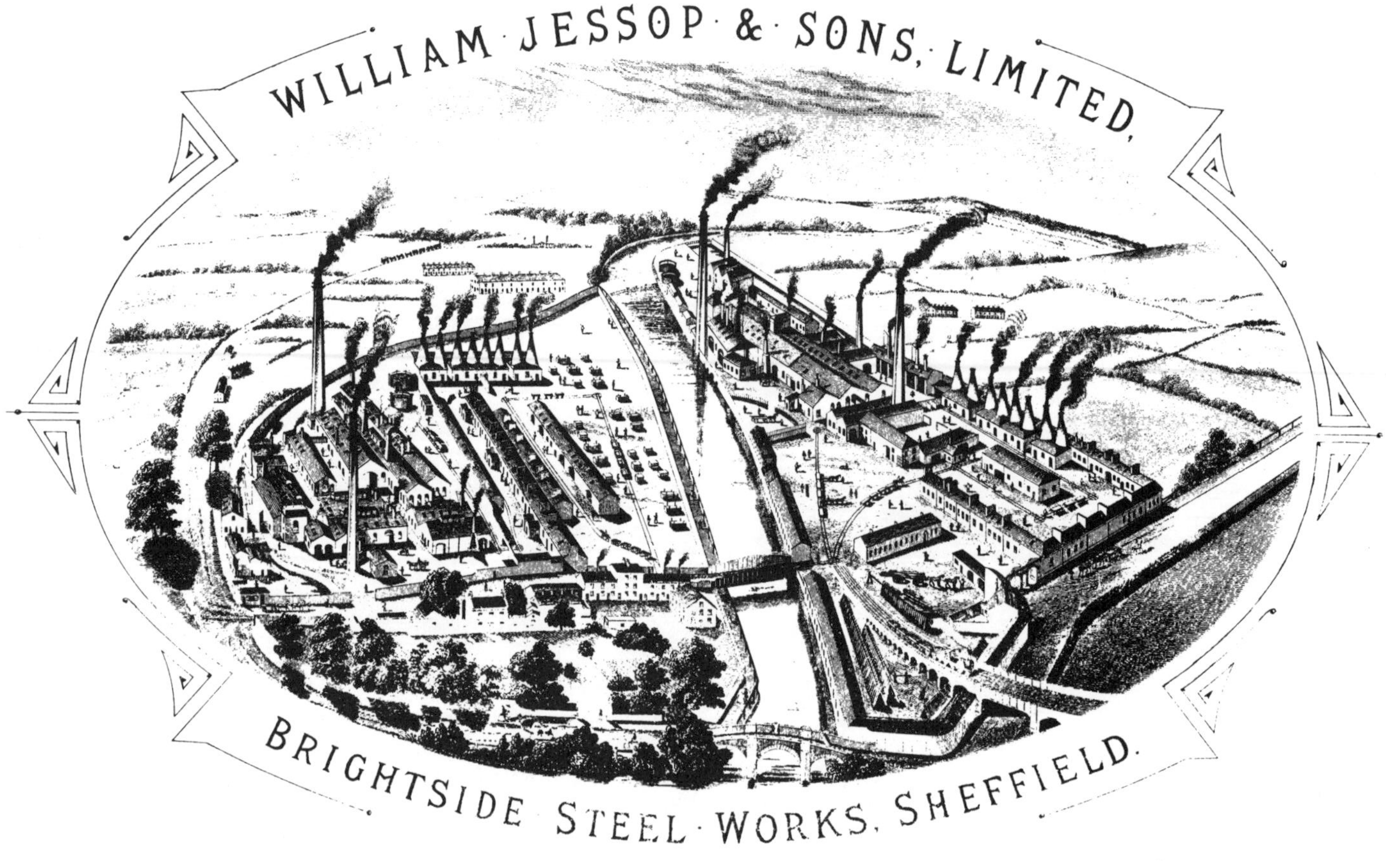

Plate VI- William Jessop & Sons, Ltd. Brightside Steel Works, Sheffield
[*Illustrated Guide to Sheffield, 1879,* adjacent p.217]

through the works, the two parts being connected by an iron bridge. The premises are of the most complete character, including extensive converting and melting furnaces, forges, rolling mills, wire mills, steel foundry, gas works, engineering shops and all the necessary appliances for the successful manufacture of steel in its various forms, from the heaviest forgings to the smallest bars, and from the largest plates to the thinnest sheets. The firm, since their establishment in 1793, have confined their attention chiefly to the production of the highest classes of steel, for which they have obtained a world-wide reputation. They make the finest qualities of steel for engineering and machinery purposes, for edge tools, cutlery, hammers, mill picks, shear blades, needles, fish hooks, watch and clock springs and other most delicate purposes. The rolling of sheet steel is carried on to great perfection, and is very interesting to witness. The company also make large quantities of sheets for circular and other saws, and an important branch of their business is the manufacture of sheet steel for making steel pens. They have long been the sole importers of several of the most noted brands of Swedish bar iron, amongst which may be named the famous brand (L) LEUFSTA every bar of which comes into their possession, and, if genuine, bears their name. The company have establishments in the principal cities on the Continent and in the United States, where they have always large stocks of steel on hand. It may be mentioned that the "gold medal" was awarded to the firm at the Paris Exhibition of 1878.

Messrs. VICKERS, SONS AND Co. Limited.--The striking development which has taken place in the steel trade of late years is well illustrated by the processes carried on at the River Don Works, Brightside, (Messrs. Vickers, Sons and Co. Limited,) of which we present an illustration, (Plate VII). The Company are the successors to Messrs. Naylor, Vickers and Co., one of the oldest steel manufacturers in the town, and the first to introduce some of the more important branches of that trade. They were the first to commence in England the manufacture of castings in steel to shape, and the first to introduce the manufacture of cast steel tyres and heavy forgings in steel. In 1854, they acquired the patent of E. Riesse (which up to that time had been used in Germany) for making cast steel bells, and, working under this patent, Naylor, Vickers and Co. gradually developed the art of steel casting which has now become a most important branch of the Sheffield trade. In our previous issue we described the casting of a steel bell by this firm for the Exhibition of 1862, which required the contents of 176 crucibles and weighed about 4½ tons, (Plate VIII). At that time this was the largest steel casting that had been made in England. The trade since then developed so rapidly that, having erected their new works in 1866, this firm was able soon after to make a cast steel marine shaft that weighed nearly 22 tons in the rough and required 672 crucibles, and was used (without forging) in the steamer "Wisconsin". The present firm is now making ingots up to 30 tons weight for heavy forgings, and could produce them of any size required.

Messrs. Burys And Co. Limited.[11] Like the firms already mentioned, Messrs, Burys and Co. Limited, of Regent Works, Penistone-road, are large manufacturers of steel, but they carry on other branches which those firms do not. Regent Works were founded over forty years ago by Mr. John Bedford, and passed some twenty years afterwards into the hands of Messrs. Burys, by whom they were transferred to a Limited Company, in August, 1865. (See Table III). They are now carried on under the management of Mr. Brown. The Company have made large additions to their premises, which cover an area of over six acres, and, as an examination of our illustration (Plate LX) will show, are among the most compact and complete works of their kind in the town, comprising converting and casting furnaces, ranges of workshops of many kinds, forges, rolling mill, grinding wheel, &c. The works include interesting specimens of the old forges and tilts worked by water power. Messrs. Burys make blister, shear, crucible and spring steel in large quantities for the general trade, as well as for their own manufactures. Among the goods they manufacture are files, saws, edge and engineers' tools of all kinds, hammers, miners' picks, steel plough plates, knives for reapers, choppers, and other agricultural implements, &c. The best evidence of the completeness of Regent Works is found in the fact that they provide for all the various processes of so many branches of trade. Buying in only fuel and raw material, Messrs. Burys are able to send out steel and manufactured goods of the most varied character without extraneous assistance of any kind. They do everything on their own premises, even to the making and repair, in the engineers' shops, of the tools and much of the machinery with which they work elsewhere. In this respect Regent Works are but examples of many factories in the town, though among the best of those factories. The advantage of such works is that the manufacturers know exactly the quality of the steel they use, and having complete control of all the operations of their trade, can regulate production in every way to the demands of their customers. Messrs, Burys and Co. moreover keep very large stocks

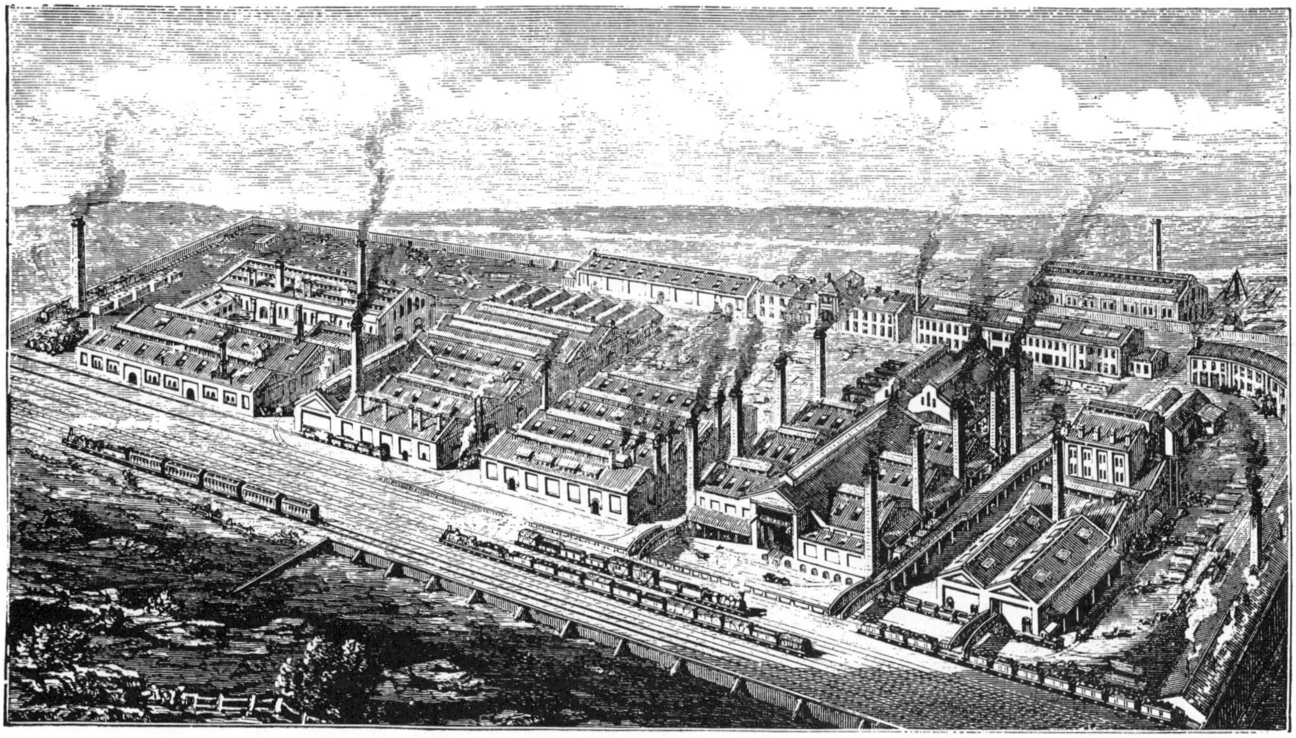

RIVER DON WORKS—MESSRS. VICKERS, SONS AND CO. LIMITED.

Plate VII - RIVER DON WORKS - MESSRS. VICKERS, SONS and CO. LIMITED
[*Illustrated Guide to Sheffield, 1879*, p.218]

STEEL BELL CASTING—RIVER DON WORKS.

Plate VIII - STEEL BELL CASTING - RIVER DON WORKS
[*Illustrated Guide to Sheffield, 1879*, p.219]

REGENT WORKS—MESSRS. BURYS AND CO. LIMITED.

Plate IX - REGENT WORKS - MESSRS. BURYS and CO. LIMITED
[*Illustrated Guide to Sheffield, 1879,* p.221]

SCOTIA STEEL WORKS—MESSRS. THOMAS JOWITT AND SONS.

Plate X - SCOTIA STEEL WORKS - MESSRS. THOMAS JOWITT and SONS
[*Illustrated Guide to Sheffield,* p.223]

of the goods they manufacture, and are usually able to consign assortments of the most varied character within a few days of the receipt of orders. They do a large business in France, Russia, Poland and other continental states, in addition to a considerable home trade, and their goods have long had a high reputation in these markets both for quality and finish. They received a silver medal at the Paris Exhibition of 1855; gold medals at Paris in 1867 and 1875; a gold medal at the Cape in 1877; and at the Paris Exhibition last year they have been awarded a gold medal.

Messrs. Thomas Jowitt and Sons.--Messrs. Thomas Jowitt and Sons of the Scotia Steel Works, Attercliffe-road, are well-known manufacturers of steel, files, engineers' tools, railway plant, heavy forgings, &c. The business was founded, in 1848, by Mr. Thomas Jowitt, the father of the present members of the firm, at Saville Works, Saville-street, and was afterwards removed to Royds Works. These proving inadequate to the requirements of a rapidly extending business, the firm erected Scotia Works in 1864. It will be seen from our illustration (Plate X) that the works are large and complete, including converting and crucible furnaces, tilt and forge hammers, rolling mill and extensive ranges of general workshops. Mr. Thomas Jowitt's first market was Scotland, where his name continues to be one of the best known Sheffield names. The firm have now extensive connections in Canada, India, Australia, Africa, America and on the continent of Europe, in addition to a good home trade. They import large quantities of Swedish and Russian iron, which they convert into fine steel of all kinds. Their trade mark is a beam engine, and "beam engine steel" is known far and wide for its uniform temper and excellent quality. Their files, made of the best steel, are all hand-cut, and in other branches of manufacture they have made quality a first consideration. First-class medals were awarded to the firm at the London Exhibitions of 1851 and 1862, and the Paris Exhibition of 1855, and they received a prize medal at the Paris Exhibition of last year.

Messrs. Thomas Turton and Sons.--Messrs. Thomas Turton and Sons are large manufacturers of steel, railway springs, files, saws and edge tools, Sheaf Works,[12] their place of business, being near the ticket platform at the Victoria Station of the Manchester, Sheffield and Lincolnshire Railway, for trains from the south and east. It will be seen from our engraving (Plate XI) that these far-famed works are of large extent, flanking the railway on both sides, and having free communication through its open arches, some of which are used as workshops. The works are also intersected by the canal which runs to Thorne and Hull. Messrs. Turton and Sons are among the oldest firms in the town, and have a very high reputation. In edge tools they succeeded the old eminent firm of William Greaves and Sons of Sheaf Works, whose reputation they have greatly enhanced. They are at the head of the file and railway spring trades. They make shear and refined steel for all purposes, and their steel is known in all markets for its excellent quality. Messrs. Turton and Sons do an enormous business in high-class goods. The quality of their productions has been acknowledged by awards of the highest honours at the International Exhibitions at which they have competed.

Messrs. Samuel Osborn and Co.--Messrs. Samuel Osborn and Co., of the Clyde Steel and Iron Works, are large manufacturers of steel and steel castings, railway springs, files, saws, reaping machine knives, tools, &c. The works at which the steel and heavier industries are

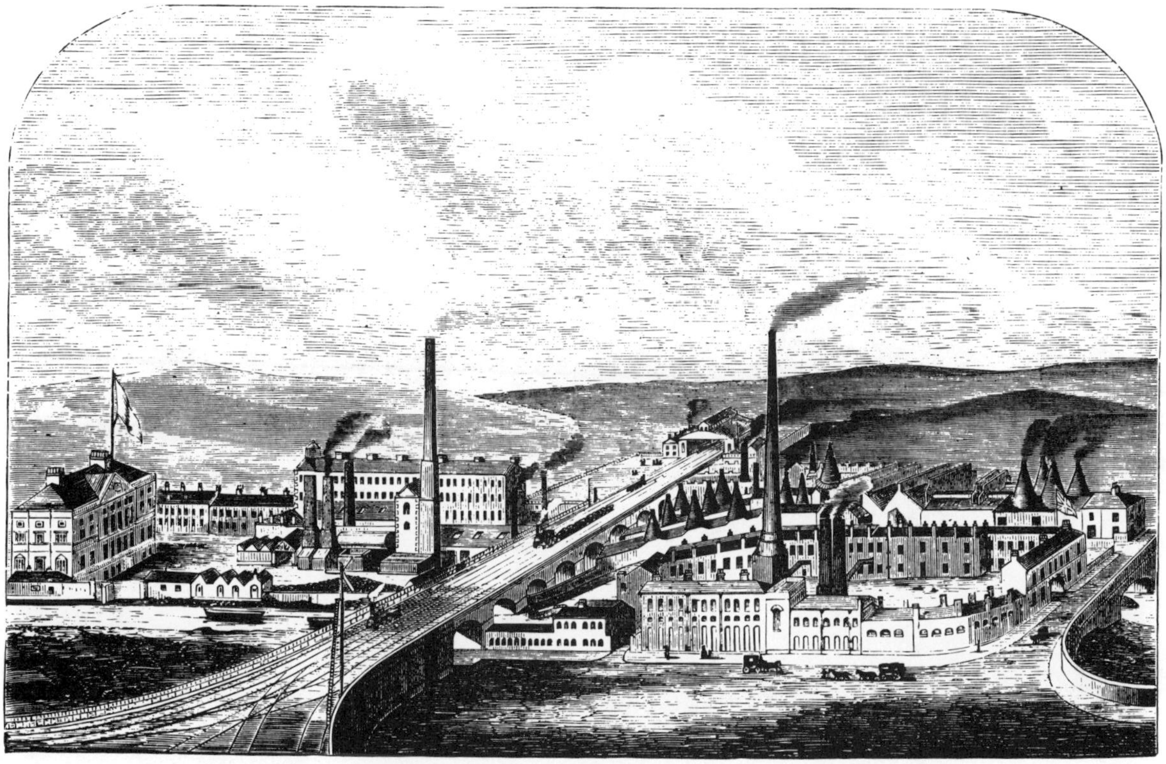

SHEAF WORKS—MESSRS. THOMAS TURTON AND SONS.

Plate XI - SHEAF WORKS - MESSRS. THOMAS TURTON and SONS
[*Illustrated Guide to Sheffield, 1879,* p.224]

carried on are in the Wicker, extending backwards to the river Don, and include converting and melting furnaces, forge hammer and spring shops, rolling mills, grinding wheel, and all the other adjuncts of large and complete premises. The firm have branch works at Brookhill, for the manufacture of files, saws and tools. Our illustration (Plate XII) shows the extent of the works. The firm have spared no expense in adapting their works to the improvement and cheapening of production, and in applying the best labour-saving machinery to all operations in which it can be advantageously used. Their steel, tools, files, and springs have long been well known in the American and continental as well as in the home markets. Of late years they have devoted much attention to the production of steel castings for engine and other purposes, and also to the manufacture of the great variety of knives and castings used by the large agricultural implement makers in the construction of reapers and other machines--the manufacture of these knives and castings being a considerable and important branch of Sheffield industry. Messrs. Osborn and Co. are also manufacturers of two important specialties--Mushet's special and Titanic steels. Mr. Robert F. Mushet is well known as one of the most skilled English metallurgists. In 1875, the Iron and Steel Institute awarded to him the Bessemer gold medal, in recognition of his great improvements in the manufacture of iron and steel. The name of Mr. Mushet is well known in commercial circles in connection with Bessemer steel. When Mr. Bessemer had invented his very remarkable apparatus for converting cast iron into steel by blowing air through it, his work was but half done. The steel when re-heated was found to lack cohesiveness. It broke into small particles under the hammer, acquiring the name of "red-short." To Mr. Mushet is due the credit of completing the invention by the discovery of a remedy for this want of cohesiveness,

CLYDE STEEL AND IRON WORKS—MESSRS. SAMUEL OSBORN AND CO.

Plate XII - CLYDE STEEL and IRON WORKS - MESSRS. SAMUEL OSBORN and CO.
[*Sheffield Illustrated Guide, 1879,* p.226]

in the addition of a triple compound of iron, manganese and carbon, in the form of spiegeleisen--a Prussian iron not commercially used in England or known up to that time. We mention this as a very interesting fact in the history of one of the most valuable metallurgical inventions. Mr. Mushet has since discovered a new process of making hard cast steel, which requires neither hardening nor tempering. This product is now well-known in the engineering world, and although the price of it--about 140 a ton--is high, its hardness and great durability make it indispensable for the best lathe and planing tools. Titanic steel, also an invention of Mr. Mushet, is cast steel manufactured with Titanium, found chiefly in Norway, in an iron ore known as ilmenite--the effect of the Titanium being to increase greatly the cohesive strength and cutting properties of steel.[13] Messrs. S. Osborn and Co. are the sole manufacturers of these special steels.

Plate XIII - TOLEDO STEEL WORKS - MESSRS. J. H. ANDREW and CO.
[*Sheffield Illustrated Guide, 1879*, p.228]

Messrs. J. H. Andrew and Co., of the Toledo Steel Works, Neepsend, are manufacturers of best crucible steel for tools and general purposes. (Plate XIII) One of their specialties is the forging of octagon rods of unusual length for rock drills; and they roll sheet, spindle and rod steel for a variety of purposes, employing the best machinery for straightening and finishing it. Their great specialty is the rolling of wire rods for cable, rope and other such purposes. Wire rolling, as carried on at Toledo Works, is an exceedingly striking and interesting process. The rods are rolled of various gauges down to about three-sixteenths of an inch and up to two inches diameter, and of extraordinary length. Perhaps the best test of the superiority of a wire mill is the length of rod that can be rolled. This depends much upon the speed at which a mill can be worked, the heat passing off so quickly as to prevent the operation being continued more than a few minutes. For the Paris Exhibition, at which they received a prize medal, Messrs. Andrew rolled an ingot of 270 lbs. into an unbroken rod of over 750 yards, being nearly half-a-mile in length. The rapidity with which coil after coil of wire rods is produced is one of the marvels of the rolling trade, and shows the perfection to which this class of machinery has been brought. In addition to every convenience in shape of the most improved and modern machinery, Messrs. Andrew and Co. have their own collieries at Dronfield. They do a large trade in the home, American, continental and other markets, in wire rods and general goods.

Among other leading steel manufacturers are Messrs. Sanderson Brothers and Co. Limited, Ibbotson Brothers Limited, Spear and Jackson, Seebohm and Dieckstahl, Wilson Hawksworth, Ellison and Co., the Hallamshire Steel and File Co., John Kenyon and Co., Jonas and Colver,

Wingfield, Rowbotham and Co., S. and C. Wardlow, Bedford and Sons, Crookes, Roberts and Co., J. R. Wright, jun., Turton Brothers and Matthews, William Cooke and Co. Limited, Makin and Son, Beardshaw and Sons, Howell and Co., Leadbeater and Scott, S. S. Brittain and Co., Edgar Allen and Co., H. Rossell and Co., W. Jackson and Sons, J. and R. Dodge Limited, Moss and Gamble Brothers, W. K. and C. Peace, Austin and Dodson, Walter Spencer and Co., J. R. Spencer and Son, Gregory and Bramall, W. Spencer and Son, Cocker Brothers, Jabez Vernon and Co., Marriott and Atkinson, Flockton, Tompkin and Co., D. H. Coupe, jun. and Co., Fenton and Sons, J. C. Clarke and Co., Vessey and Friend."

[end of transcription p.227 *1879 Illustrated Guide to Sheffield*]

Plate XIV - Thos. Firth & Sons Limited, Norfolk Works
[an engraving "entered at Stationers' Hall, c.1862, collection Firth,Brown Ltd.]
reproduced by kind permission of Firth Brown Limited,Sheffield, England

Another prominent firm producing tool steel was Thomas Firth & Sons, Ltd. This company was established by Thomas Firth, a practical steel melter, about 1842.[14] Subsequently his sons joined him in the business. An engraving, dated about 1862, (Plate XIV) illustrates their Norfolk Works on Saville Steet East, known as Norfolk Works. Adjacent to their Works, not shown in this illustration, was Spear & Jackson's Aetna Works on the left and John Brown's Atlas Works on the right. In addition to tool and ordnance steel the firm maintained a high reputation for their manufacture of files, saws and edge tools. Along with William Jessop their tool steel was highly regarded among customers in the United States.

Another steel firm that deserves mention is that of the Warrington Steel Works of Peter Stubs located near Sheffield at Rotherham. This was acquired by the Stubs' company of Warrington, Lancashire from Joshua Warner & Co. in 1826.[15] (See Plate XV). At the time of this acquisition this famous file making concern was annually consuming 50 tons of steel. Desiring to be independent from Sheffield sources, most of whom were competitors, the Stubs' company purchased this steel factory.

Plate XV - The Warrington Steel Works of Peter Stubs near Rotherham
[Photographed from an engraving in 1851 Peter Stubs Catalogue]
[Archives, File 124/1/Box 25, Central Library, Manchester]

Commercial advertisements reveal valuable prima facie historical data. Plates XVI - XXV show several advertisements that appeared in the 1879 *ILLUSTRATED GUIDE TO SHEFFIELD* noting the products of some of the prominent steel producing firms. The decorative and fancy type designs, many of which are no longer available, employed in printing these notices had attractive styles for capturing the attention of potential customers, Great pride was apparently attached to awards given at exhibitions and considered influencial promotional publicity. Engravings of medals or seals of such awards, often supplemented by corporate marks, accompanied the text. Notice of infringement to trade marks were sometime noted. (See Plate XXV).

The majority of firms producing cast steel for tool applications or converters thereof, also manufactured files and saws. Not only did these products have high marketing demands, but it was excellent recommendation that their quality was high. It was certainly advantageous for a saw manufacturer to also make files due to the large number of the latter consumed in making the former. Remelting of scrap and used files also gave such steel producers an economic advantage.

Prior to the development of machine cutting tools, the hand file late in the 18th and early in the 19th centuries was a most important tool for shaping metal in building machinery. Indeed it was the skill of Henry Maudsley in using the file and realizing its limitations for absolute precision work that led to his developing the screw cutting lathe, methods of finishing a surface plate and his micrometer in 1805.[16] Holtzappfel has presented an excellent description, classification and uses of the file.[17] Plate XXVI illustrates the standard types and various cuts in which files were manufactured.

Early in the 19th Century a market began to develop for circular saws. Richards credits the invention of this tool to Samuel Miller of Southampton, who was awarded British Patent No. 1152, 5th August 1777.[18] The 1816 Smith's *SHEFFIELD KEY* noted item #625 - "Circular Saw - 4 inches diameter to 36". Jno. Davenport in the 1821 Sheffield Directory was the first listed to make circular saws at Rockingham Street. Under Saw Makers noted in the 1825 Sheffield Directory Richard Groves & Sons at Allen Street "also manufacturers of the patent circular saws warranted to run on an improved principle". In the 1839 Sheffield Directory John Davenport noted - "Patent Lathe-turned Circular Saws".

Throughout the 19th Century hand saws remained the most important cutting tool for the carpenter, joiner and cabinetmaker. The principal forms were the cross-cut, rip, panel, tenon, sash, carcase, dovetail and compass or keyhole. These are classified by dimension and tooth form by Holtzappfel.[19] Some Sheffield hand saws are illustrated in Fig.4.

The sections concerning the manufacturing of files (pp.260-267) and saws (pp.278-282) follow transcribed from the 1879 *ILLUSTRATED GUIDE to SHEFFIELD*. This includes descriptions of the file works of William Hall, Thomas Turner & Co., J.R. Spencer & Sons and the saw works of Spear & Jackson and Moses Eadon & Sons. Plates XXXIV - LIX illustrate Sheffield advertisements of saw and file makers. Plate LX shows the Adelaide Works of the Taylor Brothers in Sheffield that appeared in an 1858 Birmingham publication. A check list of 403 firms engaged in making or selling saws in Sheffield during the 19th Century appears as an Appendix to this Chapter.

B. HUNTSMAN'S

REFINED STEEL,

ATTERCLIFFE,

NEAR

SHEFFIELD.

Plate XVI - B. Huntsman's Refined Steel Advertisement from 1865

TURTON BROTHERS & MATTHEWS,

MANUFACTURERS OF ALL KINDS OF

32

NOT EXHIBITED SINCE.

WARRANTED CAST STEEL FOR TOOLS, &c.;

SINGLE AND DOUBLE SHEAR, BLISTER AND SPRING STEEL;

Sheets, Plates, Forgings, Castings, and every description of Steel for Engineering and Agricultural purposes;

T.B.&M.

OCTAGON CAST STEEL for MINERS' DRILLS;

HAMMERS, SAWS and FILES, marked "TURTON BROS. & MATTHEWS."

RAILWAY CONICAL, VOLUTE & SPIRAL SPRINGS;

Engineers', Contractors', and Shipbuilders' Tools; and also General Merchants. FILES marked "THOS. WING," or "THOS. WING & CO."

PHŒNIX STEEL WORKS, SHEFFIELD.

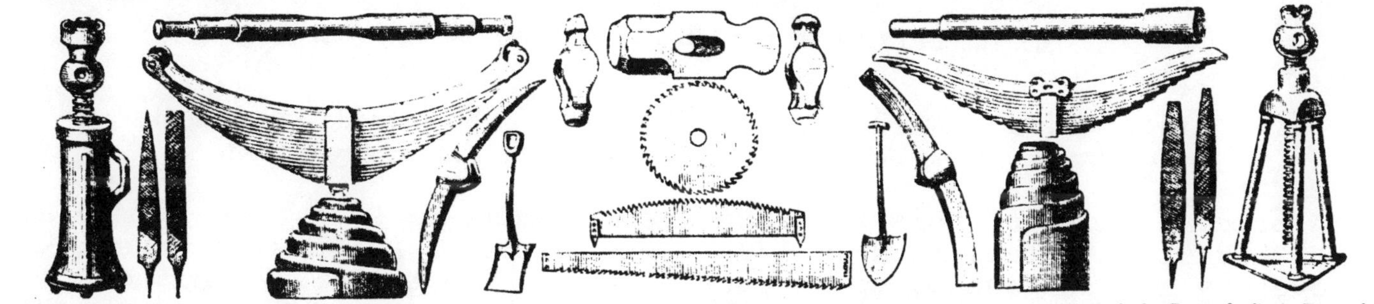

LONDON—JAMES LIVINGSTONE, 34, Gracechurch Street.
FRANCE—SOULÉ FRERES, 36, Rue du Château d'Eau, PARIS.
GERMANY—HENRY HAHN, 34, Gracechurch Street, LONDON.
AUSTRALIA—B. C. BURSTALL, MELBOURNE. / T. B. LINLEY, 111, Clarance Street, SIDNEY.

SPAIN—LUIS ATHY, Calle de Caracas, 5, Pral, da, Pasco de Santa Engracia, MADRID.
AUSTRIA—FERD. EUG. BIBA, III. Marokkanergasse 23, VIENNA.
RUSSIA—THEODOR SAZONOFF, 67, Fontanka, ST. PETERSBURG. / J. A. NUSSBAUM, Ostojenka Sawelowsky Pér, MOSCOW.
ITALY—FELIX BARBIÉ, Via Arcivescovado 6, TURIN.

Plate XVII - Turton Brothers & Matthews - Phoenix Steel Works
Page 32 Advertising Section 1879 Illustrated Guide to Sheffield

Gold Medal Awarded, Paris Exhibition, 1878.

WM. JESSOP & SONS LIMITED,

MANUFACTURERS OF

EVERY DESCRIPTION OF

TOOL STEEL,

Cast, Shear, German, Spring, Blister, and Sheet Steel for all purposes.

CRUCIBLE STEEL CASTINGS.

FILES OF FIRST QUALITY.

Steel Lagging Plates for Locomotive and Marine Engines. Boiler and Ship Plates and Forgings of all kinds.

BRIGHTSIDE STEEL WORKS,

SHEFFIELD;

ALSO AT

LONDON, MANCHESTER, PARIS, DUSSELDORF,

AND ALL THE PRINCIPAL CITIES IN THE UNITED STATES.

9

Plate XVIII - Wm. Jessop & Sons, Limited - Brightside Steel Works
Page 9 Advertising Section of 1879 Sheffield Illustrated Guide

A brief history of this firm appears on pages 23 & 24. The firm claims to have been established in 1774. The works at Brightside were taken over by Thomas Jessop, son of the founder, about 1835.[20] In the United States Thomas Jessop was known as"The Steel King", having made his first visit here in 1829. An American works was established at Washington, Pennsylvania, near Pittsburgh, about 1906 for the manufacture of sheet steel and saw plate. This followed the precedence of Thomas Firth & Sons, Ltd. who amalgamated with the Wheeler Sterling Steel Co. of Pennsylvania in 1896, establishing the Firth Sterling Co. in United States. *[History of the Firths,* Sheffield]

WILSON HAWKSWORTH, ELLISON & Co.,

CARLISLE WORKS,

SHEFFIELD.

MANUFACTURING UNION COMPANY

TRADE MARKS

REFINED CAST STEEL,

SHEAR, SHEET & BLISTER STEEL,

CAST STEEL WIRE

FOR ALL PURPOSES;

PATENT AND OTHER RAILWAY CARRIAGE, LOCOMOTIVE, WAGGON, BEARING AND BUFFING SPRINGS;

FILES, SAWS, EDGE TOOLS, HAMMERS AND ENGINEERS' TOOLS;

AND GENERAL MERCHANTS.

AGENTS AT—

AMSTERDAM	CHEMNITZ	PHILADELPHIA, U.S.
BRUSSELS	HAVANA	PARIS
BOSTON, U.S.	MONTREAL	PRAGUE
COLOGNE	NEW YORK, U.S.	ST. PETERSBURG

65

Plate XIX - Wilson, Hawksworth, Ellison & Co. - Carlisle Works
Page 65 Advertsing Section of 1879 Sheffield Illustrated Guide

Note international distribution and locations of Agents.

THOMAS JOWITT & SONS,

PRIZE MEDALS AWARDED TO THOMAS JOWITT & SONS.

PARIS, 1855. | LONDON, 1862. | LONDON, 1851. | PARIS, 1878.

TRADE MARK
TESTED

SCOTIA STEEL WORKS,

SHEFFIELD,

MANUFACTURERS OF THE CELEBRATED

BEAM ENGINE BRANDS OF

CAST & DOUBLE SHEAR STEEL

OF EVERY DESCRIPTION,

FOR TURNING TOOLS, CHISELS, TAPS, DIES, PUNCHES, DRILLS, &c.,

FOR HOME AND EXPORT;

ALSO

FILES, SAWS,

HAMMERS & ENGINEERS' TOOLS.

Agencies:—LONDON, PARIS, MONTREAL, NEW YORK and SAN FRANCISCO.

Registered Telegraphic Address:—JOWITT, SHEFFIELD.

135

Plate XX - Thomas Jowitt & Sons - Scotia Steel Works
Page 135 Advertising Section 1879 Sheffield Illustrated Guide

A brief description of the works and history of the firm appears on page 28 and an illustration of the works Plate X, page 27

Plate XXI - John Henry Andrews & Co. - Toledo Steel Works
Page 86 Advertising Section 1879 Sheffield Illustrated Guide

Note mention of international depots. A brief description of the works with sketch, Plate XIII, appears on page 31.

CORPORATE MARK.

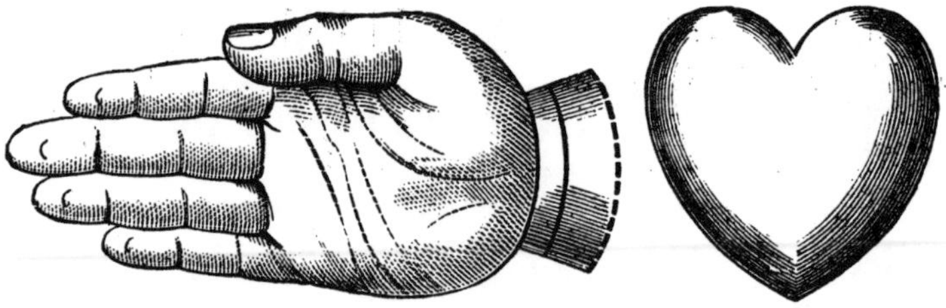

Granted according to Act of Parliament, 1770, by the Cutlers' Company, Sheffield.

SAMUEL OSBORN & CO.

MANUFACTURERS OF

BEST CAST STEEL,

FOR ENGINEERS' AND MINERS' TOOLS.

SPRINGS,

FOR LOCOMOTIVES, RAILWAY CARRIAGES AND WAGONS.

Steel Sheets, Blister, Coach and Railway Spring Steel, Single and Double Shear Steel.

CRUCIBLE STEEL CASTINGS & FORGINGS.

BEST REFINED CAST STEEL FILES.

Saws, Tools, Solid Cast Steel Hammers and Sledges.
Plough Plates, Rake Teeth Steel, Osborn's Patent Drawn Finger Bars, and all kinds of Steel for Agricultural purposes.

REAPING AND MOWING MACHINE KNIVES.

SOLE MAKERS OF

"R. MUSHET'S SPECIAL STEEL,"

FOR LATHE AND PLANING TOOLS, DRILLS AND COUNTERSINKS.

This Steel turns out at least double work by increased speed and feed, and cuts harder Metals than any other Steel. Neither hardening nor tempering required.

ALSO SOLE MAKERS OF

"R. MUSHET'S TITANIC STEEL,"

FOR ENGINEERS' AND MINERS' TOOLS, DIES, SNAPS, &c.

CLYDE STEEL AND IRON WORKS,

SHEFFIELD.

17

Plate XXII - Samuel Osborne & Co. - Clyde Steel & Iron Works
Page 17 Advertising Section 1879 Sheffield Illustrated Guide

A brief history and description of the works appears on pages 28 - 30 with a sketch of the works Plate XII, page 30. A more detailed record of this important firm has been published as: *PIONEERS for a CENTURY 1852-1952, A History of the Growth and Achievement of Samuel Osborn & Co., Ltd. Clyde Works Sheffeild* by T.Alec Speed [Sheffield, 1952]

GOLD MEDAL

PARIS UNIVERSAL EXHIBITION, 1878.

SILVER MEDAL, PARIS, 1855. | GOLD MEDAL, PARIS, 1875.
GOLD MEDAL, PARIS, 1867 | GOLD MEDAL, CAPE TOWN, 1877.

BURYS & Co. Limited,

REGENT WORKS,

AND

Philadelphia Rolling Mills and Forges,

SHEFFIELD;

CONVERTERS, MELTERS, ROLLERS & TILTERS

OF EVERY DESCRIPTION OF

STEEL

For Engineers, Machinists, Ship Builders, Agricultural Implement Makers, Rock-boring Tools, and general Mining purposes;

MANUFACTURERS OF **FILES** OF ALL KINDS;

EDGE TOOLS, ENGINEERS' TOOLS, SAWS AND HAMMERS.

CORPORATE AND TRADE MARKS.

"WM ASH & CO" "REGENT"

"BURYS & CO"

PARIS ADDRESS:—RUE CHARLOT, 62 (AU MARAIS).
LONDON ADDRESS:—17, FENCHURCH STREET, E.C.

40

Plate XXIIa - Burys & Co. Limited - Regent Works
Page 40 Advertising Section 1879 Sheffield Illustrated Guide

A brief history and description of the works appears on page 25 and a sketch of the works Plate IX, page 28. Note that the firm acquired the mark of "Wm. Ash & Co." and undoubtedly made plane irons with this imprint long after Wm. Ash ceased. (See Edge Tool List). See Table III, page 42 for development record of this firm.

Development of John Bedford, Edward Bury, John Bramall in Sheffield Edge Tool, File, Saw Businesses 1792 - 1970-

John Bedford, 1792
Grayson Fordge, Oughtibridge

↓

1852 to Sheffield at
Regent and Oughtibridge Works
Merchant & Steel Refiner and Saw and Edge Tool, cutlery Manufacturer and Tilter and Forger

↓

Edward Bury, esq.
Merchant and Manufacturer
[Cammell & Co.]

↓

Bedford, Bury & Co., 1856-1857
[John Bedford, Edward J. and William T.Bury]

↓

Bedford, Burys & Co., 1857-1862
Merchants, Steel Refiners, and Manufacturers of files, saws, edge tools &C. Regent Works and Oughtibridge Steel Works, Penistone Road

↓

Burys & Co., 1863-1970-
[Edward Jame Bury & William T.Bury]
Manufacturers of Steel Files, Edge Tools, Saws, Hammers, Vices, &c.
Regent Works & Philadelphia Works
[Acquired Mark of William Ash]

Bramall & Bedford, 1863-1872
Merchants and Manufacturers of Improved Cast Steel, files and edge tools, &c.

↓

John Bedford, 1872-1854-
Regent Works, Penistone Road

↓

John Bedford & Sons, 1856-1962-
Lion Works, Mowbray Street
Merchants & Manufacturers of Steel Files and Tools and Importers of Sweedish and Russian Iron

↓ 1970

SPEAR & JACKSON, 1760-1980-

TABLE III

TRADE MARK.

JOHN BEDFORD & SONS,

MANUFACTURERS OF

CAST STEEL

OF EVERY DESCRIPTION;

ALSO,

FILES, EDGE TOOLS, SAWS,

SPADES, SHOVELS,

AND

MACHINE KNIVES.

1855. 1878. 1855.

LION WORKS,

MOWBRAY STREET,

SHEFFIELD.

144

Plate XXIII - John Bedford & Sons - Lion Works
Page 144 Advertising Section 1879 Sheffield Illustrated Guide

Note Development Record, Table III, page 42

GOLD

MEDAL,

TRADE

MARK,

PARIS,

1878

SEEBOHM AND DIECKSTAHL,

DANNEMORA STEEL WORKS,

SHEFFIELD,

STEEL CONVERTERS;

MELTERS AND MANUFACTURERS

OF EVERY DESCRIPTION OF

CAST STEEL,

IN BARS, SHEETS AND WIRE.

SPECIALITY { TOOL STEEL AND MINERS' DRILL STEEL.

SINGLE OR DOUBLE SHEAR STEEL.

MILD CENTRED CAST STEEL FOR TAPS, &c.

SHAPED STEEL IN VARIOUS SECTIONS FOR NAIL DIES, &c.

SPECIAL DANNEMORA CAST STEEL

FOR TURNING AND PLANING HARD MATERIAL.

SPRING STEEL.

CAST STEEL FORGINGS OF EVERY DESCRIPTION.

FILES, SAWS, HAMMERS, &c.

11

Plate XXIV - Seebohn and Dieckstahl - Dannemora Steel Works

Advertising Section 1879 Sheffield Illustrated Guide

SUFFOLK WORKS,

SHEFFIELD.

THOMAS TURNER & CO.

MANUFACTURERS OF

Every description of Cutlery;

CONVERTERS OF STEEL;

**MAKERS OF BEST CAST STEEL FILES AND SAWS, WARRANTED;
ALSO, GENERAL MERCHANTS.**

THOMAS TURNER & CO. caution all parties against using their Name or Trade Mark, granted by Act of Parliament. Injunctions have been obtained from the Vice-Chancellor to restrain certain individuals who have thus infringed.—See Report in the *Times*, January 11, 1850; also March 6, 1863. All Goods bearing their Name or Mark as below are warranted.

TURNER.

THOMAS TURNER & C°
SHEFFIELD.

TURNER & C°
SUFFOLK WORKS.
SHEFFIELD.

V R
THOMAS TURNER & COMPY
SUFFOLK WORKS SHEFFIELD.

THOMAS TURNER.

THO^s TURNER.
CAST STEEL.

A CAUTION.—Whereas, certain parties have of late assumed the title of "TURNER & CO.," with the evident intention of deceiving the public and disposing of inferior goods as being those of our manufacture, we deem it advisable, both in justice to ourselves and for the protection of our friends, to notice the matter by public advertisement, and to request that all parties will kindly see that our address be written in full, and that the goods be marked with our

CORPORATE MARK.

THOMAS TURNER & CO.

SUFFOLK WORKS, SHEFFIELD.

We further beg to observe that we shall commence the usual proceedings against any person infringing upon any of our Trade Marks.

26

Plate XXV - THOMAS TURNER & CO. - Suffolk Works
Advertising Section Sheffield 1879 Illustrated Guide

Note CAUTION notice regarding fraudulent uses of Name or Trade Mark of Thomas Turner & Co.

FILES AND SAWS.

" Sheffield is popularly known principally as the world's great steel and cutlery mart, but an enormous trade has been done for many years in files, edge tools, sheep shears, and saws, these goods being in most cases produced by firms who also make steel.

The manufacture of files is very interesting, from the peculiar skill manifested in one department--file cutting. Files vary greatly in shape and size. They are made flat, round, half-round, three-square, and square; and range in length from two or three inches to the same number of feet, differing proportionately in weight and strength. (See Plate XXVI). In teething, there are the single and double cut, made with the broad chisel, and the rasp cut, with the pointed chisel; the smallest teeth--those of watchmakers' files for instance--being so find that a good naked eye is required to see them, while the teeth of the largest rasp files are half an inch apart and a sixteenth long. The work is greatly subdivided, especially in cutting, the men being kept as far as practicable to one class of work, in which they necessarily acquire very great skill.

Files to be of any value must obviously be made of the best steel, or the file would wear away instead of the metal upon which it is used. The steel is hammered or rolled in long bars of various sizes and shapes. The forger selects his bar with reference to the size of the file to be made, and forges it in the usual way, on anvils usually fixed for the sake of firmness on large blocks of stone. There are single and double-hand forging, the forger of all sizes up to ten inches in length working alone, and a "striker" being employed for the larger sizes. The bar of steel is cut into lengths, and as files are usually tapered at one end and have a tapering tang at the other, they are cut obliquely to facilitate this part of the forging. Flat files are shaped entirely with the hammer, and it is interesting to the onlooker to note the ease and rapidity with which the long tapering tang is drawn out, and the blade of the file is shaped and cut to measure. For other kinds, "swages" or bosses fixed in the face of the anvil are used to secure exactness of shape and size. The next process is annealing, the object being to soften the steel for the operations of the grinder and cutter. This is done by putting the forgings over a fire or in an oven, in piles, heating them evenly through, and cooling them gradually by letting the fire die out. Having been ground, the smooth file, if it may be so called, is taken to the cutters' shop to be roughened in that way which gives the characteristic property to the tool. File cutting is done chiefly by hand, and the process furnishes one of the most remarkable instances of manual dexterity which is to be seen in the whole round of human industry. The accompanying engraving (Plate XXVII) gives a representation of a row of file-cutters at work. Under the windows there are a number of low stone benches, and close to them the seats for the workmen. The hammer which the cutter uses weighs from one to six pounds, according to the size of the file to be operated upon; and it is constructed so as to allow the metal which composes its head to be pulled towards the workman while he is making the blows on the file. The chisel is formed of very strong, tough steel, and, like the hammer, varies with the size of the file it is used upon. When the workman proceeds to cut, the chisel is held in the left hand, somewhat in the same way as a pen, and so that the hollow of the hand is turned towards the workman. But this is only a general description: the method of working varies with the different

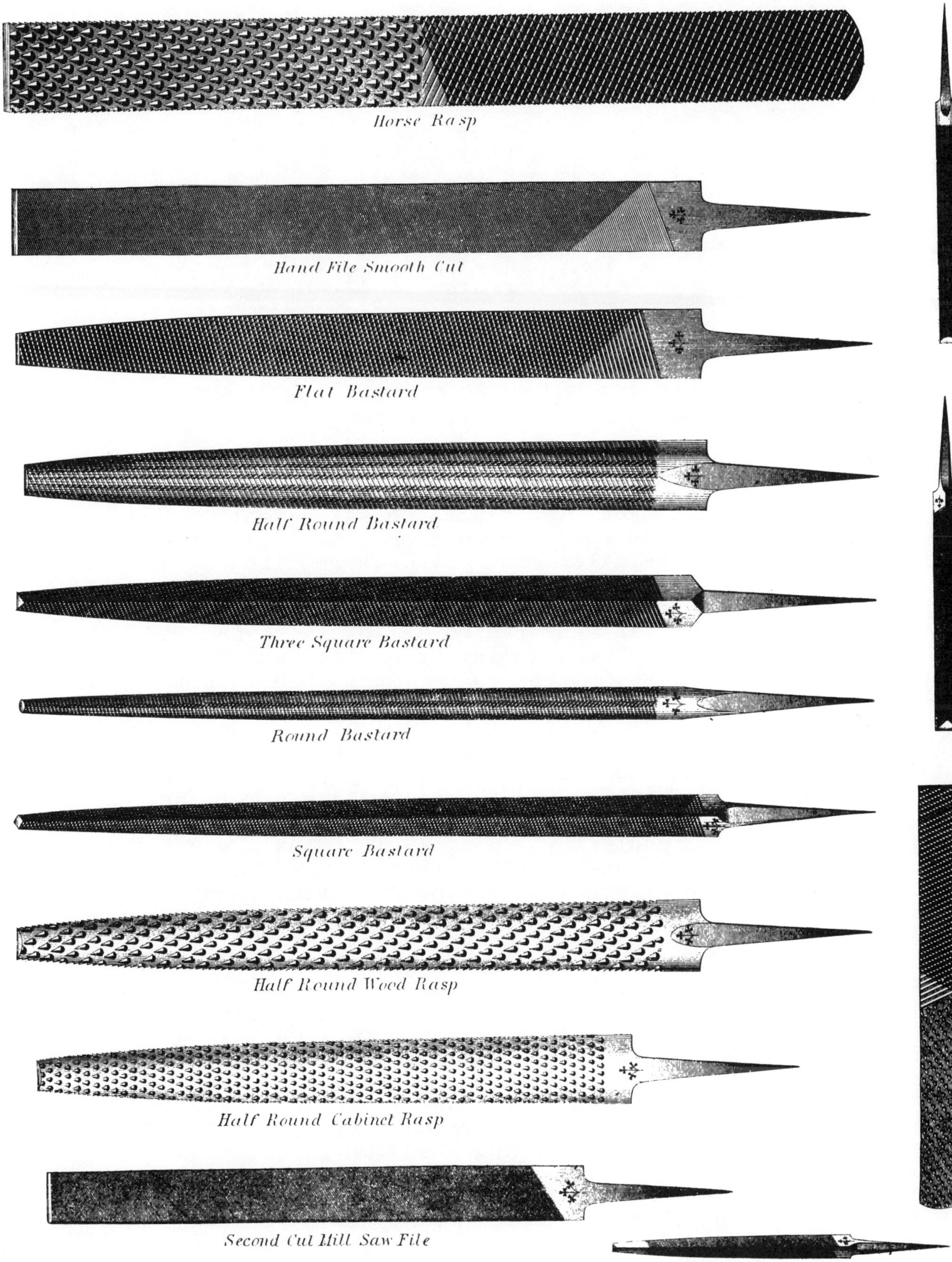

Plate XXVI - Various Cuts of Files - p.112, *Wm. Marples & Sons, Sheffield, 1891 Catalogue*

FILE CUTTERS' SHOP—MR. WILLIAM HALL, ALMA WORKS.

Plate XXVII - FILE CUTTLERS' SHOP - Mr. William Hall, ALMA WORKS
[*Illustrated Guide to Sheffield, 1879,* p.261]

kinds of files. The file is held in its place by means of a leather strap passing over each end, and going round the feet of the workman like stirrups. At each blow of the chisel a tooth is cut, and the blows are repeated in rapid succession until the whole surface of the file is covered, the file being moved as required by loosening the tread on the straps. When one side of the file is finished the workman deals with the other in the same manner; but as the finished part would be liable to injury if unprotected while the second side is cut, a flat piece of metal, being a mixture of lead and tin, is placed under it during the process. It is quite impossible to obtain an adequate notion of the skill of the workmen in cutting the various kinds of files, without an inspection of their labours. Some idea, however, of the dexterity required, may be gathered from the fact that in fine round files as many as from ten to twenty rows of cuts are required to cover the surface with teeth, that there are sometimes more than a hundred teeth within the space of an inch, and that they are cut with the most perfect regularity. Single-cut files are cut only one way--obliquely across the steel. "Double-cuts" have a second row of oblique cuts crossing the first, and forming an immense number of small diamond-shaped teeth. The rasp is cut by striking the pointed chisel more deeply into the steel, and producing a rough, pointed tooth, with a broad, rounded base. The cutting finished, hardness has to be restored to the file; and this is done by heating it well in a smithy, and then dipping it in a cistern of water strongly impregnated with salt, to increase the coldness. The formation of scale which usually follows the heating of steel having to be avoided on account of the teeth, the file is dipped in a tub of thick brewers' grounds before putting it into the fire. Some makers harden in lead, but the mode we have described is generally preferred. After hardening, the files are scoured with sand and water, and then put in lime and water for the night to take out the salt. When taken out of the lime they are dried quickly to prevent rust, oiled and cleaned; the tangs being afterwards dipped in molten lead and cooled slowly to soften them and prevent snapping in use. In the warehouse files are submitted to the final operation of a most searching examination, which not only takes into account size, shape, regularity of cutting and truth in setting, but the still more important point of temper, upon which the value of the file mainly depends. To test the temper of files, a piece of finely-tempered blade steel is carefully drawn over the whole surface of the file by a skilled hand capable of detecting in a moment the least softness. Every file is separately examined, those which prove to be defective being thrown to the waste heap.

Of late years steam hammers have come largely into use for the forging of the larger sizes of files. We give an illustration from the works of Messrs. Lockwood Brothers, of Arundel-street. (Plate XXVIII).

Plate XXVIII
Sketch of File Forging Shop at
Lockwood Brothers, Arundel Street

[Sheffield Illustrated Guide,1879]
p.275

FILE FORGING SHOP.

On the left the workman is forging the red-hot end of a bar of steel under the hammer, his assistant attending to the heating of other bars at the hearth. On the right, work at the hammer has been done, and the file having been cut off the bar and re-heated, the forger and striker are drawing out the tang and completing the forging on the anvil. The grinding of files by machinery has been several times attempted. During a protracted wages dispute some years ago, a file grinding association was formed and erected several machines. The association lasted a very short time, and file grinding in private hands has not been more successful. The work is still done almost entirely by hand. The question of file-cutting machinery is a much more interesting one owing to the far larger number of hands employed in that branch, and the peculiarly delicate character of the work. File-cutting machines were invented many years ago, and companies were formed in Manchester, Birmingham and other towns for working them. The machines were pronounced to be a great success, but the companies one after another failed, and the workmen learned to despise the machinery they had at first feared. The difficulties of cutting files by machinery have, however, been gradually surmounted to a considerable extent, and large quantities of files cut by machines are now in constant use. Many manufacturers use machines, cutting some of the files wholly by the machine; cutting others partially, and completing the work by hand. The superiority of hand-cutting is, however, stoutly maintained, and some of the best makers employ hand labour almost exclusively.

Mr. William Hall, of the Alma Works, Barker-pool, has been extensively known as a large manufacturer of steel, files, edge tools, saws, &c., for nearly a century. The firm had premises originally in Porter-street, but was compelled by the extention of their business to provide a factory on a much larger scale. In erecting the present buildings great attention was paid to sanitary requirements, which in the construction of such works were formerly altogether overlooked.

The forging and other shops, instead of being each a small separate building, as used to be the case, are open to each other for the purpose of ventilation through the whole length of the building, without anything to break the free current of air. The accompanying sketch (Plate XXIX) gives an idea of the extent of the works, and the illustrations of grinding and file cutting accompanying the descriptions of those processes are taken from the manufactory, which embraces furnaces for the conversion and casting of best steel, and long ranges of workshops for carrying on the manufacture of finished goods. The files and other manufactures of Mr. William Hall have had a high reputation at home and abroad for many years, and the present members of the firm are scrupulously careful to preserve their good name by using the best material and employing highly skilled labour. Their manufactures are all hand-forged; their files are hand-cut; and every article is rigidly tested before it is sent out. Mr. Wm. H. Brittain, the present Master Cutler, is the senior partner of the firm.

ALMA WORKS—MR. WILLIAM HALL.

Plate XXIX - ALMA WORKS - Mr. William Hall
[Illustrated Guide to Sheffield, 1879, p.264]

Messrs. Thomas Turner and Co. - The business of Messrs. Thomas Turner and Co., of Suffolk Works, was founded more than seventy years ago by the father of the present senior partner, and includes the manufacture of steel, saws, files and cutlery. The works adjoin the new station of the Midland Railway Company, and, as will be seen from (Plate XXX), include extensive melting furnaces, grinding wheel, and other adjuncts of large and complete manufacturing premises. Like other leading Sheffield houses Messrs. Thomas Turner and Co. have gradually established a wide reputation by the unvarying excellence of their goods. The first essential in the character of Sheffield goods is the quality of the steel employed in making them. Messrs. Turner make their steel from the best Swedish brands of iron, using Dannemora iron for the finer work. They trade chiefly in the home markets, in some of which--South Wales and West of England for instance--their cutlery is so well known and so highly appreciated that it is sold almost exclusively, and is found in nearly every ironmonger's stock. Their files are almost entirely hand-made, machinery being used in the minor operations only, whilst their cutlery is entirely made by hand workmanship. In the hafting of table cutlery with pointed tangs, the handle is drilled by a small borer. Into the hole thus made resin is poured, and the tang is then inserted. It is one of the approbriums of table cutlery that, owing to the carelessness of servants is putting the knives into hot water, the handles become loose, and many precautions are taken against this negligence. In the "Suffolk knife," patented by Messrs. Thomas Turner and Co., the difficulty is, we are assured, effectually overcome by a groove in the handle to which the tang is adjusted.

SUFFOLK WORKS—MESSRS. THOMAS TURNER AND CO.

Plate XXX - SUFFOLK WORKS - MESSRS. THOMAS TURNER and CO.
[*Sheffield Illustrated Guide, 1879*, p.265]

ALBION STEEL WORKS—MESSRS. J. R. SPENCER AND SON

Plate XXXI - ALBION STEEL WORKS - MESSRS. J. R. SPENCER and SON
[*Illustrated Guide to Sheffield, 1879,* p.267]

Messrs. J. R. Spencer and Son, of the Albion Steel Works, Peacoroft, is another old firm doing an extensive business in several of the staple trades we are describing. An illustration of their works will be found on Plate XXXI. The chief manufacturers of the firm are steel,files,cutlery and shovels; they are also general merchants, doing a very extensive business with home and continental customers in all kinds of hardware, as well as in their own manufactures. The business of Messrs. J. R. Spencer and Son is one of the oldest in Sheffield, having been founded in the year 1749. They have had a high reputation for several generations, especially as manufacturers of files. There have been for years, and still are, markets on the Continent in which "Spencer's" is the only file storekeepers will buy or consumers use. The firm has won high honours at several of the great international exhibitions of past years, and at Paris this year they have been awarded a silver medal for the excellence of their files and steel.

CUTLERS' SHOP.

Sketch of Cutler's Shop - Lockwood Brothers - Arundel Street
[Illustrated Guide to Sheffield, 1879, p. 275]

SAWS, &c.

The manufacture of saws is one of the oldest of the staple trades. Steel of the best quality for saws being cast into ingots of proper weight, and reduced by rolling onto sheets, is then cut into the shape required by a paring machine worked by steam power, which cuts steel plates with the ease and rapidity with which silk is cut with ordinary scissors. It is then taken to the toother, who places the sawplate on a bed. A punch is forced down by steam power and cuts out a piece of steel, leaving the edge standing of the exact size and form required for the tooth. The saw is then moved a space, and a similar process is performed until the whole saw is toothed. The movement of the saw under the toothing machine is regulated with great simplicity and accuracy by means of a cog wheel working in a place similarly toothed. It is then marked and filed, after which it is submitted to the hardening process, which is peculiar. The saw is placed in a furnace heated to a high and uniform temperature, after which it is taken out and immersed in oil, resin, &c., which have the property of hardening the saw to the proper degree. This process requires great care and experience; for, heated too highly, the saw becomes brittle, and if the heat does not reach a certain temperature it will be too soft. It is then carefully tempered, after which it is straightened. This is now partly done by machinery, the saw being subjected to pressure between large flat plates, and completed by hammering on the anvil. Great skill and judgment are required on the part of the workmen in reducing the face of the saw into a perfectly level surface. It is now prepared for the grinder, under whose hands it undergoes very elaborate manipulations. (Plate XXXII). The stones at which saws are ground are much larger than those previously described. The grinder, after shaping the back, places the saw on a "scorching" board, and brings it then across the grindstone. Unlike the grinder of cutlery or edge tools, he does not sit at his work, but, placing his knees on one edge of the scorching board and his hands at the other, works

SAW GRINDING.

Plate XXXII - SAW GRINDING [*Illustrated Guide to Sheffield,1879*,p.279]

the saw across the stone from the tooth edge to the back, moving it transversely from heel to point. Having thus ground it on both sides crossway, which is called "scorching," he then grinds it lengthway, which is called "drawing," the grinder bringing the whole weight of his body to bear upon the saw as it moves along the surface of the rapidly-revolving stone. Straight saws are usually ground thinner towards the back. The circular saw in the process of grinding used to be secured to a vertical circular-faced plate, opposite to which is a "lap," which reduced the surface to a smooth face, sliding backwards and forwards, slightly reducing it from the circumference to the centre. This is the old process of saw grinding by hand, of which we give an illustration. We should add that first-class work in saw grinding is now done chiefly by machinery; costly but very admirably contrived machinery for the purpose having been recently put down by leading houses. The next process is hammering, which corrects any irregularity that has arisen from the friction caused by grinding. This is an operation requiring the greatest care, as upon it depends greatly the straight and satisfactory working of the saw. Long saws are then stiffened or tempered by being placed over a coke fire until they assume a straw colour. The saw is next polished, and then rubbed by machinery. If a short saw, it goes to the setter and sharpener, who sets or bends the teeth alternately with a hammer, so as to enable it in actual work to cut with little friction. After setting, the teeth are made sharp by filing, and the saw is ready for the handle or frame, as the case may be.

Messrs. Spear and Jackson. - The firm of Spear and Jackson, of Etna Works, is one of the oldest as well as one of the leading Sheffield firms, having been established more than a hundred years ago in Gibraltar-street. The works have been twice removed, however, to meet the increase of business, and are now situated in Saville-street, between those of Messrs. Charles Cammell and Co. Limited and Messrs. Thomas Firth and Sons. (Plate XXXIII). The firm is engaged in the manufacture of steel and steel goods of every description, including saws, files, edge tools, machine knives, hay and manure forks, spades, shovels, &c. Of these the saw department may be considered the most important, and here have been turned out the largest finished circular saws ever yet made. Hand saws, mill saws, &c., are also made in immense quantities. By the adaption of special machinery, not only is the cost of production reduced, but the articles are more perfectly finished than was possible before; and it is thus that the firm has continued to maintain during the keen competition of prolonged periods of depression the lead acquired years ago. The garden and field tool department, though a comparatively recent addition, has developed to an extra-ordinary extent, and is now one of the most important departments of the business. Many thousands of dozens of steel hay, manure, and digging forks, spades, shovels, hoes, &c., are manufactured annually, whilst improvements in shape, temper, and finish are continually taking place. The most striking improvement, however, comparing the present goods with those of a few years ago, is shown in the steel hayforks, the highly-finished prongs of which are so light that they weigh only a few ounces, and are yet so finely tempered that it is considered almost impossible to break them. Messrs. Spear and Jackson are also large manufacturers of edge tools, files, &c., their mark being well known and in much repute in this country and in colonial and continental markets. Very ingenious machinery has been adapted by the firm to this as to other branches of their trade--machinery which does its work with far greater accuracy and precision than would be

ETNA WORKS.—MESSRS. SPEAR AND JACKSON.

Plate XXXIII- ETNA WORKS - MESSRS. SPEAR & JACKSON
[*Sheffield Illustrated Guide, 1879,* adjacent p.280]

possible by hand labour, and this is one chief cause of the remarkable success they have attained in a branch of manufacture in which it has been erroneously supposed the Americans "licked creation." No better or more handy and beautiful garden and farming tools are made than those of Messrs. Spear and Jackson. Very distinguished honours have been awarded to the firm for the excellence of their goods--the Council Medal of the Great Exhibition of 1851, the Cross of the Legion of Honour at the Gold Medal of Honour at the Paris Exhibition of 1855, the Order of Francis Joseph and the Medal for Progress at the Vienna Exhibition of 1873--decorations not held by any other house engaged in the same branches of the staple industries of the town.

Messrs. Moses Eadon and Sons are large manufacturers of steel, saws and other goods, their manufactory, President Works, being in Saville-street East, adjoining the Midland Railway. The business was established by the late Mr. Moses Eadon. The saws made by the firm have long had a very high reputation, and are largely sold in distant foreign markets as well as at home. The firm also manufacture steel, files, tools, machine knives, &c.

The pre-eminence of Sheffield in the file, edge tool and saw trades was shown by the number of prize medals awarded at the International Exhibition of 1871, and has been repeatedly demonstrated since. The number of eminent firms engaged in these industries is so large that we can only give a selection. Among the principal houses, exclusive of those already mentioned, are:--FILES--Messrs. C. Cammell and Co. Limited, Thos. Firth and Sons, John Brown and Co. Limited, Ibbotson Bros. and Co. Limited, J. Kenyon and Co., John Bedford and Sons, W. K. and C. Peace, Howell and Co., Wm. Spencer and Sons, The

Hallamshire Steel and File Co., Limited, S. S. Brittain and Co., Walter Spencer and Co., Fredk. Brittain, H. Rossell and Co., M. & J. Wing, Leadbeater and Scott, Camm, Bagshaw and Co., Wm. A. Tyzack and Co., J. Beardshaw and Son, Tyzack, Sons and Turner, S. Newbould and Co. Limited, Askham Bros. and Co., J. Baker and Sons, A. Beckett and Sons, J. and Riley Carr, Edgar Allen and Co., W. Jackson and Sons, G. Fisher and Co., Turton Bros. and Matthews, W. Makin and Sons, J. Thompson and Sons, G. Barnsley and Sons, Deakin, Reuss and Co., D. Flather and Sons, T. L. Green and Co., Gregory and Bramall, Hoole, Staniforth and Co., J. and R. Dodge Limited, J. and F. C. Wild, Moss and Gamble Bros., Marsh Bros. and Co., Bromley and Fisher, W. Turner and Son. H. Ecroyd and Co., Dickson Bros. and Co. EDGE AND JOINERS' TOOLS--Messrs. Robt. Sorby and Sons, David Flather and Sons, Marsden Bros., Turner, Naylor and Marples, F. G. Pearson and Co., Moulson Bros., T. Ibbotson and Co., Josh. Cam and Sons, W. Burkinshaw and Sons, F. Harris, A. Jinkinson. SAWS, MACHINE KNIVES, &c.--Messrs. Slack, Sellars and Co., Taylor Bros., Crownshaw, Chapman and Co., Drabble and Sanderson, Colver Bros., Wheatman and Smith Limited, Crookes, Roberts and Co. SHEEP SHEARS--Robt. Sorby and Sons, Wm. Wilkinson and Sons, Burgon and Ball, T. Ibbotson and Co., J. Linley and Sons, &c.

In the list given above we have made an attempt at classification, but as some of the firms manufacture goods in all the classes, and many in two or three, it can only be taken as a rough indication of the leading department.

SPADES, SHOVELS, &c.

Scythes, sickles and reaping knives of various kinds are manufactured in Sheffield and the neighbourhood in large quantities. Among the principal manufacturers in these branches are Messrs. S. and R. Linley, Wm. A. Tyzack and Co., Tyzack, Sons and Turner, Thomas Staniforth and Co. (Hackenthorpe), Hutton and Co. (Ridgeway), Garfit and Sons, R. and J. Linacre (Norton Woodseats), Wm. Fox and Son (Ridgeway).

Among the manufacturers of spades and shovels are Messrs. Skelton, of Heeley; E. Lucas andSons, of Dronfield; Bedford and Sons, Burys andCo. Limited, Spear and Jackson, F. G. Pearson and Co., Morton Brothers, &c.

ENGINEERS' TOOLS, HAMMERS, VICES, &c.

This class of tools is largely manufactured in the district. Amongst the principal makers are Messrs. Ibbotson Bros. and Co. Limited, Edgar Allen and Co., Askham Bros. and Co., Robert Renton, Wm. M. Macbrair, Easterbrook, Allcard and Wild, Crampton Bros., Robert F. Drury, Turner, Naylor and Marples, Senior and Swift, Thomas Linley and Sons, Turton Bros. and Matthews, Benjamin Nicholson, &c.

[end of transcription, p.282 *1879 Illustrated Guide to Sheffield*]

Plate XXIV - Thomas Wright
Highfield, Sheffield

Plate XXXV - Wright & Son
New George Street Sheffield

Advertisements Appearing in the 1828 Sheffield Directory

The diversity of products is noted from the above two engraved advertisements appearing in the 1828 Sheffield Directory. Their listings were as follows:

Wright, John and Son,	manufacturers of saws, edge and joiner tools, steel busks, elastic steel trusses, lancets, skates, table knives, &c. New George Street
Wright, Thomas,	manufacturer of saws, edge tools, steel busks, elastic steel truses, pallette knives, lancet, &c. Highfield

THOMAS TURTON & SONS,

MERCHANTS,

AND

MANUFACTURERS

OF

STEEL, FILES,

EDGE TOOLS, SAWS,

AND

CUTLERY,

LOCOMOTIVE ENGINE AND RAILWAY CARRIAGE

SPRINGS, &C.

SHEAF WORKS,

SPRING WORKS,

AND

UPPER SLACK STEEL WORKS,

SHEFFIELD;

No. 12, St. Benet's Place, Gracechurch-street.

LONDON,

AND AT

18, CLIFF STREET, NEW YORK.

N 63

Plate XXXVI - Advertisement of Thomas Turton & Sons. 1854 Sheffield Directory
[From Collection of the Author]
A brief description of the works appears on page 28, also see Plate XI.

PRIVATE MARK S J WARRANTED OF THE BEST QUALITY

SPEAR & JACKSON,

ÆTNA WORKS, SHEFFIELD;

No. 14, FISH STREET HILL, LONDON;
No. 20, RUE D'ANJOU, PARIS;
No. 32, CLIFF STREET, NEW YORK.

STEEL CONVERTERS AND REFINERS,

MANUFACTURERS OF

SAWS, FILES,

EDGE TOOLS,

LEDGER BLADES, CURRIERS' KNIVES,

MACHINE KNIVES, &c.

Great Exhibition of all Nations.—1851.

The only Great or Council Medal Of the Exhibition,

For Class 21, Awarded to Spear & Jackson, Sheffield.

EXTRACT FROM THE JURORS' REPORT.

"With respect to the United Kingdom, we find that articles in the Class of Cutlery and Edge Tools have been sent from a great variety of places. In England, from London, Sheffield, Birmingham, Warrington, Stourbridge, and a few other towns of less note; from Glasgow and Edinburgh, but chiefly from the former, in Scotland; and from Cork, Clonmel, and Limerick, in Ireland. Among these seats of the manufacture, there is none, as might naturally be expected, which for extent, variety, and excellence of collection, can compare with SHEFFIELD—its most ancient home. We here find every article, from the most exquisite Razor down to the plainest Pocket Knife, and from the finest Saw or File, to the most ordinary Chisel, displayed with various degrees of merit it is true, but with a large proportion of the highest.

"*From this collection, the Jury have thought themselves justified in awarding for one remarkable object, a Council Medal.* Messrs. SPEAR & JACKSON *have Exhibited, among an assortment of Edge Tools of great excellence, a Cast Steel Circular Saw, of the large size of five feet diameter, and of such signal beauty and perfection that it stands far above comparison with any other in the building. The mere excellence of its quality and workmanship, however, would not, the Jury are aware, have enabled them to distinguish by a Council Medal, if they had not been able to satisfy themselves that its merits is the result of a new and peculiar process of manufacture. But they entertain no doubt, from the information they have received, that mechanical ingenuity of a novel and special character has been employed by these manufacturers for the production of such articles, without which they could not be carried to equal perfection; and they therefore consider them justly entitled to the highest mark of distinction.*"

N 37

Plate XXXVII - Advertisement of SPEAR & JACKSON in 1854 Sheffield Directory
[Note detailed description concerning their Award at 1851 Great Exhibition]
Also see Plates XXXVIII and XXXIX [Author's Collection]

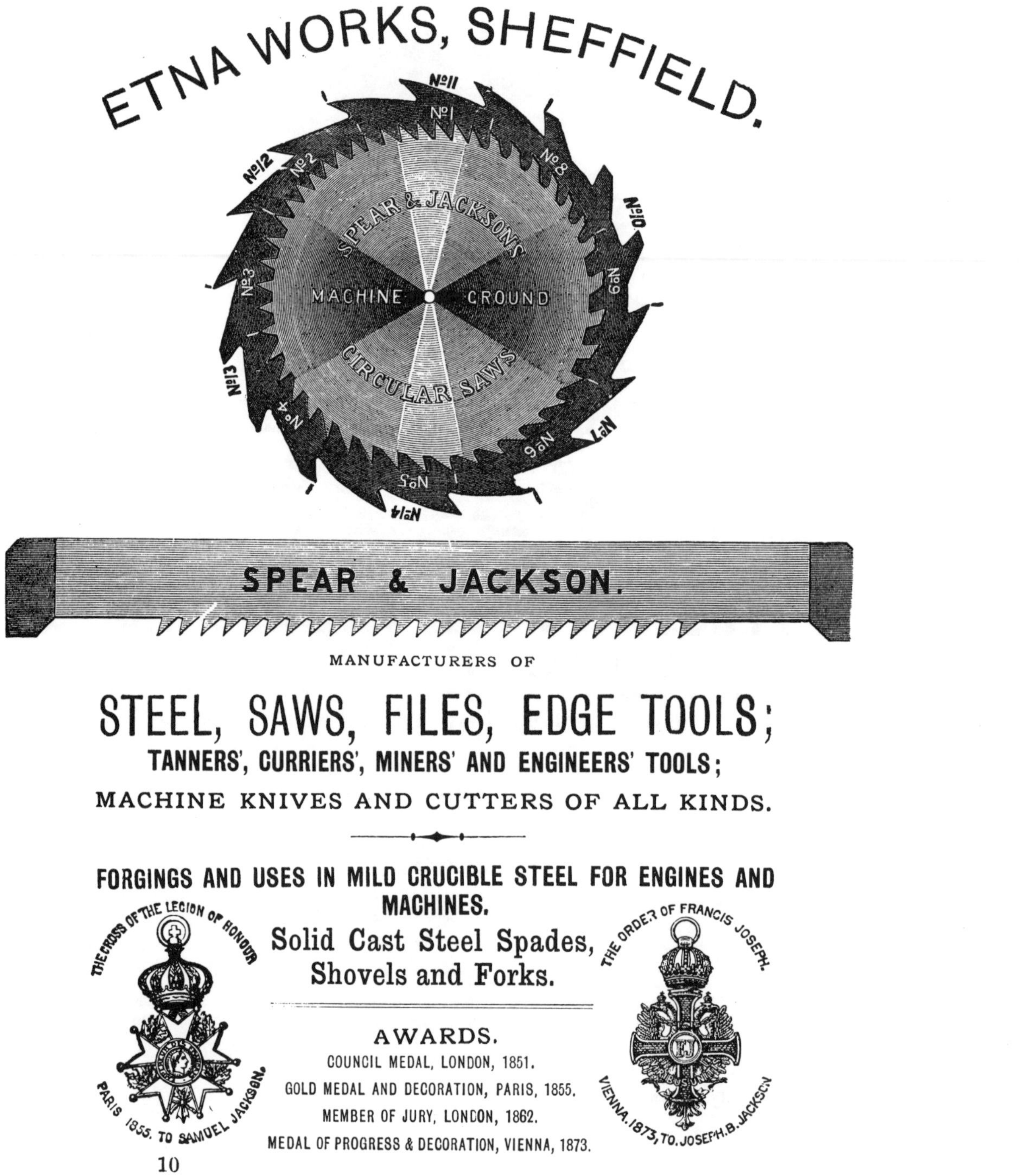

Plate XXXVIII - 1879 Spear & Jackson Advertisement from *Sheffield Illustrated Guide*

A brief description of the Aetna Works of Spear & Jackson appears on page 54 and a sketch of this works Plate XXXIII, page 55. Research from the Sheffield Directories first notes the establishment of this company in 1825 by the sawmaker, John Spear, and filemaker Nicholas Jackson. John Spear is first listed in a Sheffield Directory in 1817, while Nicholas Jackson is noted as a file maker at the Wicker in the first Sheffield Directory of 1774. He moved to Shemold Croft in 1797.

SPEAR & JACKSON,

ÆTNA WORKS, SHEFFIELD,

MANUFACTURERS OF

STEEL, SAWS, FILES, EDGE TOOLS,

TANNERS' AND CURRIERS' TOOLS,

MACHINE KNIVES OF ALL KINDS,

ELASTIC CAST STEEL HAY, MANURE, AND DIGGING FORKS,

STEEL BREASTS FOR PLOUGHS, &c.

PATENTEES of the SOLID CAST STEEL SPADES and SHOVELS which received the Special Prize at the Royal Agricultural Society's Meeting, at Canterbury, and for which First Class Prizes have been awarded at every other Show where they have been Exhibited.

44

Plate XXXIX - Advertisement from 1864 Sheffield Directory - [Note Plate XXXVII]

The beginning of the firm is traced to John Love, who founded a saw business in 1760 at Sheffield. This is noted on page 55 of *Story of Saw* by P. d'A. Jones and E.N. Simons [London, 1960] published to mark the second centenary of the world's largest sawmakers. Saws are still manufactured at their Sheffield factories along with lines of garden and mechanic's tools, formerly respectively made by C.T. Skelton & Co. and John Bedford & Sons, acquired by Spear & Jackon during this century.

IBBOTSON BROTHERS & CO.
Sheffield.
STEEL CONVERTERS AND REFINERS
AND MANUFACTURERS OF THE JUSTLY CELEBRATED
"GLOBE" FILES,
SAWS, EDGE TOOLS,
CUTLERY, &c.
AND
GENERAL COMMISSION MERCHANTS.

Plate XL - Ibbotson Brothers Advertisement, 1854 Sheffield Directory
[Author's Collection]

JOSEPH PEACE & CO.,
SHEFFIELD.

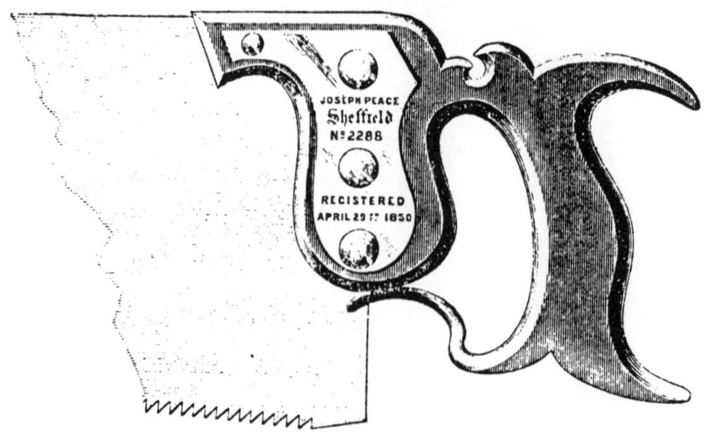

MANUFACTURERS OF
STEEL, SAWS, MACHINE KNIVES, SPRINGS, &c

Plate XLI - Joseph Peace & Co. Advertisement - 1876 Sheffield Directory

TRADE MARK.

W. K. & C. PEACE,

EAGLE WORKS, SHEFFIELD,

Manufacturers of

STEEL & FILES, SAWS, EDGE TOOLS, &C.

TRADE MARK.

SOLE MAKERS OF THE GOODS MARKED

"IBBOTSON, PEACE & CO.,"
"R. IBBOTSON," AND
"J. PATTERSON."

CORPORATE MARK GRANTED 1710.
P

LONDON—O. LOFTHOUSE, 61, GRACECHURCH STREET, E.C.

PEACE BROTHERS,

CORPORATE MARK—3593, GRANTED 1813.

CORPORATE MARK
P
WARRANTED QUALITY OF THE BEST

GRANTED 1770.

STEEL CONVERTERS AND REFINERS,

MANUFACTURERS OF

IMPROVED CAST STEEL

OF EVERY DESCRIPTION,

ENGINEERING & MACHINE FILES OF ALL KINDS, SAWS, EDGE TOOLS,

ETC., ETC.

GARDEN STREET WORKS, SHEFFIELD.

55

Plate XLII & XLIII - Advertisments of W.K. & C. Peace and Peace Brothers
1870 Sheffield Directory

SLACK, SELLARS, & Co.,

CORPORATE MARK

SPEED.

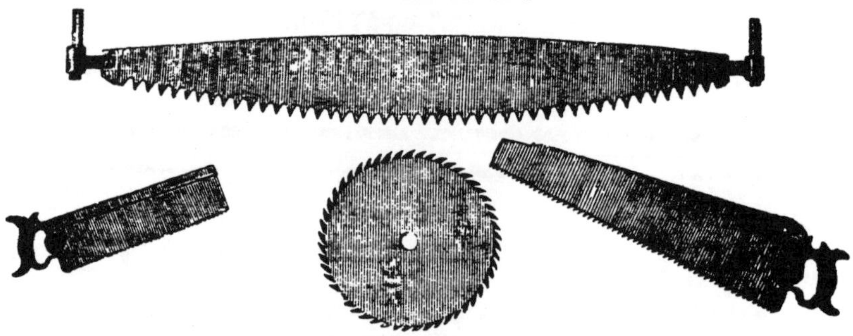

MERCHANTS,

MANUFACTURERS OF IMPROVED

SAWS, FILES, EDGE TOOLS,

LEDGER BLADES, SPIRAL CUTTERS, SPRING BEDS,

And every Description of

SAWS

TO WORK BY POWER,

☞ **STEEL CONVERTERS & REFINERS.**

N.B.—MACHINE KNIVES OF ALL KINDS.

TOWNHEAD STREET, SHEFFIELD.

S. S. & Co. were honoured with Prize Medals of the Exhibition of 1851.

Plate XLIV - Slack Sellars & Co. Advertisement, 1860 Sheffield Directory

1883.] SHEFFIELD ADVERTISEMENTS. 63

WM. JACKSON & CO.

MANUFACTURERS OF

ALL KINDS OF

CAST

Steel,

Files,

TRADE MARK

MARKS:—

"WM. JACKSON & SONS,"

"R. VERNON & SONS,"

Tools,

Saws,

AND

CUTLERY.

SHEAF ISLAND WORKS,

SHEFFIELD.

Plate XLV - Wm. Jackson & Co. - Advertisement in 1883 Sheffield Directory

WING'S FILES.

M. & J. WING,

MANUFACTURERS OF

STEEL, FILES AND RASPS,

JERICHO WORKS,

SHEFFIELD.

Plate XLVI - M. & J. Wing - Advertisement in 1878 Birmingham Directory

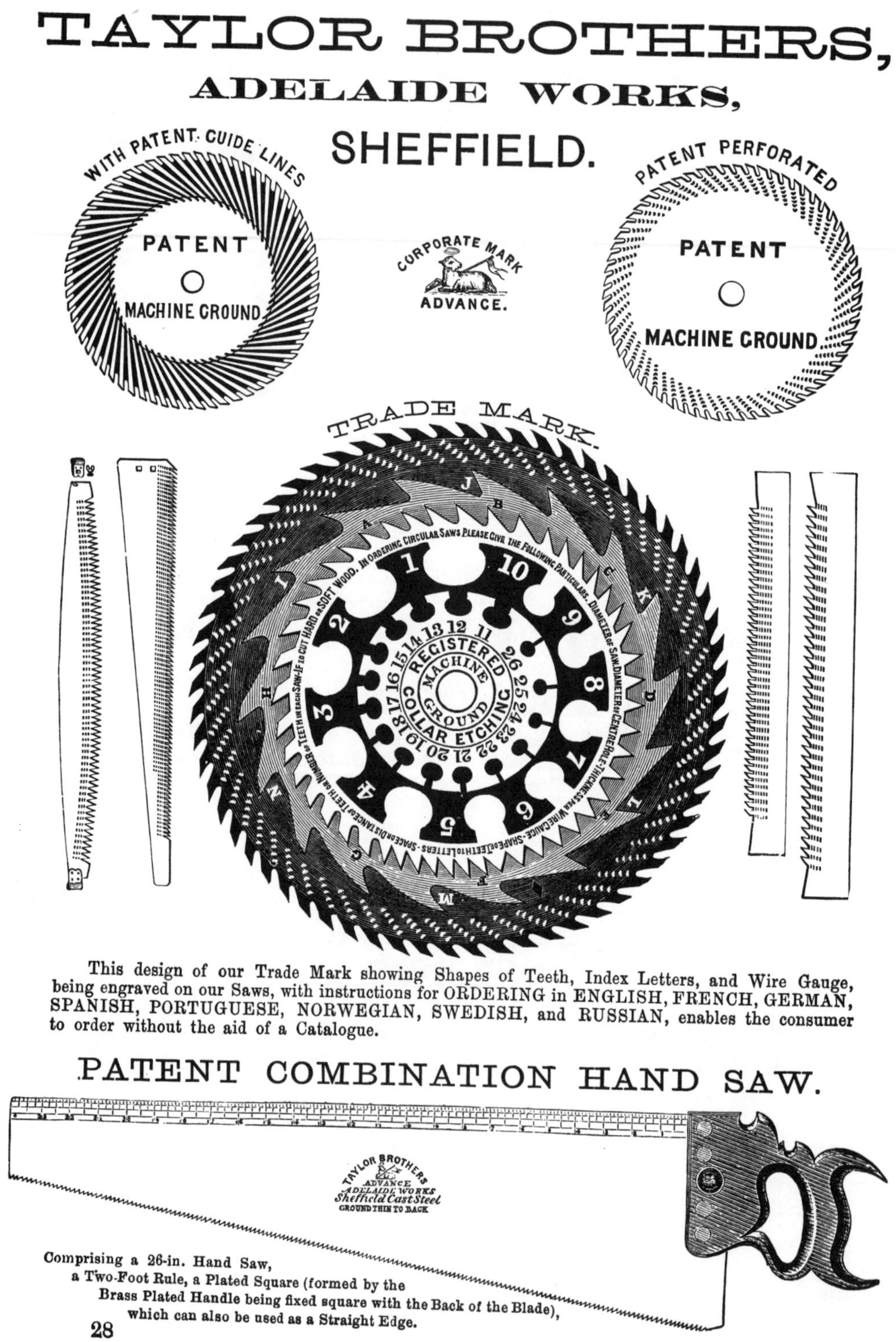

Plate XLVII - Taylor Brothers - Advertisement Section of the 1879 Sheffield Illustrated Guide - page 28

[See Note 22,page 80 regarding Perforation Patent]

PATENT PERFORATED CIRCULAR SAWS.

TAYLOR BROTHERS,

Manufacturers of Saws,

VIENNA EXHIBITION, 1873.

MEDAL OF MERIT.

STEEL, FILES,

EDGE TOOLS,

PLANING AND MOULDING IRONS,

SAW BUCKLES,

MACHINE CHAFF KNIVES,

SUGAR CANE KNIVES,

CALABOZO MATCHETS, &c.

THE HIGHEST PRIZE AWARDED

FOR CIRCULAR SAWS.

Makers to Her Majesty's and Foreign Governments.

PERFORATED SAWS.—Advantages of the same:—

(See drawing on other side.)

Great SAVING of Files. Effectually PREVENTS EXPANSION. Half the amount of FILING SAVED. Liability to FRACTURE done away with. Keeps the Teeth of UNIFORM SIZE. Serves as a GUIDE to the FILER. Perforations CUT OUT bulk of what is generally FILED away. Frequent RE-GULLETTING saved. Great saving of steam power, &c., &c.

PATENT GUIDE LINED CIRCULAR SAWS.

The lines etched on these Saws serve as guide to the filer. *(See tracing on other side.)*

GROUND BY PATENT MACHINERY.

All Saws Toothed, Ground, Glazed and Polished by the most recent improved Patent Machinery.

All our BETTER QUALITIES of long and short Saws are ground thinner towards the back ot the Saw, and being perfectly flat, they work with LESS SET, LESS FRICTION, and WASTE LESS TIMBER. VISIBLE TEMPER.

BRANDS.—"*Taylor Brothers,*" "*J. Taylor and Son,*" "*Joseph Taylor,*" "*J. & I. Taylor,*" "*Adelaide Saw Works,*" "*Melvin,*" "*Adelaide,*" "*Advance,*" *and* "*Smithson.*"

ILLUSTRATED CATALOGUES AND PRICE LISTS FREE ON APPLICATION; ALSO, ORIGINAL TESTIMONIALS AND REFERENCES IF DESIRED.

ADELAIDE WORKS,

SHEFFIELD.

29

Plate XLVIII - Taylor Brothers - Advertisement Section of 1879 Sheffield Illustrated Guide - page 29

ESTABLISHED

IN

1770.

RICHARD GROVES & SONS,

Bee Hive Works, SHEFFIELD,

MANUFACTURERS OF

PATENT MACHINE-GROUND CIRCULAR

SAWS,

Steel of every description;

SAWS, FILES, EDGE & JOINERS' TOOLS,

Machine and Chaff Cutting Knives, &c.

CORPORATE MARK

GRANTED BY THE CUTLERS' COMPANY — USE — AND REGISTERED IN ENGLAND AND GERMANY.

19

Plate XLIX - Richard Groves & Sons - Advertisement from 1879 Sheffield Directory

Plate L - Moses Eadon & Sons - Advertisement in White's Hardware Trade Marks
[Edited by Herbert Hughes, 1st Edition, 1892, Sheffield]

HOWELL & CO.,

BROOK WORKS, BROOK HILL,

SHEFFIELD.

ONLY MEDAL AWARDED
IN THE EXHIBITION OF 1862

FOR EXCELLENT QUALITY OF
CAST STEEL
IN BARS, SHEETS & ROLLS.

NOTICE.

In consequence of the decision of the High Court of Chancery in the recent cause of "Howell *v.* Lloyd," wherein it is declared that the word HOMOGENEOUS, heretofore used by me as my private Trade Mark, cannot be legally held as a private trade mark; I hereby give public notice that in future all homogeneous metal of my manufacture will in addition to the word HOMOGENEOUS or HOMO., bear upon it the following mark, viz., X·TRÉ; and that this mark will henceforward be used by me as my private Trade Mark.

(Signed) J. B. HOWELL.

TRADE MARK.
X·TRÉ
HOMO
FOR HOMOGENEOUS METAL.

TRADE MARK FOR FILES.

MANUFACTURERS OF ALL KINDS OF

Blister, Shear, Spring and Cast Steel,

FILES, SAWS, &c.

HOWELL'S SPECIAL STEEL, for Turning, Planing, and Slotting Tools, enables the Speed of the Lathe to be Increased from 25 Feet to 70 Feet per Minute.

HOWELL'S HOMOGENEOUS METAL, for Boiler and Fire-box Plates, and other purposes, combines the Toughness of Copper with the Strength of Steel.

HOWELL'S PATENT HOMOGENEOUS METAL and CAST STEEL TUBES, for Locomotive or other Boilers.

CAST STEEL ROLLS,

For Rolling Precious Metals, all Sizes, and warranted Glass Hard.

15

Plate LI - Howell & Co. - Page 15 Advertising Section, 1879 Sheffield Illustrated Guide - Notice regarding Trade Mark

Plate LII - Sanderson Brothers & Co. - Page 8 Advertsing Section
1879 Sheffield Illustrated Guide

Plate LIII - Crownshaw Chapman & Co. - Page 52 1879 Illustrated Guide to Sheffield

Plate LIV - Joseph Ashworth & Co. - Page 55 1879 Illustrated Guide to Sheffield

S. S. BRITTAIN & CO.
SAW MANUFACTURERS,
AND
STEEL CONVERTERS AND REFINERS,
ST. GEORGE'S WORKS,
SHEFFIELD.

S. S. BRITTAIN & CO., being fully aware how important it is to Merchants and Shopkeepers to be enabled to supply their Customers with SAWS of first-rate quality, have determined to manufacture an Article of such materials and workmanship as cannot be surpassed by any House in the Trade.

S. S. B. & CO. have their Saws hardened and tempered on a new and improved principle (known only to themselves), which combines strength with flexibility.

N. B.—FRAME, PIT, CROSS-CUT AND MILL SAWS, SEGMENTS, MACHINE BLADES, AND PAPER KNIVES.

Circular Saws, Patent Turned, made to any Size, Strength, Tooth or Axle.

Also Patent Black CIRCULAR SAWS, strongly recommended for use and durability.

BLISTER, SHEAR, CAST, SPRING, AND ALL OTHER DESCRIPTIONS OF

STEEL.

S. S. B. & CO. are the Manufacturers of the Superior Saws, marked "S. SWANN & CO." principally for the export trade.

Plate LV - S.S.Brittain & Co. - Advertisement from 1841 Sheffield Directory

WILLIAM SPENCER & SON,

TRADE SPENCER SHEFFIELD MARK.

MANUFACTURERS OF

STEEL AND FILES,

AND

GENERAL MERCHANTS,

KINGSTON STEEL WORKS,

MALINDA STREET, SHEFFIELD.

Plate LVI - William Spencer & Son - Advertisement from 1876 Sheffield Directory

J. R. SPENCER & SON,

MERCHANTS,

AND

MANUFACTURERS OF

STEEL, FILES,

Silver Medal

Paris, 1878.

EDGE TOOLS.

SAWS,

Silver Medal

Paris, 1878.

TABLE KNIVES, RAZORS,

SHOVELS,

STEEL WIRE, &c., &c.,

OF EVERY DESCRIPTION,

ALBION STEEL WORKS,

SHEFFIELD.

CAUTION—NONE ARE GENUINE UNLESS STAMPED WITH THE ORIGINAL CORPORATE MARK GRANTED 1749.

N SPENCER
SHEFFIELD

23

Plate LVII - J.R. Spencer & Son - Page 23 Advertising Section 1879 Illustrated Guide to Sheffield

LOCKWOOD BROTHERS,

(JOHN SORBY & SONS,
SPITAL HILL,)

74, Arundel Street, SHEFFIELD,

MANUFACTURERS OF

FILES, STEEL, CUTLERY,

EDGE TOOLS,

SHEEP SHEARS AND SAWS.

Sole Manufacturers of the

CUTLERY.

EDGE TOOLS & SAWS MARKED } I. & H. SORBY.

SHEEP SHEARS MARKED } I&H SOR BY

FILES MARKED } LOCKWOOD BROTHERS C✝X SHEFFIELD OR C✝X.LOCKWOOD.

CUTLERY MARKED } C ✝ X LOCKWOOD BROTHERS, SHEFFIELD.

ALSO,

"PAMPA" BUTCHERS' KNIVES.

25

Plate LVIII - Lockwood Brothers - Page 25 Advertising Section of 1879 Illustrated Guide to Sheffield - History of this Firm noted in Section on Edge Tools

CHARLES CAMMELL & CO.,

MANUFACTURERS OF THE PATENT

MACHINE FILES,

ON SCIENTIFIC AND PECULIAR PRINCIPLES,

PRODUCING THEIR

IMPROVED AND CELEBRATED QUALITY,

LONG KNOWN TO THE

ENGINEERING AND RAILWAY WORLD.

INVENTORS OF THE

NOW UNIVERSALLY ADOPTED

CURVILINEAR TANGED FILES,

Registered No. 665.

SOLE MANUFACTURERS OF THE

CONTINUOUS TOOTH CONCAVE AND CONVEX FILES,

FOR WHICH

The Medal of the Scottish Society of Arts & Manufactures

WAS AWARDED, REGISTERED No. 1069; ALSO

MEDALS OF THE GREAT EXHIBITION, 1851.

C. C. & Co. beg to observe, that in the Hardening and Tempering of FILES, they have adopted such decidedly superior principles as to retain and fix in the Steel the greatest possible amount of Carbon, ascertained by considerable research and numerous experiments in Chemical Science, aided by those safe and invaluable guides—long practice and great experience.

IN THE SEVERAL DEPARTMENTS OF THEIR VARIOUS MANUFACTURES

THE CYCLOPS STEEL WORKS STAND UNRIVALLED.

Plate LIX - Charles Cammell & Co. - Advertsiment from 1854 Sheffield Directory

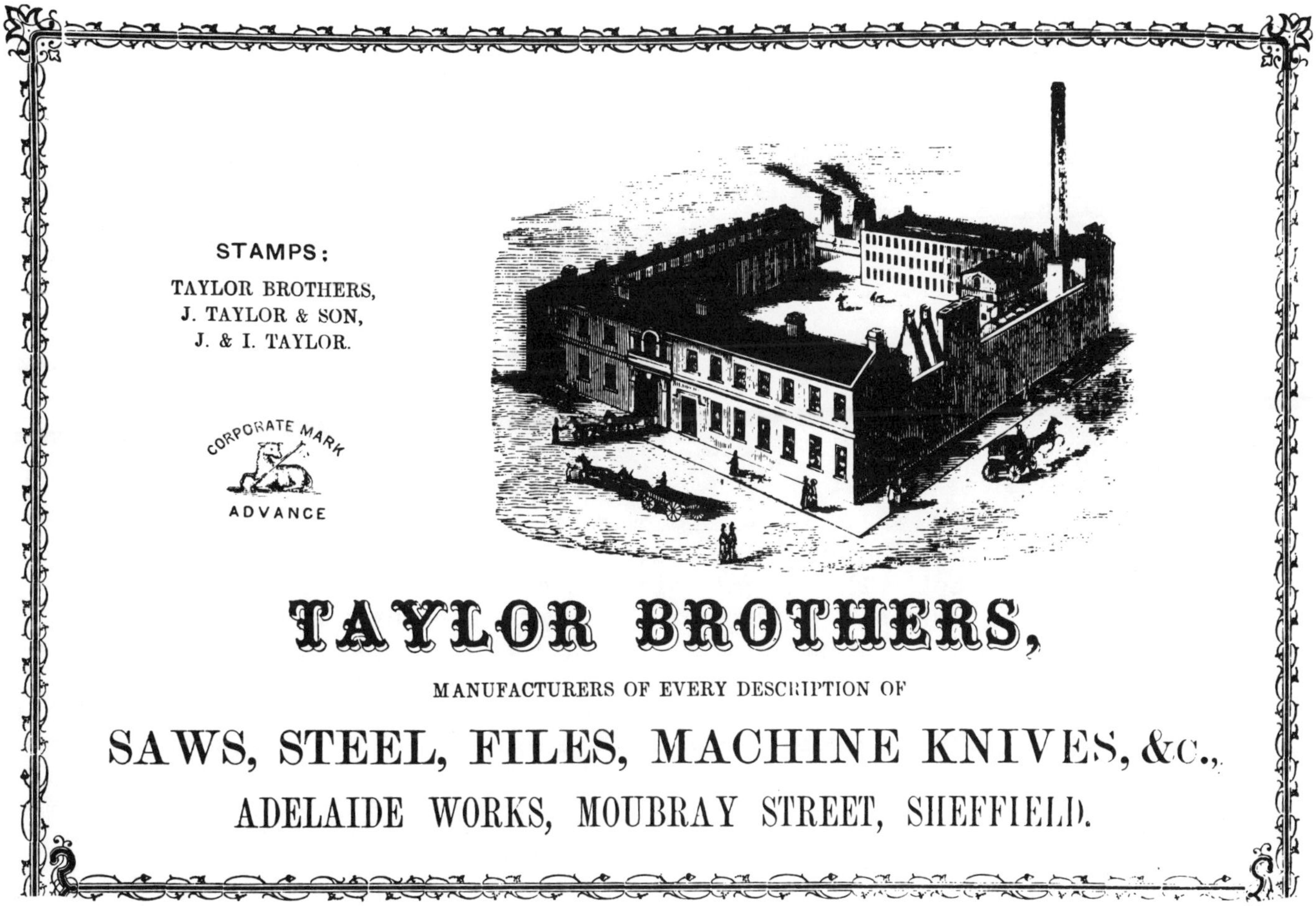

Plate LX - Taylor Brothers Adelaide Works, Sheffield - Advertisement from *A Look at Modern Things,* [Birmingham, 1858], Vol. II, p.158
Photograph Courtesy of the Local History Department, Birmingham Reference Library

A typical arrangement of a tool factory during the latter half of the 19th Century in England is seen in Plate LX of the Adelaide Works of Taylor Brothers, saw and file manufacturers of Sheffield, 1849 - 1900-. The front entrance, usually with a seal or trade mark of the works over the cornice, led to an inside courtyard, surrounded on all sides by buildings or walls. In the above illustration the two conical furnaces along the right wall were converters for making blister steel. The adjacent tall stacks were for batteries of crucible steel melting furnaces. The tall stack was for the steam engine. The offices were in the front buildings. Manufacturing was done in the left wing and rear buildings. The stacks at the rear left corner may have been for forges.

The similarities in factory layouts may be seen from the sketches of the Alma Works (plate XXIX, p.50); the Suffolk Works (Plate XXX, p.51) and the Albion Steel Works. (Plate XXXI, p.52)

Analysis of the Check List of Sheffield Saw Makers & Dealers presented in Appendix I reveals that approximately 398 firms made saws during this period. The durations of these enterprises are compiled in Table IV.

Table IV - Duration of 19th Century Sheffield Sawmakers

Years in Continuous Operation	Number of Firms	Percent of Total
Less than 5	172	43.2
5 but less than 10	77	19.3
10 " " " 20	62	15.6
20 " " " 30	28	7.0
30 " " " 40	20	5.0
40 " " " 50	14	3.5
50 " " " 60	7	1.8
60 " " " 70	10	2.5
Over 70	8	2.0

Twenty five firms or 6.3% were in business for more than 50 years, two of which, Spear & Jackson and Robert Sorby and Sons, still produce saws at this date. The eight firms in business for more than 70 years making saws in Sheffield during the 19th Century were: Ibbotson Brothers, 72 years; Jonathan Beardshaw & Son, Drabble,Sanderson & Co., and Mathius Spencer & Son, 75 years; John C. Kenyon & Co., 77 years; and Richard Groves & Sons; Samuel Newbould & Son and Spear & Jackson, 86 years. Apparently it was a stable business with a steady market. After serving an apprenticeship, not requiring a large outlay of capital, it was a relatively easy business to enter. There were numerous suppliers of steel and other necessary parts, such as saw handles and screws for attaching the blades. Many sawmakers emigrated to America, establishing saw enterprises in United States, especially to New York State.[21] The Keystone Works in Philadelphia, perhaps the most imporatnt saw firm in United States during the 19th Century, was established by Henry Disston, who was born in England.

DEAKIN, REUSS & CO.

TIGER WORKS,

SHEFFIELD,

TIGER — REGISTERED TRADE MARKS — TIGER

MANUFACTURERS OF

STEEL, FILES, TOOLS, SAWS,

SHEEP SHEARS,

AND ALL DESCRIPTIONS OF

TABLE AND POCKET CUTLERY.

Merchants for all kinds of SHEFFIELD and BIRMINGHAM Goods, and General Export Agents.

Advertising Section 1879 Sheffield Illustrated Guide - Deakin, Reuss & Co.

Fig. 4 - A Group of Sheffield Manufactured Hand Saws. [22]

1. John Brown moved from Furival Street to Savile Street where Atlas works were set up in 1857. See *Sheffield Steel*, p.60, Fig. 67.
2. Manganese functions as a de-oxidizer, having more affinity for oxygen than iron. It is also beneficial in tying up sulfur as a compound inclusion.
3. The term 'crystallized' in this instance is a misnomer. In the solid state iron is actually crstalline at all times. The gradient of carbon from the outside the inside during converting varied from 1.1% to below 0.5%.
4. This account has not been factually documented, but has become legendary. See *Sheffield Steel*, p.13.
5. Tungsten was formerly called 'Wolfram' in honor of its discoverer. This element alloyed with steel promotes hardness at high temperatures; i.e. high speed steel.
6. In this instance manganese acts as a deoxidizer,manganese oxide rising to the slag. See note No. 2 above.
7. In this instance 'temper' refers to carbon content. (see p.9, Plate I)
8. Siemens-Martin refers to the Open Hearth Process. Neither this or the Bessemer steel making or section (4) Steel Founding will be considered further in this study since they do not relate to tool steel.
9. The first successful production of crucible steel made in America was by the brothers William and John Garrard, natives of England, at Cincinnati in 1832. See Douglas A. Fisher,*The Epic of Steel*,[New York,1963,p.112]. It is claimed in the 1917 *Disston Lumberman Handbook* [p.8] that Henry Disston operated the first successful crucible steel melting plant in United States making saw steel in 1855.
10. According to James W. Swank,*History of Iron Manufacture in all Ages* [Philadelphia, 1892] the "best grades" of crucible steel were not produced in America until the Civil War period.[p.393]. In the 1892 Stanley Rule & Level Co. Catalogue, revised in 1897, it was noted that their spokeshave irons were made of "the best English steel".
11. See Table III, p.42.
12. The Sheaf Works was built in the early 1820's by Wm. Greaves & Sons and was the first integral steel mill at Sheffield where raw material was processed into finished goods. See Next Chapter concerning edge tools for further history.
13. Mushet alloy steels developed at Samuel Osborn & Co. included "RMS", tungsten bearing for drills and lathe tools which were the basis of high speed steel. See,*Pioneers for a Century,1852-1952*,[Sheffield,1952](See p.29,this text)
14. A.C.Marshall & H.Newbould,*The History of the Firths*,[Sheffield, 1924], p.1.
15. E.Surrey Dane,*Peter Stubs and the Hand Tool Industry*,[Altrincham,1973], p.181.
16. L.C.Rolt,*A Short History of Machine Tools*,[Cambridge,Mass.,1965],p.88.
17. C.Holtzappfel,*Turning and Mechanical Manipulation*,Vol.II,[London,1856],p.817-703.
18. J.Richards,*A Treatise on Construction of Wood Working Machinery*,[London,1872],p.3.
19. C.Holtzappfel, *op. cit.*,p.669 & 684
20. *Visit to Sheffield Steel Works*, [Sheffield,1913], p.3
21. K.D. and Jane W.Roberts, *Planemakers and Other Edge Tool Enterprises in New York State in the Nineteenth Century*, [Cooperstown,N.Y.,1971], p.58 & p.116.
22. The perforations on the top rip saw, made by Taylor Brothers, was their patent for guides in refiling a new set of teeth. This was By Her Majesty's Royal Letters Patent, No. 2579, under the Great Seal of Great Britain, noted on Page 142 Birmingham Directory 1872, which gave the following details:

THIS INVENTION stands prominent among the Improvements of the present age, and we therefore beg to call the **SPECIAL ATTENTION** of all persons interested to **A FEW FACTS** connected with the use of these Saws.

Our theory (confirmed by practice) is, that PERFORATING THE SAW AT THE TEETH not only SAVES MORE THAN HALF THE AMOUNT OF FILING, but also EFFECTUALLY PREVENTS THE EXPANSION of the SAW (caused by friction), partly by the air rushing through the Perforations, and so cooling it, and partly by the Perforations carrying the sawdust out of the cut, thus doing away with the cause of friction; the Perforations also become receptacles for grease for lubricating the Plate.

The Perforations save frequent regulletting, and serve as a guide to the filer, enabling him to KEEP THE TEETH OF UNIFORM SIZE.

The LIABILITY TO FRACTURE at the roots of the Teeth is also DONE AWAY WITH, the Perforation answering the same purpose as a hole drilled at the end of a crack, which prevents it extending; and enables a better and more uniform temper to be given to the teeth.

There is also A GREAT SAVING OF FILES IN SHARPENING, as the Perforations CUT OUT the bulk of what is generally *filed* away.

Where the Perforated Saws are in use they are HIGHLY APPROVED. They combine UTILITY with BEAUTY OF APPEARANCE, and will, sooner or later, force themselves into the first place in the market.

Appendix I - CHECK LIST OF SHEFFIELD SAW MAKERS & DEALERS DURING 19th CENTURY

Ashforth, Holman & Co.		-1856-
Ashforth, Joseph & Co.	Bruce Works	-1860 - 1884-
Ashton, Abraham		-1896 - 1900-
Atkin, Henry & Co.	Celtic Works	-1865 - 1872-
Atkin, Henry & Handcock		-1879 - 1884-
Atkin, William C. & Co.	Reliance Steel Works	-1879 - 1900-
Atkin & Peace		-1856-
Atkinson Brothers	Milton Works	-1879 - 1884-
Bagshaw, J.J. & Co.	Thames Steel Works	-1865 - 1872-
Barker, Mitchell & England	Havelock Works	-1879-
Barber, Genn & Co.		-1797 - 1817-
Barker & Lord		-1825-
Barnes, F. & Co.	Corn Hill Works	-1865 - 1900-
Barnsley, George & Sons	Cornish Works	-1872 - 1900-
Beardshaw, George		-1839-
*Beardshaw, Jonathan & Son	Baltic Steel Works	-1825 - 1900-
Beckett, Alfred	Brooklyn Works	-1845 - 1865-
Beckett, Alfred & Sons	Brooklyn Works	-1868 - 1900-
Bedford, John		-1849 - 1856-
Bedford, John & Sons	Lion Works	-1879 - 1900-
Bedford, Bury & Co.		-1856 - 1860-
Beet, Jonathan & Sons		-1839 - 1849-
Beet, Jonathan Sons & Griffith		-1852 - 1854-
Biggin, S. & Sons		-1845 - 1860-
Blackford, Wm.		-1845 - 1854-
Blake, Joseph & Son		-1814 - 1817-
Blake, Thomas & Son		-1821 - 1828-
Blake & Parkin		-1845 - 1854-
Blake & Thomsom		-1837 - 1845-
Boothby, Groves & Co.		-1814 - 1823-
Bramall, Edward		-1849-
Bramall, John		-1860-
Bramall & Bedford		-1868-
Brittain, Frederick	St. George's Works	-1865 - 1884-
*Brittain, Samuel Swann & Co.	St. George's Works	-1841 - 1900-
Brittain, Hoole & Co.		-1837 - 1839-
Broadhurst, William		-1839 - 1841-
Broadhurst, William & Son		-1845 - 1849-
Brookes, William & Sons		-1860 - 1865-
Brookes Brothers	Brunswick Works	-1865-
Brooksbank, Abraham	Malinda Works	-1849 - 1900-
Brown, Charles		-1854 - 1865-
Brown, Henry & Sons (D)		-1879 - 1884-
Brown, William		-1841 - 1852-
Brumby & Middleton		-1841 - 1860-
Burgon, Green & Co.		-1817 - 1828-
Burys & Co.	Regent Works	-1865 - 1900-
Butcher, William & Samuel		-1837 - 1865-
Butcher W. & S. (D)		-1896 - 1900-
Butcher, Brown & Butcher		-1825-
Buxton, John		-1837 - 1860-
Cadman, Peter & Sons		-1839 - 1845-
Cam, James		-1823 - 1825-
Cam, James (Marsh Brothers D)		-1896 - 1900-
Carnall, George & Co.		-1818-
Carr, George		-1797 - 1825-

Carr, George & Co.		-1837 - 1841-
Carr, William		-1845 - 1856-
*Carr, John & Riley		-1837 - 1900-
Carr, Woodhouse & Carr		-1817 - 1828-
Cawthorn, John		-1865 - 1879-
Cawthorn, W.		-1884 - 1896-
Cawthorn & Naylor		-1854 - 1860-
Chambers, Chas. & Co.	Castle Hill Works	-1865 - 1884-
Clegg & Barton		-1879 - 1890-
Cock Brothers		-1865 - 1884-
Cocker, Grayson & Co.		-1865 - 1884-
Colver Brothers	Pilot Works	-1879 - 1900-
Constantine, A. & Co.		-1841-
Constantine, Fras.		-1845 - 1849-
Constantine, Henry & Charles		-1852-
Constantine Brothers		-1854 - 1900-
Cook, William & Sons		-1884 - 1900-
Cooper, Edward & Co.	Astley Steel & File Works	-1879 - 1884-
Crookes, Henry & Co.		-1896 - 1900-
*Crookes, Robert & Co.	Argus Works	-1837 - 1900-
Crowshaw, Chapman & Co.	Don Saw Works	-1879 - 1900-
Cuthberson, Edward		-1823-
Cutler, Hiram G.		-1837 - 1845-
Cutler, John		-1823 - 1839-
Cutler, John jr.		-1845 - 1852-
Cutler, Son & Chambers		-1849 - 1860-
Cutts, Wm. & John		-1828-
Davenport, James		-1852 - 1900-
Davenport, John Alfred		-1821 - 1865-
Davenport Brothers		-1845-
Davenport & Fletcher		-1817-
Deakin, George		-1860-
Deakin, Reuss & Co.		-1879 - 1890-
Dodge, Joseph & Robert	Continental Works	-1872 - 1884-
Donovan, Wm.		-1896 - 1900-
Drabble, Benjamin		-1823-
*Drabble & Sanderson	Ebenezer Works	-1825 - 1900-
*Eadon, Moses	President Works	-1837 - 1900-
Eadon, Sheldon & Co.		-1828-
Edley, John		-1837-
Edley and Rutherford		-1825 - 1828-
Edleys & Carter		-1841-
Ellis, Joseph		-1828-
Ellis, Thomas		-1821 - 1823-
Eyre, Edward		-1823 - 1825
Fearnhough & Co.		-1821-
Fearnhough, Joseph		-1823 - 1837-
*Fearnhough, William		-1837 - 1900-
*Fearnhough & Son		-1839 - 1900-
Fenton, Joseph & Sons	Sykes Works	-1879 - 1884-
Thomas Firth & Sons	Norfolk Works	-1852 - 1900-
Fisher, George & Co.	Hoyle Steel Works	-1860 - 1872-
Fisher, Thomas		-1821-
Flather, David & Sons (D)		-1868 - 1900-
Fletcher, Charles		-1849 - 1852-
Flockton, Tomkins & Co.		-1884 - 1900-
Fretwell, Isaac		-1849 - 1852-
Furniss, Cutler & Stacey		-1821 - 1828-

Garfit & Adsetts		-1849-
Garlick, Charles & Sons		-1872 - 1900-
Garlick & Haslam		-1868-
Glauert, L. & Co.	Burgess Works	-1879 - 1900-
Glossop, Herbert		-1879 - 1896-
Goodwin, Barker & Co.		-1814 - 1823-
Graves, R. & Sons		-1865 - 1879
Gray, G.		-1865 - 1872-
Gray, Geo. & Thomas		-1849 - 1860-
Gray, John	Phoenix Works	-1879-
Gray, Jonathan & Sons		-1823 - 1854-
Gray & Son		-1821-
Grayson, Samuel		-1837-
Greaves, Isaac		-1825 - 1828-
Greaves, John & Son		-1839-
Greaves, Nathaniel		-1841-
Greaves & Walker		-1837-
Green, Jane & Sons		-1814 - 1817-
Green, Thomas		-1823-
Gregory, W. & Sons		-1879 - 1884-
Gregory & Bramall	Soho Steel & File Works	-1879 - 1890-
Gregory & Redfern		-1845-
Groves, Richard & Son		-1817 - 1824-
*Groves, Richard & Sons	Beehive Works	-1825 - 1900-
Groves & Sons		-1814-
Hauge, John & Co.		-1879-
Hauge, Clegg, & Barton		-1868 - 1872-
Hall, T.		-1865-
*Hall, William	Alma Works	-1839 - 1900-
Hallamshire Steel & File Works		-1879 - 1900-
Hammond, Anthony		-1837-
Hanbey, Pearson & Green		-1821 - 1825-
Harrison, Thomas		-1817-
Haywood, Hinchcliffe & Co.		-1825 - 1828-
Hearnshaw Brothers	Sorby Street Works	-1879 - 1884-
Herring, John & Sons	Navigation Works	-1879 - 1884-
Hill, Abraham		-1845 - 1879-
Hill, Anthony		-1821 - 1825-
Hill, Samuel		-1837 - 1852-
Hincliffe, G.W.		-1839 - 1849-
Hitchen & Wharton		-1823 - 1825-
Hobson, Francis		-1841 - 1860-
Hobson, Jno.		-1841-
Hobson, Wm.		-1852 - 1865-
Holland, Robert		-1814-
Holy, Daniel & George		-1837 - 1852-
Hoole, Wm. & Co.		-1839-
Hoole, Staniforth & Co.	Minerva Works	-1841 - 1890-
Hoole, Staniforth & Gray		-1845-
Hoole & Brooksbank		-1845-
Hoole & Pinder		-1825-
Horn, Wm. Ibbetson		-1837 - 1868-
Howarth, J. & Son (D)		-1865 - 1879-
Howell & Co.	Brook Works	-1879 - 1896-
Ibbetson, William & George		-1797 - 1828-
*Ibbotson, Richard	Shoeham Works	-1839 - 1900-
Ibbotson, Robert		-1841-
Ibbotson, Thomas & Co.		-1837 - 1879-

Ibbotson, Peace & Co.		-1845 - 1852-
*Ibbotson Brothers	Globe Works	-1828 - 1900-
Ibbotsons & Roebuck		-1821-
Ibbotson & Smith		-1828-
Jackson, William	Sheaf Island Works	-1856 - 1900-
Jackson, Newton & Co.		-1865-
Joel, G. & Sons	St. Mary's Works	-1865 - 1872-
Johnson, Samuel		-1825-
Johnson & Co.		-1828-
Johnson, Cammel & Co,		-1839-
*Kenyon, John C. & Co.		-1823 - 1900-
Kenyon & Co.		-1821 - 1823-
*Kenyon, Ellison & Co.		-1839 - 1900-
Kenyon, Sykes & Co.		-1814 - 1817-
Kitchen, Richard		-1845 - 1856-
Kitchen, Robert		-1839-
Leadbeater, John & Co.		-1823 - 1828-
Linley, Samuel & Robert	Clough Works	-1868 - 1872-
Lockwood Brothers		-1854 - 1900-
Lomas, Joseph		-1872-
Long, H.G. & Co.		-1849 - 1860-
Long, Gregory & Co.		-1845-
Long, Hawksley & Co.		-1868 - 1900-
Long, Hawksley & Marples		-1852 - 1865-
Loukes & Russell		-1845 - 1849-
Low, George		-1868-
Low, G. & W.	George Washington Works	-1879 - 1900-
Low, Phillips & Son		-1828 - 1849-
Makin, Thos.		-1845-
Makin, Thos. & Son		-1849 - 1884-
Makin, Wm. & Sons		-1884 - 1900-
Marples, Wm. & Sons (D)		-1884 - 1900-
Marples & Evill	Victorica Works	-1865 - 1872-
Marriot & Atkinson		-1841 - 1860-
Marshall & Co.		-1837-
Marshall, Charles		-1856 - 1860-
Marshall, Martin		-1814 - 1837-
Marshall, J. & W.		-1849 - 1852-
Marshall, Addy & Co.		-1854-
Marshall, Henry & Co.		-1849-
Marshall & Cheetham		-1839-
Marsh Brothers (D)	Pond Works	-1852 - 1900-
Marshes & Shepherd (D)	Pond Works	-1837 - 1849-
Marsden, E.J. & Co.		-1828-
Marden, Wm.		-1825 - 1828-
Marsden Brothers (D)		-1856 - 1872-
Marsden Brothers & Silverwood (D)		-1849 - 1852-
Marsden & Downend		-1837 - 1841-
Maw & Staley		-1823-
Memmott, Wm.		-1839 - 1856-
Millington, James & Co.		-1814 - 1849-
Millington, John		-1823-
Mitchell & Co.		-1821-
Mitchell, Robert		-1837 - 1845-
Mitchell, S. & Co.		-1823 - 1825-
Mitchell, W.B. & Co.		-1837-
Mitchell Brothers		-1828-
Mitchell, Wreaks & Co.		-1814 - 1817-

Mort, Henry & Co.		-1837 - 1872-
Morton, Matthew		-1845-
Moss, Joshua & Gamble Brothers	Franklin Works	-1854 - 1900-
Mottram,Thomas		-1837-
Mottram, Thomas & Son		-1839-
Mottram, Thomas & Sons		-1841 - 1868-
Mottram, Taylor & Co.		-1852 - 1860-
Moulson, J.		-1823-
*Moulson Brothers		-1828 - 1884-
Naylor, G. & Co.	Reliance Works	-1865 - 1900-
Neaves, Thomas		-1837 - 1839-
New, Jeremiah		-1837 - 1854-
*Newbould, Samuel & Son	Bridgefield Works	-1814 - 1900-
Newton, Francis & Sons		-1849 - 1860-
Nicholson, Hoole & Co.		-1837 - 1841-
Oldham, Daniel		-1868 - 1900-
Oldham, William		-1849-
Osbaldiston, Louis & Co.		-1865 - 1900-
Osborn, Samuel & Co.	Clyde Steel Works	-1854 - 1900-
Outram, Henry	Canal Works	-1860 - 1865-
Outram, Wm. & Son	Delhi Works	-1860 - 1879-
Pacey, Elizabeth		-1884 - 1890-
Pacey, John & Co.		-1860 - 1872-
Pacey & Co.	Burnt Tree Lane Works	-1854 - 1856-
Pacey & Armitage		-1852-
Padgin, John		-1828-
Pagdin, John & Sons		-1837 - 1841-
Pagdein, Wm. W.		-1845 - 1872-
Parker, Ebenezer & Sons	Ecclesall Works	-1865 - 1884-
Parkin, John,		-1860-
Parkin, Wm. & Co.	Crown Steel Works	-1865 - 1896-
Parkin, Parker & Clough		-1817-
Parkin & Marshall	Telegraph Works	-1865 - 1900-
Parkin & Riley		-1860-
Peace Brothers	Garden Steel Works	-1854 - 1879-
Peace, C. & S.	Eagle Works	-1828 - 1841-
*Peace, Joseph & Co.	Merchant Works	-1849 - 1900-
Peace, Samuel & Co.		-1814 - 1823-
Peace, W.H. & Co.		-1852-
Peace, Wm. K. & Chas.	Eagle Works	-1860 - 1900-
Peace, Spafford & Co.		-1854-
Peace, Ward & Co.	Agenor Works	-1860 - 1868-
Pearson, Robert		-1837 - 1845-
Pearson, F. G. & Co.	Hope Works	-1879 - 1900-
Penestone, Wheatcroft & Co.		-1814-
Peneston, Wheatcroft & Singleton (Penestone & Co.)		-1817 - 1828-
Pinder, John		-1841 - 1864-
Pinder, Wm. & Sons		-1837 - 1841-
Potts, Baxter & Brumby		-1837-
Potts, Baxter & Brown		-1839-
Rawson Brothers		-1879-
Reaney & Wood		-1849-
Renton, Robert		-1879-
Reuss & Co.		-1896 - 1900-
Rhodes, William Chas. & James		-1868-
Rhodes Brothers	St. James' Steel Works	-1872-
Robinson, Jonathan		-1839 - 1845-
Rodgers, Russell & Horsfield		-1849 - 1852-

Royston, Benjamin		-1823-
Russell, Thomas & Co.	Canada Works	-1860 - 1900-
Russell & Horsfield		-1854 - 1856-
Russell, Horsfield & White		-1879-
Rutherford, John		-1825 - 1837-
Rutter, Alexander		-1814 - 1823-
Sanderson, James		-1833-
Sanderson, John		-1821 - 1852-
Sanderson & Kirk		-1817-
Scott, Pacey & Tasker		-1849-
Seebohn & Dickenstall		-1896 - 1900-
Senior & Fernebough		-1817-
Shaw, James	Dauntless Works	-1860 - 1884-
Shaw & Co.		-1821-
Shaw, Marshall & Co.		-1797 - 1825-
Shaw & Cutler		-1814 - 1817
Sheffield Steel & Manucaturing Co.		-1879-
Sheldon & Son		-1853 - 1865-
Sheldon, Thomas		-1837 - 1839-
Sheldon, Thomas & Son		-1841 - 1849-
Sheldon, Wragg & Brown		-1856-
Slack, Sellars & Co.	Townhead Steel Works	-1860 - 1900-
Slack, Sellars & Grayson		-1837 - 1856-
Slimm, Robert		-1823-
Smith, John		-1823 - 1825
Smith, John & Son		-1814 - 1828-
*Sorby, John & Son		-1821 - 1884-
Sorby, Robert & Co.		-1841 - 1845-
*Sorby, Robert & Son		-1849 - 1900-
Southern, William		-1879-
Spafford, A. & Co.	Imperial Works	-1868 - 1900-
Spear, John		-1814 - 1823-
*Spear & Jackson	Aetna Steel Works	-1825 - 1900-
*Spencer,Mathias & Son		-1825 - 1900-
Spencer, J.R. & Son	Albion Steel Works	-1865 - 1879-
Spencer & Sons		-1823-
Stacey, James		-1868-
Stacey & Handley		-1865-
Staniforth & Co.		-1821 - 1823-
Stanifield, Newbould & Baildon		-1849-
Stansfield & Lowe		-1865-
Stanfield Brothers		-1868-
Stones, Frederick		-1821 - 1845-
Stone. J.C.		-1865-
Stuart, Jackson & Co.		-1828-
Sykes, Wm.		-1823-
Sykes, Wm. & Sons		-1825-
Tasker, Henry		-1852-
Taylor, Edward		-1828 - 1839-
Taylor, Joseph & Jno.		-1841 - 1845-
*Taylor Brothers,	Adelaide Works	-1849 - 1900-
Thompson, John		-1817-
Thorpe & Fearncclogh		-1817-
Tingle & Co.		-1821-
Tingle, Drabble & Co.		-1823-
Turner, James		-1841-
Turner, John		-1821-
Turner, Jonathan		-1839 - 1845-

Turner, Naylor & Co.		-1872-
Turner, Naylor & Marples		-1879 - 1884-
*Turner, Thomas & Co.		-1841 - 1900-
Turner, Davies & Co.		-1837 - 1845-
Turner & Swift		-1814 - 1817-
Turton, Thomas & Son		-1837 - 1839-
Turton, Thomas & Sons	Phoenix Steel Works	-1841 - 1896-
Turton Brothers		-1865 - 1872-
Turton Brothers & Mathews		-1879 - 1896-
Tyzack, Ebenezer		-1841-
Tyzack, Thomas		-1837 - 1845-
Tyzack, Thomas & Sons		-1849 - 1896-
Tyzack, T. & N.		-1817 - 1821-
Tyzack, T & W. Little		-1823-
Tyzack, William		-1825 - 1841-
Tyzack, Wm. & Sons		-1845 - 1872-
Tyzack, Wm. Sons & Turner		-1879 - 1900-
Tyzack Wm. A. & Sons	Stella Works	-1872 - 1900-
Vickers, James & Co.		-1814 - 1828-
Wade, Wingfield & Robotham		-1845 - 1849-
Wales, Geirge		-1837-
Wall & Cooper		-1845-
Ward, David		-1821 - 1823-
Ward, David & Son		-1825 - 1828-
Ward, Wm. & Sons		-1879 - 1896-
Ward & Payne (D)		-1849 - 1900-
Webster & Johnson		-1839 - 1845-
Weldon & Furniss		-1814 - 1817-
Wharton, George & Co.		-1845 - 1849-
Wheatman, John		-1845-
Wheatman & Smith	Russell Works	-1849 - 1900-
Whetham, Charles		-1828-
Whetham & Co.		-1821 - 1823-
Wilcock, Benjamin & John	Bolton Works	-1868 - 1900-
Wilson, Hawksworth, Elison & Co.	Carlisle Works	-1856 - 1884-
Wilson, Hawksworth & Moss		-1845-
Wilson, Hawksworth, Moss & Elison		-1849 - 1852-
Wilson & Schofield		-1845-
Winder Brothers	Royd Works	-1872 - 1879-
Wing, M. & J.		-1860 - 1884-
Wingfield, Robotham & Co.		-1852 - 1890-
Woodhouse & Caterer		-1837-
Woodroofe, Wm.	Whitley Wood Works	-1872 - 1890-
Woolhouse Brothers		-1839 - 1856-
Wreaks, John & Co.		-1821 - 1823-
Wright & Son		-1814-
Wright, James		-1837 - 1839-
Wright, John		-1823 - 1856-
Wright, John & Son		-1817 - 1828-
Wright, Samuel		-1836-
Wright, Thomas		-1828 - 1852-
Wright, Boothroyd & Co.		-1841-
Yates & Wood		-1856 - 1900-
Yeomans, Yates & Stabdfield		-1837 - 1839-
Youle, John	Phoenix Works	-1860 - 1900-

* = in business over 50 years; (D) Dealer, not Manufacturer

Chapter III

Edge Tools

Ten makers of edge tools were listed in the first published 1867 directory of Birmingham. The accompanying note claimed that at that date Birmingham was the leading center of this production.

EDGE TOOL MAKERS

There is considerable Trade in these articles carried on here, and it may be said with Justice, that this is the First Town in the kingdom for these Goods; they are the makers of Axes, Chissels, Plane Irons, Adzes, Hooks, Bills, &c. Great Quantities of which are exported to North America, and various other Parts.

With the increasing availability of cast steel at Sheffield leadership in edge tool manufacturing soon shifted here. Twelve firms making such products were listed in both the 1787 (see Table IV) and 1797 Sheffield directories. This surpassed the half dozen firms then listed at Birmingham. Unfortunately a third Sheffield directory was not published until 1814, which then noted 30 firms in this trade. This gap of seventeen years in directories makes it difficult to trace industrial developments during this period, 1798 - 1813.

Table V - List of Edge Tool Makers [*1787 Sheffield Directory*]

EDGETOOLS.

Manufacturers in SHEFFIELD.	*Marks*
BISHOP George, Brinſworth's Orchard	G. BISHOP
Biſhop George, jun. do.	LIBERTY
Gillot Malin, Far-gate - -	
Green Hannah, and Son, Burgeſs-ſtreet	JOHN GREEN
Green Jane, and Sons, Scotland-ſtreet -	GREEN
Hall Thomas, Pea Croft; *alſo draws Steel*	BOSTON
Law Philip, Carver-ſtreet -	P. LAW
Maxfield Thomas, Barker Pool -	HOW +
Mitchell Joſeph, and Co. Shudehill -	BOLSOVER
Newbould Samuel, Sheffield Moor	NEWBOLD
Weldon William, Colſton Croft	WELDON
Wilde John, Church-lane -	WILD

Manufacturers in the NEIGHBOURHOOD.	
FIELD John, Marſh Lane -	
Hutton William, do. - -	HUTTON
Parkin George, Brightſide -	+ AMO
Reaney James, Unſtone - -	REANEY
Shaw Thomas, Bridgehouſes; *Gravers and Turning Tools* -	SHAW
Wilkinſon Paul, Grimeſthorpe -	PAUL WILKINSON

The practice of making edge tools combined the edge of cast steel which was fire welded to less expensive wrought iron. The two pieces of different materials were beveled (scarf cut), heated to the proper temperature in a charcoal or coke fire, withdrawn from the forge, put on an anvil, fluxed with borax and hammered welded together. The demarkation between steel and iron is apparent on surviving plane irons chisels, axes, draw knives and other edge tools. This technique had been followed in the cutlery and sickle making trades for several centuries. Smaller chisels and gouges were usually forged entirely from steel. Several alternate methods for making such edge tools were patented around the turn of the 19th Century. None of these were practical nor economical, but are noted for historical record.

Arnold Wilde, sawmaker from Sheffield, was granted Patent No. 2023, 19th January 1795; "making and manufacturing all sorts of Plane, Irons, Scythes, Sickles, Drawing Knives, Hay Knives, and all other Kinds of Edge Tools, from a preparation of Cast Steel and Iron United and Incorporated Together by means of Fire". His idea was to cast molten steel into a mould shaped similarly to the finished tool and inject a heated iron handle to form a weld of the two materials during solidification. This appears to be impractical at that date due to insufficient temperature control and limited equipment.

William Bell, engineer from Derby, was granted Patent No. 2997, 22nd December 1806; "new improved Method of Making Smoothing Irons, Sad Irons, Plane Irons and various Edge Tools". Bell proposed soft soldering a piece of hardened and tempered steel into a shaped iron blank or stock. This required added finishing expenses and precision machining, not generally available at this date.

George Babb, engineer of Bordsley, Birmingham, was granted Patent No. 3526, 23rd. January 1812; - "A New Method or Principle of Producing Files, Plane Irons and Turning Chisels". He proposed to case carburize "British or other iron" using powdered charcoal thus converting it into steel. The problem was length of time required to build up adequate depth of case, temperature control and subsequent grinding during sharpening would remove such a case.

Thomas Dyson, scythe manufacturer from Sheffield, was granted Patent No. 4512, 11th Novemeber 1820; - "Improvements in Manufacture of Plane Irons and Turning Chisels". His idea was to make a cutting edge from cast steel and insert this in between two plates clamped together by a single screw. Adjustment was much the same as a slotted double plane iron. Such an assembly must have formed an objectionable bulky arrangement, particularly for chisels.

James Boydell, iron master from Dudley, was granted Patent No. 9607, 26th January 1843; - "Improvements in the Manufacture of Metals for Edge Tools". Boydell's method was to make a composite material of cast steel and iron much the same as Boulsover's Sheffield Plate of silver to copper. The problem at the time was lack of control of the high temperature required to weld large pieces together by rolling.

Two other English Patents concerning edge tools deserve passing mention. Albert Newton, mechanical draughtsman of London, was granted Her Majesty's Royal Letters Patent No. 11,380, 24th September 1846 for a furnace designed for insuring more accurate temperature control during hardening and tempering. Henry Wilson, foreman at William Greaves & Sons, Sheaf Works at Sheffield, was granted Patent No. 12,272, 21st September 1848 for combining a socket and tang in the manufacture of chisels.

Previous to the 19th Century all trades producing cutting instruments from steel at Sheffield had been associated with the Cutler's Company. Specialization soon led to sectional trade societies into nine groups, as classified by Lloyd in Table VI.[2]

Table VI - The Cutlery Trades

I.	Table Knives
II.	Steel Forks
III.	Spring Knives
IV.	Razors
V.	Scissors
VI.	Scythes and Sickles
VII.	Files
VIII.	Saws
IX.	Edge Tools

Actually the saw and edge tool makers were exempt from the original Cutlers' Company - "their articles not being among the ancient manufacturers of the town".[3] The File-smiths' Society was first to establish as a benefit club in 1784 apart from the Cutler's Society.[4] Soon after each of the trades had at least one, some as many as four, different societies leading to a type of labor union for each group.

Edge tool manufacturing differed from the general cutlery trades. It was not a cottage industry, generally requiring machinery, while cutlery was essentially a hand skilled occupation. Edge tools were classified as heavy or light. The former included such products as axes, adzes, hatchets and chopping bills. The latter included chisels, gouges, engraving tools, draw knives and plane irons

With the deterioration of the Cutler's Company as a protective society early in the 19th Century, numerous labor unions were instrumental in securing high prices. In an effort to demonstrate solidarity the Master manufacturers united March 28, 1814 to form the Sheffield Mercantile and Manufacturing Union. Among the resolutions stated by the Chairman of this Committee was:[5]

> "The Peace and good order of the Town at this juncture particulary demand Union and energy among all Manufacturers, as observation tends forcibly to demonstrate that exhorbitant Wages do not promote industry and sobriety; but on the contrary, produce idleness and dissipation, with concomitant evils".

Twelve trades, including that of the Edge Tool Manufacturers, formed a Committee, two from each trade, to adopt a policy beneficial to their mutual interests. From this may have come an agreement for both standardization of products and sale prices, which continued throughout the remainder of the 19th Century. It is also possible that the *Key to the Manufactories of Sheffield,* published and sold by Joseph Smith in 1816 was an outcome of this organization.[5] Joseph Smith, an engraver in Sheffield, had previously prepared *Peter Stubs' 1801 Pattern Book.*[6]

An earlier pattern book containing 37 Plates, printed from copper plate engravings with some pages printed on paper watermarked 1803, has been observed in Sheffield. Three Plates of this are idendtical with those in the *1816 Smith's Key*. One page of this contains a crown mark "J.S.", which may stand for J.Smith. The Plate illustrating the plane irons, noted *"Smith's Warranted"*. This may refer to Joseph Smith, edge tool maker at Townhead Street listed in the 1814-1815 Sheffield Directory, who may possibly have been one and the same Joseph Smith the engraver. It is possible that this pattern book might be the one to which Dane referred as *"Sheffield List,1803"*[7]

Nineteen plates (LXI - LXXIX) from an unidentified pattern book having 23 frames and printed on paper watermarked 1799, are herein reproduced.[8] It is possible that these might be those referred to by Dane. In any event they are probably of Sheffield origin and predate those of the *1816 Smith's Key*.

The earliest known price list of Sheffield edge tools is dated July 25, 1823, issued by Spencer & Son.[9] The heading on the title page is shown in Plate LXXX.

Spencer and Son

STEEL REFINERS AND CONVERTERS, AND

MANUFACTURERS OF FILES, RASPS, EDGE TOOLS, SAWS, &c.

***JULY* 25, 1823.**

PRICES OF FILES AND RASPS.

Plate LXXX - Title to Spencer & Son Edge Tool Price List

This extensive three page list included:files, cast steel chisels and gouges, engravers and printcutter's carving tools, firmer chisels and gouges, mortise chisels, turning chisels and gouges, plane irons, augers, drawing knives, wheelers' bruzzes, matchets, mincing and cheese knives, hatchets, choppers, ship scrappers, axes, adzes,hoes, trowels, garden and turnip hoes, rakes, bills and saws.

The earliest known agreement of prices among the Sheffield Edge Tool Manufacturers is dated November 12, 1824. This is shown as Plates LXXXI - A,B,& C.Note the announcement on the heading: - "At a Meeting of the Edge Tool Manufacturers, held this day, (Sheffield, Nov. 12th, 1824) it was Resolved that in consequence of the very great advance on Materials, the following Prices shall take place upon all orders received on and after the 15th instant". An extant price list of Mitchell, Brothers, & Co. confirmed these prices.[10] The title to this is shown as Plate LXXXII.

MITCHELL, BROTHERS, & Co.

(LATE SAMUEL MITCHELL AND Co.)

SHEFFIELD, MANUFACTURERS OF

Saws, Fenders, Edge Tools, Patent Cast Steel Scythes, Hay and Straw Knives, Hoes, &c. &c.

SHEFFIELD, APRIL 1, 1827.

Plate LXXXII - Mitchell, Brothers,& Co. 1827 Price List

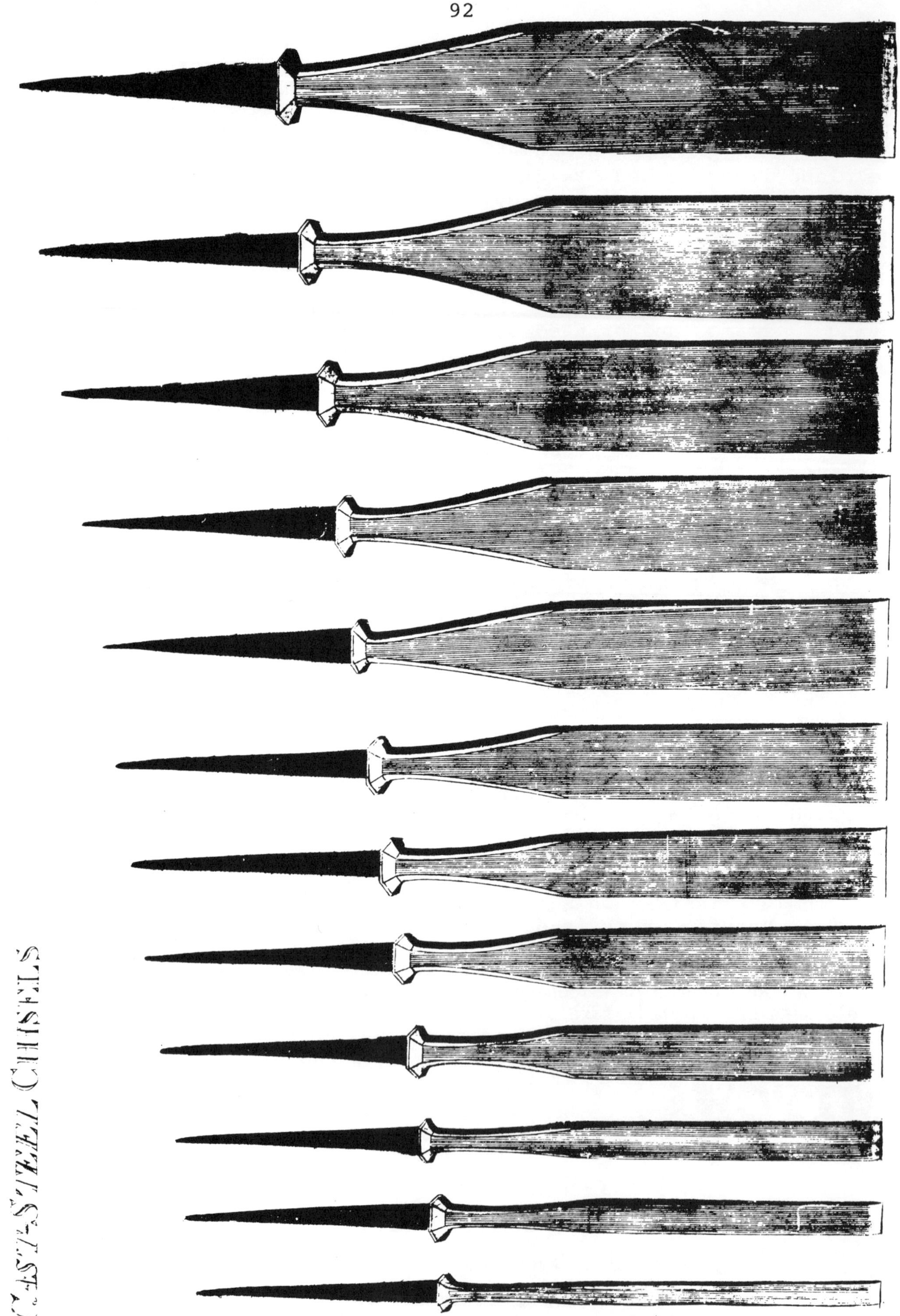

Plate LXI Firmer Chisels 3/16" - 1" widths

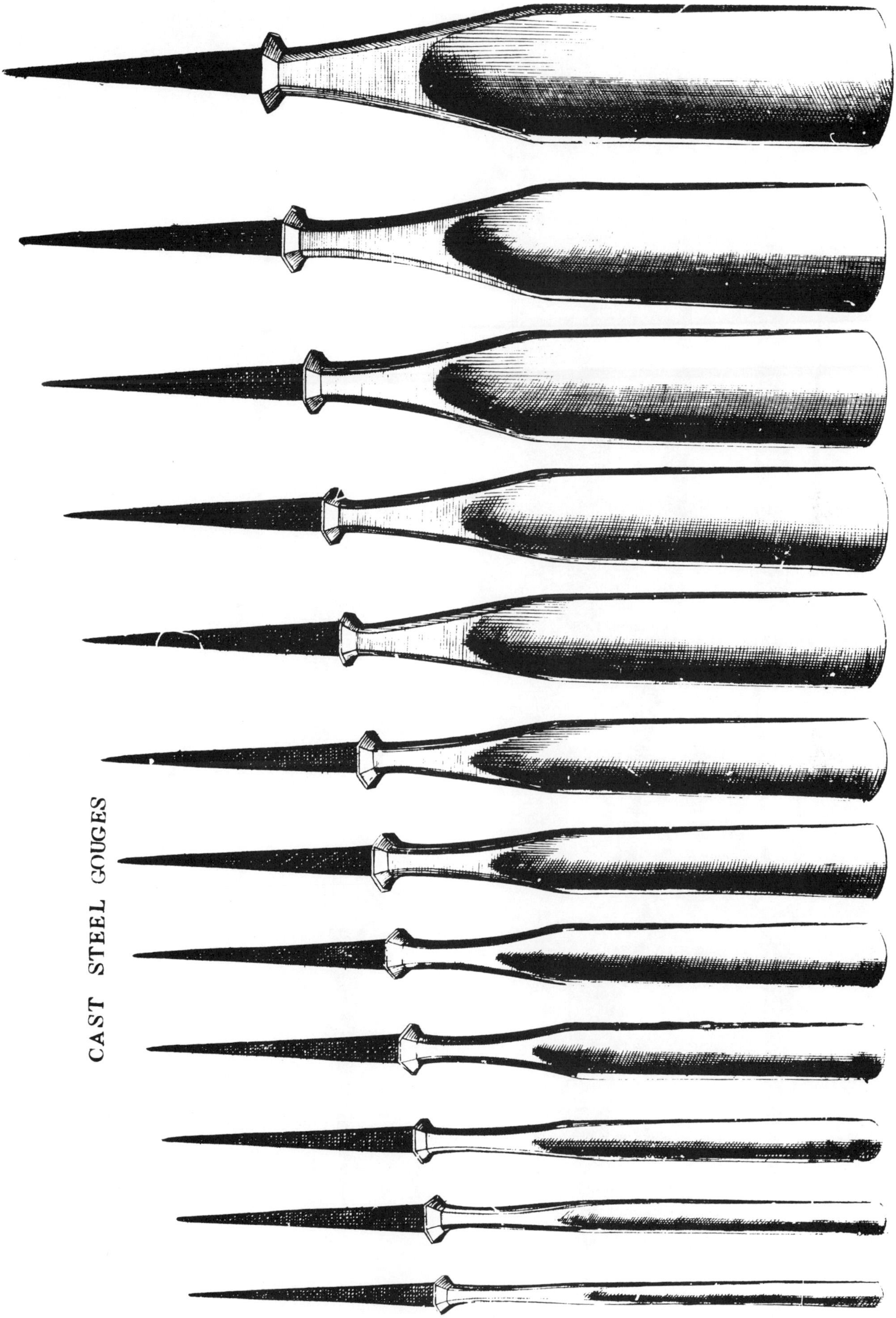

Plate LXII - Firmer Gouges 3/16" to 1" widths

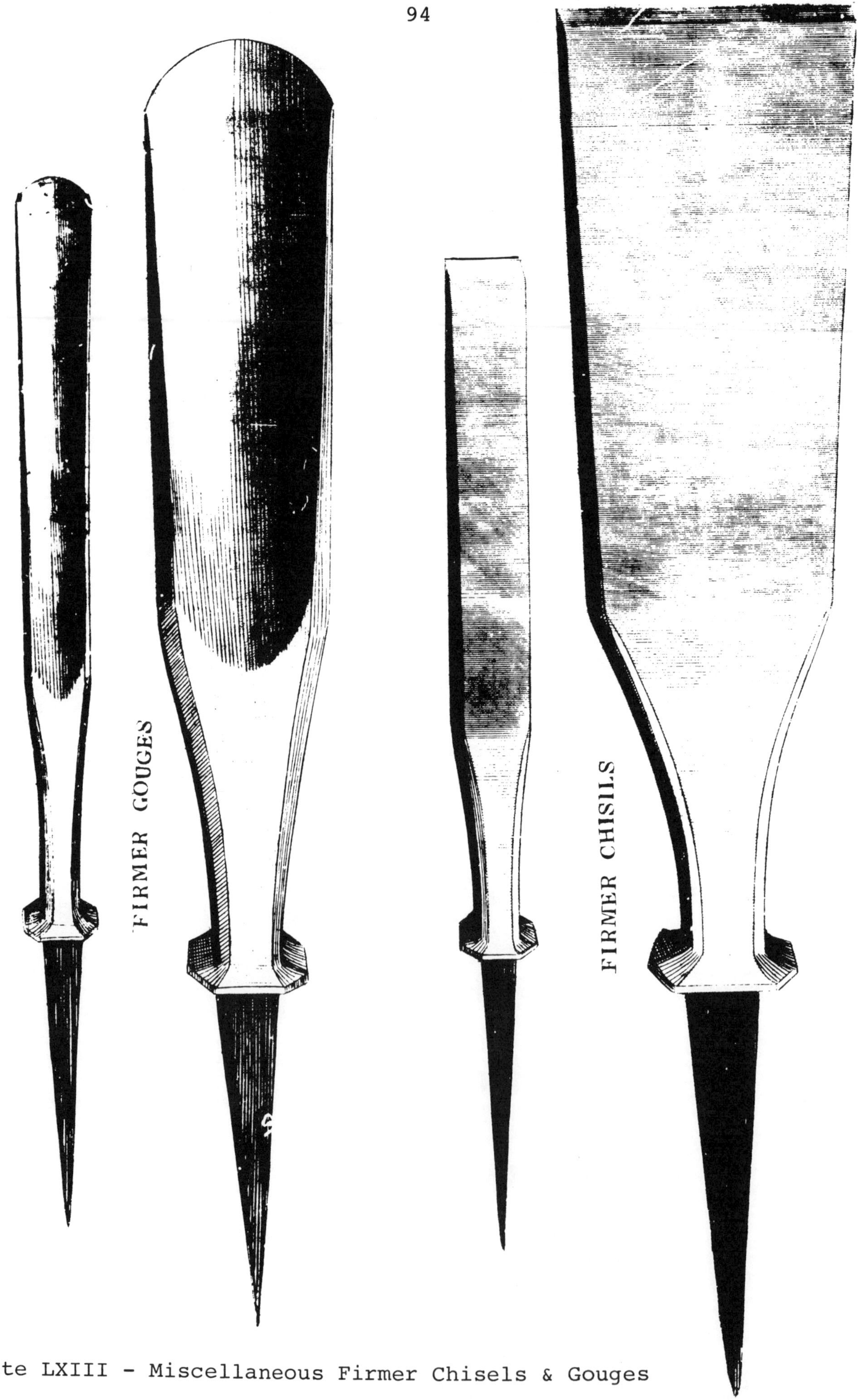

Plate LXIII - Miscellaneous Firmer Chisels & Gouges

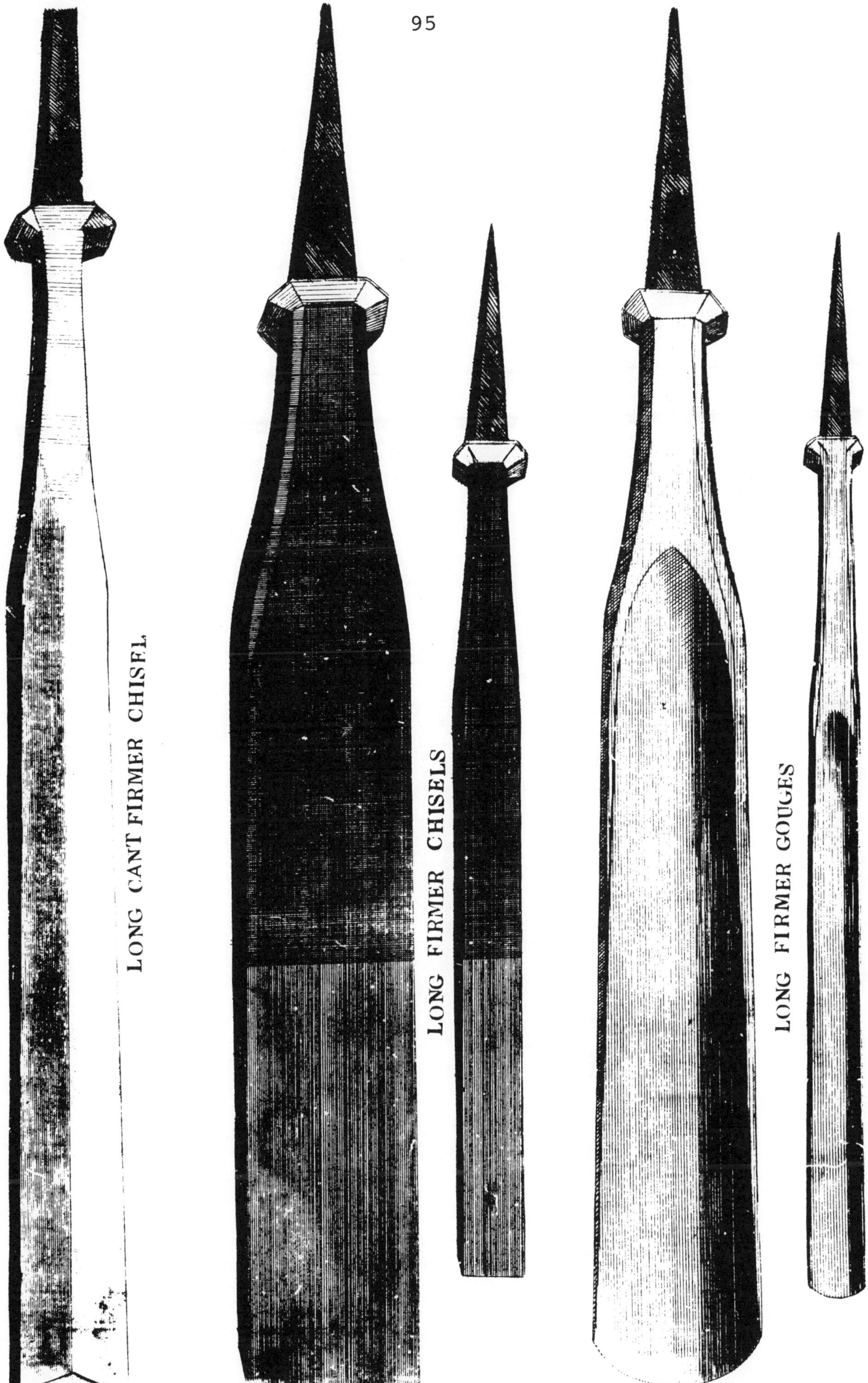

Plate LXIV - Miscellaneous Firmer Chisels and Gouges

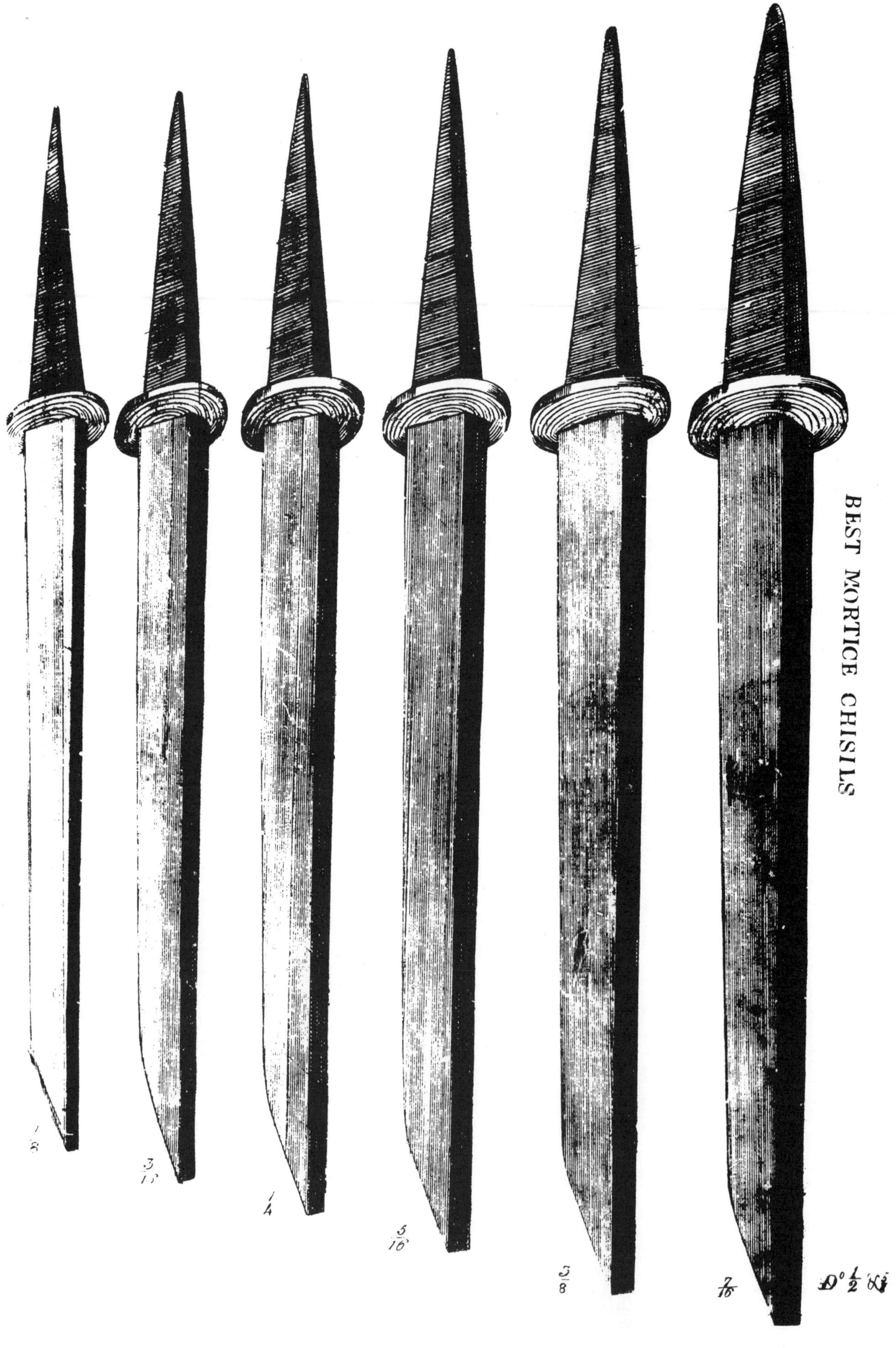

Plate LXV - Mortice Chisels 1/8" - 7/16"

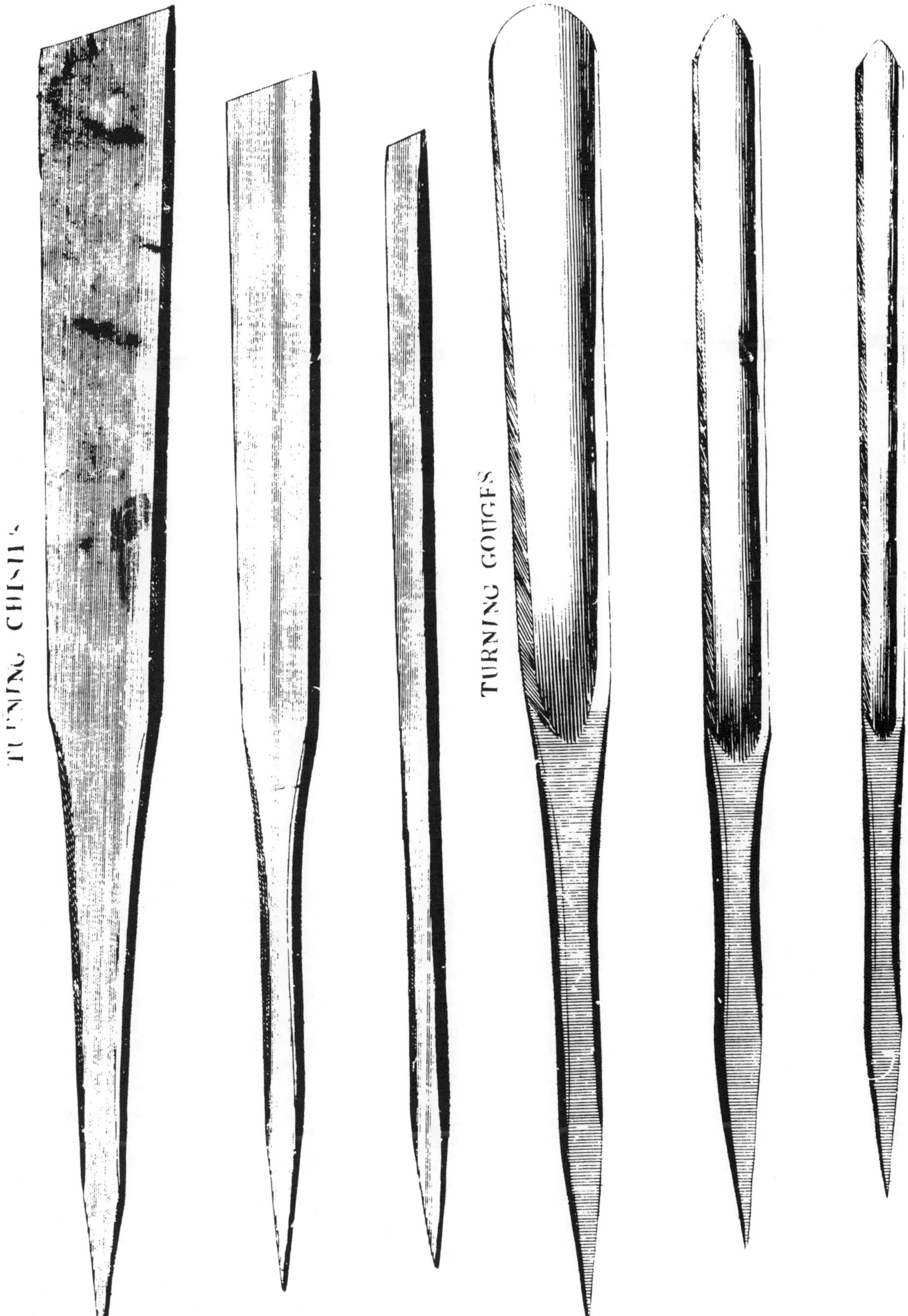

Plate LXVI - Turning Chisels and Gouges

Plate LXVII - Socket Chisels: 1-3/4", 1-1/4", 7/8" & 1/2"

Plate LXIII

Socket Gouges

1-3/4",1-1/4",7/8" & 1/2"

Short Plane Iron

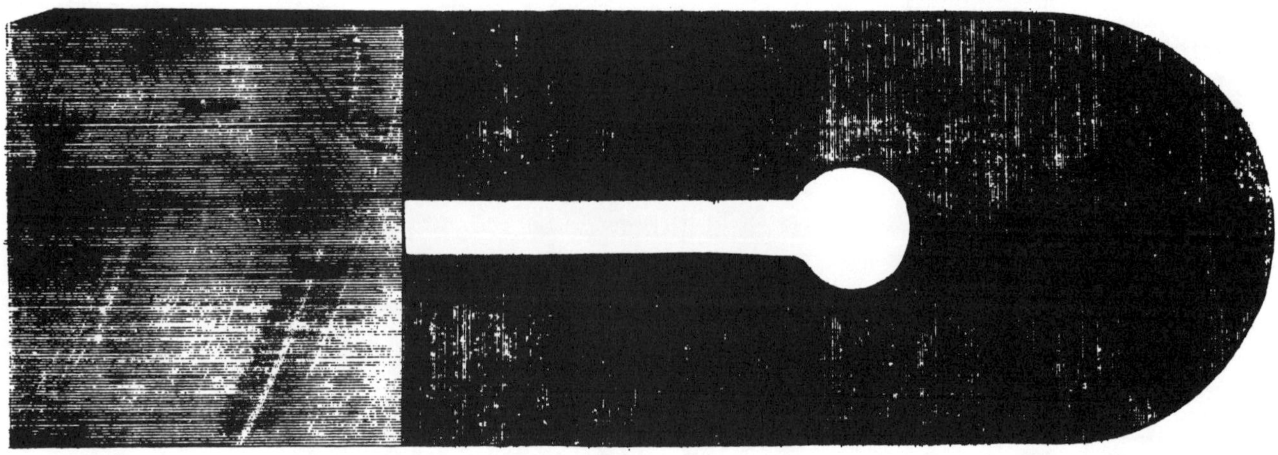

Cut Plane Iron

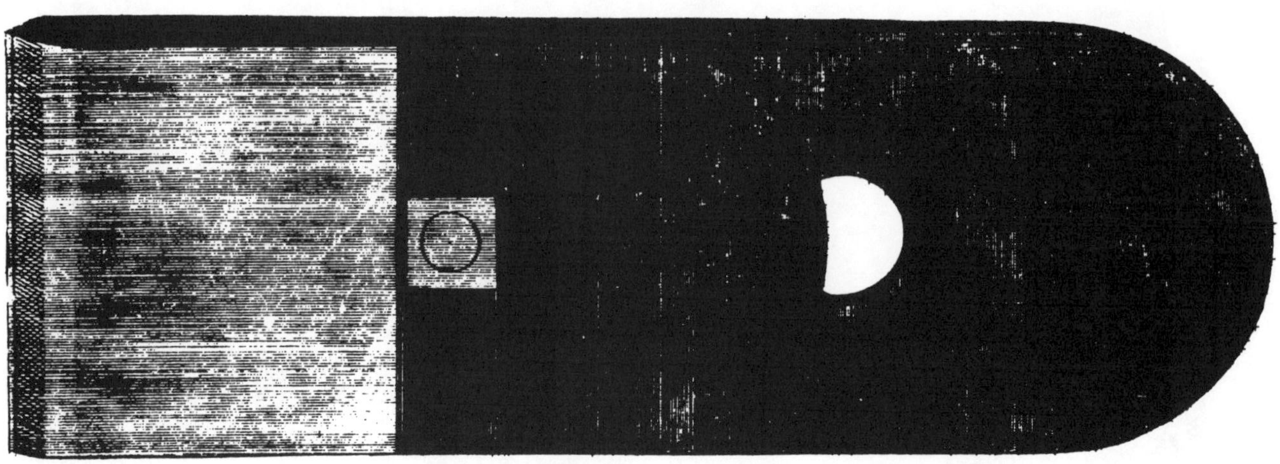

Double Plane Iron

Plate LXIX - Plane Irons

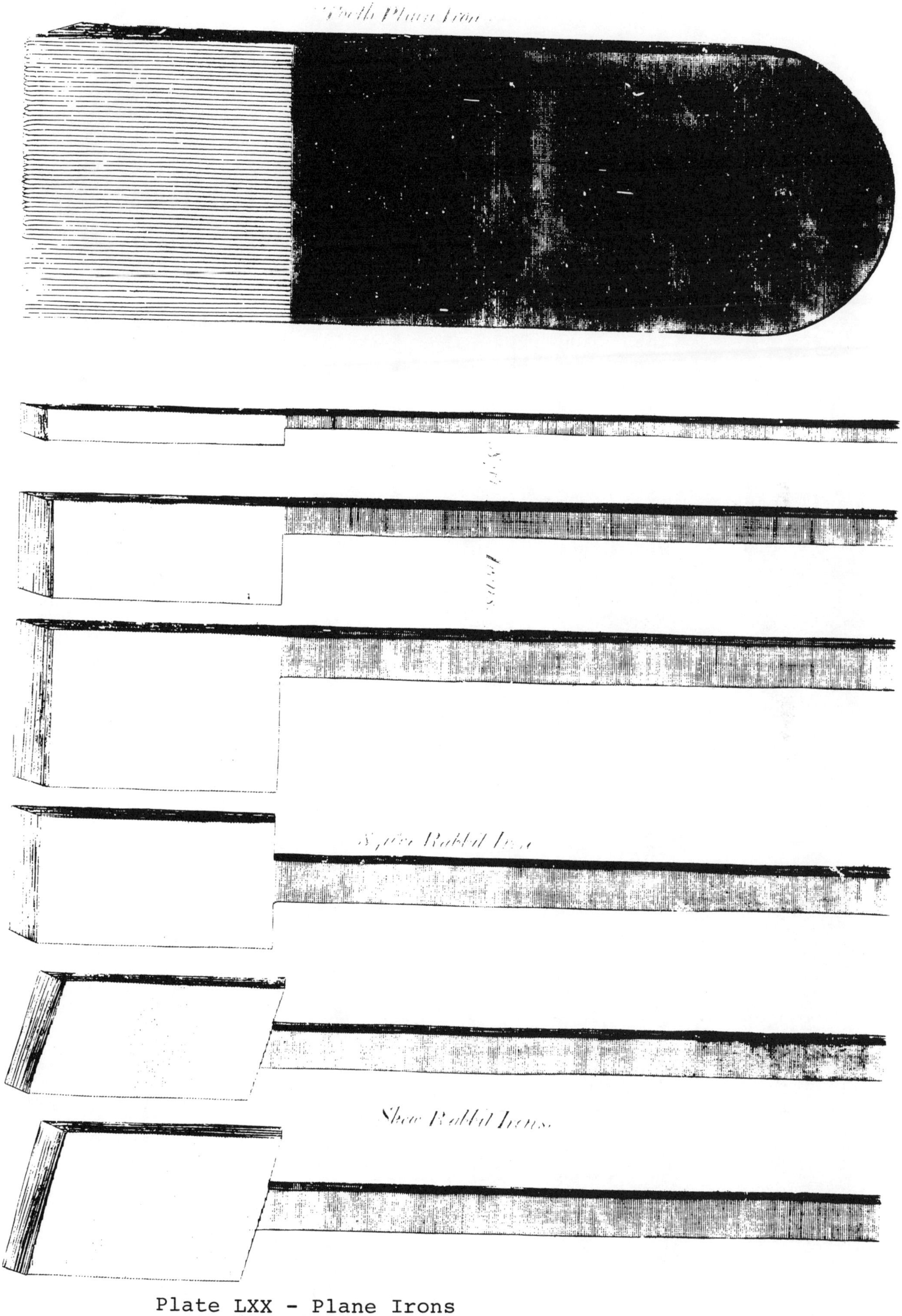

Plate LXX - Plane Irons

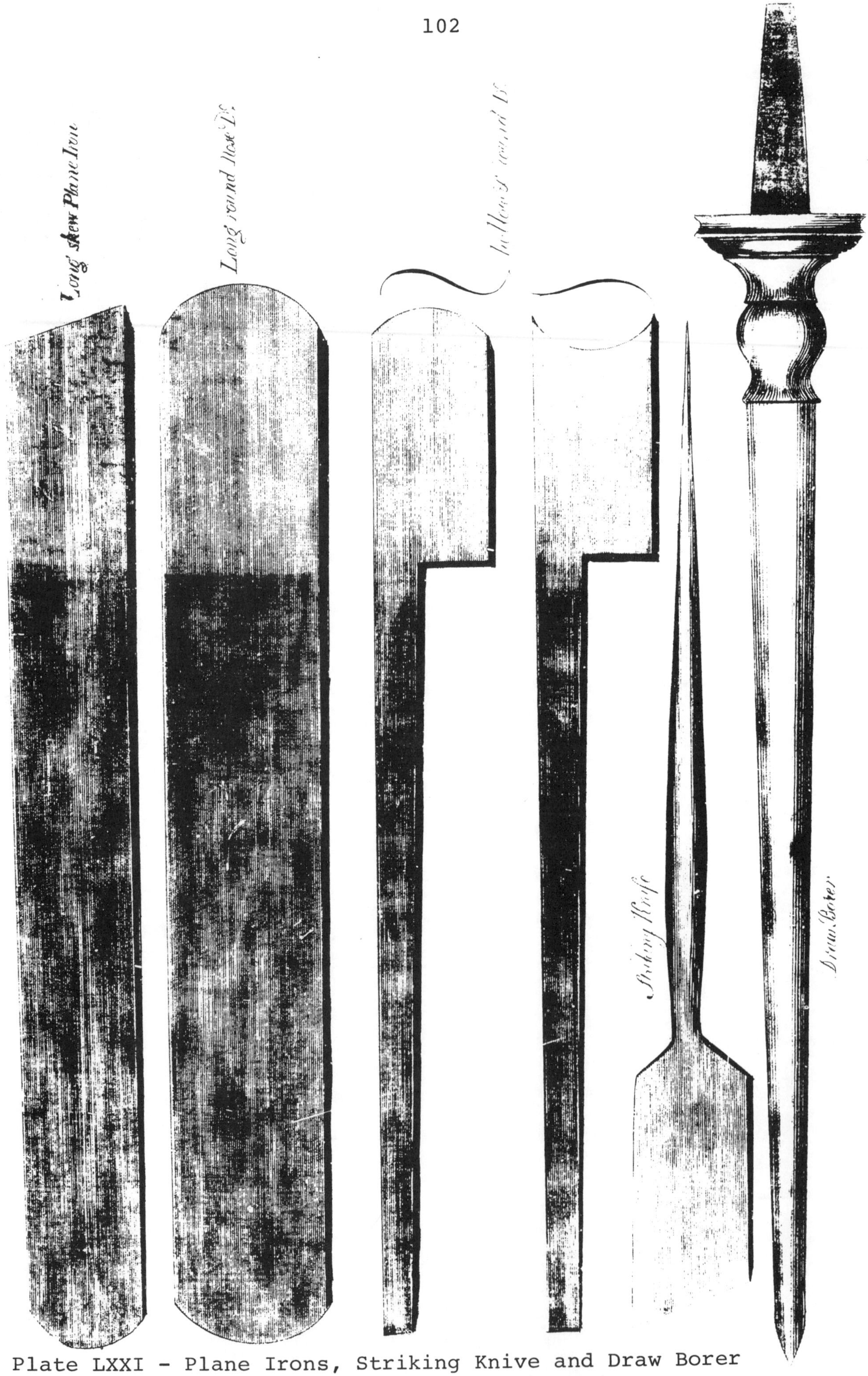

Plate LXXI - Plane Irons, Striking Knive and Draw Borer

Plate LXXII - Plow and Grooving Plane Irons

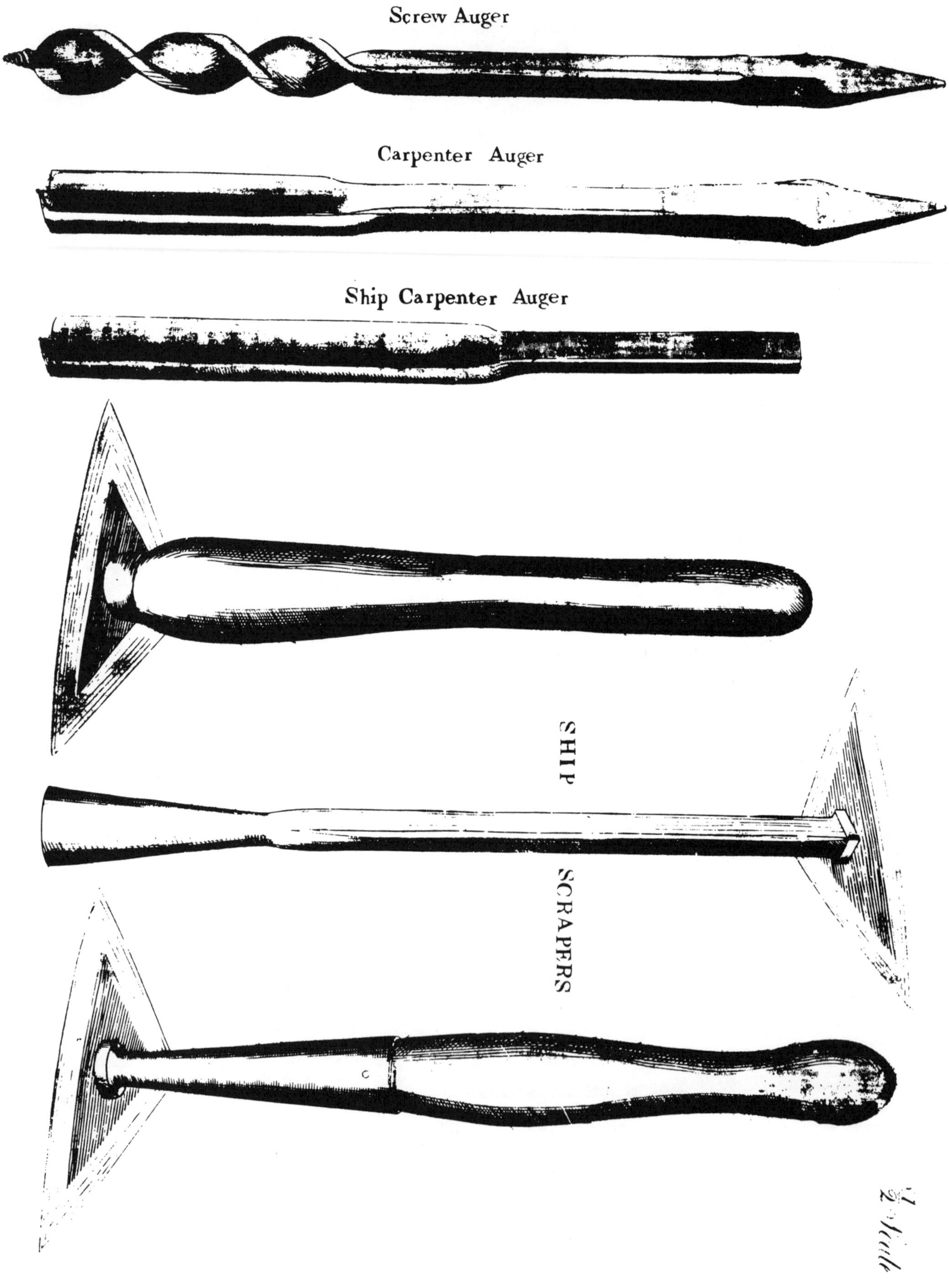

Plate LXXIII - Augers and Ship Scrapers

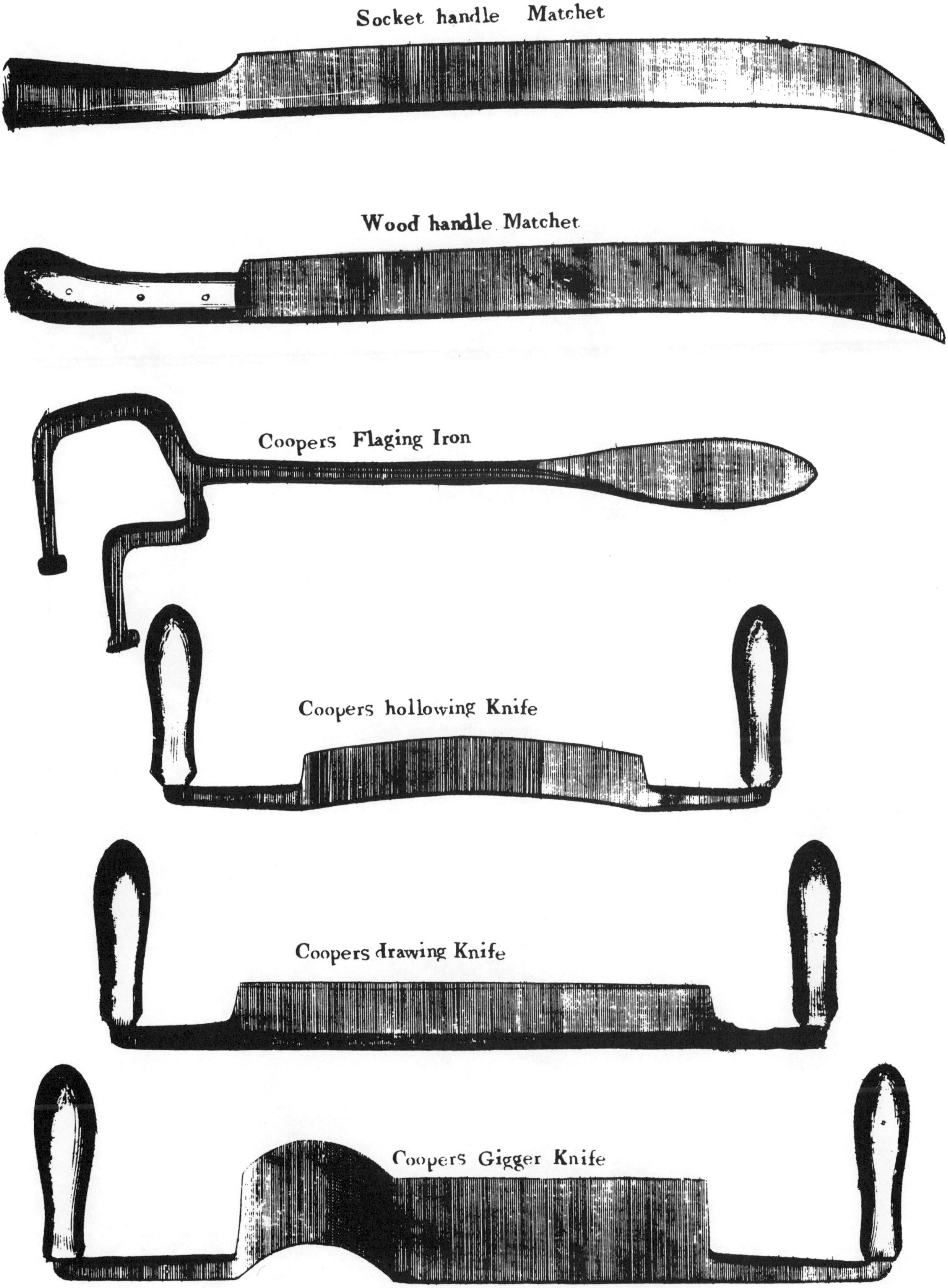

Plate LXXIV - Draw Knives, Cooper's Tools and Matchets

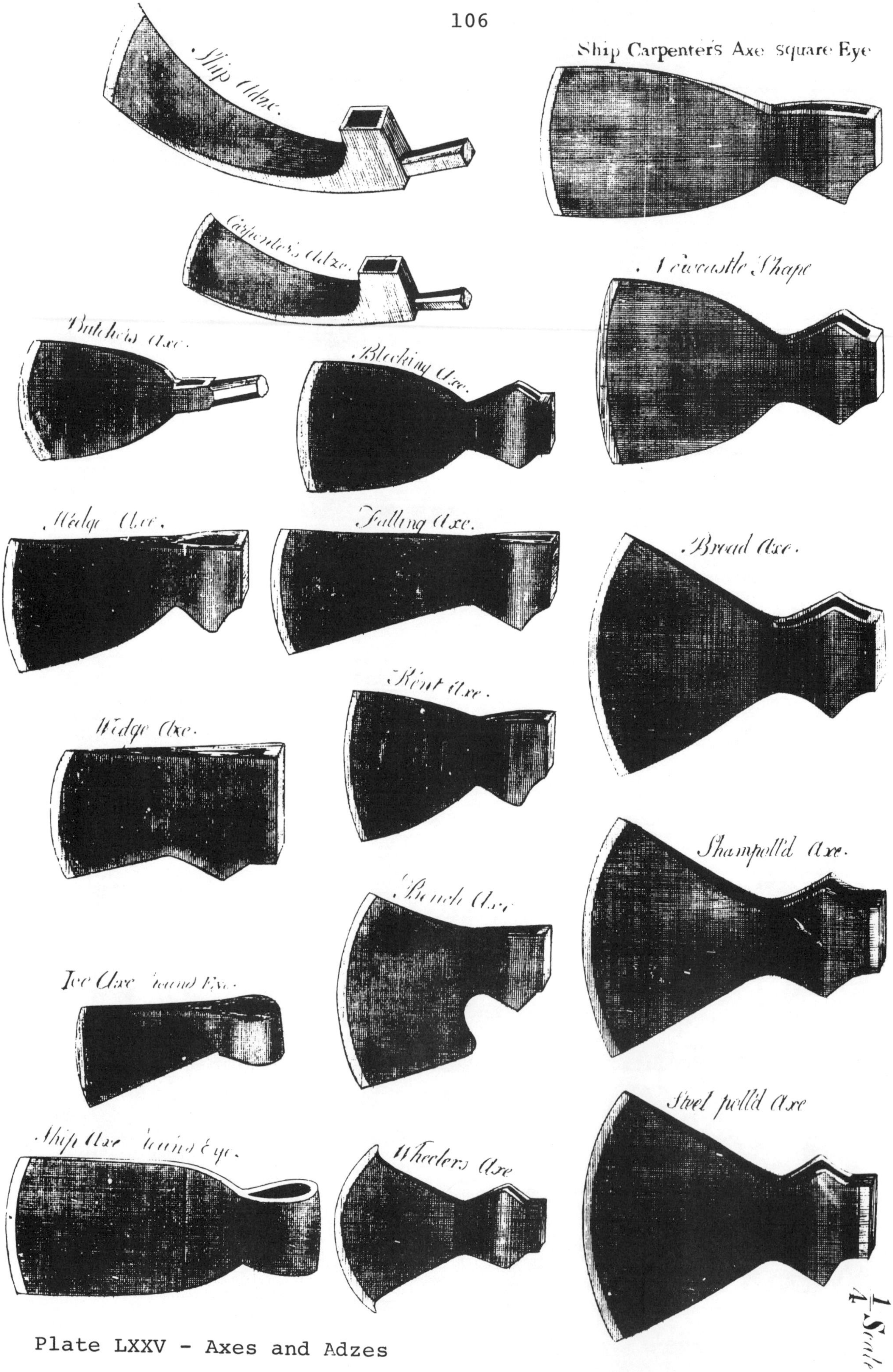

Plate LXXV - Axes and Adzes

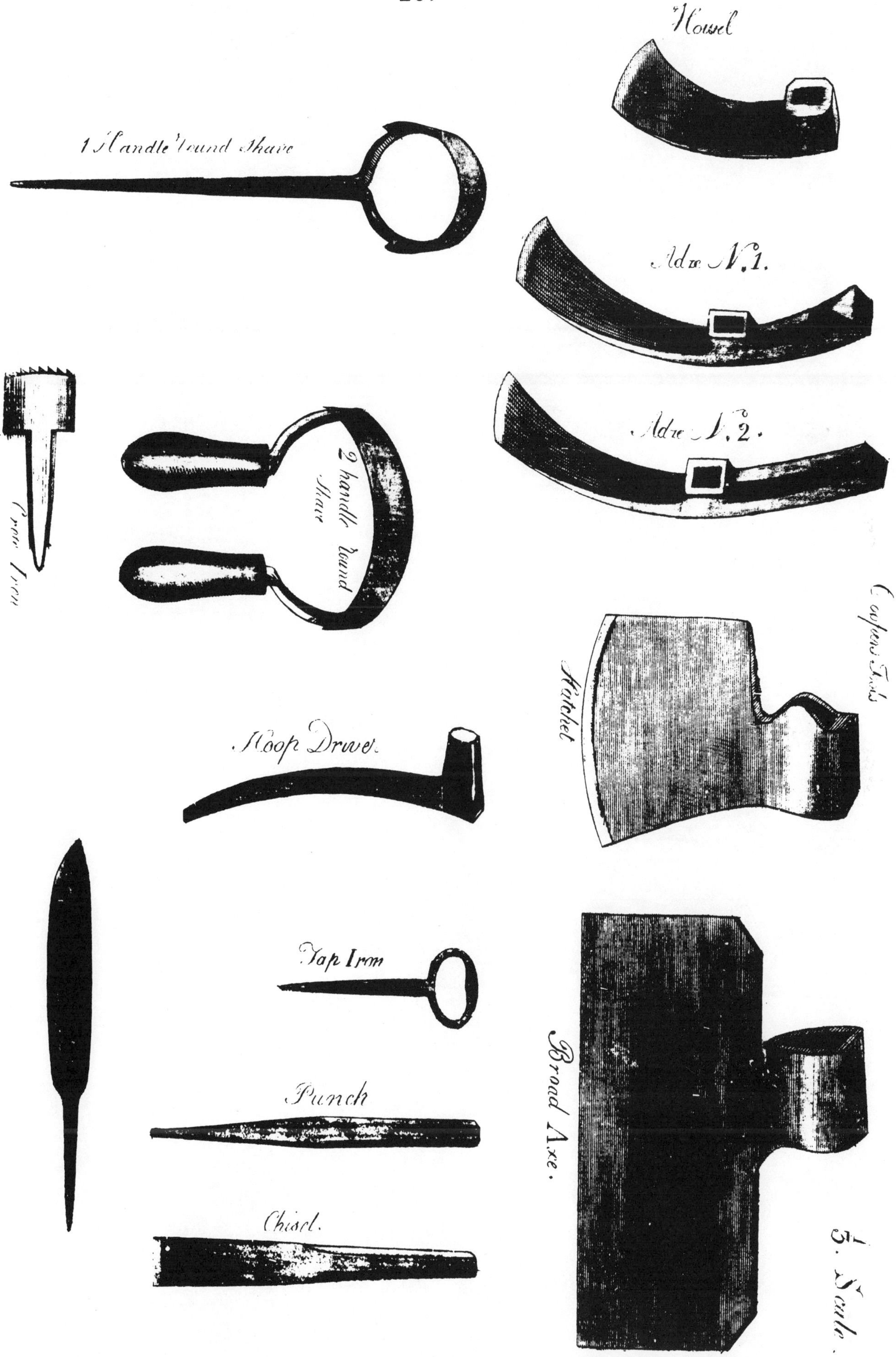

Plate LXXVI - Cooper's Tools

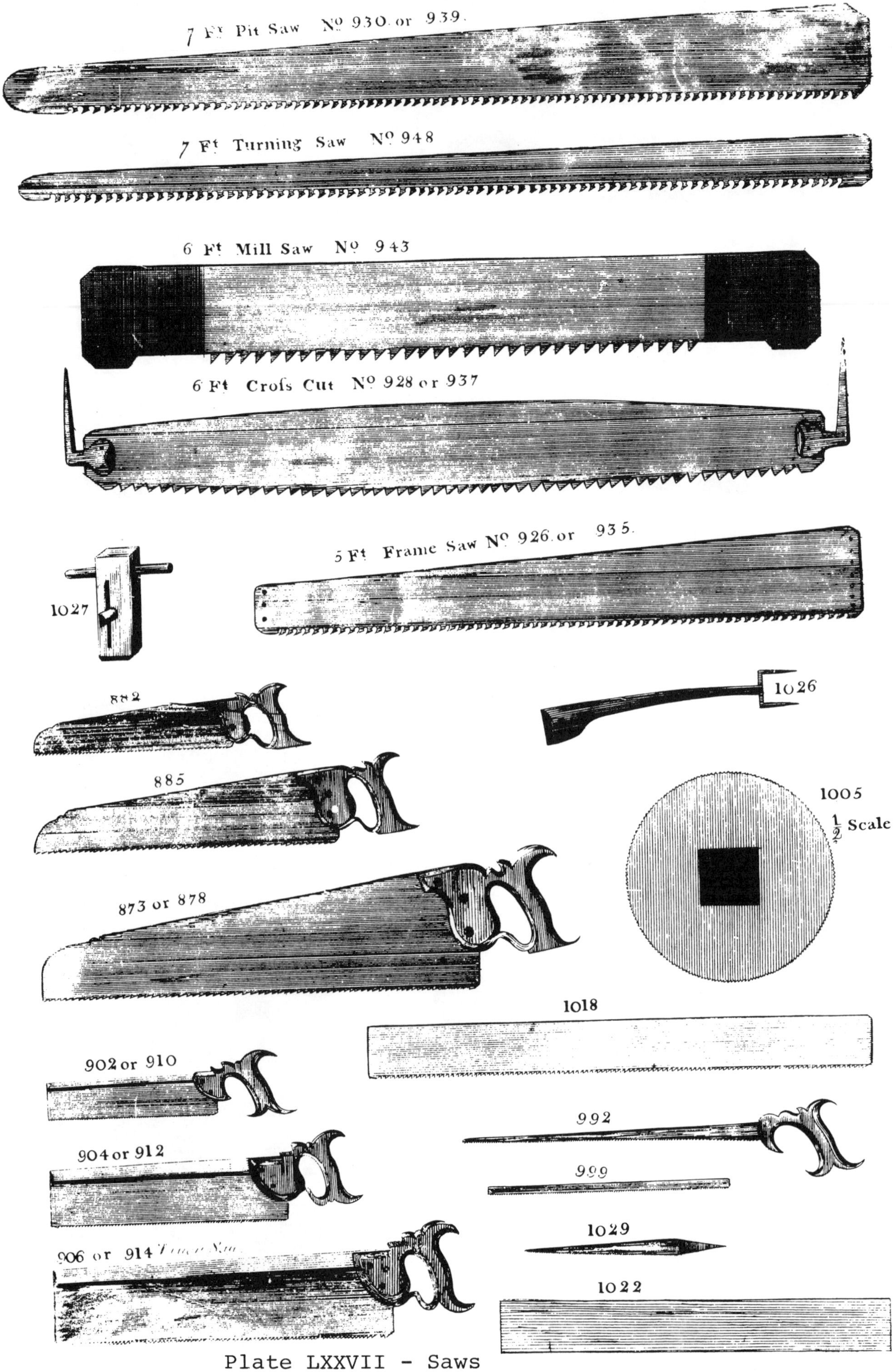

Plate LXXVII - Saws

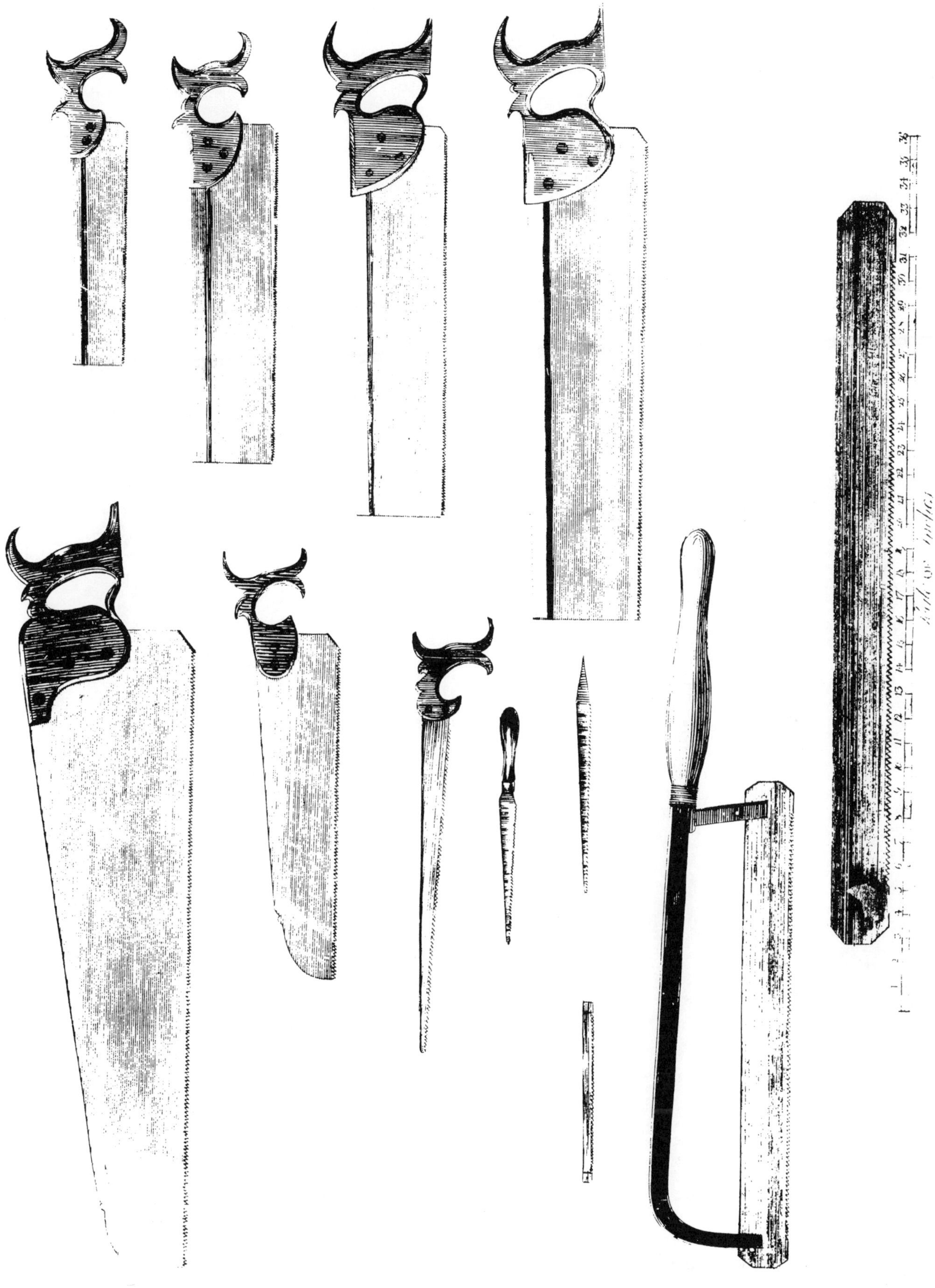

Plate LXXVIII - Saws

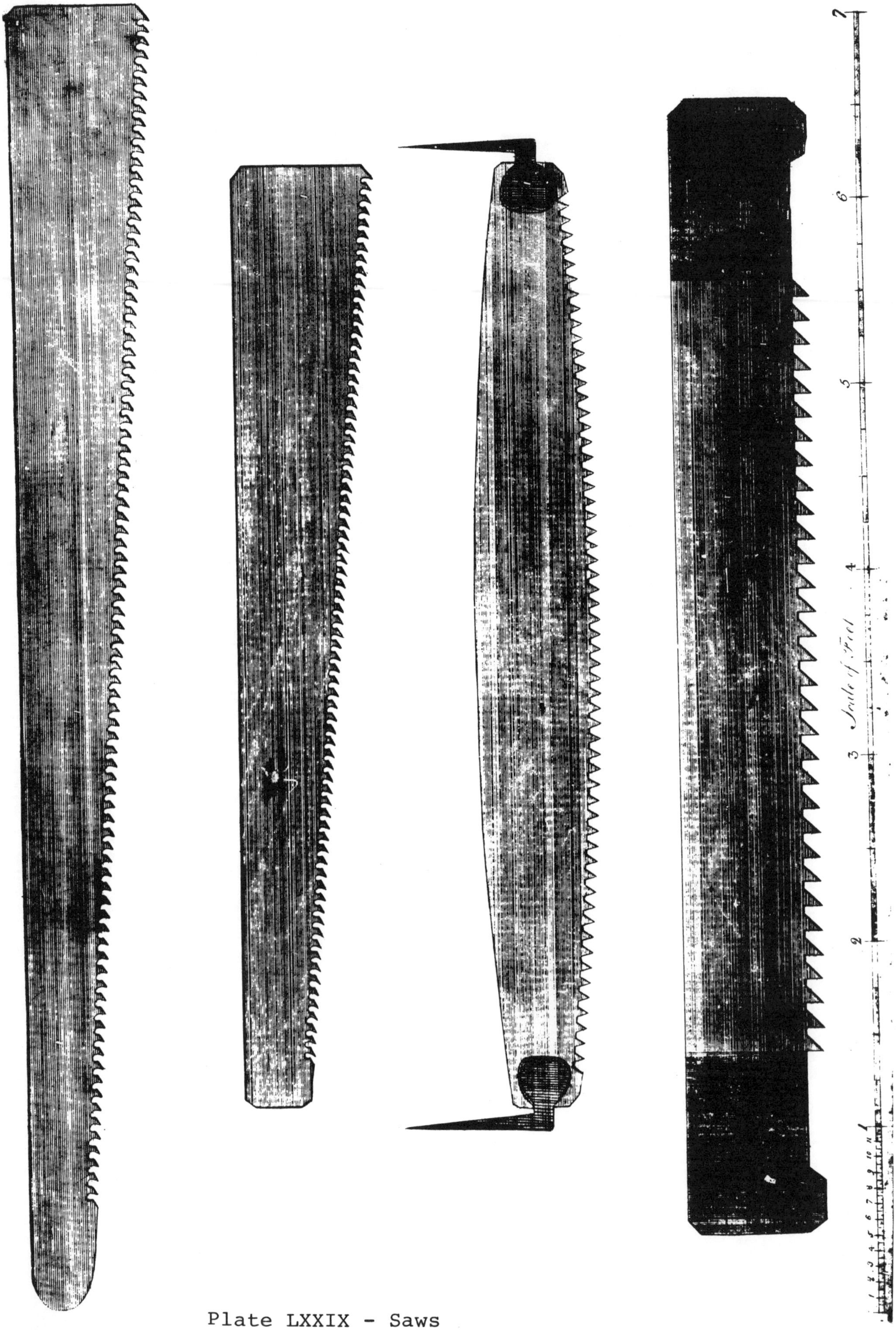

Plate LXXIX - Saws

PRICES OF
LIGHT EDGE TOOLS.

Sheffield, November 12, 1824.

At a Meeting of the Edge Tool Manufacturers, held this day, it was Resolved, that in consequence of the very great advance on Materials, the following Prices shall take place upon all Orders received on and after the 15th instant

CAST STEEL CHISELS AND GOUGES.

Cast Steel Firmer Chisels,	1-16th	1/8	3-16ths	1/4	5-16ths	3/8	7-16ths	1/2	5/8	3/4	7/8	1	1 1/8	1 1/4	1 3/8	1 1/2	1 5/8	1 3/4	2	2 1/4	2 1/2
	3s6d	3s6d	3s9d	3s9d	4s	4s	4s3d	4s3d	4s6d	5s	5s9d	6s6d	7s9d	9s	10s	11s	12s6d	14s	16s	19s	24s

Cast Steel Firmer Gouges	1/8	3-16ths	1/4	5-16	3/8	7-16ths	1/2	9-16ths	5/8	3/4	7/8	1	1 1/8	1 1/4	1 3/8	1 1/2	1 5/8	1 3/4	2	2 1/4	2 1/2	Inch
	4s	4s3d	4s3d	4s6d	4s6d	4s9d	4s9d	5s	5s	5s6d	6s3d	7s	8s3d	9s6d	11s	12s6d	14s	17s	20s	24s	30s	doz.

Strong Cast Steel Firmer Chisels	1/4	3/8	1/2	5/8	3/4	7/8	1	1 1/8	1 1/4	1 1/2	1 3/4	2	2 1/4	2 1/2	Inches.
	5s6d	5s9d	6s	6s9d	7s6d	8s6d	9s6d	11s6d	14s	16s6d	20s	24s	30s	36s	per dozen.
Do. Do. Gouges	6s	6s3d	6s6d	7s3d	8s0d	9s0d	10s	12s6d	15s	18s	24s	29s	36s	42s	ditto

Long Cast Steel Millwrights' Chisels	1/4	3/8	1/2	5/8	3/4	7/8	1	1 1/4	1 1/2	1 3/4	2	2 1/4	2 1/2	2 3/4	3	Inch
	6s6d	7s	7s6d	9s	10s6d	12s6d	15s	19s	24s	29s	34s	40s	48s	57s	67s	doz

Cast Steel Paring Chisels	1/8	1/4	3/8	1/2	5/8	3/4	7/8	1	1 1/8	1 1/4	1 1/2	1 3/4	2	2 1/4	2 1/2	2 3/4	3	Inch
	5s3d	5s6d	5s9d	6s3d	6s9d	7s6d	8s6d	9s6d	11s6d	14s	17s	20s	24s	30s	36s	43s	50s	doz.
Do Long thin do......	6s	6s6d	6s6d	7s	8s	9s	10s	11s	13s	15s6d	19s	22s	27s					
Cast Steel Coach Makers' Chisels	5s3d	5s6d	5s9d	6s3d	6s9d	7s6d	8s6d	9s6d	11s6d	14s	17s	20s	24s	30s	36s	42s	48s	doz.

Cast Steel Turning Chisels	1/8	1/4	3/8	1/2	5/8	3/4	7/8	1	1 1/8	1 1/4	1 1/2	1 3/4	2	2 1/4	2 1/2	2 3/4	3 in.
	5s	5s	5s6d	6s6d	7s6d	8s6d	9s6d	10s6d	12s	14s	17s	20s	24s	30s	36s	48s	60s do.
Do. Do. Gouges		6s	7s	8s	9s	10s6d	12s	14s	16s	18s	24s	28s	33s	39s	46s		

Long Cast Steel Turning Chisels	1/2	3/4	1	1 1/4	1 1/2	1 3/4	2	inch
	9s6d	12s	16s	20s	24s	28s	33s	doz

Do. Do. Gouges	1/4	3/8	1/2	5/8	3/4	7/8	1	1 1/4	1 1/2	Inch
	13s6d	15s	17s	19s	22s	25s	28s	34s	40s	per dozen.

Cast Steel Gunstocker's Chisels	2 1/4	2 1/2	2 3/4	Inch
	16s	18s	20s	doz.

Cast Steel Gunstocker's Gouges	5/8	3/4	7/8	1	Inch
	7s6d	8s	9s6d	11s	doz

Cast Steel Firmer Chisels sorted to 1 inch, 4s6d per set
Ditto Ditto Gouges sorted to 1 inch 5s0d per set
Cast Steel Turning Chisels assorted 1/4 to 1 1/2 inch, 9s6d dozen
Ditto Ditto Gouges assorted 1/4 to 1 inch 9s 6d per doz.
Ditto Ditto Thick Backs assorted to 1/4 inch 10s do.

PRINT CUTTERS' CARVING TOOLS AND ENGRAVERS.

Cast Steel Fluting Gouges 6s per set. Veaning Tools 5s6d per set. Parting Tools 8s per doz. Double-bent Fluting Gouges 7s per set. Flat and Middle Sweep Print Cutters' Gouges 3s3d per doz. Quick Sweep do. do. 3s6d per doz. Shouldered Print Cutters' Gouges Flat and Middle Sweep 6s per doz. Quick do. do. 6s per doz. Print Cutters' Chisels 2s per doz. Print Cutters' Blades 3s per doz. Print Cutters' Flat Shank Dog Legs 2s6d per doz. Ditto Round Shank 3s per doz. Black Carving Chisels 4s3d per do. Black Skew Carving Chisels 4s3d per doz. Moulding Gouges 7s per doz. Beading Gouges 7s per doz. Entering Chisels 6s per doz. Entering Gouges 7s per doz. Square and Lozenge Gravers 18s per gross. Scorpers ditto 18s per gross. Spit Stickers 20s. Knife Gravers 20s per gross. Three and Four Square Fluted Scrapers 15s6d per doz. Double Burnishers 10s. Single ditto 6s6d per doz.

Plate LXXXIA - Prices of Light Edge Tools - November 12, 1824
Sheffield Edge Tool Manufacturers' Resolution

FIRMER CHISELS AND GOUGES.

	1/8	1/4	3/8	1/2	5/8	3/4	7/8	1	1 1/8	1 1/4	1 3/8	1 1/2	1 5/8	1 3/4	2	2 1/4	2 3/8	2 1/2	Inch
Short Firmer Chisels	3s	3s	3s3d	3s3d	3s6d	3s9d	4s	4s6d	5s	5s9d	6s6d	7s6d	8s6d	9s6d	11s	13s	14s6d	16s	per dozen
Do. Do. Gouges..	3s6d	3s6d	3s9d	3s9d	4s	4s3d	4s6d	5s	6s	6s9d	7s6d	8s6d	9s6d	11s	13s	15s	17s	19s	do.

	1/4	1/2	5/8	3/4	7/8	1	1 1/8	1 1/4	1 1/2	1 3/4	2	2 1/4	2 1/2	Inch
Strong Firmer Chisels	4s	4s3d	4s9d	5s	5s6d	6s	6s9d	7s6d	8s6d	10s6d	13s	16s	20s	per dozen
Do. Do. Gouges	4s6d	4s9d	5s3d	5s6d	6s	6s6d	7s9d	8s6d	9s6d	12s	15s	18s	22s	do.

	1/8	1/4	3/8	1/2	5/8	3/4	7/8	1	1 1/8	1 1/4	1 3/8	1 1/2	1 5/8	1 3/4	2	2 1/4	2 1/2	inch
Long Firmer Chisels	3s9d	3s9d	4s	4s3d	4s6d	4s9d	5s	5s6d	6s6d	7s6d	8s3d	9s	10s	11s	13s	15s	18s	doz
Do do Gouges	4s3d	4s3d	4s6d	4s9d	5s	5s3d	5s6d	6s	7s	8s6d	9s3d	10s	11s	12s	14s	16s	19s	do.

	1/4	3/8	1/2	5/8	3/4	7/8	1	1 1/8	1 1/4	1 3/8	1 1/2	1 5/8	1 3/4	2	inches
Long Cant Firmer Chisels	8s	8s3d	8s6d	9s	9s6d	10s6d	11s6d	13s	14s6d	16s	18s	20s	22s	25s	per dozen

Short Firmers assorted 1/8 to 1 inch, 3s6d per doz. Do Gouges 1/8 to 1 inch 4s per doz.

MORTICE CHISELS.

	1/8	3-16ths	1/4	5-16ths	3/8	7-16ths	1/2	9-16ths	5/8	3/4	7/8	1	inch
Mortice Chisels	4s9d	5s	5s6d	6s	7s	8s	9s6d	10s6d	12s	15s	20s	25s	per dozen
Socket Mortice Chisels			8s	8s6d	9s6d	10s6d	12s	14s	16s	19s	24s	30s	per doz

Tanged Mortice Do..do. sorted 1/8 to 1/2 inch 6s6d per doz.

	1/8	3-16ths	1/4	5-16ths	3/8	7-16ths	1/2	9-16ths	5/8	3/4	7/8	1	inch
Best Joiners' Mortice Chisels	11s6d	11s6d	12s	12s6d	13s	13s6d	14s6d	16s6d	20s	25s	34s	42s	per doz
Best Cabinet do do	11s	11s	11s6d	12s	12s6d	13s	13s6d	15s6d	19s	24s	33s	40s	do

Best Joiners' Mortice Chisels 9s per set of 8, sorted to 9-16ths—Do. to 5-8ths 9s4d—Do. Cabinet 8s6d per set of ditto.

	1/8	3-16ths	1/4	5-16ths	3/8	7-16ths	1/2	9-16ths	5/8	3/4	7/8	1	inch
Cast Steel Joiners' Mortice Chisels	13s6d	13s9d	14s	14s6d	15s	15s6d	16s6d	18s6d	22s	27s	36s	44s	per dozen

Cast Steel Cabinet Mortices 9d per doz. less than Cast Steel Joiners' do.
Cast Steel Joiners' Mortice Chisels 10s6d per set. Ditto Cabinet 10s per set.

TURNING CHISELS AND GOUGES.

	1/4	3/8	1/2	5/8	3/4	7/8	1	1 1/8	1 1/4	1 3/8	1 1/2	1 3/4	2	inch
Turning Chisels	4s	4s3d	4s6d	4s9d	5s	5s6d	6s	6s6d	7s	7s6d	8s6d	10s	12s6d	per doz
Ditto..Gouges	4s9d	5s	5s3d	5s9d	6s	6s6d	7s6d	8s6d	9s6d	10s	11s6d	14s	17s	do.

Turning Chisels assorted to 1 1/2 inch 6s per dozen......Ditto Gouges to 1 inch 6s per dozen.

SOCKET CHISELS AND GOUGES.

	1/2	5/8	3/4	7/8	1	1 1/4	1 1/2	1 3/4	2	2 1/4	2 1/2	Inch
Socket Chisels	7s	7s6d	8s	8s9d	9s6d	10s	10s6d	11s6d	13s	15s6d	19s	per dozen
Do. Gouges	7s6d	8s	8s6d	9s3d	10s	11s	11s6d	12s6d	14s	17s6d	21s	do.

	1/2	5/8	3/4	7/8	1	1 1/4	1 1/2	1 3/4	2	2 1/4	2 1/2	2 3/4	3	inches
Strong Socket Chisels	8s	8s9d	9s6d	10s	11s	12s6d	14s	16s	18s	22s	27s			
Cast Steel Socket Chisels ..	10s6d	11s	12s	12s6d	13s	14s	16s	18s	21s	26s	32s			per dozen
Ditto....ditto....Gouges	11s	11s6d	12s6d	13s	13s6d	15s	17s	19s	22s	28s	34s			per dozen
Stalking Chisels	10s6d	11s6d	12s6d	13s6d	14s6d	16s	18s	21s	25s	29s	33s	38s	42s	per dozen

Socket Chisels assorted 1/2 to 1 1/2 inch 8s6d per dozen.—Do. Gouges do. to 1 inch 8s6d per dozen.

Plate LXXXIB - Prices of Light Edge Tools - November 12, 1824
Sheffield Edge Tool Manufacturers' Resolution

2

PLANE IRONS.

Short Plane Irons to	1¼	1⅜	1½	1⅝	1¾	1⅞	2	2⅛	2¼	2⅜	2½	2⅝	2¾	2⅞	3	inches
	3s9d	4s	4s	4s3d	4s6d	4s9d	5s	5s6d	6s	6s6d	7s3d	8s	9s	10s6d	12s	per dozen
Cut Plane Irons	4s3d	4s6d	4s6d	4s9d	5s	5s3d	5s6d	6s	6s6d	7s	7s9d	8s6d	9s6d	11s	12s6d	do
Top Plane Irons			3s4d	3s7d	3s10d	4s1d	4s4d	4s10d	5s4d	5s10d	6s7d	7s4d	8s4d	9s10d	11s4d	do

With solid nuts 2s per dozen more.

Toothed Plane Irons double the price of common Plane Irons.

Double Plane Irons to	1¾	1⅞	2	2⅛	2¼	2⅜	2½	2⅝	2¾	inches
	11s6d	12s	12s6d	13s6d	14s6d	15s6d	17s	18s6d	20s	per dozen

Long Plane Irons to	1⅛	1¼	1⅜	1½	1⅝	1¾	1⅞	2	2⅛	2¼	2⅜	2½	inches
	4s	4s3d	4s6d	4s9d	5s	5s3d	5s6d	6s	6s6d	7s6d	8s6d	10s	per doz

Warranted Cast Steel Plane Irons to	1¾	1⅞	2	2⅛	2¼	2⅜	2½	2⅝	2¾	inches
	7s	7s6d	8s	8s6d	9s	10s	11s	12s6d	14s	per dozen.
Do......Cut......... ...	7s6d	8s	8s6d	9s	9s6d	10s6d	11s6d	13s	14s6d	per dozen.

WARRANTED DOUBLE IRONS.

Warranted Cast Steel Double Plane Irons, to	1¾	1⅞	2	2⅛	2¼	2⅜	2½	2⅝	2¾	inch
	15s6d	16s	16s6d	17s	18s	19s	20s6d	22s	25s	per dozen

Common Double Irons with Solid Nuts, 2s per dozen extra.

Soft Moulding Irons assorted to 1½	do to 1¾	& do to 2 inch
27s	30s	33s per gross

Common Plough Bits assorted 3s per dozen
Best.....ditto.....ditto.... 3s4d per set of 8
Groving Irons assorted 1½ to 1¾ at 5s6d per dozen
Strong....do....do. 1½ to 1¾ inch at 19s per dozen pair

Hollows and Rounds 7s per dozen pair
Square Rabbit Irons assorted to 1½ inch at 3s6d per dozen
Skew....ditto....ditto....ditto to 2 inch 4s6d per dozen

Coopers' Jointer Irons	3	3⅛	3¼	3⅜	3½	3¾	4 inches
10 Inch Long....	15s6d	16s	17s	18s	19s	21s	23s per dozen
11 Inch Do......	16s6d	17s	18s	19s	20s	22s	25s do
12 Inch Do......	17s6d	18s	19s	20s	22s	24s	27s do.

Coopers' Jointer Double Irons 3s per dozen more than double the price of Coopers' Jointers.

DRAWING KNIVES.

Drawing Knives	8	9	10	11	12	13	14	15	16	inches
	12s6d	13s6d	14s6d	15s6d	16s6d	18s	20s	22s6d	26s	per dozen
Brass hoops & rivetted Cast Steel	19s	20s	21s	22s6d	25s	28s	33s			

Coopers' Hollowing Knives 8 Inches 14s per dozen.
Do......do......do.... 9 Inches 16s do.

Coopers' Broad Hollowing Knives 9 inch, 18s per dozen
Do..1 and 2 handled Round Shaves... 12s6d per dozen

Broad Coopers' Spokeshave Irons, to	5	5½	6	6½	7	inches
	7s6d	8s	8s6d	9s	9s6d	per dozen

Small Joiners' Spokeshave Irons to 3½ inch, 5s9d per dozen, and Middle from 3½ to 4½ inch, 6s9d per dozen.

Coopers' Whittles, 4s per dozen.

Cast Steel Joiners' Spokeshave Irons	2½	3	3½	4	4½	5	6	inches
	5s3d	5s9d	6s3d	6s9d	7s3d	7s9d	8s6d	per dozen

Bookbinders' Plough Knives 15s per dozen............Ditto Cast Steel 18s per dozen.

Draw Borers,	8	9	10	inches
	12s	13s	14s	per dozen

H. A. BACON, PRINTER, INDEPENDENT OFFICE, ANGEL STREET, SHEFFIELD.

Plate LXXXIC - Prices of Light Edge Tools - November 12, 1824
Sheffield Edge Tool Manufacturers' Resolution

The Edge Tool Manufacturers met and revised prices again in 1836, 1846, 1853, 1864 and finally in 1872. These were strictly adhered by all manufacturers. Such an agreement on February 2, 1836 appeared on the following notice.[11] (Plate LXXXIII)

AT A MEETING

Of the Edge-Tool Manufacturers, held this 2nd Day of February, 1836,

at the Cutlers' Hall, in Sheffield;

IT WAS UNANIMOUSLY RESOLVED,

That, in consequence of a considerable advance on Materials, &c., the Discount on Edge-Tools should be reduced 10 per Cent. upon all Orders received on and after the 8th instant.

WILLIAM ASH & CO.
WILLIAM & SAMUEL BUTCHER.
SAMUEL BISHOP.
JAMES CAM.
JOHN CORFIELD.
JOHN CUTLER.
GEORGE CARR & CO.
WILLIAM GREAVES & SONS.
SAMUEL GREEN.
ISAAC GREAVES.
GEORGE HATTERSLEY.
THOMAS HAYWOOD.
PHILIP LAW.
THOMAS MOTTRAM.
MOULSON BROTHERS.
HENRY MORT.
GEORGE MOSELEY.
SAMUEL NEWBOULD & SONS.
WILLIAM OAKES.
J. W. PARKIN & CO.
FRANCIS PEARSON.
JOHN SORBY & SONS.
FREDERICK STONES.
WILLIAM STACEY.
JOSEPH SMITH.
JOHN TURNER.
DAVID WARD & SON.

Plate LXXXIII - Meeting of Edge Tool Manufacturers, February 2,1836

Expansion of the edge tool business resulted with the completion of the Canal from Tinsley into Sheffield in 1819. Coal became more available, resulting in increased steel production. William Greaves, a successful factor, probably realizing the opportunities in making edge tools had the Sheaf Works built in the early 1820's along this canal. (See Plate LXXXIV) This is reported to have been the first factory where raw materials of coal and iron entered at one end and finished goods of edge tools came from the other end.[12] Numerous other edge tool factories were soon to follow: Pond Works, Castle Hill Works, Park Works, Cyclops Works, Globe Works, Spital Hill Works, etc.

A sketch of the Sheaf Works is also shown in Plate XI, page 29, and a brief account appears on page 28.

Plate LXXXIV - THE SHEAF WORKS

The earliest extant Sheffield trade catalogue of edge tools yet reported was by Cutler & Co., Castle Hill Works, c.1833 - 1837.[13] This is in the collection of the Victoria & Albert Museum in London. The Plate of Gentleman's Tool Chest from this has been previously published.[14] The frontispiece from this book with its interesting quotation: *"It is well known that a great proportion of Tools are manufactured ONLY to Sell; and, an INFALLIBLE TRUTH That those are the Cheapest which do the most Work".* is shown in Plate LXXXV.[15] It is not known whether Hiram Cutler or the predecessor firm of Furniss, Cutler and Stacey used available engravings from an existing Sheffield list or had these specially prepared for their own use. Plate LXXXVb illustrated a plate from this book showing some chisels made from the best German steel and warranted by Weldon.

William Weldon was noted in the 1774 Sheffield Directory at Colston Court as a wool shear and edge tool maker. In 1814 he was in partnership with Furniss at Castle Hill. After Weldon's death this firm became Furniss, Cutler and Stacey in 1821 "merchants, factors, edge tool, saw and steel manufacturers". The latter implied that they were "converters of bar steel and cast steel refiners", specifically mentioned in the 1828 Directory. In 1837 Hiram G. Cutler acquired sole proprietorship, but in 1849 the name changed to H.Cutler, Son and Chambers, continuing until 1865.

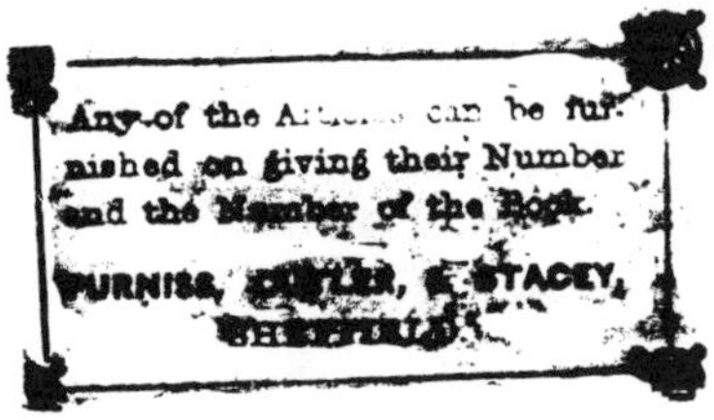

"*Any of the Articles can be furnished on giving their number and the Numbers of the Book*"
FURNISS, CUTLER & STACEY
SHEFFIELD

c.1837

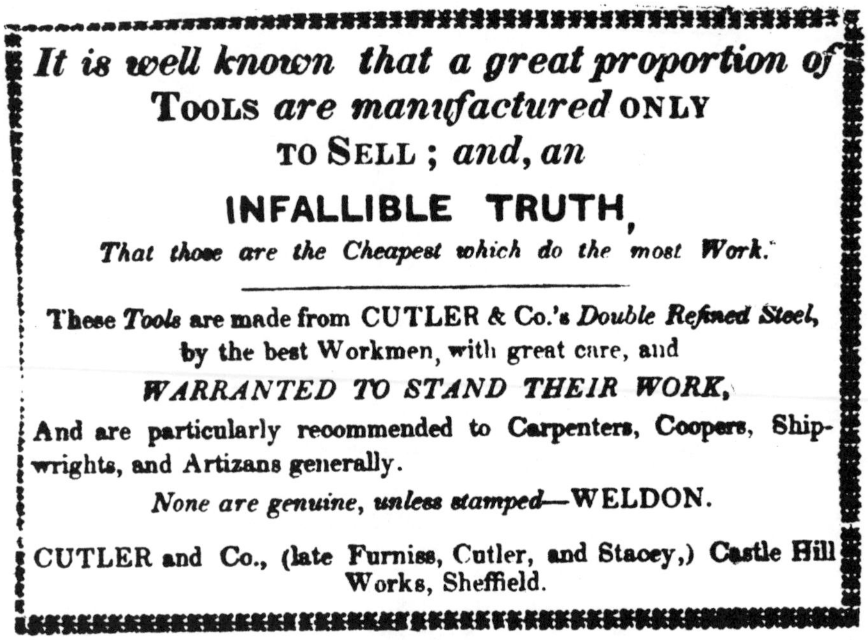

Plate LXXXV - Frontispiece to *Cutler and Co. Pattern Book of Tools*
[From Book 87, Collection of Victoria & Albert Museum, London, Item No. M63.1]
[Crown Copyright, Plates CXXV & CXXVI reprinted courtesy of V. & A. Museum]

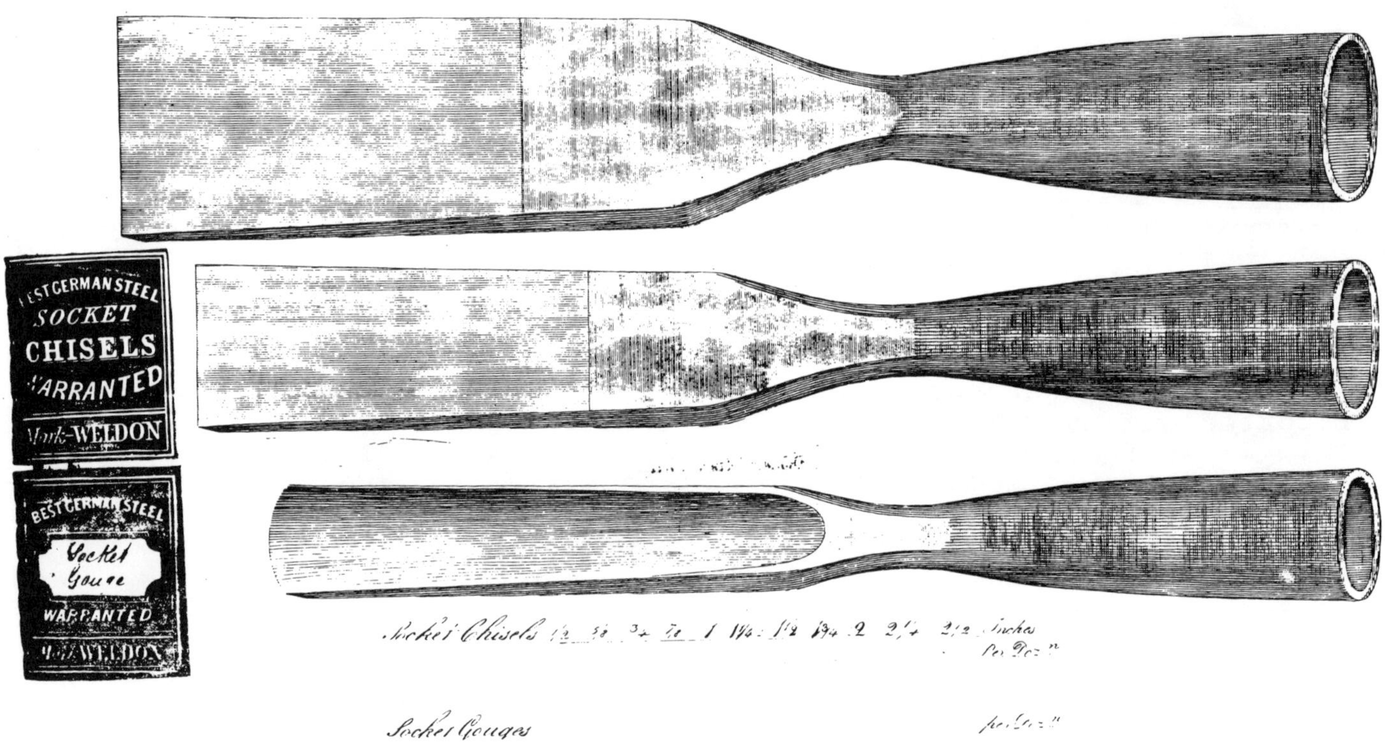

Plate LXXXV b - Plate of Chisels from Above Pattern Book c.1837

The catalogue was 9 in. x 14-3/4 in. and contained 56 plates with no watermark or date. The tools illustrated included a full line of joiners' and carpenters' chisels (firmers,gouges and mortise); engravers' tools;plane irons; draw knives; spokeshaves; bevils; squares; saws; coopers' tools; tools for skinners, tanners, weavers, book binders; ship builders, shoemakers, tailors; files and rasps; garden tools, picks, shovels, mattocks and spades. Apparently the mark of Weldon, also used by the former parent firms, was acquired and used long after his death.

Surviving records of the firm Marshes & Shepherd (1824-1848) and their successors, Marsh Brothers & Co.(1849-1970-) provide valuable study. These include several price lists and some dated trade catalogues.[16] Pollard has written a history of the Marsh family in Sheffield over three centuries.[17] They had been cutlers since 1631. James Marsh & Co. was organized in 1810 when James took in his three sons and son-in-law, a Shepherd, into partnership. The firm soon became known as Marsh & Shepherd. In 1816 they rented quarters at Castle Hill, but by 1822 their business had expanded. They moved to Porter Street to larger quarters. After 1824 the firm became Marshes and Sheperd. Their business increased to a point that it became desirable to melt and fabricate their own steel. Again it was necessary to move. In 1828 they acquired steel producing facilities at Pond Works. Sales agencies were established in America at New York and Philadelphia. Export sales for their products increased. In combination with sales in America of their own products other Sheffield edge and joiner tools were offered on commission. In 1838 with the firm of Nayloy & Vickers they acquired ownership of the established edge tool business of the late James Cam. Cam had been in business at Norfolk Street since 1817. This acquisition permitted them to the right of the mark *"James Cam"* and *"JAMES CAM & CO."* which they then stamped on their own manufactured files and edge tools. An arrangement was made with Samuel Peace, a small edge tool manufacturer, to run this part of the business at the former works of James Cam so that Peace could eventually earn ownership of this factory.[18] Accordingly tools were produced by Peace with the marks: *Marshes & Shepherd;James Cam; and Samuel Peace & Co.*

Sometime after July 1846 and before 1849 Marshes & Shepherd issued an extensive trade catalogue (77 Plates, probably using the engraving blocks from an existing Sheffield List) noting their products as: general merchants; importers of Sweedish and Russian iron; converters and cast steel refiners; makers of single and double shear steel and other Blister Steel; Rollers, Tilters and forgers; Table Cutlery. pen and Pocket Knives; fine razors; superior files; circular and other saws; *sole manufacturer of"JAMES CAM's" celebrated edge tools* ; also manufacturers of bearing and buffer springs for Locomotive engines, Tenders, Passenger carriages; &c.[19] Additionally the catalogue illustrated a full line of joiner and garden tools. This was accompanied by a price list of 18 pages.[20] This catalogue can be dated bewteen the above limits as the prices conformed to those of an existing Trade List of Prices of Edge Tools and Joiner Tools Manufactured by James Howarth, Fitzwilliam Street, Sheffield, which stated: "At a General Meeting of Edge and Joiners' Tool Manufacturers, held at Cutlers' Hall, Shefield, *July 3,1846*, it was unanimously resolved, on Account of the Advances of Wages and Materials, that the following List of Prices should be adopted by the Trade."[21] In 1849 the firm name of Marshes & Shepherd was changed to Marsh Brothers & Co.

Soon after the formation of Marsh Brothers & Co. in 1849, this trade catalogue was re-issued with a new title page (See Plate LXXXVI). Note that this is dated 1849 and reference is made to *LATE MARSHES & SHEPHERD*. The 18 page price list accompanying this Catalogue tabulated the same prices as the previous by Marshes & Sheperd. Plate 26 from this 1849 Catalogue illustrated Plane Irons with the imprint of James Cam (Plate LXXXVII) , who had been deceased 12 years. As will be noted in subsequent Chapters, the planes offered in this Catalogue were made by Thomas Turner, the braces by Brown & Flather, the edge tools by Samuel Peace & Co., and the saw works by Messrs. Millington & Co. at Shemeld Court.[22] Undoubtedly several other items of joiner and other

MARSH BROTHERS & CO.,

(LATE MARSHES AND SHEPHERD,)

POND WORKS, SHEFFIELD,

GENERAL MERCHANTS,

IMPORTERS OF SWEDISH AND RUSSIAN IRONS,

CONVERTERS AND CAST STEEL REFINERS;

MAKERS OF SINGLE AND DOUBLE SHEAR STEEL, GENUINE Ⓛ, AND OTHER BLISTER STEEL,

WARRANTED COACH AND RAILWAY SPRING STEEL;

ROLLERS, TILTERS, AND FORGERS

Of all descriptions of Cast, Shear, and Bar Steel in uses,

SHEETS, RODS, &c., FOR MECHANICAL AND MANUFACTURING PURPOSES.

MANUFACTURERS OF

TABLE CUTLERY, PEN AND POCKET KNIVES,

FINE RAZORS,

SUPERIOR FILES, CIRCULAR & OTHER SAWS,

AND SOLE MANUFACTURERS OF "JAMES CAM'S" CELEBRATED EDGE TOOLS.

ALSO, MANUFACTURERS OF BEARING AND BUFFER SPRINGS, &c.

FOR LOCOMOTIVE ENGINES, TENDERS, PASSENGER CARRIAGES, LUGGAGE WAGONS, AND OTHER RAILWAY PURPOSES.

CORPORATE MARK.

ENTERED AT STATIONERS' HALL.

1849.

Plate LXXXVI - Title Page of 1849 Marsh Brothers & Co. Trade Catalogue
[Archives, Marsh Collection #117, Sheffield Central Library]

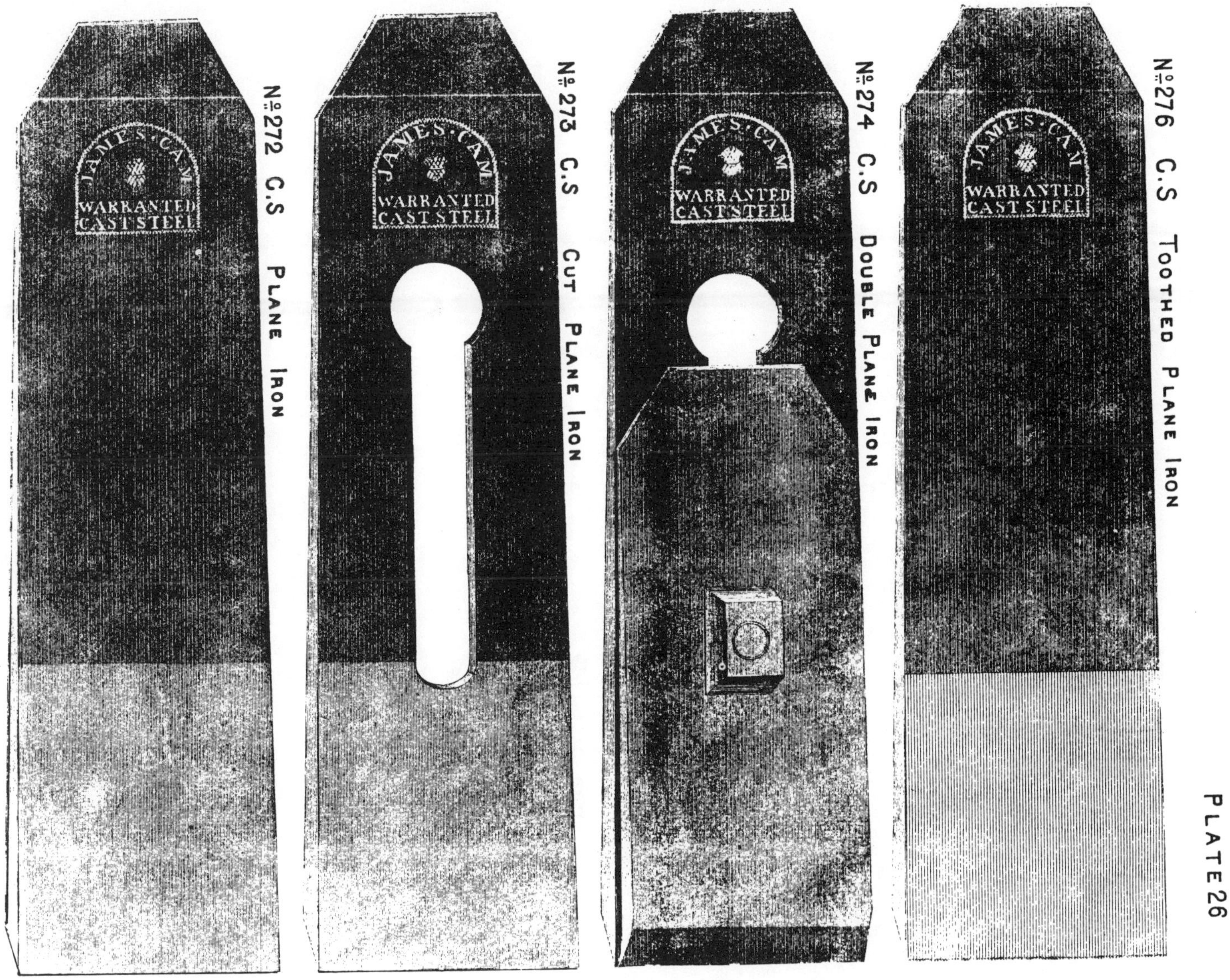

Plate LXXXVII - Plate 26 from Marsh Brothers 1849 Trade Catalogue
[Archives - March Collection #117 Central Library Sheffield]

Plane irons with the imprint of "JAMES CAM" sold by Marsh Brothers.

CHARLES CAMMELL & CO.'S

Illustrated List

OF

FILES & RASPS, SAWS,

LIGHT AND HEAVY EDGE TOOLS,

JOINERS' TOOLS, ENGINEERS' TOOLS,

ANVILS,

BELLOWS, LANCASHIRE TOOLS,

SPADES & SHOVELS, &c.,

CYCLOPS STEEL & IRON WORKS, SHEFFIELD,

AND 5, BARGE YARD, BUCKLERSBURY, LONDON, E.C.

Plate LXXXVIII - Title Page CHARLES CAMMEL & CO. Trade Catalogue, c1858 - 1865
[Collection of Central Library, Sheffield]

Plate LXXXIX - Cyclops Works of Charles Cammell & Co. [Advertisement from *A Look at Modern Things*, Birmingham, 1858]
Photograph Courtesy of Local History Department, Birmingham Reference Library

RAILWAY VIEW
OF
THE CYCLOPS STEEL WORKS,
SHEFFIELD.

CHARLES CAMMELL & CO.,

MANUFACTURERS OF

SPRINGS,

For Locomotive Engines, Tenders, Railway Carriages and Waggons; viz. Bearing, Buffing, Spiral, Patent Bow, Elliptic, Disc, and every Description of Springs in use.

In this Department the Proprietors respectfully invite Consumers to visit their Works, and inspect the Beautiful Machinery they have erected for producing exactness and uniformity.

MANUFACTURERS OF

VICES, HAMMERS, SMITHS'

AND SHIP BUILDERS' TOOLS,

RAILWAY COUPLING SCREWS,

BUFFER RODS, WAGGON AND CARRIAGE IRON WORK,

AND ALL USES IN STEEL AND IRON

FOR RAILWAY WORKS;

ALSO EVERY DESCRIPTION OF IRON WORK FOR GUN CARRIAGES & FIELD PIECES.

Plate XC - Advertisement of Charles Cammel & Co. *[1854 Sheffield Directory]*

Also see Plate LIX - page 76

tools appearing in this catalogue and price list were not actually manufactured by Marsh Brothers & Co, but were sold on a commission or taken in exchange for steel or other goods produced by this firm. This practice of marketing products of other firms along with goods made by the concern issuing a trade catalogue is believed to have been widespread among Sheffield tool and steel companies. This same arrangement in issuing pattern books of goods had been followed by Birmingham manufacturers and merchants since the turn of the 19th Century.[23] Also in 1801 Pater Stubs of Warrington, Lancashire had issued a pattern offering goods other than his own manufacture.[24]

A comparable situation developed with Charles Cammel & Co., who published an extensive trade catalogue of 71 pages (9 in. x 12 in.) c.1858 - 1865.[21] The title page of this is shown as Plate LXXXVIII. Charles Cammel was first listed in the 1837 Sheffield Directory as a commercial traveller (salesman). Visualizing the opportunies and fortune to be made with steel, he combined with another merchant, Henry Johnson in 1839 and formed Johnson, Cammel & Co. They advertised as: steel converters, refiners and manufacturers of files, saws and general merchants. After building the Cyclop Works adjacent to the Rotherham Railway in 1845, the partners separated in 1854. Henry Johnson moved to Napier Street, continuing as a screw and bolt maker. Charles Cammel & Co. remained at the Cyclops Works (see Plate LXXXIX) continuing with the same steel products. (see Plate XC) Subsequently this firm developed into among the largest producers of steel in England, becoming Cammel, Laird & Co. in 1909 and part of the English Steel Corporation in 1924. The Trade Catalogue noted above was published for the firm by Pawson & Brailsford using printing cuts which they had also used in printing the Sheffield List of the period. The only tools actually manufactured by Charles Cammel & Co. that appeared on this list were the files and saws. The others were either commission items or obtained from other Sheffield concerns in exchange for steel.

The following excerpt from pages 268 - 278 of the *1879 Illustrated Guide to Sheffield* presents details concerning the manufacture of edge tools and brief insights into four prominent edge tool and joiner firms: Messrs. Howarth & Sons; John Sorby & Sons; Ward & Payne; William Marples & Sons.

"The manufacture of light and heavy edge tools, and other joiners' tools, is also a considerable industry in Sheffield. The variety of these tools is very great, including chisels, gouges, augers, gimlets, plane irons, braces and bits, squares, levels, spoke shaves, hammers, axes, adzes, hedging and other bills, hatchets, &c., &c. Some of the tools, moreover, are made of many sizes and shapes. In gouges, for instance, there are no less than seven different "sweeps" or bends, and each bend is made of twenty different sizes, making 140 varities of one simple tool.[25] The amount of detail in the trade is consequently enormous, and the sub-division of labour is proportionately great. The forgers work in pairs, as in the table knife and the heavier branches of the file trade--a forger and striker in each shop--and the two are kept as far as practicable to one class of work--often to a few sizes of one tool--in which they acquire great skill. In the shaping of tools, "prints" and "bosses" are much more largely used than for table cutlery or files, and in much greater variety of size and shape. These articles are, in fact, moulds, into the recesses of which the pliant metal is crushed by the hammer while red hot, instead of being poured while in a liquid state, as in the making of castings; and the process is called "mooding" (moulding). As previously mentioned, one side of the dies or moulds, known as "prints" or "bosses," consists of a small square block of metal, and fits into the anvil; but the other side, which is held in the hand during use, is hammer-shaped, and has a very primitive handle, formed by twisting a twig of hazel round it, and fastening the two ends together with a ring. Special dies being required for every variety in the size or shape of the tool to be made, prints and bosses form an important part of the furniture of an edge tool forger's workshop. The walls of workshops in which the smaller tools are made are often almost concealed by rows of these implements, carefully arranged according to size and pattern, the set for a single shop costing £40 to £50. To this explanation of the simple means by which apparently difficult ends are attained, we must add that though the forging of edge tools does not differ in principle from the forging of cutlery and files, the greater number and complication of the processes occasions a much larger amount of labour. Take, for instance, a common joiner's chisel, which is one of the most simple of edge tools. The forger heats the end of a rod of steel selected to size; roughly shapes the blade; cuts it off with sufficient for the bolster and tang; re-heats; shapes the bolster with prints, and draws out the tapering tang with the hammer. The blade is then re-heated, and carefully "plated" (hammered out) to the size and shape required, water being thrown on at intervals during the brief process to remove the scale. Gouges are forged in the same way, the blades being afterwards rounded in "hollowing bosses." The forgings then pass into other hands to be hardened, tempered, ground, glazed, and "finished," ready for packing. The more complicated socket chisel passes through the forger's hands no less than six times; is then hardened, tempered, set, ground, glazed, japanned, and finished, passing three times through the hands of the grinder in the course of these operations. The processes of axe forging are also numerous, the body of the axe being iron, to which steel is welded for the pole or hammer as well as for the blade. The first operation is to weld a small piece of the end of a flat steel rod across the middle of a short length of bar iron. The ends of the iron are then plated out, lapped over, and welded together to form the eye, the centre of the bar with the adhering steel becoming the pole, or hammer, of the axe. The eye is afterwards smithed on a tapering wedge, put in first on one side and then on the other, the effect being not only to secure uniformity of size, but to leave the eye narrower in the middle than at the sides, so that the handle, when properly wedged at the end, cannot loosen with use. Steel for the blade is next inserted between two laps of iron, left open for the purpose in forming the eye, and is welded to them. The mooding and plating of the blade follow, after which the axe is ready for hardening, tempering, grinding, &c.

These examples sufficiently illustrate the general processes of edge tool forging, though far from exhausting their almost endless variety. The forging of edge tools at Sheffield is done chiefly by hand, the great difficulty in the way of the general application of machinery being the immense variety of size and shape required--a variety which is not, as in table cutlery, a mere matter of taste, but is a necessity of the varied work for which the tools are used. There is also the undoubted fact of the superiority of hand labour in the manipulation of the fine steel used for the blades of the best edge tools. Boring, drilling, and other such operations are, however, done by machinery, while steam hammers and other machinery are being more and more applied to the rougher and heavier operations of the forger. One feature in the edge tool trade is the quickness and dexterity with which the forger performs the more minute and delicate operations by a few rapid strokes. The observant visitor will note also the extreme care with which he watches the heating of the steel, lest it should be rendered worthless by over-heating. Another operation requiring the greatest care is the setting of many of the more important tools--a part of the work which is very carefully scrutinized by the employer or his managers before the tools are allowed to leave the premises. The grinding and glazing are also very carefully examined. The more extensive edge tool manufactories include grinding wheels. The accompanying illustration (Plate XCI) shows the grinding wheel at the works of Messrs. James Howarth and Sons, of Bath-street. Hardening and tempering are essentially the same processes in the edge tool as in the cutlery trades. The tools are hardened by being drawn through a cistern of cold water when red hot; and the extreme brittleness produced by hardening is removed by heating the tools slowly on a furnace plate, and allowing them to cool gradually. Many of the tools have afterwards to be taken back to the grinding wheel to be glazed or polished.

The framing and hafting of edge tools is for the most part a subsequent process, though some few tools are roughly hafted before they are ground. For many tools ebony and other expensive woods are used. These are cut up in lengths, and stored for periods of five or ten years before they are used, in order to ensure dryness, without which the handles would rust the steel. The preparation of handles and the wood framework of tools is a separate department, in which the use of steam machinery is very fully developed, and many of the operations are very interesting to strangers.

FORGING AND GRINDING AT MESSRS. JAMES HOWARTH AND SONS'.

Plate XCI Forging and Grinding at Messrs. James Howarth & Sons
[*Illustrated Guide to Sheffield, 1879*, p.270]

MESSRS. LOCKWOOD BROTHERS, ARUNDEL-STREET

MESSRS. JOHN SORBY AND SONS—SPITAL-HILL WORKS.

Plate XCII - Lockwood Brothers, Arundel Street (top)
John Sorby & Sons, Spital Hill Works (bottom)
[Sheffield Illustrated Guide, 1879, p.274]

Messrs. Howarth and Sons.--Among leading manufacturers of edge tools and joiners' tools are Messrs. James Howarth and Sons, of Broomspring Works, Bath-street. The business was commenced in 1835, by the present senior partner, who was joined by his sons in 1863. Messrs. Howarth manufacture light and heavy edge tools of all kinds, including every variety of joiners' tools, hammers, skates, augers, &c. Restricted in the first instance to the home markets, the trade of the firm has rapidly extended, and they have now direct business connections with leading continental states, Canada, Australia, China and other distant markets, as well as with home consumers. They were exhibitors at the London Exhibitions of 1851 and 1862, and at Paris in 1855, receiving on each occasion prize medals of the first class. The Universal Society for the encouragement of Arts and Industry, founded in London in 1851, reported on the goods exhibited at Paris four years later. This society, awarded its first-class prize medal to Mr. Howarth for the excellence of his tools--an honour paid to no other British manufacturer--and elected him an honorary Vice-President of the Society. Messrs. Howarth are all practical men, and personally superintend all the departments of their manufactory, giving the closest attention to the general workmanship as well as to setting and finishing. Edge and joiners' tools, though made for the use of the same class of workmen, are separate and distinct trades. The edge tool, which is mostly a cutting tool, is made of iron and steel combined, or of solid steel as already described; while joiners' tools, such as squares, bevels, spokeshaves, gauges, &c., are made chiefly of ebony and other expensive woods combined with brass or steel. The manufacture of skates is also one of the special features of Messrs. Howarth's business. They make hammers from one ounce to 25 lbs. each in weight, and of almost infinite variety of shape to suit the most varied requirements. They also manufacture large quantities of steel "toys," including pincers, pliers, &c.--a branch in which they are almost the only Sheffield house engaged.[26] Using the best material and employing the most skilled labour, Messrs. Howarth produce goods of a very high class, and have made a reputation in the most distant markets. They are also largely engaged in the steel, file, saw and other kindred trades.[27]

Messrs. John Sorby and Sons.--Among the best known and most celebrated manufacturers of edge tools in Sheffield are Messrs Lockwood Brothers, of Arundel-street, and John Sorby and Sons, of Spital-hill. (Plate XCII).[28] These two names have long represented one and the same firm, and the two factories, though in distant parts of the town, are

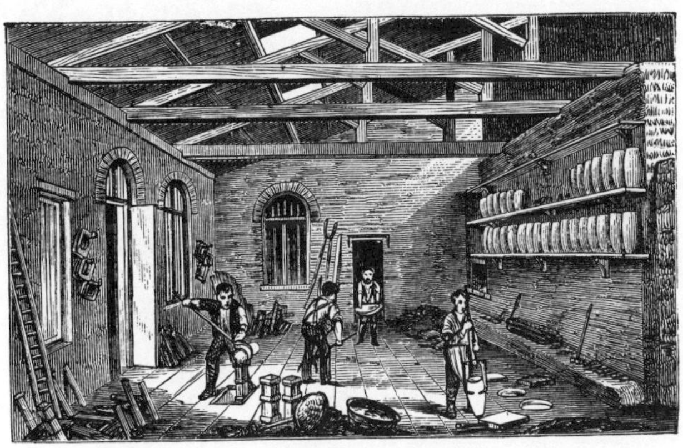

STEEL MELTING FURNACE.

SHEAR BENDING AND FINISHING ROOM.

Plate XCIII - Steel Melting and Shear Bending at John Sorby & Sons
[*Sheffiled Illustrated Guide, 1879*, p.279]

are in immediate communication by telephone. The firm now consists of Mr. Joseph Lockwood, his son, and nephews. The ancestors of the firm began business at Ecclesfield at a date not precisely known, but more than a century ago, records of business they did in 1767 being still in existence. Nor is the time of the removal to Sheffield precisely known; but apparatus was discovered at a recent stock-taking, showing that the old firm carried on a considerable business here in those early days when retail shops were few, and manufacturers provided in a much more general way than is now permitted or necessary for the multifarious wants of their workpeople. We have already referred to the firm in connection with the manufacture of cutlery. They are also considerable manufacturers of steel, files and saws as well as edge tools. Our illustrations of steel melting (Plate XCIII) and file forging (Plate XXVIII) are both taken from their works in Arundel-street. From the latter it will be seen that the enterprize of the firm had led them to adopt steam machinery in the forging of files. While availing themselves of the latest appliances of machinery so far as they can be used with advantage, they adhere rigidly to hand labour in all operations affecting the quality of the goods. The chief markets of the firm are Scotland, Australia, New Zealand, South America, Canada and the continent of Europe. In France and neighbouring States their tools have a specially high reputation. Messrs. Lockwood have not greatly cultivated internationsl exhibitions, but their important continental connections induced them to compete at Paris in 1855, when they received a first-class medal for excellence. The great specialty of the firm is the manufacture of sheep shears, which they carry on at the works on Spital-hill. The manufacture is long and complicated, but some of the numerous operations are very interesting. The forger is supplied with flat rods of iron cut into lengths of about fifteen inches; this he "moods" by broadening the two ends for the blades, and then laying the centre edgewise on the anvil and flattening it transversely for the bow and shanks. The blades are them "tagged," i.e., strips of steel are welded on for the cutting edge. These preliminaries involve five successive heatings and beatings, the blades being mooded and steeled or tagged separately. The work thus roughly begun is completed by a still more elaborate series of operations on the anvil. The blades are successively plated. The shanks are plated and afterwards hammered at a low heat to impart spring. The next process is "kneeing," as the formation of the slight bend at the junction of the bow and shanks is called. The blades are then "smithed" and stamped with the mark and name of the firm. This long succession of processes, during which the forging has to be repeatedly heated, completes the work of the forger. The shear is next passed to the hardening and tempering rooms, and then to the wheel to be ground and glazed. Two important operations remain--bending and setting. The shear reaches the bending and setting room in the shape of a long dagger with a tapering blade at each end, curved shanks, and a straight flat centre. The business of the workman is now to bring the two blades and shanks together, and this is done by bending the shear in the middle. The process is shown in one of the illustrations on Plate XCIII. The tools used are a "break" and pair of tongs. The break consists of a flat piece of iron fixed upright on a wooden frame, and terminating at the top in two round blunt prongs nearly an inch apart. The tongs have round fingers similar to the prongs of the break. The flat centre of the shear is put between the prongs of the break, and the bender then seizes it with his tongs, and draws it round, producing a perfect bow in a few moments. All that remains is to adjust the shears so that in use the edges of the blades close together perfectly from shank to point--an operation requiring much care--and the shear is ready for rubbing and packing. The old firm of John Sorby and Sons were the first manufacturers of sheep shears in Sheffield, and, we suppose, have exported more of those very useful articles to the sheep farmer than any other house in the trade. They still export largely to all quarters of the world, especially to the Australian and South African colonies and to South America.

Messrs. Ward and Payne.--Mr. David Ward, the present mayor of Sheffield, is a leading manufacturer of edge tools and sheep shears, and has works in West-street. The business was established in 1803 by Mr. Edward Ward, with whom Mr. Henry Payne was afterwards associated.[29] The edge tools of Ward and Payne acquired a very wide reputation, and the business is still carried on in the name of the old firm. Mr. David Ward, the present proprietor, succeeded to the business on attaining his majority, and though still a young man, has added two important branches to the old trade--the manufacture of carving tools and sheep shears. Mr. Ward purchased the carving tool business of Mr. S. J. Addis, of London, and succeeded in transplanting the trade in Sheffield, in spite of many difficulties thrown in the way by the regulations of the edge tool unions.[30] Abandoning old, worn-out designs, and adopting styles and designs more acceptable to the taste of a somewhat fastidious class of customers, Mr. Ward has made a reputation as a manufacturer of carving tools, which he exports largely to the protected markets of America, as well as to European, colonial, and other markets. He displayed equal enterprise in entering upon the manufacture of sheep shears. The scarcity of cotton caused by the American civil war brought wool greatly into favour, and led to the investment of large capital in the rearing of sheep in Australia, the Cape, New Zealand, and South America. The demand for shears became so great that leading manufacturers had orders on their books to last them for two or three years. It was under these circumstances Mr. Ward entered upon the manufacture of sheep shears. The high reputation of the old makers, and the excellence of the goods they supplied, were a great difficulty; but by going direct to the consumers, and supplying a first-class article, Mr. Ward created a demand for his goods, which compelled factors and dealers to stock them. His success in Australia was decisive, and he is now successfully cultivating valuable markets at the Cape, &c. Messrs. Ward and Payne hold many distinguished honours, including the London Exhibition Prize Medal, the Vienna Medal of Progress, the Philadelphia Medal, and the Diploma of Merit and Gold Medal of the recent Cape Exhibition, having been successful wherever they have exhibited. We have seen how numerous and complicated are the operations of shear-forging by hand. Mr. Ward believes that these operations may be performed to a large extent by machinery--which has already been introduced to a limited extent--and he has erected machinery by which they can be manufactured at a minimum of cost without any deterioration of quality.

HIBERNIA WORKS—MESSRS. WILLIAM MARPLES AND SONS.

Plate XCIV - Hibernia Works - Messrs. William Marples & Sons.
[*Illustrated Guide to Sheffield, 1879*, p.277]

Messrs. William Marples and Sons.--Hibernia Works, the factory of Messrs. William Marples and Sons, edge tool manufacturers and merchants, is in Westfield-terrace, extending backwards to Rockingham-street. The business was founded nearly fifty years ago by the father of the present proprietors, who early established a reputation for excellence of quality and superiority of finish which enabled him to build up a large business.[31] The business has continued to increase under the energetic management of his sons, who have now one of the most extensive edge and general tool trades in Sheffield. The large extent of their manufactory in Sheffield is shown in Plate XCIV. They have branch establishments in London, Liverpool, Germany, Canada, and the United States of America, and their goods are well known in all our colonies and in most other countries. The specialty of the firm is the very large stock of goods they have always on hand. Their warehouses are filled with an assortment of tools of every known description, cutlery and general hardware of all kinds. They have, probably, the largest stock of tools of any house in the world, it being the endeavour of the firm to supply tools for every purpose for which they can be required. Their enormous stock enables them to execute the largest orders at a few hours' notice--this special feature of the business having contributed materially to the wide extension of their connections. In addition to the stock of goods of their own manufacture, and of general English hardware, the firm import tools and hardware of all kinds from the United States, Germany, France, and other manufacturing countries, supplying from their extensive stores every diversity of tools and hardware that customers can possibly require.

[end of transcription p. 278, *1879 Illustrated Guide to Sheffield*]

Table VII

Development of William Marples & Sons from William Marples Jr. 1828 into Record Ridgway, Ltd. 1972

Wm. Marples, Sr., 1794
Sheffield Moor
Edge Tools

→ 1819-1837
Rockingham Street
1837-1844
Solly Street

→ William Marples, Jr., 1828-1832
Broomhall Street
Joiner Tools

→ 1832 Broomspring Lane

→ Wm. Marples & Sons, 1859
25/27 Westfield Terrace
Hibernia Works

→ 1869, Wm. Marples Retired

→ Edwin Henry Marples
William Kent Marples 1869-1883
Albert Marples

→ Albert Marples 1883-1888
Edwin Henry Marples

→ Edwin Henry Marples with Sons
Henry Edgar and Edward Albert
Marples 1888

Thomas Ibbotson, 1821
Charles Street
Joiner Tools

→ Thomas Ibbotson & Co., 1833-1905
Charles Street, Paternoster Row

Isaac Sorby
1810

→ Turner, Naylor & Co.
1871-1876

→ Turner, Naylor &
Marples 1876-1893

→ Turner, Naylor & Co., 1893-1909
Northern Tool Works, John Street
(owned and operated by Wm. Marples & Sons, 1909-1963)
Dissolved in 1963

John Mosley & Son, 1730-1892
London, purchased in 1892
Moved to Sheffield, 1902

Turner, Naylor & Co.
John St. Northern Works
1909

C. & J. Hampton, 1898-1972
owned 50% of
Wm. Marples & Sons
in 1962

William Ridgway 1871-1898
William Ridgway & Sons, 1898-1972
owned 50% of
Wm. Marples & Sons
in 1962

→ Combined & Formed
RECORD RIDGWAY, LTD. 1972

ACQUISITIONS AND COMBINING OF SHEFFIELD EDGE TOOL FIRMS INTO
WILLIAM MARPLES & SONS

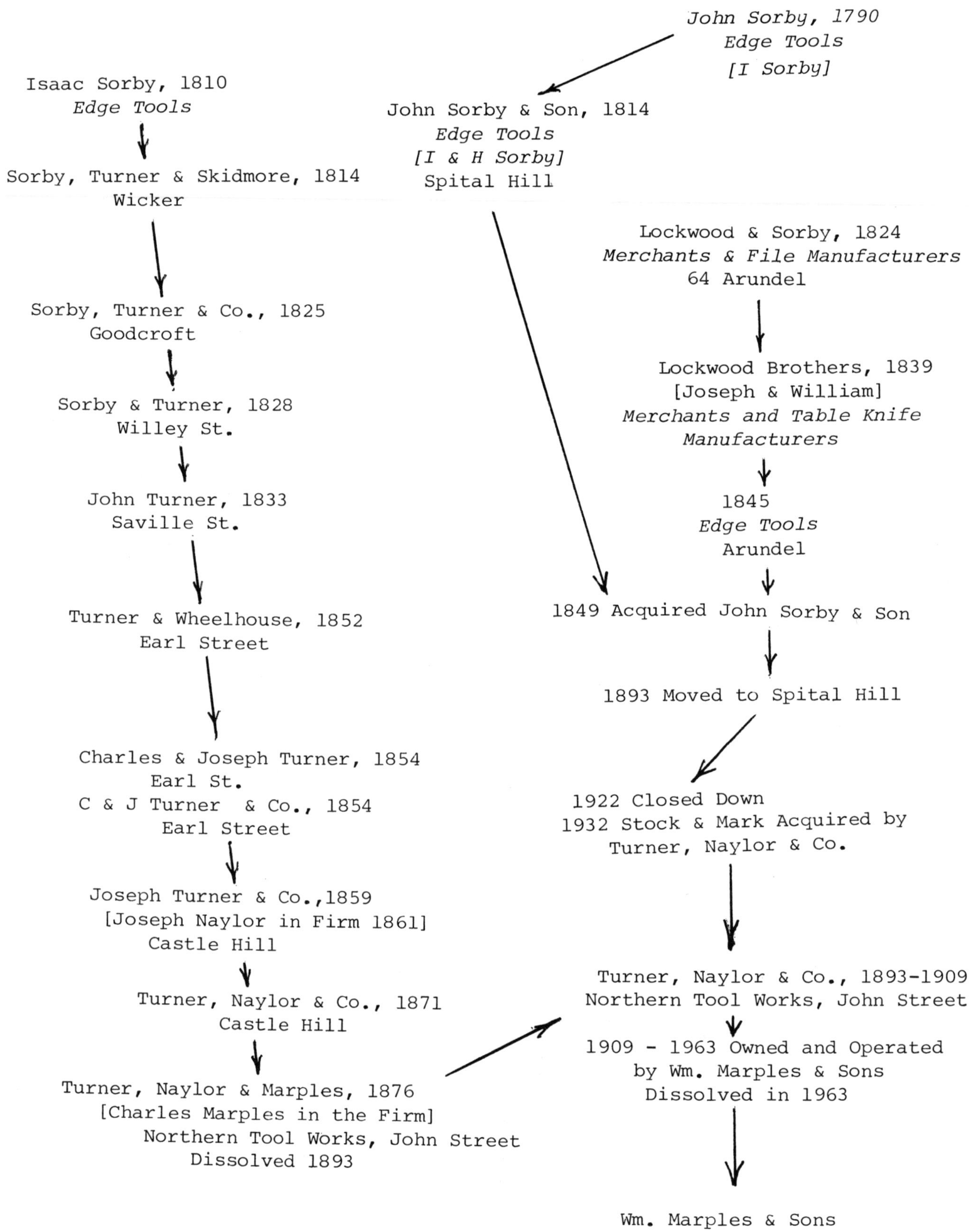

Actually the firm of William Marples commenced business as joiner tool manufacturers, as will be discussed in the next chapter. The development of this concern is traced in Table VII. It was not until 1871 that this firm dealt in edge tools. At that date William Marples bought an interest in the firm Turner, Naylor & Co. (see Table VIII) for his youngest son, Charles.[32] The firm name was then changed to Turner, Naylor & Marples. In 1893 although Marples still held a controlling interest, the firn name reverted to Turner, Naylor & Co. Apparently this was for sales purposes, as that mark and name had established an excellent reputation for certain foreign markets. Wm. Marples & Sons continued to operate Turner, Naylor & Co. until 1963, having in the meantime also acquired the marks of John Sorby and Lockwood Brothers.

At the end of this chapter a Check List of 19th Century Sheffield Edge Tool Makers and Dealers is presented as Appendix II. This records 346 firms doing business during this period. The durations of these enterprises are compiled in Table IX. The data for this check list was obtained from the following Sheffield and Yorkshire Directories: 1797; 1814; 1817; 1821; 1825; 1828; 1830; 1833; 1834; 1837; 1839; 1841; 1845; 1849; 1852; 1854; 1856; 1857; 1859; 1860; 1861; 1863; 1865; 1868; 1871; 1872; 1876; 1879; 1881; 1884; 1887; 1888, 1890; 1892; 1893; 1895; 1896; 1898; 1900. (all known to exist during this period) The names were recorded from the listings of "Sheffield Trades", and not from the alphabetical resident listings. Names appears in only one directory after 1850 were not included, unless subsequently noted as part of another firm. It should be pointed out that such listings are not infallible, as ommisions did occasionaly occur. Also since there are gaps for years that directories were not published some firms were in business a year or two earlier and later than indicated from these sources. It should also be cautioned that these dates do not necessarily establish the tenure of a firm, but the range that they dealt in edge tools. Many had previously manufactured saws or joiner tools and continued with these products after discontinuing edge tools If either a (J or S) is noted after a firm listing, Appendix III or I should also be checked.

Some imprints have been observed on tools for which firm listings do not appear in the directories. For example, the imprint BARTON BROTHERS has been noted on plane irons, chisels and bit braces with no listing in the Sheffield Trades. By checking the alpabetical listing, a single entry in the 1852 Sheffield Directory noted Edward Barton, merchant and tool manufcaturer, 231 Glossup Road with in Barton Brothers with his brother William listed at New York.

Table IX - Duration of 19th Century Sheffield Edge Tool Firms

Years in Continuous Operation	Number of Firms	Percent of Total
Less than 5	113	32.7
5 but less than 10	70	20.2
10 " " " 20	54	15.6
20 " " " 30	47	13.6
30 " " " 40	17	4.9
40 " " " 50	26	7.5
50 " " " 60	9	2.6
60 " " " 70	3	0.9
Over 70	7	2.0
	346	100.0

Analysis of Table IX shows that approximately 1/3 of these firms were in business less than five years, 1/3 more than five but less than 30 years and about 1/3 more than 30 years. While 19 firms (3.5%) were in business for more than 50 years, more than half this number were saw manufacturers as their principal business. The majority of firms in business from more than 30 years also marketed joiner and also saws along with their lines of edge tools. These were the more successful enterprises who engaged in extensive trading with other manufacturers and in effect became tool factors and merchants along with their own production of tools.

Among the firms in edge tool business for more than 50 years who almost exclusively manufactured edge tools, at least during their early years, were: Wm. & Sam. Butcher; Philip Law; Robert Sorby; Henry Taylor and Ward & Payne. John Sorby & Son, 1814-1849, were also principally edge tool manufacturers, but their mark and business name was continued through the remainder of the 19th Century by others, including Lockwood Brothers and then Turner, Naylor & Marples, and then into the mid-20th Century as subsidiaries of Wm Marples & Sons. James Howarth began as an edge tool maker, but after the firm became James Howarth & Sons, they became tool factors and general merchants of edge and joiner tools and saws, continuing to manufacture some of their products. This firm failed about 1913 and was acquired by Robert Sorby & Sons about 1920. The same situation developed with Thomas Ibbotson & Co., who claimed in their 1903 Catalogue to have been established in 1814. The firm failed in 1905 and was acquired by William Marples & Sons. After 1872 Wm. Marples & Sons became the most prominent general tool dealer in Sheffield.[33]

Among the obvious reasons for changes in ownership and management of these firms were disagreements among partners, hardships during difficult financial times or strikes, old age of proprietors with no successors or death of the owners. In a few instances widows continued to manage until their sons became of age or they could arrange an equitable sale of the firm. An interesting dissolution was that of Wm. Greaves & Sons, who established at 12 Division St as merchants and manufacturers of table knives and razors in 1823. Their development and building of the Sheaf Works has been related on page 28. The extent of their products is revealed in their advertisement which appeared in the 1849 Sheffield Directory, shown in Plate XCV. Apparently they had successfully built up an export trade to America. In any event this was their final advertisement and for reason not known,the firm dissolved in 1850. Their cutlery line was taken over by Eyre, Ward & Co., noted in the 1852 Sheffield Directory as merchants and cutlery manufacturers at the Sheaf Works (late Wm. Greaves & Sons). The steel and edge tool line was taken over by Thomas Turton & Sons noted in the 1852 Directory as "steel, file, saw, railway springs" at the Sheaf Works and also at Russell Street and as forgers and tilters with forge at Upper Slack.

As evidenced from surviving trade catalogues and though undergoing changes in ownership and management, the firm Thomas Turton & Sons continued producing a full line of light and heavy edge tool throughout the remainder of the 19th Century and well into the 20th. During this time they continued the use of the imprint *Wm. GREAVES & SONS* [frequently also with the imprint *ELCTRO BORACIC STEEL*]. Plate XCVI shows plane irons with these imprints from a Thos. Turton & Sons 20th Century trade catalogue.

Plate XCV - Wm.Greaves & Sons Advertisement in 1849 DIRECTORY

An extract from the *Illustrated Guide to Sheffield, 1862* noted: -

> "Thomas Turton & Sons are large manufacurers of files, saws and edge tools. Their place of business is called the Sheaf Works, and the works are very large. Thomas Turton & Sons are amongst the oldest firms in the town, and possess a high reputation for all their various manufactures. They are especially noted for files and edge tools. In In edge tools they succeeded the eminent house of William Greaves & Sons, of the same works; and they have greatly increased the former reputation of this firm. Their *"BORACIC"* edge tool are made by a peculiar process, and are of such repute that they command a higher price than any other goods of the same kind"

A notice that appeared in an 1880 circular of Thomas Turton & Sons, shown in Plate XCVII called to attention that Wm. Greaves & Sons had announced the development of Boracic Steel on May 1, 1839.

Actually the term "Boracic Steel" was probably more advertising propoganda than a metallurgical alloying accomplishment.[34] It is possible that borax was used as a composite of the flux during melting with sincere belief that an improvement in steel quality resulted. The prefix "elctro" may have referred to the amber color of the slag, but most certainly had no relation to electricity.

Several advertisement of edge tool manufacturers which appeared in various directories and some from the *1879 Illustrated Guide to Sheffield* are shown in Plates XCVIII - CXIII, pages 138 - 152. These attest to the diversity of products, manufacturing facilities and international marketing of some of the more prominent Sheffield firms in this field.

Plate CXIII, page 152, shows the advertisement of Thomas Turner & Co. at the Suffolk Works. Thomas Turner was first listed in the 1823 Sheffield Directory as a "gentleman", with no business connection. He established his firm Thomas Turner & Co. at 38 Norfolk St. in 1825 as a "merchant and manufacturer of table & pocket knives, files, tools, &c." This was an example of a person who entered the manufacturing field as an investment rather than an apprenticed tradesman. The firm expanded and in 1837 occupied the Suffolk Works, adding steel refining and a line of edge tools. The firm continued through 1900. There was another Thomas Turner, planemaker, 1841-1879, Orion Works at Queen St.

2

WM. GREAVES & SONS,

SHEAF WORKS, SHEFFIELD.

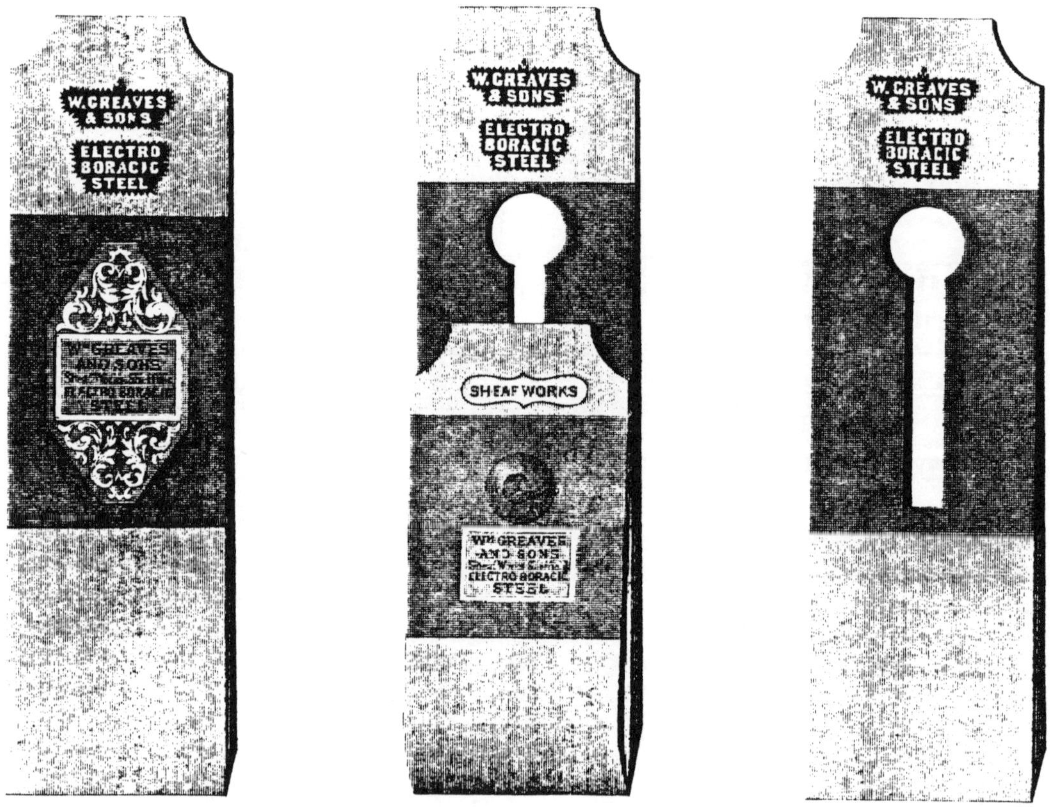

No. 7.	No. 8.	No. 226.
E.B. IMPROVED SINGLE PLANE IRON.	E.B. IMPROVED DOUBLE PLANE IRON.	E.B. IMPROVED CUT PLANE IRON.

For Prices see pages 3 and 15.

Plate XCVI - Plane Irons from Thos. Turton & Sons Trade Catalogue, c.1910
[from Collection of K.W. Hawley, Sheffield]

THOMAS TURTON AND SONS,

SUCCESSORS TO

WM. GREAVES AND SONS,

SHEAF WORKS, SHEFFIELD,

SOLE INVENTORS OF THE

ELECTRO BORACIC STEEL.

WM. GREAVES AND SONS

Are aware that in introducing to the Publie a NEW ARTICLE, it is usual to say much in explanation of its superiority—so much that the whole is often viewed with distrust. In order to avoid this evil, W. G. & SONS deem it best to say nothing either in explanation or praise of their

ELECTRO BORACIC EDGE TOOLS,

Leaving the entire question of merits to be decided by the Consumer.

These merits cannot be known without a trial; and a trial is earnestly and respectfully asked.

W. G. & SONS would remark in conclusion, that their ELECTRO BORACIC STEEL, (as is usual with all experiments in Metals and Alloys) is the result of much labor and investigation. May 1st, 1839.

SHEAF WORKS,
May 1st, 1880.

Plate XCVII - Notice Regarding Electro Boracic Steel
Thos. Turton & Sons Trade Catalogue, c.1910, Sheffield Central Library

[Collection of Central Library, Sheffield]

ESTABLISHED 1835.

JAMES HOWARTH & SONS,

CORPORATE MARK
HJ
Granted by the Cutlers' Company.

BROOMSPRING WORKS,
BATH STREET,
SHEFFIELD,
MERCHANTS & MANUFACTURERS
OF
LIGHT AND HEAVY

JAMES HOWARTH
HJ
BROOMSPRING
WORKS
SHEFFIELD
TRADE MARK—REGISTERED.

EDGE TOOLS & JOINERS' TOOLS,

In every Variety;

SAWS, FILES, HAMMERS, PINCERS,

AUGERS, GARDEN TOOLS, SHEARS, TOOL CHESTS, SKATES,

CUTLERY, &c.

LIGHT EDGE TOOLS, viz. :—
CAST STEEL CHISELS AND GOUGES—Firmer, Coachmakers', Model Makers', Millwrights', Turning, Socket, & Mortising.
PLANE IRONS—Single, Cut, Double, and Tooth.
DRAWING KNIVES—Carpenters', Wheelwrights', Coopers', &c.

CARVERS' and ENGRAVERS' TOOLS in all their Varieties.

JOINERS' TOOLS, viz.:—Braces and Bits of all kinds, together or separate; Squares, Bevels, Spokeshaves, Gauges, Levels, Turnscrews, Saw Frames, Saw Sets, Saw Pads, &c.

TOOL CHESTS, in Oak and Deal:—Youths', Gentlemen's, Emigrants', and Carpenters'; containing Tools of First Quality.

HAMMERS, viz.:—Kent, Scotch, Garrick, Canterbury, Garden, Lath, Shoe, Ladies', Gentlemen's, Saddlers', Upholsterers', Coachtrimmers', Grocers', Farriers', Coal, Geologists', &c.

HAMMER HEADS–London, Warrington, Manchester, Bath, Roofing, Flooring, Clinch. Sheathing, Bricklayers', Boiler Makers', Engineers', Machinists', Hand, Sledge, and Stonebrakers'.
Slaters' Axes, Hammers, Stakes, & Rippers. FIREMEN'S AXES, &c.

PLANES of all kinds. SKATES of every Variety.

AUGERS—Carpenters', Ship, Shell, and Screw Augers; Screw Bits, Ship Gimlets, Brad Awls, and Gimlets.

SAWS, viz.:—Hand, Panel, Rip, Compass, Dovetail, Carcase, Sash, Tenon, and Fret Saws; Frame, Pit, Cross-Cut, Mill, Tarning, Billet, and Canada Webs. CIRCULAR SAWS, Saw Benches, and Spindles.

HEAVY EDGE TOOLS, viz.:—Axes and Adzes, for Joiners, Carpenters, Coopers, Wheelers, Coachmakers, and Shipwrights; Hatchets, Choppers, Mincing and Cheese Knives, Bills, single and double-edged Switching Hooks, &c. Brick, Masons', Pointing, Gauging, and Plastering Trowels.

GARDEN TOOLS, viz.:–Trowels, Hoes, Rakes, Spades, Forks, Reels, Syringes; Sets of Ladies' Garden Tools.
CARRIAGE and WAGON BUILDERS' TOOLS.

SHEARS, viz.:—Garden, Pruning, Lopping, and Grass; Sheep, Fellmongers', Woolstaplers', Thatchers', Rag, Horse, and Docking Shears.

FILES, viz.:—Flat, Half round, 3-Square, Square, Hand, Equalling Slotting, Mill, and Cotter; Frame and Hand SAW, Gulleting, Warding, and Topping Files.

RASPS, viz.:—Horse, Shoe, Bread, Wood, Cabinet, and Farriers'.

ENGINEERS' and LANCASHIRE TOOLS, viz.:—Screw Stocks, Dies and Taps, Screw Plates and Taps, Ratchet Braces, Coach Wrenches, Hand, Bench, Staple, and Parallel Vices; Chipping Chisels, Spring Dividers, Compasses, Callipers, Pliers, Cutting Nippers, &c.

GAS FITTERS' TOOLS, viz.:—Pipe Tongs, Pliers, Pipe Cutters, Screw-Cutting Tools, &c.

CRAMPS—Joiners'. Bench, Floor, and Sash Cramps; Bench Screws, Holdfasts, &c.

SCYTHES, SICKLES, REAPING HOOKS, HAY AND STRAW KNIVES.

CAST STEEL HAY, DIGGING, SPADING, and MANURE FORKS;

Rules, Measuring Tapes, Oil Stones and Slips, Tool Basses, Table and Pocket Cutlery.
N.B.—Cast Steel, Single and Double Shear, Blister and Spring Steels, to any Size or Temper.

FIRST-CLASS PRIZE MEDALS, as under:—

LONDON, 1851. PARIS, 1855. SOCIETY OF ARTS, 1856. LONDON, 1862

58

Plate XCVIII - James Howarth & Sons' Advertisement
[page 56 Advertising Section - 1879 ILLUSTRATED GUIDE TO SHEFFIELD]

REGISTERED **GOLD MEDALS** for EDGE TOOLS, SAWS, FILES, and STEEL awarded at Calcutta, 1884, and Edinburgh Exhibition, 1886.

TRADE MARK.

ADZES & AXES.

AUGERS AND AUGER BITS.

EDGE TOOLS,

Light and Heavy. JOINERS' and COOPERS' TOOLS.

KANGAROO BRAND

SAWS,

HAND, RIPPING, and BACK SAWS
MACHINE GROUND.
CROSS-CUT, PIT, FRAME and CIRCULAR SAWS

FILES, WARRANTED HAND-CUT.

PRIZE MEDAL: LONDON, 1851.

DIPLOMA, New Zealand, 1865

HIGHEST AWARD–
SILVER MEDAL & DIPLOMA
At the Forestry Exhibition, Edinburgh, 1884,
For FORESTRY TOOLS.

SCYTHES,

PATENT and SOLID SCYTHES,
REAPING HOOKS and SICKLES,
HAY, STRAW, and MACHINE KNIVES.

SHEEP SHEARS,

Patentees of the "DUPLICATE BOW."

CRUCIBLE STEEL,

For MINING and ENGINEERS' TOOLS, &c.

Double and Single Shear, Blister and Spring Steel.

REGISTERED
TRADE MARK.

ROBERT SORBY & SONS,

CARVER STREET,

SHEFFIELD,

ENGLAND.

(SOLE MANUFACTURERS.)

Plate XCIX - Robert Sorby & Sons' Advertisement
[page 67 Advertisements - 1887 SHEFFIELD DIRECTORY]

FENTON & MARSDENS,

MANUFACTURERS OF

EDGE TOOLS,

JOINERS' TOOLS,

BRACES AND BITS,

SKATES,

Screw Boxes, &c. &c.

AND

GENERAL MERCHANTS,

STEEL CONVERTERS & REFINERS.

MARSDEN BROTHERS & SILVERWOOD,

SKATE MANUFACTURERS,

BY APPOINTMENT,

To His Royal Highness Prince Albert,

BRIDGE-STREET WORKS

SHEFFIELD.

N (15)

Plate C - Fenton & Marsdens' Advertisement
Calling to Attention Appointment as Skate Manufacturers to His Royal Highness Prince Albert to their previous firm MARSDEN BROTHERS & SILVERWOOD.
[page 15 Advertising Section - 1854 SHEFFIELD DIRECTORY]
(From Collection of the Author)

Prize Medal, 1851,

AWARDED TO

MARSDEN, BROTHERS, & CO.,

FOR MECHANICS' TOOLS.

BRIDGE STREET WORKS, SHEFFIELD.

MARSDEN, BROTHERS, & CO.,

(LATE FENTON & MARSDEN,)

MANUFACTURERS OF

EDGE AND JOINERS' TOOLS, BRACES AND BITS, AUGERS, SCREW BOXES,

Saws, Files; Pen, Pocket, and Table Cutlery, &c., of superior quality.

STEEL CONVERTERS, REFINERS, AND GENERAL MERCHANTS.

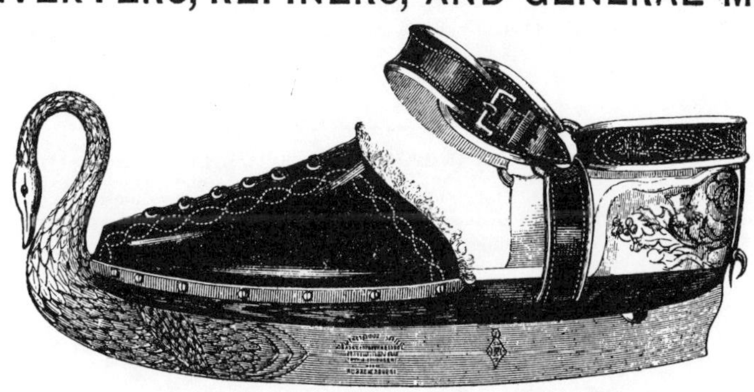

Skate Manufacturers by Special Appointment to The Queen and Royal Family.

Plate CI - Marsden Brothers & Co. Advertisement
[*A Look at Modern Things*, [Birmingham,1858] Photocopy courtesy of Local History Department, Birmingham Reference Library

PRIZE MEDAL CARVING TOOLS !

Manufacturers to the Royal Family.

J. B. ADDIS & SONS,

IN returning best thanks to their numerous Friends and the Public generally for the liberal patronage and support extended to them in their Business of

CARVING TOOL MANUFACTURERS,

would beg to assure them that no pains shall be wanting on their part to send out an Article, in Quality, Style, and Finish, immensely superior to all others; and trust to be honoured still further with their obliging commands, which it will be their constant and most earnest endeavour respectfully to deserve. At the same time, J. B. A. & SONS would especially caution intending Purchasers to beware of spurious articles vended as of their manufacture, and marked with the same name. These they emphatically disown; such goods being at best but a contemptible imitation, and unworthy of use by any first-class artist or workman. The Genuine Article is Manufactured only at

ARCTIC WORKS, ROCKINGHAM STREET, SHEFFIELD,

where all Orders addressed, will be executed with prompt attention and in their usual superior style, for which they obtained PRIZE MEDALS in 1851, 1862, 1870, 1871, and 1873; and also a GOLD MEDAL, subscribed for by the Merchants, Manufacturers, and Workmen of Sheffield, for superior workmanship.

Messrs. J. B. ADDIS & SONS also beg to say that they have ceased to have any connection with the firm of WARD & PAYNE.

Extract from the SHEFFIELD DAILY TELEGRAPH, Monday, July 18, 1870.

An interesting collection of Carvers' Tools, contained in three cases, is exhibited by Mr. JAMES BACON ADDIS, of the ARCTIC WORKS, SHEFFIELD, and attracted the special attention of the Prince. In one case alone, that devoted to carvers' tools for wood and stone, there are fifteen dozen instruments, each differing from the other, and each intended to perform some distinct mechanical process. In another case are tools appropriate for sculptors; a third is devoted to a series of wonderfully delicate and yet effective gouges and tools for the use of ladies, amongst whom carving in wood has grown to be a fashionable art. The Prince of Wales was pleased to accept a chisel from Mr. Addis, who had been introduced to His Royal Highness by the Mayor; and the good-humoured Princess Mary accepted another specimen of Mr. Addis's handicraft—a gouge so finely wrought that, without a close examination, the eye could hardly detect the exact indent upon the ivory or the wood which would be made by it. Many ladies are extremely dexterous in the use of these delicate but somewhat formidable steel implements, and amateur wood-carving has taken its place in many town and country houses as a pleasant resource and change of occupation in moments that it would otherwise be hard to fill up. Mr. Addis showed the Prince how finely-tempered were some of the tools, and how they would bend almost double without injury; and what a beautiful colour or burnish could be given to others by the simple application of heat.

The Prince was pleased to hear Mr. Addis's explanation of the art of carving tool manufacture, and complimented him as being a very practical artist.

From the SHEFFIELD and ROTHERHAM INDEPENDENT, Tuesday, May 27, 1873.

Mr. J. B. ADDIS, of the ARCTIC WORKS, SHEFFIELD, exhibits a case of sets of the most difficult Tools used by carvers. Mr. Addis claims to have discovered the secret of producing chisels and gouges that will, with the greatest ease, cut the hardest stones. The tools are very accurately finished; and deserve the careful attention of sculptors and other artists, for whose use they are designed.

Plate CII - J.B. Addis & Sons' 1872 Advertisement SHEFFIELD DIRECTORY, p.158

J.B. Addis had been in partnership with his brother S.J. Addis in London. (See 1881 Advertisement, Plate CIII). After the latter's death, J.B. removed to Sheffield, establishing a connection with Ward & Payne to continue their reknown line of carving chisels. By 1872 this relationship with Ward & Payne was apparently cancelled.

238 SHEFFIELD ADVERTISEMENTS. [1872.

UNDER THE DISTINGUISHED PATRONAGE OF

H.R.H. the Prince of Wales. The Princess of Teck. Lord Lichfield, &c., &c.

CARVING TOOLS.

J. B. ADDIS,

ARCTIC WORKS, SHEFFIELD,

(Brother to the Late Mr. S. J. ADDIS, of 68 & 70, Worship Street, Finsbury, London,)

CARVING TOOL MANUFACTURER,

Begs most respectfully to inform the Merchants and Manufacturers of Sheffield and the neighbourhood that he is the only manufacturer of the name of ADDIS in any part of the United Kingdom making Carving Tools, &c.

Mr. J. B. ADDIS also begs to say that he has ceased to have any connection with the firm of Ward and Payne, Sheffield.

All orders entrusted to Mr. J. B. ADDIS will be executed with prompt attention, and in his usual superior style, for which he obtained Prize Medals in 1851, 1862, 1870, and 1871, and also a Gold Medal subscribed for by the Merchants, Manufacturers, and Workmen of Sheffield for superior workmanship. All orders to be addressed as above.

N.B.—All Tools made by J. B. ADDIS are marked "J. B. ADDIS, Prize Medals, 1851, 1862, 1870, and 1871."

NONE OTHER GENUINE.

J.B. Addis' Advertisement in 1872 London Directory

1881.] ADVERTISEMENTS. 43

PRIZE MEDAL CARVING TOOLS.

Under the Distinguished Patronage of

H.R.H. THE PRINCE OF WALES, THE PRINCESS OF TECK, LORD LICHFIELD,

ETC., ETC.

CARVING TOOLS.

J. B. ADDIS & SONS,

ARCTIC WORKS, SHEFFIELD.

(*Brother to the late* Mr. S. J. Addis, *of* 68 *&* 70, *Worship Street, Finsbury, London,*)

CARVING TOOL MANUFACTURERS.

BEWARE OF SPURIOUS IMITATIONS.

We beg most respectfully to inform the Merchants and Manufacturers of Sheffield and the Neighbourhood that we are the only Manufacturers of the name of Addis in any part of the United Kingdom making Carving Tools, &c.

Messrs. J. B. Addis & Sons also beg to say that they have ceased to have any connection with the firm of Ward & Payne, Sheffield.

The only Prize Medals that are awarded in the United Kingdom have been to J. B. Addis & Sons. A Gold Medal was also subscribed for by the Merchants, Manufacturers and Workmen of Sheffield for superior workmanship.

All Orders to be addressed to J. B. ADDIS & SONS, Carving Tool Manufacturers, Arctic Works, Sheffield, England.

N.B.—All Tools made by J. B. Addis & Sons are marked "J. B. Addis & Sons 9 Prize Medals, 1851 and 1862." Protected by Act of Parliament.

Plate CIII J.B. Addis & Sons Advertisement in 1881 Sheffield Directory

Note the distinguished patronage of "Gentlemen Carvers" noted in the above advertisements. Apparently soon after 1891 this firm dissolved and Ward & Payne again took over their business. A group of Addis brand carving chisels are illustrated in Fig. 11. (page 168).

96 ADVERTISEMENTS.

F. G. PEARSON,

(Formerly Manager at Messrs. Spear and Jacksons, and late Manager at Messrs. Robert Sorty and Sons),

CASTLE WORKS,

105, Edward Street, SHEFFIELD,

Manufacturer of every description of

EDGE TOOLS

AUGERS,

CAST STEEL BRIGHT ELASTIC

HAY, MANURE, AND DIGGING FORKS,

CARVERS', TANNERS', AND SKINNERS' KNIVES.

Every Tool warranted of superior Quality and Workmanship, and on the most improved principle.

CORPORATE MARK. ACUTE.

Plate CIV - F.G. Pearson's Advertisement - 1858 Sheffield Directory

Francis G. Pearson was in the edge tool business 1821-1837 before managing at Spear & Jackson and then at Robert Sorby and Sons. In 1854 he went into his own firm which continued through 1900.

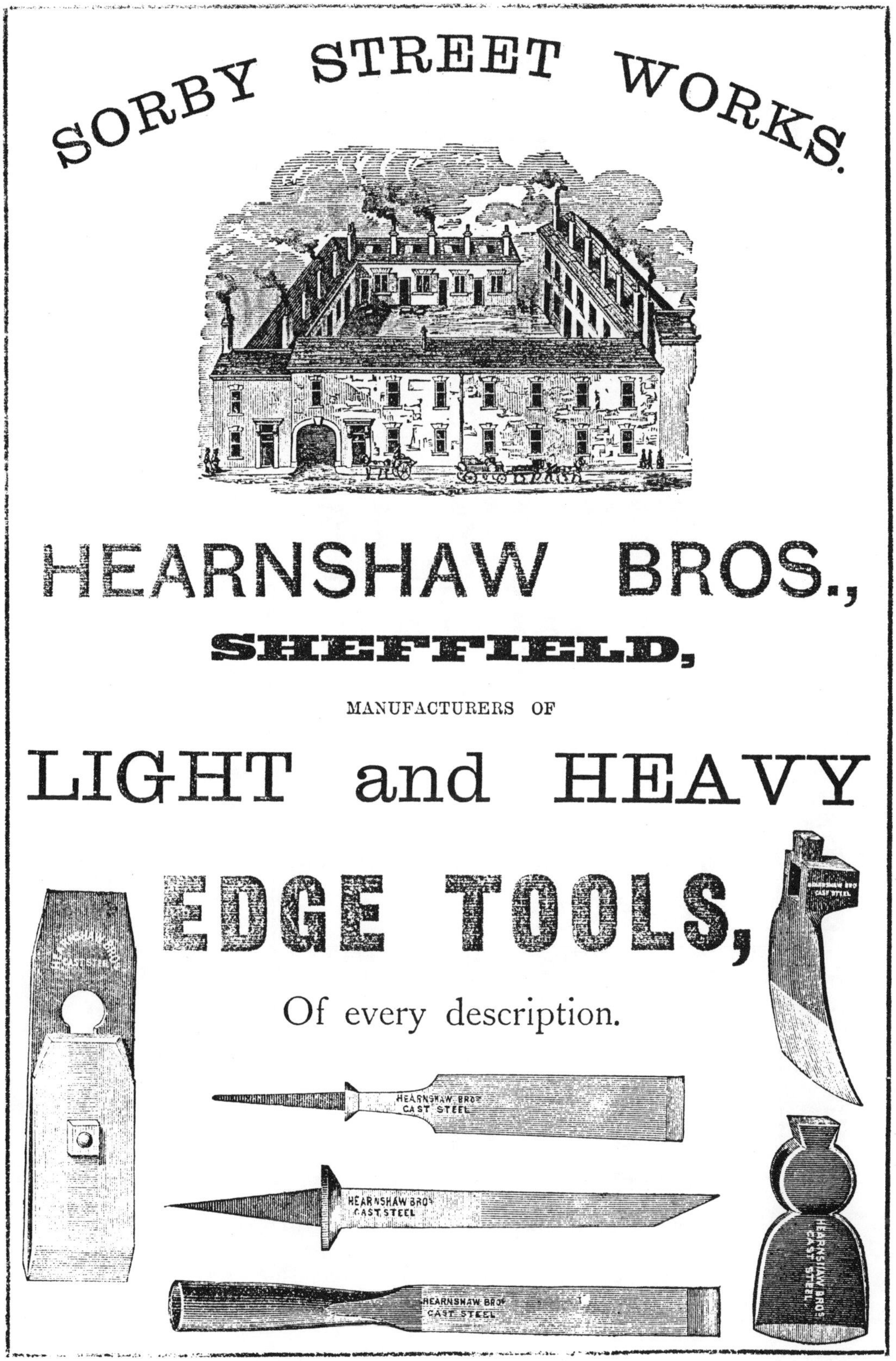

Plate CV - Hearnshaw Brothers' Advertisement - 1876 Sheffield Directory

WM. MARPLES & SONS,

HIBERNIA WORKS, **SHEFFIELD.**

LONDON OFFICES: 14, IVY LANE, E.C.

First Award at both Sydney & Melbourne Exhibitions for Edge Tools.

MANUFACTURERS OF EVERY DESCRIPTION OF

TOOLS AND SKATES,

AND GENERAL FACTORS OF ALL KINDS OF

SHEFFIELD, BIRMINGHAM, & LONDON GOODS.

We wish to direct

SPECIAL ATTENTION

to our

EDGE TOOLS.

HIBERNIA.

TRADE MARK.

Using only the best material, and employing the most skilled labour, we warrant them to be equal to any in the Market.

They are stamped with our own Name and Trade Mark, which is a guarantee of the

QUALITY.

Importers of AMERICAN, GERMAN & FRENCH

TOOLS & HARDWARE

Ironmongers' Novelties are a specialty with us.

Offices also in Deep River, Conn., U.S.A., New York, Montreal, Remscheid, Cape Town and Liverpool

ILLUSTRATED PRICE LISTS SENT ON APPLICATION.

New Customers are requested to send Business Card and Trade Reference.

Plate CVI - Wm. Marples & Sons' Advertisement - 1881 Sheffield Directory
[Note Foreign Offices and Mention of being Importers of Foreign Tools]

THOMAS IBBOTSON & CO.

PATERNOSTER ROW, CHARLES ST., SHEFFIELD,

STEEL REFINERS,

AND MANUFACTURERS OF

EDGE AND JOINERS' TOOLS,

SAWS, FILES, SHEARS, SCYTHES,

Cast Steel, Hay and Digging Forks,
Engineers' Tools, Coach and Railway Wrenches, Spanners, Hammers, Pincers, Compasses, &c.
Screw Stocks and Dies, Ratchet Braces and Lifting Jacks.

GARDEN TOOLS, TOOL CHESTS, SKATES, & CUTLERY.

LIGHT EDGE TOOLS, viz.:—

Cast Steel Firmer Chisels and Gouges.
" " Coachmakers' ditto
" " Long Thin Paring "
" " Millwrights' "
" " Turning "
" " Socket "
" " Mortice "
" " Spokeshave Irons.
" " Cut and Double Irons.
" " Tooth Plane Irons.
" " Skew and Square Rabbet Irons.
Soft Moulding Plane Irons and Plough Bits.
Drawing Knives, Carpenters', Coopers', Wheelers', &c.

CARVING & ENGRAVING TOOLS of every description.

JOINERS' TOOLS, viz.:—

Squares, Bevils, Spokeshaves, Gauges, Turnscrews, Levels, Saw Pads, Saws and Frames, Saw Sets, Scrapers, &c., &c.

BRACES and BITS, together or separate, all kinds.
Gimblets, Bradawls, Punches.

TOOL CHESTS:—Youths', Gentlemen's, Carpenters', and Emigrants', containing Tools of first quality.

SKATES, in every variety.

PLANES, all kinds.

HEAVY EDGE TOOLS, viz.:—

Carpenters' Shell and Screw Augers, and Auger Bits.
Ship Augers and Wheelers' Bruzzes.
Axes and Adzes for Carpenters, Coopers, Wheelers, Coachmakers, Shipwrights, &c.
Hatchets, Choppers, Mincing, and Cheese Knives.
Bills and Switching Hooks, single and double-edged of various patterns.
Brick, Masons', Plasterers', Pointing & Gauging Trowels.
Garden and Field Hoes, Rakes, Spuds and Weed Hooks.

HAMMERS, viz.:—

Kent, Scotch, Canterbury, Carrick.
Garden, Lathing, Shoe, Ladies' and Gentlemen's.
Saddlers', Upholsterers', Coachtrimmers'.
Grocers', Farriers', Coal, &c.

HAMMER HEADS, London, Manchester & Warrington.

Joiners', Roofing, Flooring, and Clench.
Bricklayers', Sheathing, Hand, Sledge, & Boilermakers'.

SAWS, viz.:—

Hand, Panel, Ripping, Compass, Dovetail.
Carcase, Sash, Tenon, Table, and Fret Saws.
Frame, Cross-cut, Pit and Stone Saws.
CIRCULAR SAWS, Saw Benches and Spindles.
Mill Saw, Turning, Iron Saw, Billet and Canada Webs.

FILES, viz.:—

Flat, Half-round, Round, 3-Square, 4-Square.
Hand, Equalling, Slotting, Mill, and Cotter.
Hand & Frame Saws, Gulleting, Warding & Topping.

RASPS, viz.:—

Horse, Shoe, Bread, Wood, Cabinet, and Farriers'.

ENGINEERS' and LANCASHIRE TOOLS, viz.:—

Screw Stocks, Dies and Taps, Screw Plates and Taps.
Ratchet Braces, Coach or Screw Wrenches.
Drill Stocks, Braces, and Drilling Cramps.
Hand, Bench, Staple, and Parallel Vices.
Chipping and other Hammers, Chipping Chisels.
Lathe Carriers, Spanners, and Screw Keys.
Steel Squares and Straight Edges, Iron Saw Frames.
Spring Dividers, Compasses, Callipers, Pincers.
Cutting Nippers, Bellhangers' and other Plyers.
Wire Gauges, Beck Irons, Goldsmiths' Shears, &c.

GAS FITTERS' TOOLS, viz.:—

Screw Stocks, Dies & Taps for Iron or Brass Pipe, Pipe Tongs, Plyers, Pipe Cutters, Screw Cutting Tools, &c.

JOINERS' BENCH, FLOOR, and SASH CRAMPS.

SCREW LIFTING JACKS, HOISTING CRABS and PULLEY BLOCKS.

LATHES, Amateurs' & Mechanics', of every description.

DRILLING, PUNCHING & SHEARING, PLANING, SHAPING, SLOTTING, WHEEL-CUTTING, FORGING, and BOLT and NUT SCREWING MACHINES.

SHEARS, viz.:—

Garden, Pruning, Branch, and Grass Shears.
Sheep, Fellmongers', Woolstaplers', Thatchers', Rag, Horse, and Docking Shears.

GARDEN TOOLS of every description; also SYRINGES.

GARDEN TOOLS, in Sets, in Polished Oak Chests.

SCYTHES, SICKLES, and REAPING HOOKS.

HAY and STRAW KNIVES.

CAST STEEL HAY, DIGGING & MANURE FORKS.

STEEL, Cast, Single & Double Shear, Blister & Spring.

RULES and MEASURING TAPES.

JOINERS' TOOL BASKETS.

OIL STONES and SLIPS.

48

Plate CVII - Thomas Ibbotson & Co.'s Advertisement - 1876 Sheffield Directory

LOCKWOOD BROTHERS,

Manufacturers,

74, *ARUNDEL STREET*,

SHEFFIELD.

JOINERS' TOOLS, EDGE TOOLS AND SAWS MARKED. } I. & H. SORBY. FILES MARKED. { LOCKWOOD BROTHERS C‡X SHEFFIELD OR C‡X. LOCKWOOD.

SHEEP SHEARS MARKED. } I&H SORBY CUTLERY MARKED. { C‡X LOCKWOOD BROTHERS SHEFFIELD.

JOHN SORBY & SONS,

Manufacturers,

SPITAL HILL,

SHEFFIELD.

JOINERS' TOOLS, EDGE TOOLS AND SAWS MARKED. } I. & H. SORBY. FILES MARKED. { LOCKWOOD BROTHERS C‡X SHEFFIELD OR C‡X. LOCKWOOD.

SHEEP SHEARS MARKED. } I&H SORBY CUTLERY MARKED. { C‡X LOCKWOOD BROTHERS SHEFFIELD.

N 29

Plate CVIII - Lockwood Brothers' Advertisement Noting Mark of John Sorby & Sons
[1854 SHEFFIELD DIRECTORY]

TURNER, NAYLOR & MARPLES

(LATE SORBY & TURNER),

MANUFACTURERS OF

LIGHT & HEAVY EDGE TOOLS,

AUGERS, JOINERS' TOOLS

SAWS, FILES, SHEARS AND ENGINEERS' TOOLS,

Hammers, Screw Stocks, Taps and Dies,
Screw Wrenches, Spanners, Ratchet Braces,
Pulley Blocks, Lifting Jacks;
Joiners', Flooring, and Ship Cramps;
Hand, Bench, Parallel, Smiths' and Portable Vices;

TABLE AND SPRING CUTLERY;

PLUMBERS', GLAZIERS', and TINMEN'S TOOLS, &c., &c.

Trade Mark, "I. SORBY." Established 1810.

Corporate Mark, Granted by the Cutlers' Company.

THE NORTHERN TOOL WORKS,

JOHN STREET,

SHEFFIELD.

Plate CIX - Turner, Naylor & Marples' Advertisement
[page 49 Advertising Section 1879 ILLUSTRATED GUIDE to SHEFFIELD]

[see Table VIII, page 132 noting Historical Development of this Firm]

Plate CX - Ward & Payne Advertisement - 1851 STAFFORDSHIRE DIRECTORY

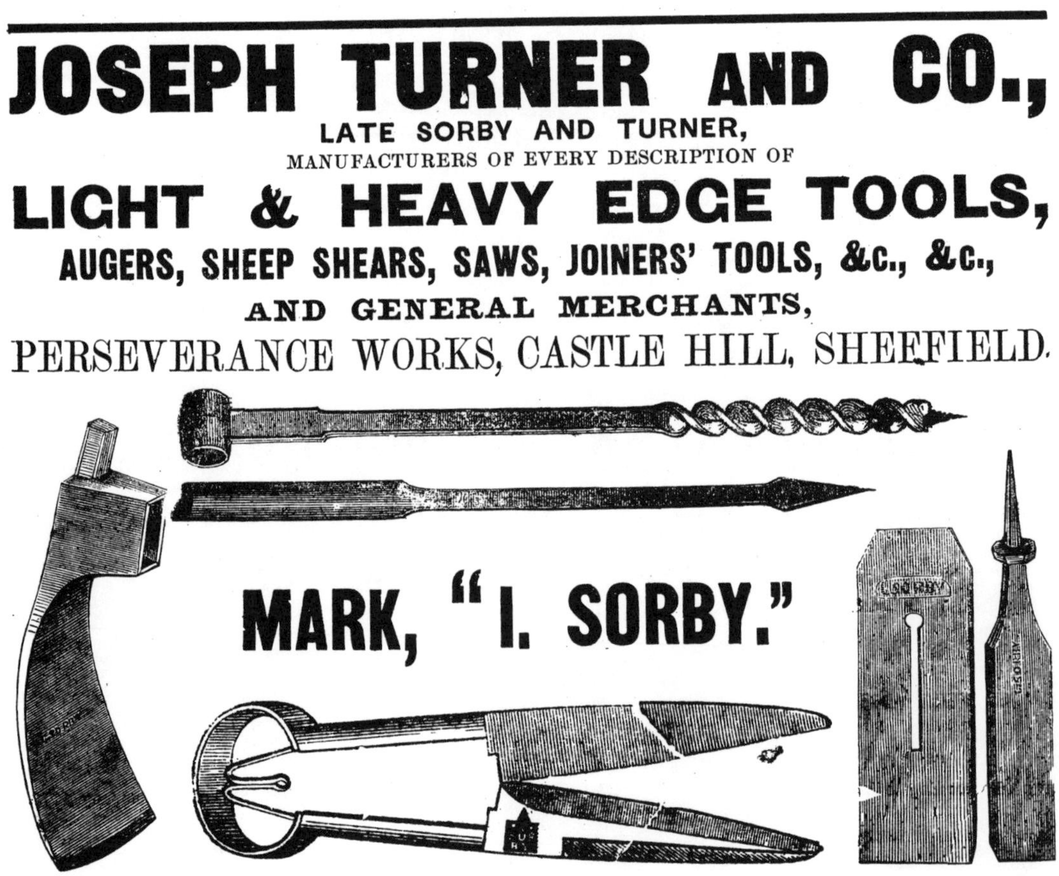

Plate CXI - Joseph Turner & Co. Advertisement - 1862 LONDON DIRECTORY, page 2478

HENRY THOMPSON,

MANUFACTURER OF ALL KINDS OF EDGE & JOINERS' TOOLS, BRIGHT SCOTCH AUGERS, BRACES, BITS, &C.

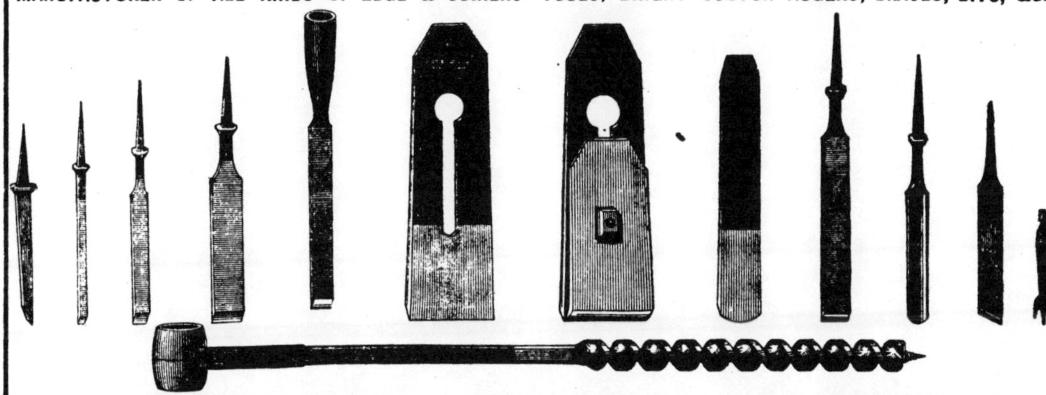

BROOMHALL WORKS, BROOMHALL ST., SHEFFIELD.

Henry Thompson's Advertisement - 1876 Sheffield Directory

JOHN CUTLER, JUN.,

MANUFACTURER OF

EDGE TOOLS,

42, CHARLOTTE-STREET,

SHEFFIELD.

John Cutler, Jr. Advertisement - 1849 Sheffield Directory

AARON HILDICK,

MANUFACTURER OF

PLOW BITS, RABBET, MOULDING,

AND ALL KINDS OF

PLANE IRONS, AND LIGHT EDGE TOOLS,

OF THE VERY BEST QUALITY;

57½, BRIDGE STREET, SHEFFIELD.

Aaron Hildick's Advertisement - 1879 Sheffield Directory
[Principally known for manufacturing of plane irons]

Plate CXII - Miscellaneous Edge Tool Manufacturers' Advertisements

SUFFOLK WORKS, SHEFFIELD.

THOMAS TURNER AND Co.,

MANUFACTURERS OF

Table Knives & Forks,

SILVER & PLATED TABLE KNIVES & FORKS,

Silver and Plated Desserts, Fish Carvers, Fish Eating Knives & Forks,
Butter Knives, Pickle Forks, Fruit Knives, & Scissors,

PEN & POCKET KNIVES, GARDENERS' KNIVES,

AND RAZORS; ALSO,

SAWS, FILES, EDGE TOOLS & STEEL.

Prize Medal of the

Great Exhibition, 1851,

THOMAS TURNER & Co. caution all parties against using their Name or Trade Mark, granted by Act of Parliament. Injunctions have been obtained from the Vice-Chancellor to restrain certain individuals who have thus infringed.—See Report in the "Times," January 11, 1850, also March 6, 1863. All Goods bearing their Name or Mark as below are warranted.

TURNER & CO.
SUFFOLK WORKS,
SHEFFIELD.

V R
THOMAS TURNER & COMPY.
SUFFOLK WORKS, SHEFFIELD.

THOMAS TURNER & CO.
SHEFFIELD.

TURNER.

TURNER & CO.
SHEFFIELD.

THOMAS TURNER.

THOS. TURNER.
CAST STEEL.

A CAUTION.—Whereas, certain parties have of late assumed the title of "TURNER & CO.," we deem it advisable, in justification to ourselves, to notice the matter by public advertisement, and otherwise to caution our friends, lest they be imposed upon by the deception thus practised. All parties will therefore kindly observe that our address be written in full, and that goods be marked with our corporate mark.

Awarded to

Thomas Turner & Co., Class 21.

THOMAS TURNER & CO., SUFFOLK WORKS.

We further beg to observe that we shall commence the usual proceedings against any person infringing upon any of our Trade Marks.

London House—19, ELY PLACE, HOLBORN.

Plate CXIII - Thomas Turner & Co. Advertisement - LONDON DIRECTORY, 1865, p. 2750

Some typical Sheffield edge tools are illustrated in Figs. 5-11.

Sales of edge tools were principally made wholesale through agents, exporters in London, importers in United States and tool stores in principal cities in England. An example was the advertisement placed in the 1854 Sheffield Directory by Stanley, Morton & Davis of Sheffield, merchants, manufacturers and AUSTRALIAN COMMISSION AGENTS who called to attention: -

"Having a thorough knowledge of the SHEFFIELD MARKET, and also of the class of Goods most in demand for Australia, beg to offer their services as Agents to Brokers, Shippers and Merchants, and whilst being chiefly employed by the First-class London and Birmingham Houses in collecting goods for Australia and the Home Market, they are open for a few engagements with parties requiring SHEFFIELD GOODS, for those or other Markets.

TABLE KNIVES, POCKET KNIVES, PENKNIVES, BOWIE KNIVES, RAZORS, SCISSORS
Britannia Metal, Electro and British Plate Goods
SAWS, FILES, JOINERS' TOOLS, EDGE TOOLS, &C. &C.

N.B.- Book of Standard Sheffield Lists, and Net Lists of Selected Sheffield Goods forwarded on receipt of Six Postage Stamps.

From time to time illustrated price lists of light and heavy edge tools, joiner tools, saws, files and cutlery products were printed and sold announcing agreed prices. Sometimes such lists were accompanied by a note stating: "The Standard Prices of the Cutlers' Company have been adopted". This merely means that the principal manufacturers who set these prices all being Members of the Cutlers' Company met there and agreed on selling at such prices. It was apparently an unwritten agreement that they would all maintain these prices, but there was no official relation to the Cutlers' Company. Discounts were offered to wholesalers that probably varied with the extent of purchases. During the period of the late 1860's through the remainder of the 19th Century these "SHEFFIELD ILLUSTRATED LISTS" were drawn, engraved and "entered at Stationers' Hall (copyrighted) by Pawson and Brailsford, publishers of *The Illustrated Guide to Sheffield*. The title page to the 1870 LIST in shown as Plate CXIV. Six plates of various edge tools from this LIST follow: B-1, Firmer chisels and firmer gouges with various sweeps; B-2, turning, paring and other miscellaneous chisels; B-3, other miscellaneous chisels; B-5, mortice chisels, plane and plough irons; B-15 Draw knives and augers; and F-3, carving chisels.

Most chisels were sold as forged to the individual smaller firms who did not have steel melting or forging facilities. These were then ground, hardened and hafted (handled). Larger firms such as Robert Sorby & Sons, Thomas Turnton and Sons, Lockwood Brothers and Ward & Payne made the complete line from cast steel to final product.

Any manufacturer could have Pawson & Brailsford prepare a trade catalogue by using printing cuts of the tools appearing on the LIST. Plate 2 from the 1870 Wm. Marples & Sons Catalogue is shown as Plate CXXI. Note that items Nos. 141 C.S. Socket Chisel, 142 C.S. Socket Gouge and 129 Best Mortice Chisel are identical to Nos. 450, 458 and 494 of B2 (Plate CXVI) from the 1870 SHEFFIELD ILLUSTRATED LIST. Similarly Marples' items Nos. 149 Carpenters' Drawing Knife and 148 C.S. Drawing Knife with Egg Handles are respectively identical with Nos. 620 and 621 on B 15 (Plate CXIX) of this same LIST.

THE SHEFFIELD ILLUSTRATED LIST,

COMPRISING

ENGRAVED PATTERNS AND STANDARD PRICES.

UPWARDS OF
TWO THOUSAND ENGRAVINGS.

Note.—The Standard Prices of the Cutlers' Company have been adopted.

ENTERED AT STATIONERS' HALL.

SHEFFIELD:
DRAWN, ENGRAVED, AND PRINTED BY PAWSON AND BRAILSFORD, HIGH-STREET AND MULBERRY-STREET.
1870.

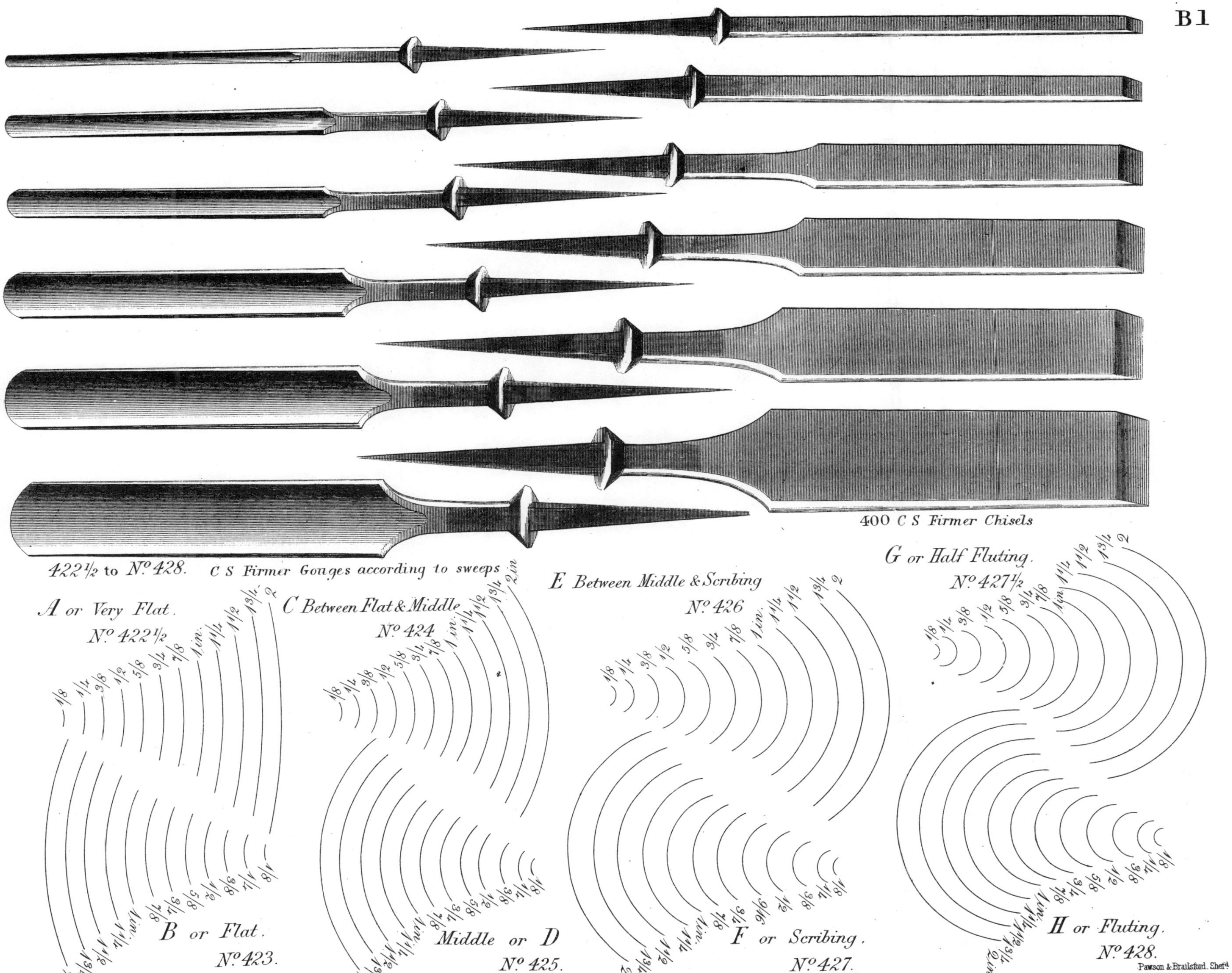

Plate CXV - Firmer Chisels and Firmer Gouges, showing Sweeps, B-1, 1870 Illustrated Sheffield List

Plate CXVI - Turning, Paring and Other Miscellaneous Chisels, B-2, 1870 Illustrated Sheffield List

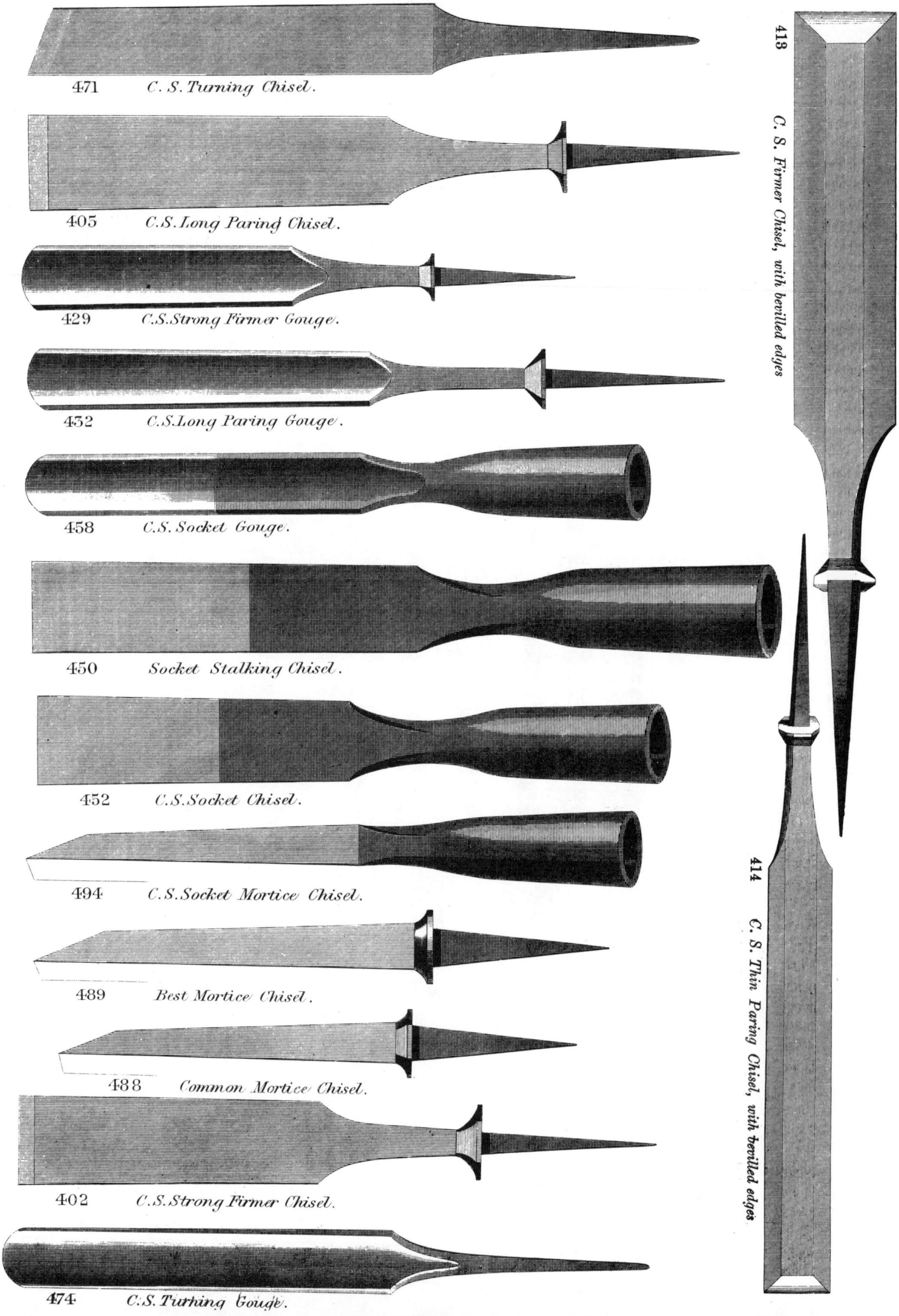

Pawson & Brailsford Lithographers Sheffield

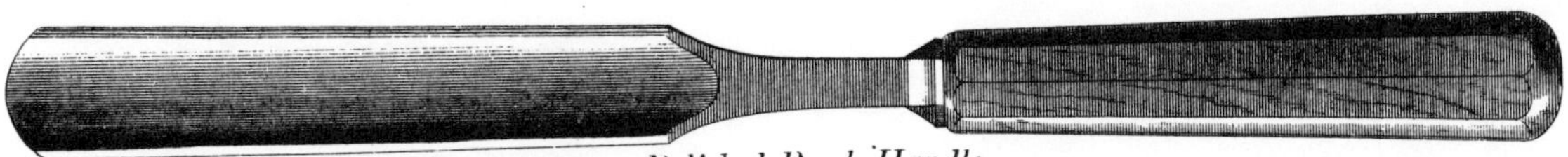

439 C.S. Firmer Gouge, Octagon Polished Beech Handle

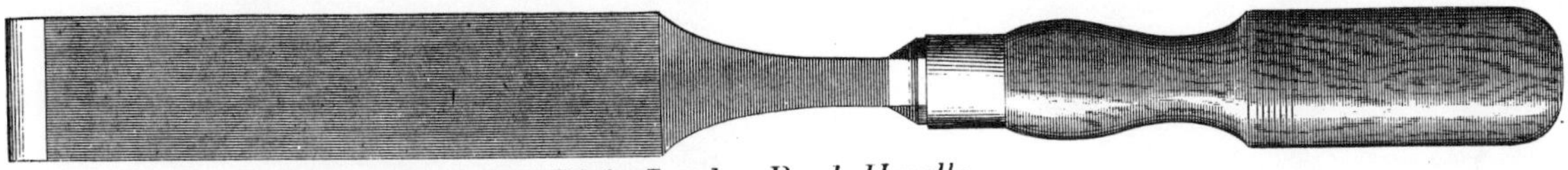

418 C.S. Firmer Chisel, best Plain London Beech Handle

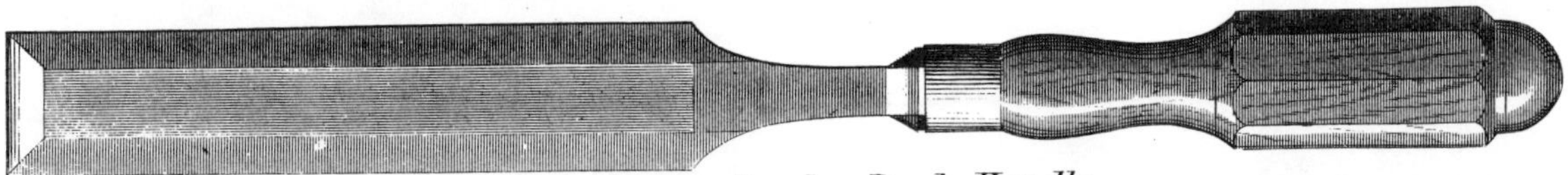

422 C.S. Bevelled Firmer Chisel best Octagon London Beech Handle

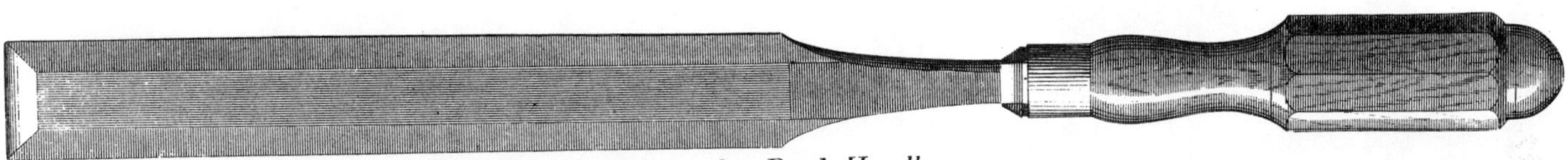

422½ C.S. Bevelled Firmer Chisel, best Octagon London Beech Handle

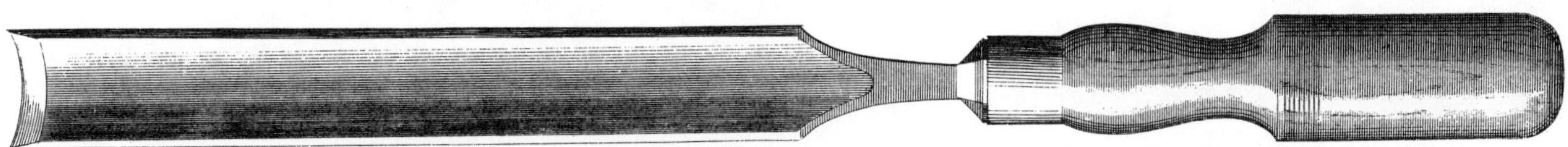

444 C.S. Long Thin Paring Gouge, cannelled inside, Plain London Beech Handle with Long Brass Hoops

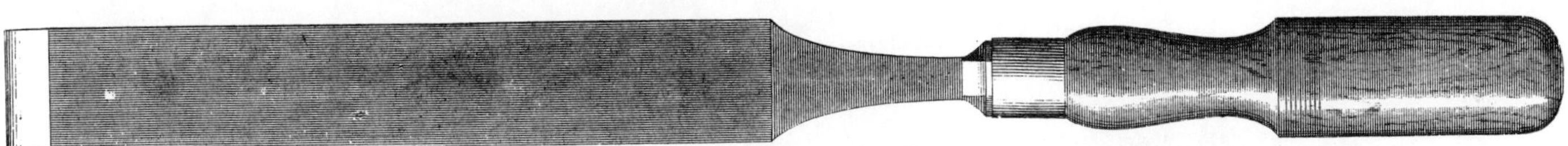

420 C.S. Long Thin Paring Chisel, best Plain London Beech Handle

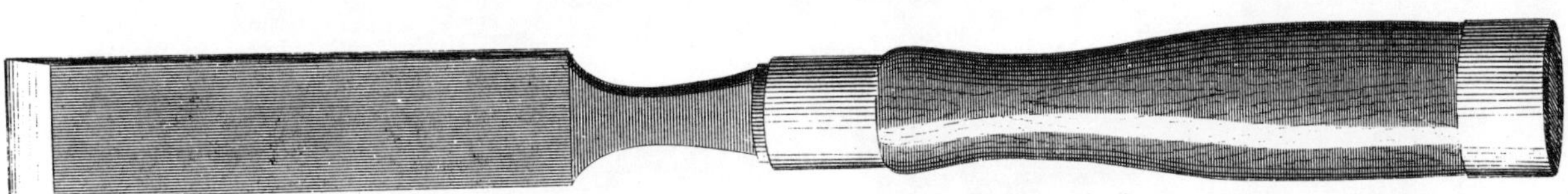

399 C.S. Registered Chisel, Handled

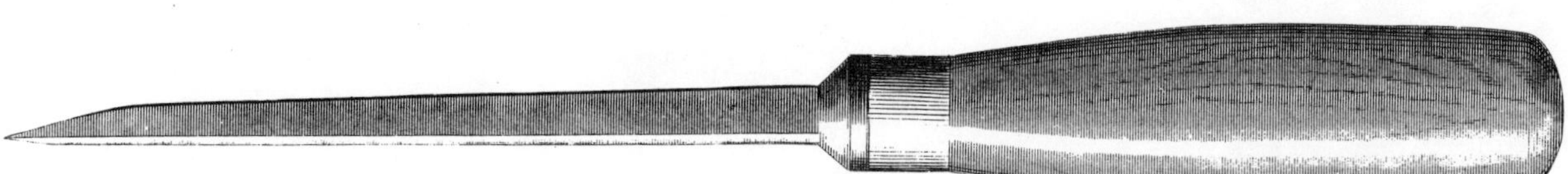

500 C.S. London Sash Mortice Chisel, Brass Hooped, Beech Handle

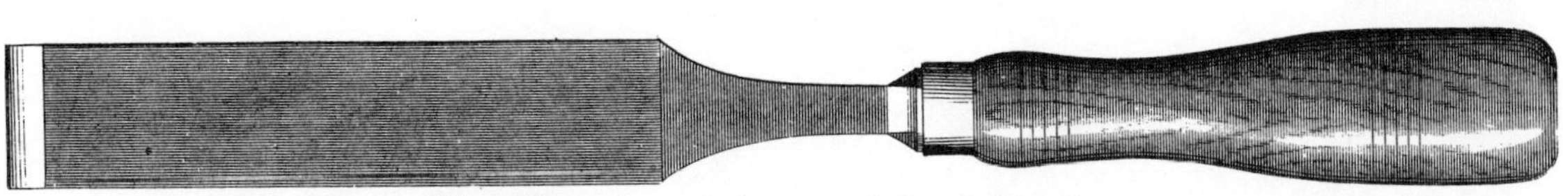

417 C.S. Firmer Chisel, Brass hooped, best round Beech Handle

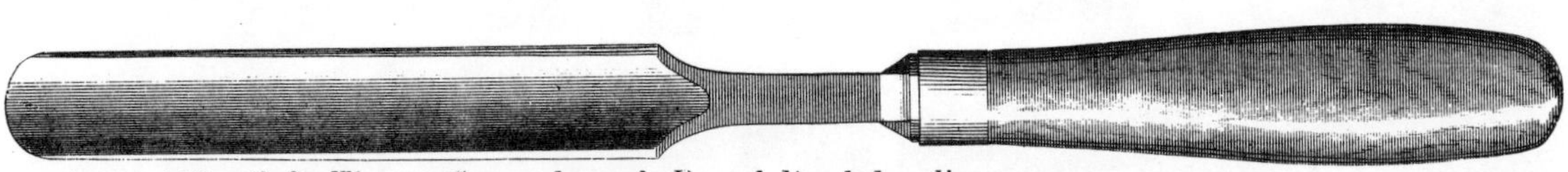

440 C.S. Firmer Gouge, hooped, Round Beech handle

Plate CXVII - Firmer, Paring and Sash Chisels; Firmer Gouges, B-3, 1870 Sheffield Illustrated List

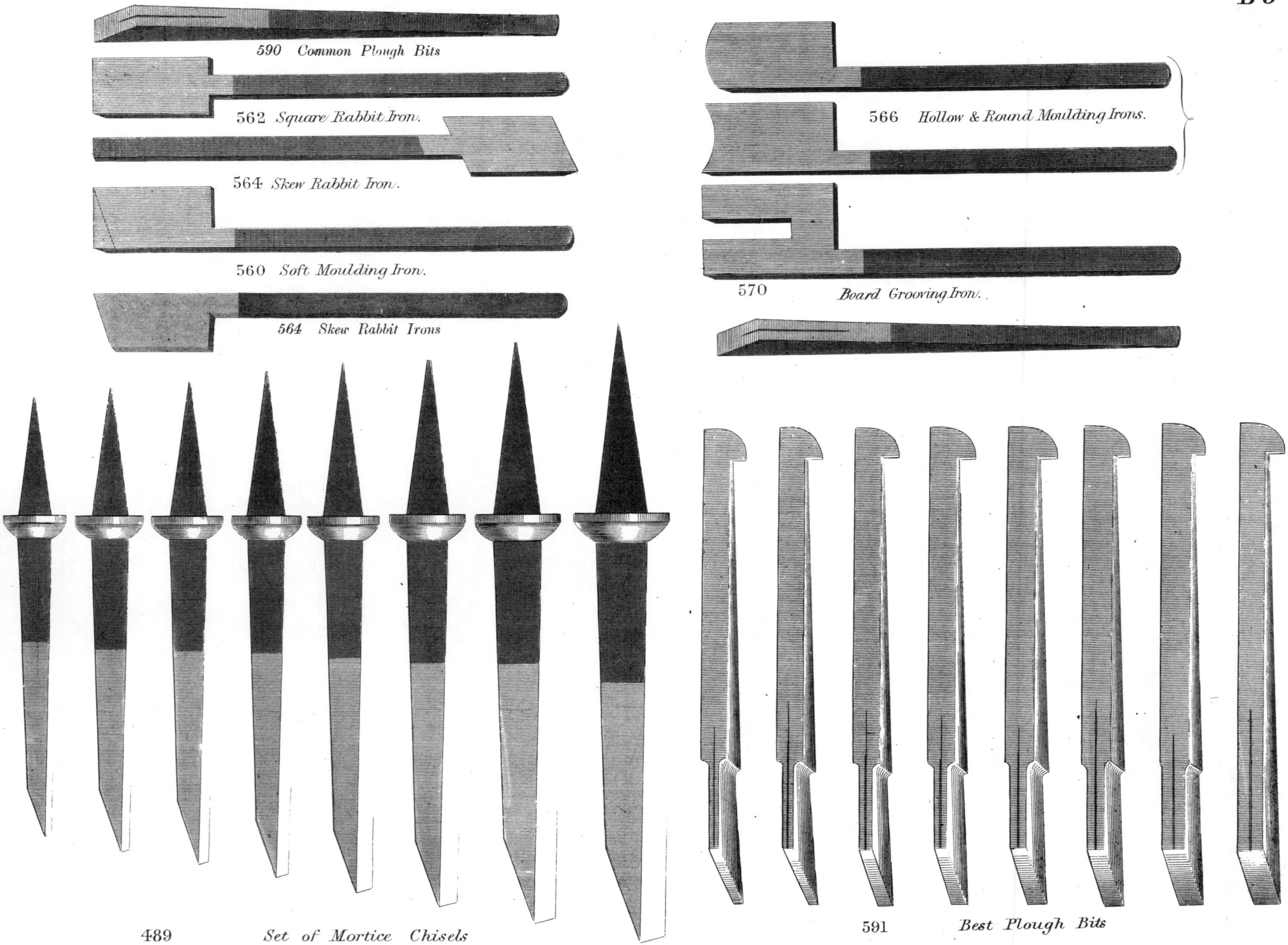

Plate CXVIII - Mortice Chisels, Plane and Plough Irons, B-5, 1870 Sheffield Illustrated List

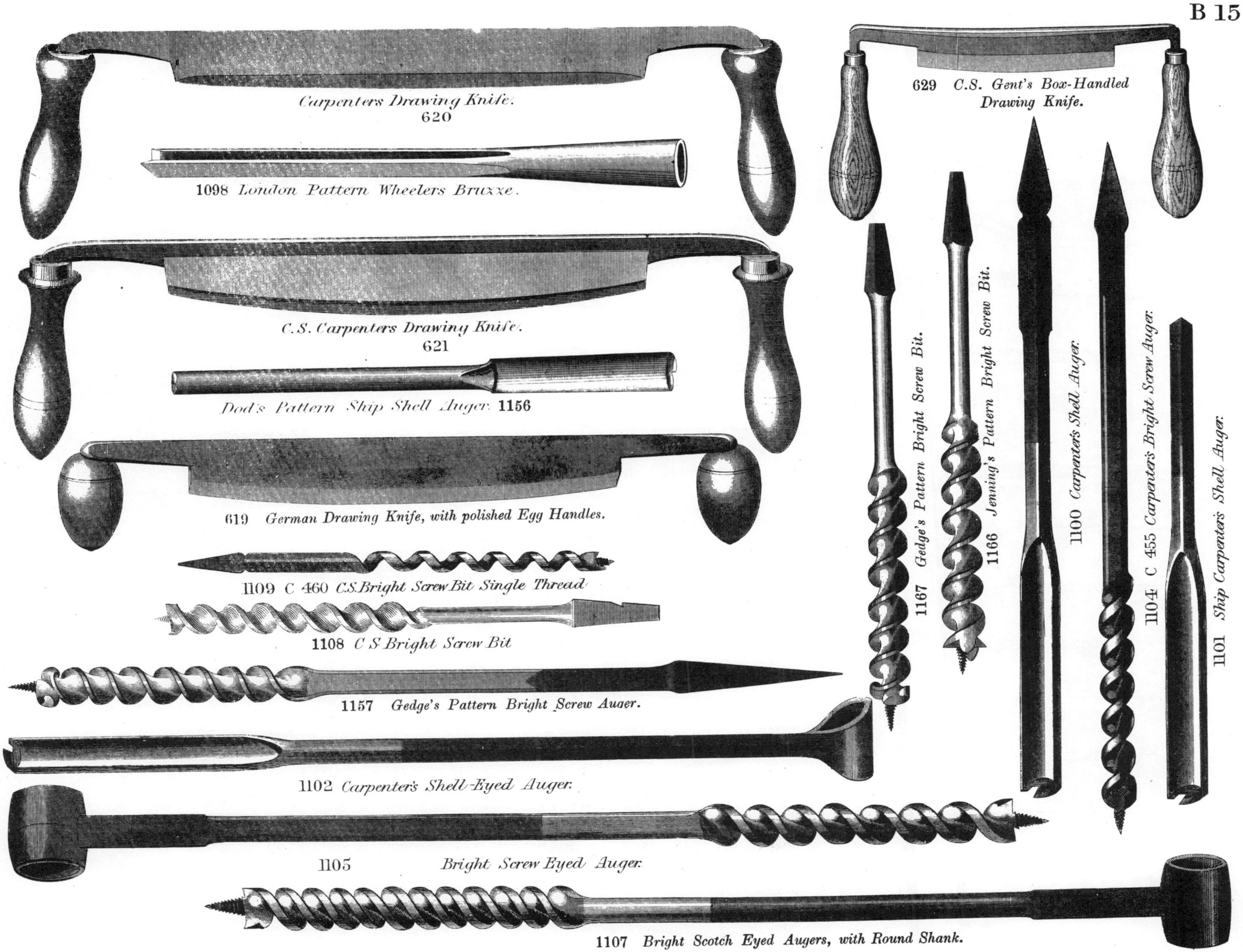

Plate CXIX - Draw Knives and Augers, B-15, 1870 Illustrated Sheffield List

Plate CXX - Carving Chisels, F-3, 1870 Illustrated List

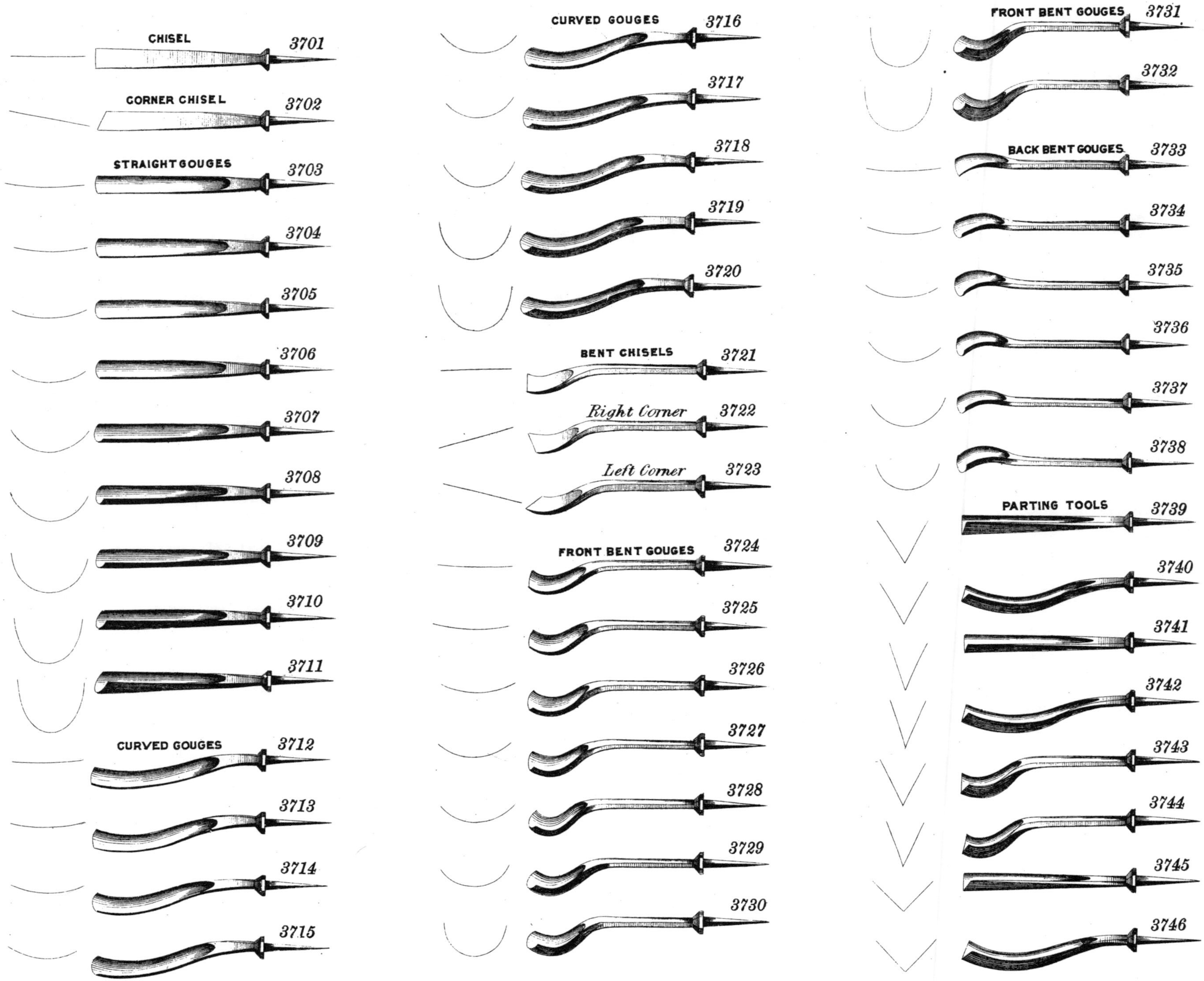

Pawson & Brailsford, Shef^d.

WILLIAM MARPLES & SONS, SHEFFIELD.

2

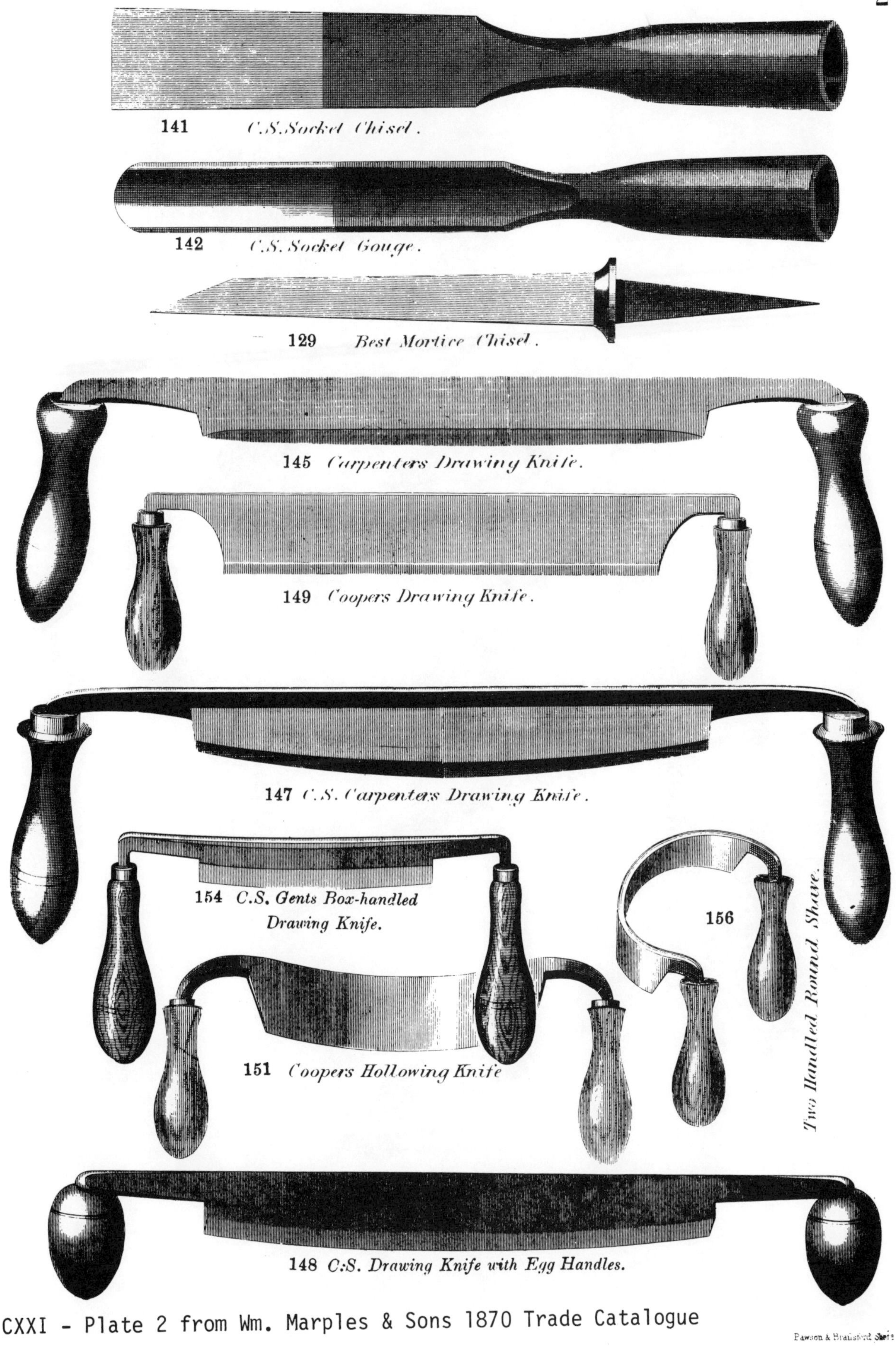

Plate CXXI - Plate 2 from Wm. Marples & Sons 1870 Trade Catalogue

Pawson & Brailsford Sheffield

Fig. 5 - Mortice Chisels by I. Sorby

Fig. 6. - Short Firmer Chisels by Robert Sorby & Sons

Fig. 7. - Gouges by Thos. Turton & Sons (bearing Mark of Wm. Greaves & Sons)

Fig. 8. - Pattern Makers Gouges and Chisels by Reaney & Wood

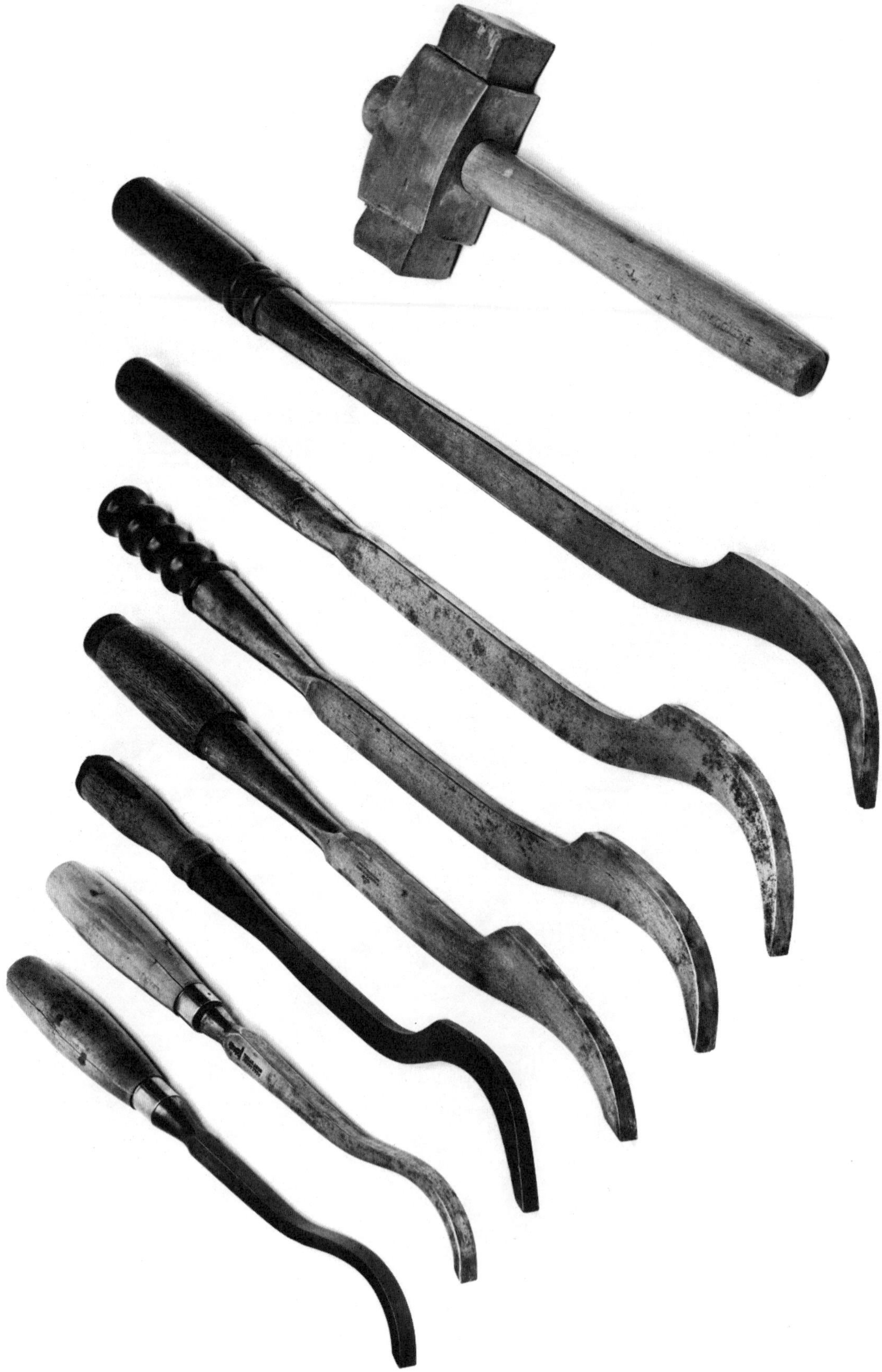

Fig. 9. - Lock Mortice Chisels by James Howarth & Sons, Robert Sorby & Sons, Thomas Ibbotson & Co., William Marples & Sons and Ward & Payne

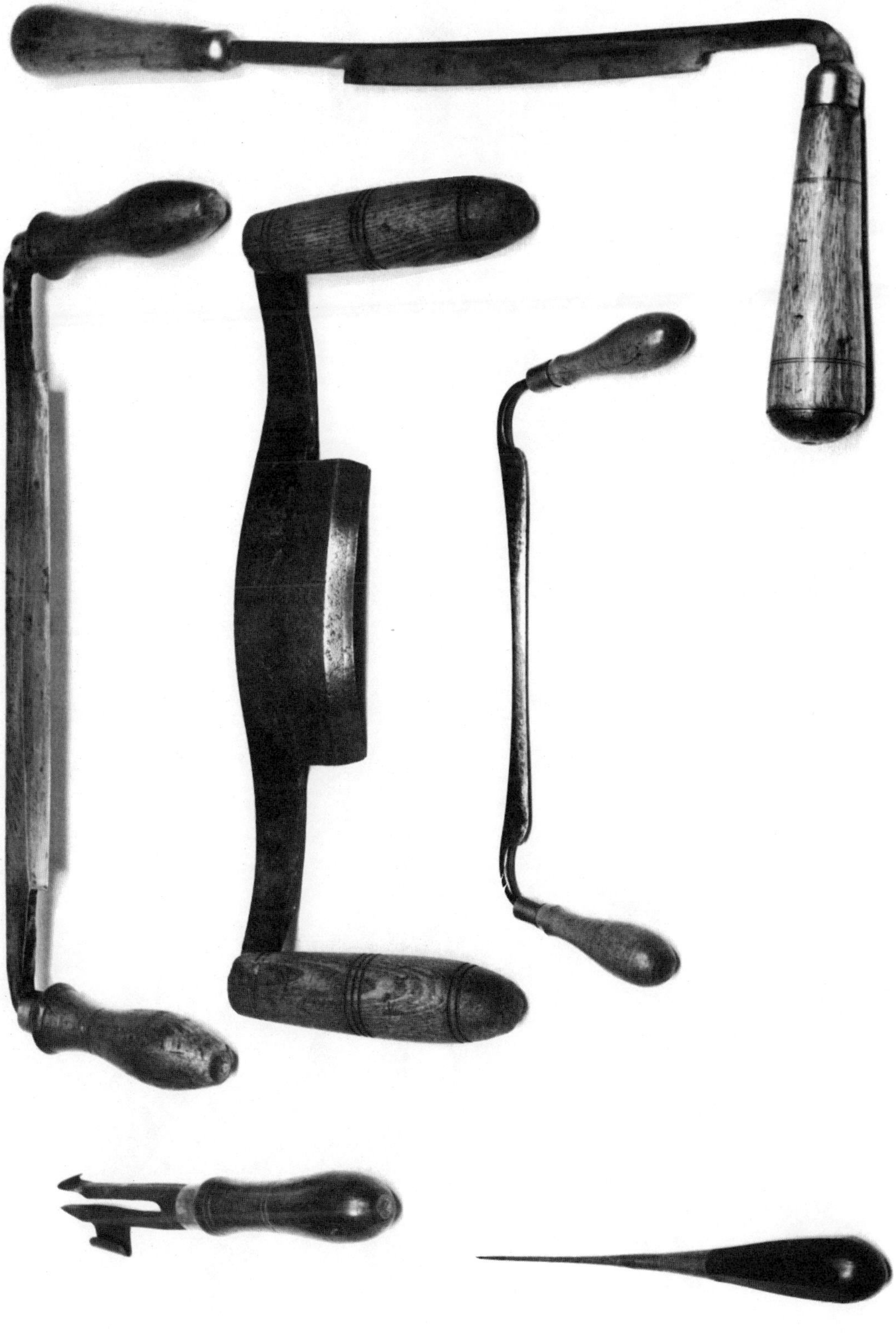

Fig. 10. - Draw Knives, Race Knife and Scriber

Fig. 11 - Addis' Carving Chisels

1. Sketchley's *Birmingham, Wolverhampton and Walsall Directory*, 1767
2. G.I.H.Lloyd, *op. cit.*, p.158
3. *1837 Sheffield Directory*, Introduction, p.33
4. *Rules, Orders, Regulations of Society Sawmakers of Sheffield*, December 1798
   P.d'A.Jones and E.N.Simons, *Story of the Saw* [London, 1961], p.26
5. *Smith's Key* [Central Library, Sheffield] Republished 1975 by EAIA
6. E.Surrey Dane, *op. cit.*, p.155
7. *ibid*, p.267, footnote #9
8. Collection of author.
9. Archives Collection of Peter Stubs, Central Library, Manchester, England
10. *ibid*, File #122
11. *ibid*
12. G.I.H.Lloyd, *op. cit.*, p.182
13. C.H.Hummel. *Winterthur Portfolio*, Vol II, 1965, p.46
14. K.D.Roberts, *Tools for the Trades & Crafts*, p.14
15. *ibid*, p.33
16. Archives, Local History Department, Central Library, Sheffield
17. S.Pollard, *Three Centuries of Sheffield Steel Making*, [Sheffield, 1965]
18. *ibid*, p.20
19. Archives, Local History Department, Central Library, Sheffield, File #114
20. *ibid*, File #115
21. S.Pollard, *op. cit*, [Note #17 above], p.20
22. *ibid*
23. *Tools for the Trades & Crafts*, p.6
24. *ibid, p.6*
25. See Plate CXV, p.155
26. for definition of "toys", see *Tools for the Trades & Crafts*, p.17
    "pincers" were generally made in Liverpool or Birmingham.
27. see Plate XCVIII, p.138
28. See Table VIII
29. Additional history of Ward & Payne appears in the next Chapter.
30. See Addis advertisements, pages 142 & 143 and discussion therein.
31. Wm. Marples began as a joiner tool maker in 1833. Wm. Marples & Sons added edge tools to their line in 1871.
32. correspondence John Sainsbury, Sheffield, England
33. A detailed history of the firm William Marples & Sons appears in the preface to the Mid West Tool Collector's Association Reprint of the 1909 Wm. Marples & Son Trade Catalogue. [1979]
34. K.C.Barraclough, Firth-Brown Ltd., Sheffield and author of *Sheffield Steel* had a plane iron with this imprint analyzed for boron content and reported only a trace present, insufficient to create an effect.

EXHIBITION OF ALL NATIONS, 1851.

PRIZE MEDAL,

CLASS 21.

Awarded to J. HOWARTH,

For Superior Quality of

EDGE TOOLS.

JAMES HOWARTH,

FITZWILLIAM STREET, SHEFFIELD,

Manufacturer of

EDGE TOOLS OF EVERY VARIETY;

Also Calico Block Cutters', Engravers', Silversmiths', Die Sinkers', Carvers', and all kinds of Turners' Tools, &c. [*Corporate Mark,*

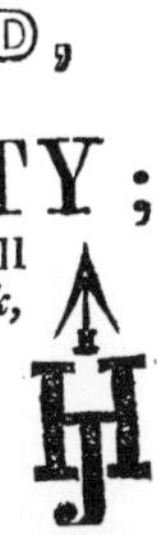

ALSO MANUFACTURER OF HAND-DRAWN CAST STEEL.

N. B.—Maker of Cork, Curriers', Skinners', Tanners', and Shoe Knives, &c. upon an improved principle.

1854 Advertisement

Sheffield Directory

p.42

Check List of 19th Century Sheffield Edge Tool Makers [See Notes at End of List]

ADDIS, J. B. & Sons	Rockingham St.		-1872-1891-
ALLEN, Edgar & Co.	Wellmeadow St.	*Wellmeadow St. Works*	-1871-1900-
ALLSEBROOK, James	Hallcar St	*Hallcar Works*	-1860-1863
	Trafalgar St.		-1871-1872-
	South St.		-1889-1884-
ASH, William [J,P]	Norfolk Lane		-1825-1852-
ASHFORTH, Joseph & Co.	Bridge St.		-1852-
	Mowbray St.	*Bruce Steel Works*	-1860-1887-
ATKINSON, Charles [J]	West St.		-1854-
	Trippet Lane		-1856-1859-
ATKINSON Brothers [J]	Milton St.	*Milton Works*	-1879-1900-
BAKER, John & Sons [J]	Monmouth St.	*Monmouth Works*	-1876-1879-
	Harmer Lane		-1881-1900-
BARNES, F.& Co. [J]	Solly St.	*Corn Hill Works*	-1865-1900-
BARNSLEY, George & Sons [S]	Cornish St.	*Cornish Works*	-1891-1980-
BATTIE, John & Son [J]	Mowbray St.	*Pacific Works*	-1871-1879-
	Bernard St.		-1881-1893-
BEARDSHAW, J. & Son [S]	Garden St		-1845-
* " " & Sons [S]	" "		-1849-
	Attercliffe		-1856-1860-
	Effingham Rd.	*Baltic Steel Works*	-1862-1900-
BEARDSHAW, J. & T.S.	Hollis Croft	*Diamond Works*	-1876-1879-
BEDFORD, John	Portobello St.		-1852-
	Peniston Road	*Regent Works*	-1854-
" " & Sons [S]	Mowbray St.	*Lion Works*	-1876-1935-
BEDFORD, BURY & Co. [J,P,S]	Peniston Rd.		-1857-1935-
BEET, Jonathan & Sons [J,P,S]	Broad Lane	*Angenoria Works*	-1841-1849-
BEET, J. & Sons & GRIFFTH	Savile St.		-1852-1857-
BELK, George	Smithfield		-1817-
BELL, Lazarus	Division Lane		-1871-1872-
BIGGIN, Elliot & Co.	Penistone Rd	*North British Steel Works*	-1883-1887-
BINNEY Brothers & Co.	Hartshead		-1833-
BISHOP, James	Young St.		-1854-1888-
" " & Sons	" "		-1890-1900-
BISHOP, Samuel	Young St.		-1837-1845-

BLOOMER & PHILLIPS	Rockingham Lane		-1845-
	Bailey Lane		-1852-
BOOTH, William	Dronfield		-1854-1879-
BRADBURY, James	Union Lane		-1833-
BRADBURY & TAYLOR	Norfolk Lane		-1821-1825-
BRAMALL & BEDFORD [S]	Mowbray St.	*Savile Works*	-1865-1872-
BRITTAIN, Frederick [S]	Wellington St.	*St. George's Works*	-1862-1865-
	Shoreham St.		-1876-1891-
BRITTAIN, Samuel S. & Co. [S,J,P]	Shoreham St.	*St. George's Works*	-1856-1900-
BROOKES, William & Son [J,S]	Rockingham St.	*Howard Works*	-1845-
*	Carlisle St.		-1849-
" " & Sons	" "		-1856-1865-
	Attercliffe		-1868-1900-
BROOKES BROTHERS [S]	Eldon St.	*Malinda Works*	-1856-1865-
*BROOKSBANK, Abraham	Malinda St.		-1849-1900-
BROWN, Henry & Son [J,S]	Rockingham St.	*Western Works*	-1876-1879-
BROWN, William [S]	Trippet Lane		-1854-1856-
BRUMBY & MIDDLETON [S]	Howard St.		-1856-1863-
BURGON & CALMANS	Pardise Square		-1833-
BURYS & Co. [J,P,S]	Penistone Rd.	*Regent Works*	-1863-1935-
BUSH, Alfred	Union Lane		-1841-
BUSH & BIRKHEAD	Rockingham Lane		-1845-
BUTCHER, William [J]	Eyre Lane		-1821-1825-
*BUTCHER, William & Samuel [J,P]	Eyre Lane		-1833-1900-
BUTCHER, BROWN & BUTCHER [J,S]	Eyre Lane		-1828-
BUTLER, James & Co.	Pond Hill		-1837-1841-
CAM, James [J,P,S]	Norfolk St.	*Pond Works*	-1787-1837-
CAM, James [Marsh Brothers] [J,P,S]	Shude Lane		-1888-1900-
CAM, Joseph & Sons [J,P]	Union St.		-1856-1863-
	Button Lane	*Lawrence Works*	-1871-1900-
CAMMEL, Charles & Co.	Savile St.	*Cyclops Works*	-1865-
CARR, George & Co. [S]	Osborn St.		-1837-1841-

CARR, J. & RILEY [S]	Bailey Lane	*Bailey Lane Works*	-1833-1837-
CARR, WOODHOUSE & Co.	Bailey Lane		-1814-1815-
CARR, WOODHOUSE & CARR	Bailey Lane		-1825-
CHAMBERS, Charles	Castle Hill		-1865-1887-
COCKER BROTHERS	Nursery St.		-1854-1879-
COLQUHOUN & CADMAN [J,P]	Arundel St.		-1883-1900-
COOK, William & Son	Attercliffe		-1883-1900-
CORNFIELD, John	Smithfield		-1833-1845-
COWLISHAW, Richard	Orchard St.		-1849-1857-
CROOKES, Charles	Marsh Lane		-1817-
CROWNSHAW, CHAPMAN & Co.	Penistone Rd.		-1891-h895-
CUTLER, Hiram G.	Castle Hill		-1837-1845-
CUTLER, H. & Son & CHAMBERS	Castle Hill		-1849-1863-
CUTLER, John	Stanley St. Charlotte St.		-1841- -1845-1854-
CUTLER, William & John	Nursery Lane		-1821-1823-
DEAKIN, George [S]	North St. Gill St.		-1825-1860- -1865-
DEAKIN, REUSS & Co. [S]	Green Lane	*Tiger Works*	-1879-1891-
DODGE, J. & R. [J,P]	Eyre Lane Bridge St.		-1865- -1876-1879-
DRABBLE & SANDERSON	Russell St.		-1879-1900-
DYSON, Joseph & Co.	Attercliffe	*Millbank Steel Works*	-1879-1884-
EADON, Moses [J,S] " " & Sons [J, S]	Norfolk St. Savile St.	 *President Works*	-1837-1857- -1860-1900-
EADON, George, William & Thomas	Park St.		-1833-
EADON, JESSOP & FOX	Park St.		-1828-
ELLIN, Thomas	Sylvester Graden		-1884-1899-
EYRE, Edward [S]	Green Lane		-1825-
FENTON, Joseph & Sons	Scotland St.	*Sykes Works*	-1871-1872-
FENTON & MARSDEN [J]	Rockingham St.		-1841-1845-
FIRTH, Thomas & Sons	Savile St.	*Norfolk Works*	-1852-1900-

FLATHER, D. & Sons [J,S]	Savile St. Love St. Solly St.	*Solly Works* *Love Works*	-1857-1859- -1862- -1863-1900-
FLETCHER, John Carr [J]	Pond Hill		-1860-1862-
FURNACE, CUTLER & STACEY	Castle Hill		-1825-1833-
GLAUERT, L. & Co. [J,S]	Cambridge St. Green Lane Furnival St	*Burgess Works* *Wallace Works*	-1879- -1884-1897- -1898-1900-
GRAY, C. & Sons [S]	Watery St.		-1865-1891-
GRAY, Edward	Mowbray St.	*Moscow Works*	-1871-1872-
GRAY & Sons	New Meadow		-1828-
GREAVES, Isaac [J.P,S]	Scotland St. Broad Street Lane	 *Park Works*	-1825-1833- -1837-1900-
GREAVES, Wm & Sons [J,P]			-1833-1849-
GREAVES & WALKER [S]	Portobello Rd.		-1833-1849-
GREEN, Alfred	Pond St.		-1833-1872-
GREEN, Jane & Sons [S]	Arundel St.		-1817-1825-
GREEN, John [P,J]	Pond Lane		-1814-1815-
GREEN, Samuel	Orchard St.		-1833-1841-
GREEN, BRADLEY & FRITH	Burgess St.		-1825-1828-
GREGORY & BRAMALL [S]	Arundel St.	*Soho Steel Works*	-1879-1893-
* GROVES, Richard & Sons [J.P.S]	Snow Lane Gibraltar St.		-1849- -1871-1890-
HALE BROTHERS [J]	Allen St. Snow Lane		-1884-1887- -1891-1900-
* HALL, William [J,P,S]	Porter St. Pool Square	*Alma Works*	-1833-1854- -1856-1890-
HAND, Thomas W.	Union Lane	*Equitable Works*	-1876-1887-
HARGREAVES, SMITH & Co.	Eyre Lane		-1184-1891-
HARRISON, John " " & Sons [J]	Dronfield Dronfield & Sheffield		-1865- -1879-1900-
HASLAM, C. & T.	Dronfield		-1865-1879-
HASLAM, Thomas	Dronfield		-1883-1893-
HASLAM BROTHERS	Dronfield & Sheffield		-1881-1883-
HATTERSLEY, George	Eyre Lane Plum St.		-1833- -1837-1841

HAVENHAND, John	Dronfield		-1854-1893-
HAWKIN, William	South St.		-1825-1828-
HAYWOOD, Thomas	Coal Pit lane		-1833-
	Rockingham St.		-1837-1841-
HAYWOOD, CLARK & Co.	Rockingham St.		-1849-
HEARNSHAW BROTHERS [J,P]	Solly St.		-1876-1900-
HERRING, Ann	Blast Lane		-1833-
HERRING, John	Broad St.		-1837-1856-
" " & Sons	" "	*Navigation Works*	-1857-1900-
HERRING, Joseph	Brightman St.		-1884-
" " & Sons [J,S]	" "		-1888-1893-
HERRING, Thomas	Coalpit Lane		-1828-
HILDICK, Aaron [J,P]	Wicker Lane		-1863-1868-
	Bridge St.		-1871-1900-
HILL, Anthony [J,S]	Hoyles		-1825-
HILL, Charles	Brown Lane	*Spring Lane Works*	-1865-1881-
	Broomspring Lane		-1883-1899-
HINCHCLIFFE	Eyre St.		-1845-1849-
HIRST, Jospeh [grinder]	Bishop St.	*Jail Wheel*	-1883-1888-
HOBSON, Benjamin	Fitzwilliam St.		-1857-1863-
" " & Son	" "		-1871-1899=
HOBSON, Francis	Savile St.		-1841-1863-
HODGSON, Chas. & Sons	Pond Hill	*Crown Works*	-1814-1833-
HODGSON, George	South St.		-1817-
HODGSON, Thomas & Co.	Norfolk St.		-1825-
HODGSON & HAWKINS	South St.		-1821-
HOLLINGSFIELD, J. & B.	Pond St.		-1821-
HOLLY, Daniel & George	Norfolk St.		-1833-
	Bridge St.		-1837-
	Eyre Lane		-1841-1852-
HORN, William	Bridge St.		-1833-1872-
* HOWARTH, James [J,P,S]	Fitzwilliam St.	*Broom Spring Works*	-1845-1857-
	Gloucester St.		-1859-
	Bath St.		-1860-1862-
" " & Sons [J,P,S]	" "		-1863-1913

HOWELL & Co.	Brook Lane		-1891-1895-
HUNTER, Michael & Son	Andrew St.		-1854-1900-
*IBBOTSON, Thomas	Charles St.		-1821-1823-
" " & Co. [J,P,S]	Paternoster Row		-1825-1905-
IBBOTSON, William & George [J,P,S]	Bridge St.		-1814-1825-
*IBBOTSON BROTHERS [S]	Penistone Rd.	*Globe Works*	-1828-1900-
IBBOTSON & Co.	Spring Court		-1821-1828-
IBBOTSON & ROEBUCK [S]	Bower Spring		-1821-1825-
IBBOTSON & SAVILLE	Broad St.		-1828-1830-
IBBOTSON, PEACE & Co. [S]	Russell St.		-1845-1852-
JACKSON, William & Co. [J.S]	Pond Hill	*Sheaf Island Works*	-1856-1900-
JACKSON, NEWTON & Co. [S]	Pond Hill		-1863-1865-
JESSOP, James	South St.		-1833-
JINKINSON, Amos [J]	Rockingham St.		-1862-
	Division St.	*Cardiagn Auger Works*	-1876-1899-
JOHNSON, Christopher	Portobello St.	*Western Works*	-1876-1935-
JOWETT, John [J,P,S]	Arundel St.	*Albion Works*	-1856-1900-
KAY BROTHERS	Eyre Lane	*York Works*	-1881-1900-
KAYSER, ELLISON & Co.	Carlisle St.		-1888-1893-
KENYON, John & Co. [S]	Willey St.		-1837-1841-
	Sheldon St.	*Brianza Works*	-1845-1865-
	Millsands	*Millsands Works*	-1881-1900-
KIRK, Peter [j]	Burgess St.		-1854-1860-
	St. Mary's Road		-1862-1876-
" " & Sons [J]	Norfolk Lane		-1841-1845-
*LAW, Philip [J,P]	Carver Lane		-1797-1856-
LEE, Richard [J,P]	Carver St.		-1814-1821-
LEON, Abraham	Solly St.		-1884-1891-
LINLEY, George	Coalpit Lane		-1828-
LINLEY, Joseph	Norfolk Lane		-1857-
	Carver St.		-1859-
LINLEY, Samuel & Robert	Clough Lane	*Clough Works*	-1854-
	Carver St.		-1856-

LITTLEWOOD, John	Dronfield		-1883-1884-
	Clough St.		-1887-1900-
LOCKWOOD BROTHERS [J,P,S]	Arundel St.		-1845-1891-
	Spital Hill		-1893-1922-
LOMAS, Joseph [S]	Solly St.		-1862-1879-
LONG, HAWKSLEY & MARPLES [S]	Rockingham St.		-1865-
	Solly St.	*Hallamshire Works*	-1876-1879-
LOWE, George & Washington	Arundel St.		-1871-1899-
LUCAS, E. & Son	Blackman St.		-1865-
" " & Sons	Dronfield		-1881-1899-
MANNERS & CAM	Hollis Croft		-1817-
MARRIOT & ATKINSON [S]	Smithfields		-1841-
	Attercliffe		-1845-1857-
	Effingham Rd.	*Fizalan Works*	-1876-1887-
MARPLES, R. & Son [J]	Hermitage St.	*Hermitage Works*	-1865-1900-
MARPLES, Wm. & Sons [J,S]	Westfield Terrace	*Hibernia Works*	-1871-1972-
MARSDEN BROTHERS [J,S]	Bridge St.	*Bridge Street Works*	-1852-1895-
MARSDEN BROTHERS & SILVERWOOD [J,S]	Bridge St.		-1849-1852-
MARSDEN & DOWNEND [S]	West St.		-1833-1841-
MARSDEN, Joseph	Ekington		-1814-1815-
MARSDEN, Mark	Duke Lane		-1854-
	Spital St.		-1857-1863-
	Bridge St.		-1888-1891-
MARSH BROTHERS & Co. [J,P,S]	Forge Lane	*Pond Works*	-1852-1900-
MARSHES & SHEPHERD [J,P,S]	Forge Lane	*Pond Works*	-1841-1849-
MARSHALL, Charles [S]	Norfolk Lane		-1852-
MARSHALL, J. & W.	Globe Works		-1849-1852-
MARSHALL, HADFIELD & Co.	Gibraltar St.		-1828-
MARSHALL & CHEETHAM [S]	Gibraltar St.		-1833-
MARSON, James & Son	Watery St.		-1876-1887-
MARTIN, John & Co.	Wentworth St.	*Wentworth Works*	-1854-1860-
MAW & Co. [J]	Rockingham St.		-1821-
MAW & STALEY [J,P,S]	Rockingham St.		-1828-

MAWHOOD BROTHERS	Pond Hill	*Palm Tree Works*	-1879-1938-
MITCHELL, Henry	Rockingham St.		-1860-
MITCHELL, Joshua	Bernard St.		-1837-
	Duke St.		-1862-1872-
MITCHELL, Robert [S]	Orchard Lane		-1841-
	Portobello St.		-1845-
	Carver St.		-1849-1852-
MITCHELL, R.B.	Earle St.		-1863-
MITCHELL BROTHERS & Co. [J,P,S]	Furnival St.		-1828-1833-
MITCHELL, S. & Co.	Norfolk St.		-1821-1825-
MITCHELL, SEAMAN & Co.	Eyre St.		-1821-
MITCHELL, WREAKS & Co. [S]	Norfolk St.		-1814-1817-
MITCHELL & LAW	Carver St.		-1854-1860-
MOORE, William & Co.	Leverson St.		-1884-1887-
	Effingham St.		-1898-1899-
MORT, Hnery [J,P,S]	Castle Hill		-1833-1854-
MORT, Henry & Co. [J,P,S]	Carver St.		-1857-1859-
	Solly St.		-1860-1872-
MOSLEY, George	South St.		-1817-
	Waingate		-1825-
	Workhouse Lane		-1828-1833-
	Spring Lane		-1837-
" " & Son	Workhouse Lane		-1841-1849-
MOSLEY, John & Co.	Rockingham St.		-1859-1866-
	Norfolk Lane		-1871-
	Stancer St.		-1881-1893-
MOSS, Joshua and GAMBLE BROTHERS [S]	Russell St.	*Russell Works*	-1856-1900-
MOTTRAM, Thomas [S]	Eyre Lane		-1825-1837-
" " & Sons [S]	" "		-1841-
	Arundel St.		-1845-1849-
	Regent Terrace		-1860-1871-
MOTTRAM & Co. [P]	Eyre Lane		-1821-1833-
MOTTRAM & GREAVES	Furnival St.		-1814-1817-
MOTTRAM, TAYLOR & Co.	Eyre Lane		-1852-1856-
	Arundel		-1857-
MOULSON, John [S]	Green Lane	*Globe Works*	-1871-1879-
* MOULSON BROTHERS [J,S,P]	Divison St.	*Union Works*	-1833-1900-

*NEWBOULD, Samuel & Co. [J,P,S]	South St. Newhall Rd.	*Bridgefield Works*	-1817-1876- -1879-1900-
NEWTON, Francis & Sons [S]	Portobello St.	*Portobello Works*	-1852-1865-
NICHOLSON, Austin	Victoria St.		-1890-1893-
OAKES, William [J]	Spitalfield	*Spitalfield Works*	-1837-1857-
OAKES & TURNER	Spitalfield		-1833-
OSBALDISON, Louis	Eldon St.	*Challenge Works*	-1871-1879-
OSBALDISON, L. & Co. [S]	Moor Head	*Glasgow Place*	-1865-1900-
OSBORN, Samuel & Co. [S]	Broom Hill	*Clyde Steel Worrks*	-1854-1887-
OUTRAM, William [S]	Gibraltar St. Henry St.		-1852- -1854-1859-
" " & Son [J,P,S]	" "	*Delhi Works*	-1860-1891-
PARKIN, Charles [J] " " & Son [J]	West Street Lane Charolette St.	*Clarendon Works*	-1863-1881- -1883-1900-
PARKIN, Henry [J]	Grindlegate St. Mary's Road		-1854-1856- -1879-
PARKIN, John & Co. [S]	Arundel		-1833-1837-
PARKIN, Joseph	Fulerton St.	*Ickles Forge*	-1898-1900-
PARKIN, Timothy	Briad St.		-1837-1841-
PARKIN, William & Co. [S]	Furnival St.		-1876-1891-
PARKIN, PARKER & Co.	Arundel		-1821-1825-
PARKIN & BLACKWELL	Cadman St.	*Canal Works*	-1854-1857-
PARKIN & MARSHALL [J,P,S]	Furnival St.		-1852-1891-
PARKER, POTTS & DENTON [J]	Arundel St.		-1828-
PEACE, Charles & Co.	Eyre Lane		-1833-
PEACE BROTHERS [S]	Garden St.		-1856-1879-
PEACE, SPAFFORD & Co. [S]	Russell St. Green Lane	*Eagle Works*	-1854-1856- -1857-
PEACE, WARD & Co.	Davile St.		-1860-1863-
PEACE, Wm. K. & Chas. [S]	Russell St.	*Eagle Works*	-1860-1900-
*PEARSON, Francis G. [J,P,S]	Hollis Croft Chapel Yard Edward St. Furnace Hill	*Hope Works*	-1821-1828- -1837- -1854-1865- -1868-1900-

PINISTER & Co.	Newfield		-1821-
PENESTON & Co.	Arundel		-1821-
PENISTON, WHEATCRAFT & Co. [S]	Arundel		-1825-1828-
POTTS, BAXTER & BRUMBY [S]	Arundel		-1833-1837-
RAINEY, John & Henry	Dronfield		-1814-1817-
REANEY, William [J] " " & Son [J]	Bernard St. " "		-1856-1863- -1865-1872-
REANEY & WOOD [J,S]	Bernard St.	*Bernard Works*	-1837-1854-
REANEY & HEATHCOTE	Bernard St.		-1876-1883-
RILEY, Henry	Leary Greave St.		-1881-1884-
ROBERTS & ASH	Norfolk St.		-1814-1817-
RODGERS, Thomas	Ridgeway		-1854-1893-
RODGERS BROTHERS	Bridge St. Solly St.		-1881- -1891-1899-
ROEBUCK, Jonathan	Broad St.		-1828-
ROGERSON, Henry	Matilda St. Margaret St.		-1881- -1883-1900-
SANDERSON BROTHERS	Attercliffe	*Attercliffe Steel Works*	-1887-1900-
SAVILLE, Charles [J]	South St.		-1828-1833-
SCHOFIELD, George [J]	Sheldon Row		-1845-1857-
SELWOOD, Joseph	Pyebank		-1862-1879-
SHAW, James [S]	Morpeth St.		-1884-1887-
SHAW, JOBSON & Co.	Roscoe Place	*Dauntless Works*	-1817-
SHAW, MARSHALL & Co. [S]	Bridgehouse Newfield		-1814-1817- -1821-1828-
SHAW & CUTLER [S]	Stanley St.		-1817-
SHAW & OATES [P]	South St.		-1833-
SHELDON, Thomas & Sons [J,S]	Trafalgar St.		-1856-1860-
SIDDALL, Samuel	Dronfield		-1865-1891-
SILCOCK, Daniel	Dronfield	*Beech Works*	-1854-1900-
SILCOCK, Elijah	Dronfield		-1854-1865-

SILCOCK, Henry	Dronfield		-1854-1865-
SILVERWOOD & HORTON	Duke Street Lane		-1856-1857-
SKELTON, Charles Thomas & Co.	Hackenthorpe		-1868-1900-
SLACK, Robert [J]	Broomspring Lane		-1833-1900-
SLACK, SELLERS & Co. [J,P,S]	Townhead St. Willey St.		-1871-1872- -1881-1900-
SLAGG, Thomas & Co.	Ford		-1814-1817-
SMITH, John	Coalplit Lane		-1814-1815-
SMITH, J. " " & Son [J,P,S]	Carver St. " "		-1825- -1828-1833-
SMITH, J. & Co.	West St.		-1837-1841-
SMITH, Joseph	Townhead Rockingham Lane Carver		-1814-1816- -1817- -1828-
SMITH & HALL	Carver St.		-1871-1872-
SMITH & STRAY	Spital Hill		-1857-
*SORBY, John & Sons [J,P,S]	Spital Hill		-1814-1900-
SORBY, Robert [J,P,S] * " " & Son [J,P,S]	Union St. Carver St. " " Trafalgar St. Chesterfield Road	 *Kangaroo Works*	-1828-1830- -1834-1839- -1845-1890- -1891-1932- -1933-1980
SORBY & TURNER	Goodcroft		-1825-1828
SORBY, TURNER & SKIDMORE	Wicker		-1814-1821-
*SPEAR & JACKSON [S]	Savile St.	*Aetna Works*	-1845-1980-
SPENCER, Mathius & Son [S]	Peacroft		-1825-1900-
SPENCER, John R. & Son [S]	Peacroft	*Albion Works*	-1854-1881-
SPOFFARD, A. & Co.	Moorfield	*Imperial Works*	-1863-1884-
STAGG, T. & W.H.	Hollis St.		-1860-1862-
STALEY, William [J,P,S]	Rockingham St.		-1833-1837-
STEVENSON, MAWHOOD & Co. [J,P,S]	Little Pond St.		-1871-1872-
STONES, Frederick [J,P,S]	Arundel St. Orchard St.		-1817-1841- -1845-
STRINGER, Arthur	Wadsley Bridge	*Niagra Works*	-1898-1900-

SWINCOW & MANUEL [J,P]	Fitzwilliam St. Coalpit Lane		-1841- -1845-1852-
SYKES, William J. [S]	Backfields		-1898-1900-
*TAYLOR, Henry [J]	Fitzwilliam St.		-1849-1900-
TAYLOR, William	Radford St.		-1860-1863-
TAYLOR BROTHERS [S]	Mowbray St.	*Adelaide Works*	-1860-1900-
TAYLOR & HOWARTH	Fitzwilliam St.		-1837-1841
THOMPSON, Henry [J]	Broomhill St.	*Broomhill Works*	-1865-1890-
TILLOTSON, John & Son	Coalpit Lane		-1828-
TORR, J. & Co.	Radford St.		-1860-1863-
TOWNEND, George R.	Radford St.		-1887-1900-
TURNER, Charles & Joseph	Earl St.		-1854-1857-
TURNER, John [S]	Broad St. Sheldon Row Savile St.		-1825- -1833- -1837-1845-
TURNER, Joseph & Co.	Castle Hill		-1859-1868-
TURNER, J.M. & Co.	North St.	*Orion Works*	-1865-
TURNER, Thomas & Co.	Norfolk St. Suffolk Road		-1825-1828- -1837-1864-
TURNER, NAYLOR & Co. [J,P,S]	Castle Hill		-1871-1872- -1895-1963-
TURNER, NAYLOR & MARPLES [J,P,S]	John St.	*Northern Works*	-1876-1893-
TURNER & WHEELHOUSE	Earl St.		-1852-
TURTON, Thomas " " & Sons [S]	Sheaf Works " "	*Sheaf Works* *Sheaf Works*	-1852-1868- -1871-1980-
TURTON BROTHERS & MATHEWS [S]	Wentworth St.	*Phoenix Works*	-1876-1881-
TYZACK, W. & Sons [S]	Abbeydale Valley Road	*Little London Works* *Meersbrook Works*	-1887- -1888-1900-
WALL, James	Dronfield		-1876-1900-
WARBURTON, Charles	Rockingham St.	*Broomspring Works*	-1857-1863-
WARBURTON, John & Co. [J]	Rockingham St.		-1856-1863-
WARBURTON BROTHERS	Furnival St. Eyre Lane		-1845-1849- -1852-1856-

WARD, David [J,P,S]	Trippet Lane		-1814-1821-
" " & Son	" "		-1825-1833-
WARD, David & Co. [J,P,S]	Portobello St.		-1837-1841-
*WARD & PAYNE [J,P,S]	West St.		-1845-1965-
WATTS, Amos	Bailey St.		-1828-
WATTS, William	Shepherd St.		-1856-1863-
WELDON & FURNISS	Castle Hill		-1814-1817-
WHARTON, George	Cadman St.		-1845-
WHARTON, George & Co.	Canal Work		-1849-
WHEATCROFT, John & Son	Arundel St.		-1833-
WHEATMAN & SMITH [S]	Kelham Island	*Russell Works*	-1856-1900-
WIDDOWSON, John	Dronfield		-1879-1899-
WILSON, John [J]	West St.	*Portland Works*	-1890-1898-
WILSON, Thomas & Co.	Nursery Lane	*Telegraph Works*	-1849-1856-
WILSON, HAWKSWOORTH, ELISON & Co. [S]	Arundle Lane		-1856-1884-
WING, M. & J. [S]	Jerico St.	*Jerico Works*	-1871-1872-
WOOD, D. & Co.	Effingham Road		-1863-1865-
WOODCOCK, James [J]	Trafalgar St.		-1863-1876-
" " & Sons [J,P]	Smithfield		-1879-1884-
	Young St.		-1887-1900-
WORSLEY, Daniel & Co.	Smithfield	*Worsley Works*	-1879-1888-
WREAKS, J. & Co. [P]	Norfolk St.		-1821-
WRIGHT, John & Son [S]	Highfield		-1814-1833-
WRIGHT, Samuel [J,S]	New George St.		-1837-
YOULE, John [J,S]	North Church St.	*Orion Works*	-1884-1887-

Notes: J also made or sold joiner tools * In business for more than 50 years.
P also made Plane Irons
S also made or sold Saws.

A date noted after 1900 simply means the firm was in business up to that time, but not necessarily engaged in making edge tools. Ward & Payne remained in business until about 1970, but is believed to have given up manufacturing their line of edge tools during World War II. The date 1935 was obtained from *Registry of Trade Marks 1935*.

Chapter IV

Joiner's Tools

Included within this classification are: squares, bevils [sic, bevel], mortice, marking and slitting gauges, spoke shaves, some saws [mitre, back, bow, pad & frame turning], saw sets, turnscrews [screw drivers], levels, screw boxes, tool chests, braces and bits and planes. The latter two items, braces and planes, will be taken up in separate chapters, and will therefore not be considered further in this chapter. Most joiner tools combine wood as a handle with a functional steel blade. It is this class of tool that resulted in the highest degree of art, associated with utility of purpose. Combinations of exotic woods, such as rosewood, ebony and boxwood, with higly polished cast brass led to pleasing contrasting colors with symmetry of designs. Some typical joiner and edge tools are illustrated in Plate CXXII.

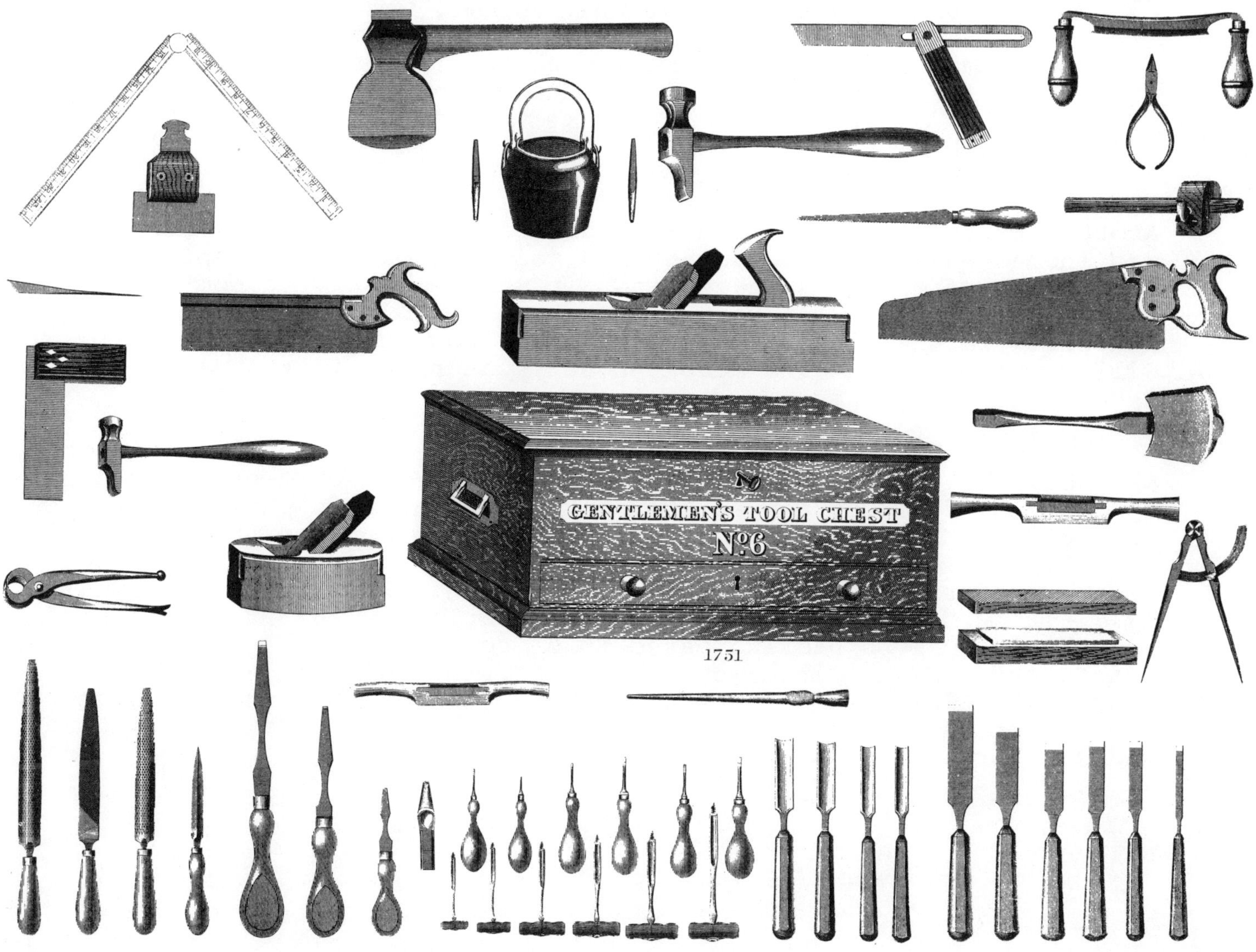

Plate CXXII - Gentlemen's Tool Chest and Contents - Plate F-5, 1870 SHEFFIELD LIST

With first consideration it may seem strange that an important side product of most joiner and some edge tool makers was the item of ice skates. Sheffield steel for the hardened blade had long been supplied to London for this purpose. After King Charles II returned to England to assume the thrown in 1660, his Court introduced the sport of ice skating. This had become popular with the King while in excile in Holland, during the time of Oliver Cromwell. Note the advertisements of Fenton & Mardens and Marsden Brothers, Plates C & CI, pages 140 & 141. Several more examples will become evident in this section.

As evidenced from several surviving pattern books, many of these joiner tools originated in the Birmingham area around the turn of the 18th Century. Such items were included in the group of artifacts known as "toys".[1] A set of interchangeable tools with the same hand pad, inscribed Rich. Timmins & Sons, Birmingham, is illustrated in Fig. 12. These are thought to have been made c1810.[2]

The earliest published Directory of Sheffield, 1774, had a single classification for *SUNDRY MANUFACTURERS OF IRON, STEEL, BRASS, ETC.:*

> George Bishop; wool shears, edge tool, spade and shoe knife maker.
> Bland's; edge tool, hammer, spade and locks, bottom of Hickstilefield
> Thos. Boulsover & Co.; manufacturer of saws, fender, edge tools cast steel and emory, Sycamore Street.
> John Gallimore; whitesmith and toolmakers, Milsands
> Malin Gillot & Sons; sheep shears and joiner tool makers, Fargate
> William Weldon; wool shears and edge tool maker, Fargate
> Joseph Wilde & Son; edge tool and shear makers, Colsoncroft
> John Wilson & Son; sheep shears and edge tool maker, Norfolk St.

Another person listed in the alphabetical section of this Directory was Thomas Maxwell, wood turner and joiner tool maker. Apparently in most cases of metal manufacturing, it was necessary to combine more than one occupation in order to make a living. In any event Gillot and Maxwell were the first documented joiner tool makers in Sheffield. Gillot, with a mark of a double heart, was noted in the next Dierctory as an edge tool maker.

As was noted earlier, unfortunately the next Sheffield directory was not published until 1814, and it is not known what progress developed in the field of joiner tool making during this period. However the 1814-15 Sheffield Directory listed ten persons in this occupation and thirty as edge tool manufacturers. Among the former group who became leaders were: James Bee, James Dixon, also listed as a Skait [sic - skate] Maker, Kergarth Maw and William Marples. By 1825 the number of joiner tool makers had increased to 21 compared to 37 edge tool manufacturers. Eight firms were on both lists: William Ash & Co., Chumbley & Edley, Thos. Ibbotson, George Linley, Penistone, Wheatcroft and Singleton, Frederick Stones (See Plate CXXIII), David Ward & Son, and John Wright.

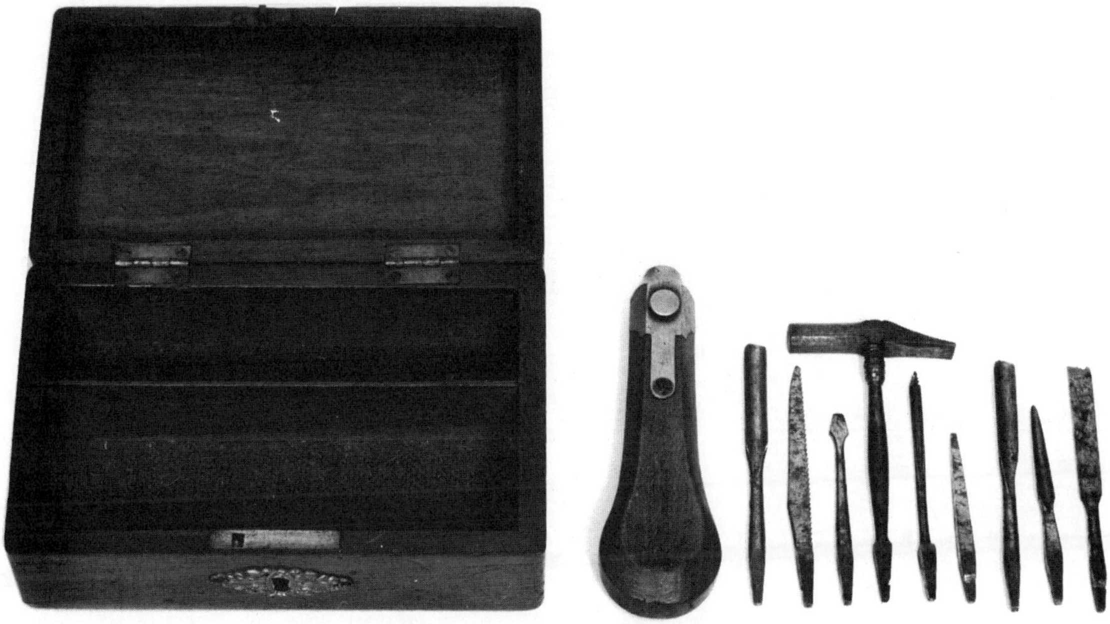

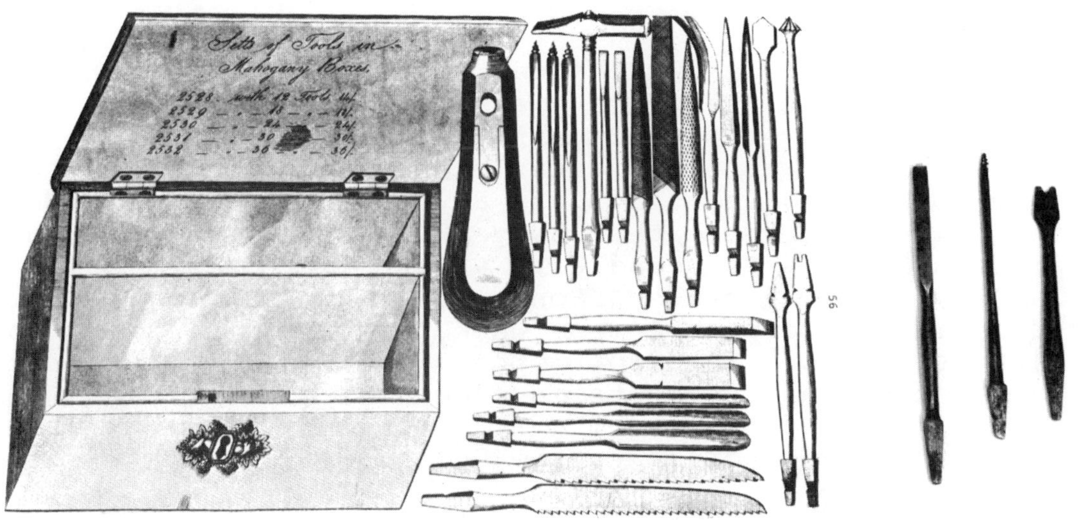

Plate 8. (Reduced 25% in size) Note spelling "Setts of Tools" and fancy brass escutchion on "Mahogany Box". Both are indications of a much earlier engraved plate than this printing. These combination sets of tools, used interchangeably in the same hand pad,date late 19th.C.

Fig. 12 - "Setts of Tools" by Rich. Timmins & Sons, Birmingham, c.1810

Plate CXXIII - Earliest Sheffield Directory Joiner Tool Manufacturer Advertisement

A Check List of Joiner Tool Makers working in Sheffield during the 19th Century is presented as Appendix III at the end of this Chapter. While this shows 260 firms, there were probably many more who were in and out of business during periods when directories were not published, some of whom may have not lasted as long as a single year. As with saw makers and edge tool makers, firms after 1850 who only appeared in a single directory were not included unless part of some other later firm. The durations of these firms are compiled in Table X.

TABLE X - DURATION OF 19TH CENTURY JOINER TOOL MAKERS

Years in Continuous Operation	Number of Firms	Percent of Total
Less than 5	67	25.8
5 but less than 10	74	28.5
10 " " " 20	49	18.5
20 " " " 30	34	13.1
30 " " " 40	18	6.9
40 " " " 50	10	3.8
50 " " " 60	2	.8
60 " " " 70	3	1.2
Over 70	3	1.2

Analysis of this data shows that approximately 25% of these firms were in business for less than five years, 60% more than 5 but less than 30, and only 15% more than 30 years. While this seems to indicate that this business was not as stable as edge tool manufacturing, it should be considered that this was a younger profession. Only 8 firms were in business for more than 50 years. Three of these; Robert Sorby & Sons; John Sorby & Sons and Ward & Payne were principally edge tool makers, and Wm Brookes & Sons, saw manufacturers. Robert Marples & Son and William Marples & Sons were principally joiner tool makers. Thoms Ibbotson & Co., Marsden Brothers and Moulson Brothers appear to have had about equal interests in edge and joiner tools. The most prominent of these firms, from a stanpoint of volume of goods produced, many of which were exported to the United States, are listed in Table XI.

TABLE XI - Most Prominent 19th Century Sheffield Joiner Manufacturers

Firm	Dates of Operation
William Ash	-1825-1841-
Brown & Flather	-1828-1849-
Henry Brown & Sons	-1852-1938-
David Flather & Sons	-1852-1954-
James Howarth & Sons	-1856-1913-
Thomas Ibbotson & Co.	-1821-1905-
Amos Jinkinson & Sons	-1862-1900-
Robert Marples & Son	-1833-1931-
William Marples & Sons	-1833-1972-
Marsden Brothers	-1854-1894-
Moulson Brothers	-1828-1898-
Robert Sorby & Sons	-1849-1980-
Ward & Payne	-1803-1913-

[Note: Some of the above firms were in business earlier offering edge tools] Brief histories of these firms follow.

William Ash & Co. - First listed as an edge and joiner tool maker on Norfolk Street in 1825, also a brace and bit maker. This listing was continued until 1841. After that date and until 1852, William Henry Ash was listed only as an edge tool amker. The firm was noted for their excellent brand of plane irons. Sometime after 1863 Burys & Co. acquired the mark of Wm. Ash & Co. and believed to have continued making plane irons with this mark. (See Plate XXIIa, page 41)

Brown & Flather; Henry Brown & Sons; David Flather & Sons. See next Chapter concerning braces.

James Howarth & Sons: James Howarth, sr. was an edge tool maker and with Henry Taylor in 1835 established the firm Taylor & Howarth. In 1832 he became sole proprietor. His award at the 1851 London Exhibition of All Nations is shown in an advertisment appearing on page 169. In 1857 joiner tools were added to their line after moving to Bath Street at Broom Spring Works. In 1863 the firm becam James Howarth & Sons An account of their forging and grinding shops is presented on page 125 (See Plate XCI). A brief history of the firm appears on page 127. The diversity of products is noted in their advertisement shown as Plate XCVIII, page 138. A steel division was established on Eyre St. where a high reputation as "melters and refiners, manufacturers of every description of cast steel, shear, blister, mining, spring and sheet steel".[3] Numerous medals awarded at various exhibitions attest to the excellency of their product line.[4] Their first class award at the Universal Society for Encouragement of Arts and Industry at the 1855 Paris Exhibition was not accorded to any other British manufacturer of edge tools (See Plate CXXIV). Upon the death of James Howarth, sr. , the firm was managed by his four sons James, Samuel, Edwin and John Charles Howarth. The firm was discontinued in 1913, and their mark was acquired by Robert Sorby and Sons about 1922.

Plate CXXIV - 1865 Sheffield Directory Advertisement of James Howarth & Sons

Thomas Ibbotson & Co. - While Thomas Ibbotson was first listed in the 1821 Sheffield Directory as a joiner tool manufacturuer, their 1903 Illustrated Price List noted the firm was established i . 1814. His enterprise became Thomas Ibbotson & Co., Paternoster Row, Charles St. as noted in the 1833 Directory. This was one of the few Sheffield firms that remained at the same location throughout its entire tenure. Edge tools were added to the line in 1845 and the firm refined its own steel. A sketch of the works is noted on the title page of their 1903 Illustrated Price List (Plate CXXV). Note that the firm address of Paternoster Row, Charles Street has been crossed out and replaced by Hibernia Works, Westfield Terrace, the address of Wm. Marples & Sons, who acquired this concern in 1905. The diversified line of products is noted on their 1876 Advertisement. (Plate CVII)

Amos Jinkinson & Sons - Amos Jinkinson was in partnership with his brother, William, at Eldon St. as edge tool manufacturers, 1854-1860. In 1862 Amos was listed alone as both an edge tool and joiner tool manufacturer. The firm's 1883 Sheffield Directory Advertisment (Plate CXXVI) noted the Cardigan Auger Works at 126 Rockingham St., calling attention to the fact that the firm was established in 1848. About 1890 the firm became Amos Jinkinson and Sons. They won an award at the 1892 Yorkshire Industrial Trade Exhibiiton held at Sheffield.

AMOS JINKINSON,

MANUFACTURER OF EVERY DESCRIPTION OF

AUGERS, GIMLETS, TOOL CHESTS, JOINERS' TOOLS, EDGE TOOLS,

AND GENERAL MERCHANT,

Cardigan Auger Works, 126, Rockingham St., & 88, Division St.,

SHEFFIELD, ENGLAND.

ESTABLISHED 1848.

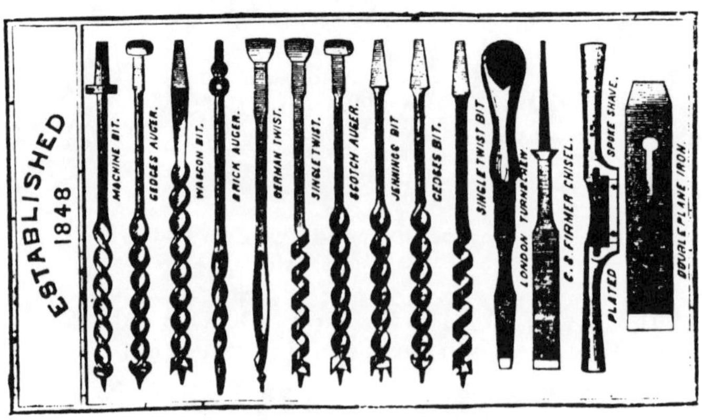

Plate CXXVI - 1883 Sheffield Directory Advertisement of Amos Jinknison

THOMAS IBBOTSON & COMPY.

REGISTERED TRADE MARK.

REGISTERED TRADE MARK.

ESTABLISHED 1814.

~~PATERNOSTER ROW, CHARLES STREET,~~

HIBERNIA WORKS, WESTFIELD TERRACE,

SHEFFIELD,

STEEL REFINERS

AND MANUFACTURERS OF

LIGHT AND HEAVY EDGE TOOLS,

JOINERS' TOOLS, SAWS, FILES,

Shears, Engineers' and Machinists' Tools,

HAMMERS, GARDEN TOOLS, TOOL CHESTS, SKATES.

Prices subject to alteration without notice. **All previous Lists cancelled.**

Plate CXXV - Title Page to 1903 Thomas Ibbotson & Compy. Trade Catalogue
[Collection of Donald Wing]

Robert Marples & Son - Robert Marples began making joiner tools in 1828 at Sheffield at Sharrow Head in company with William, assumed to be his brother. As far as can be determined there was no relation to William Marples, sr. or his son William Marples, jr., both of whom also gained a high reputation in the field of joiner tools. Robert moved to Hermitage St. in 1837, where listed alone. In 1861 he took in his son Henry Dixon Marples, establishing the Hermitage Works and adding edge tools to their line. As will be noted in the next Chapter, Robert Marples became a leader in the manufacture of braces. Their trade mark, a bust of a hermit, was granted by the Cutler's Co. in 1887. Other trade marks for this firm were: "Robert Marples, Hermitage Works"; "H.D.Marples, Hermitage Works"; "Henry Dixon Hermitage Works"; "Robert Henry, Hermitage Works" and "Henry Greaves, Hermitage Works'.

ROBERT MARPLES & SON,

MANUFACTURERS OF

JOINERS' TOOLS,

AND

LIGHT AND HEAVY EDGE TOOLS,

COMPRISING

BRACES AND BITS,

CARPENTERS' SQUARES, BEVILS, GUAGES,

SPOKESHAVES, SPOKESHAVE IRONS,

TURNSCREWS, SPIRIT LEVELS, SCREW BOXES,

GENTLEMEN'S TOOL CHESTS,

JUVENILE TOOL CHESTS,

Augers, Gimlets, Saw Hand-frames, Saw Pads,

SAW SETS, HAND PADS, PRICKER PADS,

SCRAPERS, SKATES, HAMMERS,

Cast Steel Firmer Chisels and Gouges, Drawing Knives,

PLANE IRONS, SOCKET CHISELS AND GOUGES, AXES, ADZES, &c.,

HERMITAGE WORKS,

HERMITAGE STREET,

SHEFFIELD.

Plate CXXVII - Robert Marples & Son Advertisement in 1864 Sheffield Directory

WM. MARPLES & SONS,

HIBERNIA WORKS, SHEFFIELD.

LONDON OFFICES: 14, IVY LANE, E.C.

GOLD MEDAL at MELBOURNE and

FIRST AWARD AT SYDNEY FOR EDGE TOOLS.

WE WISH TO DIRECT SPECIAL ATTENTION TO OUR

EDGE TOOLS,

AND

JOINERS' TOOLS.

USING ONLY THE BEST MATERIAL, AND EMPLOYING THE MOST SKILLED LABOUR, WE WARRANT THEM TO BE EQUAL TO ANY IN THE MARKET.

TRADE MARK

HIBERNIA

THEY ARE STAMPED WITH OUR OWN NAME AND TRADE MARK, WHICH IS A GUARANTEE OF THE QUALITY.

The "Hibernia" Amateurs' Turning Lathe.

MANUFACTURERS OF EVERY DESCRIPTION OF **TOOLS** AND **SKATES,** AND GENERAL FACTORS OF ALL KINDS OF SHEFFIELD, BIRMINGHAM, AND LONDON GOODS.

IMPORTERS OF AMERICAN, GERMAN, & FRENCH **TOOLS** AND **HARDWARE.**

IRONMONGERS' NOVELTIES ARE A SPECIALTY WITH US.

FRETWORK TOOLS AND MATERIALS OF EVERY DESCRIPTION

Illustrated Price Lists sent on Application.

New Customers are requested to send Business Card and Trade Reference.

Plate CXXVIII - Wm. Marples & Sons Advertisement in 1883 Sheffield Directory

<u>William Marples & Sons</u> - William Marples, jr. was born in 1807 and learned the trade of joiner tool making under his father. Upon reaching his majority in 1828, he is reported to have opened his own shop, on Broomhall St., where he is first noted in the 1833 Directory. In 1837 he is listed at Broomspring Lane, where he remained until 1856 moving to Hibernia Works at Westfield Terrace. In 1860 he took his three sons: Edwin Henry, William Kent and Albert into the firm. He retired in 1869 and these three sons purchased the business from him in 1872. The 1871 Directory noted that edge tools had been added to their line. In 1871 William bought an interest in Turner, Naylor & Co. at Northern Tool Works for his fourth son, Charles. This continued as Turner, Naylor and Marples until 1893. (see page 133 and Table VIII, page 132) Their 1879 and 1881 Advertisements in the Sheffield Directories called attention to their awards at Australian Exhibitions and noted that they were now also importers of "*AMERICAN, GERMAN, FRENCH HARDWARE*". After William Kent Marples' death in 1877 and Albert's retirement in 1888, Edwin Marples became the sole owner. He brought his two sons, Edgar and Edward Albert, into the business. Various other acquisitions are shown in Table VII, p.131, In 1962 William Ridgway & Sons and C. & J. Hampton acquired William Marples & Sons on a fifty-fifty basis. The old firm was maintained until all three firms combined in 1972 under the new name of Record Ridgway Ltd. A sketch of Hibernia Works (Plate XCIV) and a brief discription of the works as it was in 1879, appears on page 130. Advertisements of the firm appear on Plates CXXI, p.161, & CXXVIII, p.191.

The name of MARPLES was indeed very prominent in Sheffield joiner tool manufacturing. There appear to be at least ten different firms in this activity during the 19th Century. More than that if a firm is considered separate after later being joined by sons. These firms will be briefly traced and designated by I - X.

In addition to the three firms already mentioned: I. Robert Marples (& Sons); II. Wm. Marples, sr.; III. Wm. Marples & Sons, there appear to be two others (IV & VIII) in which another William was involved. In 1854 the firm of Marples & Mansfield (IV) operated only that year. This was composed of William Marples and Richard Mansfield. The following year William continued alone at this address, 41 Arundel St. through 1856, but was not noted in 1858. (V) This may have been the William who was first noted in partnership with Robert Marples in 1828. In 1854 Joseph Marples began at 130 Carver St. and in 1854 this firm became Joseph Marples & Son (VI). Also in 1854 Isaac Marples (VII) first appeared at 71 Broomhall St. working alone until joined by his brother, Albert, continuing through the remainder of the 19th Century as I. & A. Marples. In 1876 a Wm. Marples appeared at the address, 130 Carver St., formerly of Joseph Marples & Son (VI). This William may have been a son of Joseph (VI). In 1879 continuing at this address are Wm. & Joseph through 1884 (VIII) Joseph dropped out in 1877, Wm. continued alone, being last listed that year at this address. Another Joseph, jr. appeared at Cambridge St. in 1868, working through 1884. (IX). This Joseph is believed to be a specialist in making planes. A Charles Marples appeared at 24 Button Lane in 1876. He was different from the Charles, son of Wm. jr., who was associated with Turner, Naylor & Marples as both are listed in 1876 at different resident addresses. Charles (X) was joined by his son as Charles Marples & Son, also at Carver St. through 1891. In following the above discussion, it is easy to lose your Marples!

WILLIAM MARPLES & SONS

HIBERNIA WORKS,

WESTFIELD TERRACE,

SHEFFIELD,

Manufacturers of

EVERY DESCRIPTION OF TOOLS,

SUITABLE FOR

JOINERS, CARPENTERS,

BUILDERS, ENGINEERS,

GENTLEMEN, & EMIGRANTS;

— AND —

GENERAL FACTORS OF ALL KINDS OF SHEFFIELD, BIRMINGHAM & LONDON GOODS.

Sole Manufacturers of Wm. Marples'

PATENT ULTIMATUM METALLIC FRAMED BRACE,

WITH REVOLVING-HANDLE AND IMPROVED PAD.

PAWSON & BRAILSFORD, SHEFFIELD.

Plate CXXVI – Title Page of the 1870 Wm. Marples & Sons Trade Catalogue
[Note - Printed by Pawson & Brailsford, printers of the 1870 Sheffield List]

<u>Marsden Brothers.</u> - Kirjarth Maw established Maw & Co., edge tool makers on Button Lane in 1817. William Staley, listed in 1817 as a carpenter and builder, joined with Maw in 1825 forming Maw & Staley, advertising they made "all kinds of edge and joiner tools, braces and bits and steel refiners". In 1833 Staley became sole proprietor. In 1839 Joseph Fenton and George Marsden formed Fenton & Marsden (late W.Staley) establishing Ecceshall Works on Rockingham St. and noting products "joiner & edge tools, skates, braces & bits, screw boxes,&c." In 1845 the firm moved to Bridge Street. According to the Sheffield Rate Books Fenton's interests were purchased by the Marsden brothers (George & Robert) and Samuel Silverwood in 1846. The firm was renamed Marsden Brothers and Silverwood. (See Plate C,p.140) Their 1854 Sheffield Directory advertisement noted them as "Manufactures of Edge Tools, Joiner Tools, Braces & Bits, Skates, Screw Boxes, &c.,&c. and General Merchants. Also as Skate Manufacturers by appointment to His Royal Highness Prince Albert". (also Plate C,p.140) Sometime after 1852 and before 1854 the Marsden brothers purchased the interests of Silverwood and the firm name was changed to Marsden Brothers. A sketch of their factory, Bridge Works, appears in Plate CI, p.141. In 1896 John Wilson acquired the firm and continued as John Wilson, Marsden Brothers until 1901,when these firms were combined with Robert Sorby and Sons. (See Table XII) Imprints of Marsden Bros. have been noted on wood moulding planes, but it likely that these were made by outside workers, possibly in exchange for the firm's product of plane irons. A listing of the Marsden Bros. & Co.'s extensive products is noted on their 1876 advertisement in the Sheffield Directory. (Plate CXXIX) It appears possible that the Mardsen connection with edge or joiner tools may predate the 1817 Maw establishment. An advertisement of George R. Marsden, Argyle Works, in 1879 noted "established in 1790". This was the only listing of this firm, but the proprietor is believed to have been either a former proprietor of Marsden Bros. or a son of one.

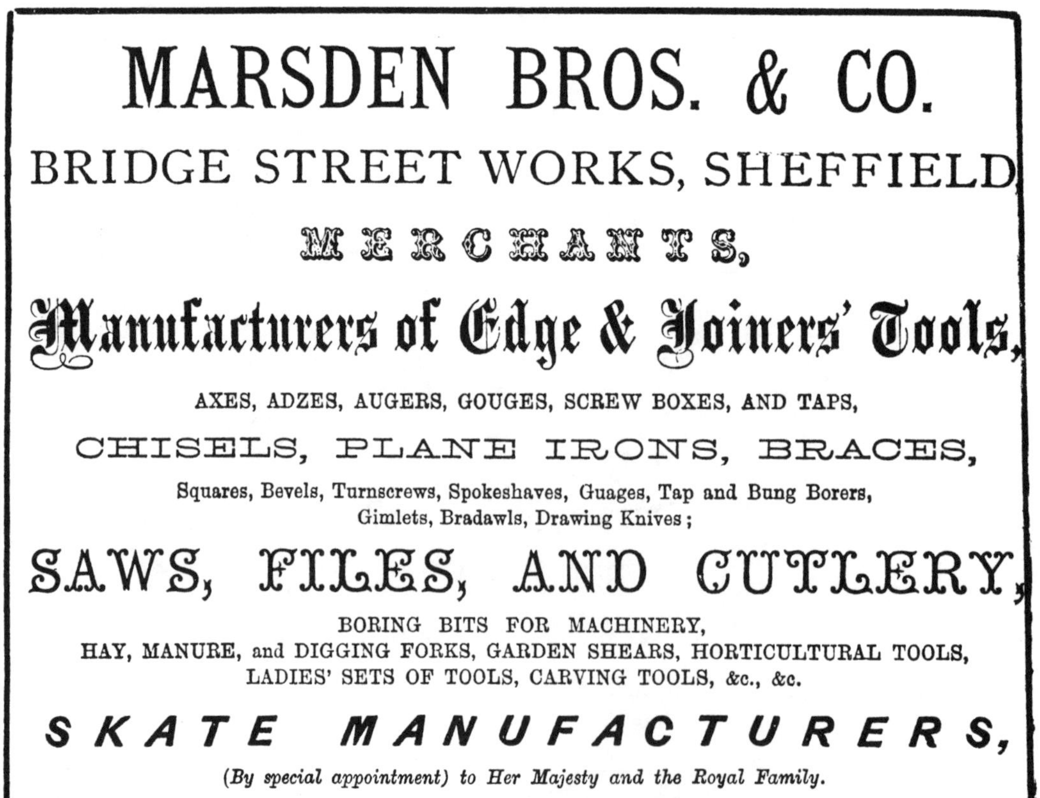

Plate CXXIX - Marsden Bros. & Co. 1876 Sheffield Directory Advertisement

Table XII

Development of Marsden Brothers, James Howarth & Sons and Robert Sorby & Son.
Sheffield Edge and Joiner Tool Manufacturers

Maw & Co., 1821
Rockingham St.
Edge Tools

↓

Maw & Staley, 1825
Manufacturers of All Kinds of Edge & Joiner Tools, Braces & Bits and Steel Refiners.

↓

Wm Staley, 1833
Joiners' & Edge Tools

↓

Fenton & Marsden, 1839
Ecceshall Street
Joiner & Edge Tools, Braces & Bits, Screw Boxes, Skates, &c.
[Joseph Fenton & George Marsden]

↓

1841 Rockingham Street
1845 Bridge Street

↓

Marsden Brothers & Silverwood, 1849
[Samuel Silverwood, Robert & George Marsden]
Edge & Joiner Tools, Brace & Bits, Skate Manufacturers for Prince Albert.

↓

Marsden Brothers & Co., 1853
Bridge Works
Edge & Joiners' Tools, Skate Manufacturers for Her Majesty & the Royal Family

↓

Taylor & Howarth, 1836
Fitzwilliam St.

↓

James Howarth, 1845
Broomspring Works, Fitzwilliam St.
Edge Tools

↓

James Howarth & Sons, Bath Street, 1863-1913 → (Robert Sorby & Son)

Robert Sorby, 1820
Edge Tools
Union Street

↓

Robert Sorby & Son, 1863
Carver Street
Edge Tools

↓

Jabez Shaw, 1857-1863-
Westfield Terrace

↓

John Wilson, 1868-1896-
Edge Tools & Joiner Tool Maker
Portland Works, West Street

↓ 1897 (joins Marsden Brothers & Co.)

John Wilson, Marsden Brothers & Co, 1898
Portland Works

↓ 1901

Robert Sorby & Son
1980

Moulson Brothers. - Thomas and William Moulson commenced as joiner tool makers in 1828 at Tudor St. Subsequently, they were joined by another brother, John. In 1830 they moved to Division Street at Union Works, adding edge tools to their line in 1833. In addition to their fine reputation as manufacturers of joiner tools, they produced high quality plane irons, continuing in business until 1912.

Robert Sorby & Sons. - Robert Sorby was first noted as an edge tool maker in 1828, becoming Robert Sorby & Son in 1841 and Robert Sorby & Sons in 1845. Their family had long previous association with the cutlery trade, Robert Sorsby (former spelling of Sorby) having been elected the first Master Cutler of the Company of Cutlers in Hallamshire in 1624.[5] They added joiner tools to their line in 1849, but from Sheffield Directory listings appear to have discontinued these after 1865. When the firm acquired John Wilson, Marsden Brothers in 1901, they continued the former concerns' line of skates. The sons of Robert Sorby the founder were Robert Arthur and Thomas Heathcote Sorby. Their Kangaroo Brand of edge tools gained worldwide reputation at Carver Street. After 1896 they moved to Trafalgar St. and in 1933 the Kangaroo Works established at Chesterfield Road, Woodseats. In addition to full lines of joiner and edge tools, the firm manufactured saws, scythes, sheep shears, files and skates. The firm is still in business and therefore has the longest surviving tenure of any edge tool manufacturer in Sheffield.[6] Plate CXXX shows a photograph of the works at Trafalgar St. from their 1907 Trade Catalogue, noting the acquisitions of Marsden Bros. and John Wilson. Incidentally, after acquiring John Wilson,Marsden Bros. in 1901, joiner tools were again added to the line of edge tools and saws. The 1907 Trade Catalouge noted a line of wooden bench and joiner planes, which were likely made by Robert Sorby & Sons before these acquistions of other firms.

Ward & Payne. - David Ward, the founder, has been reported to have commenced as an edge tool manufacturer in 1803. His firm became David Ward & Son in 1837 and David Ward & Co. in 1841 at Portobello Rd. Henry Payne became a junior partner in 1845. The firm then became Ward & Payne, moved to West St. and joiner tools were added to their line of products. Mr. Payne died in 1850, but the name continued, even though subsequently managed by the three grandsons of David Ward; David, Edwin and Herbert Ward.[7] Plate CX (p.150) shows an 1860 advertisement of the firm. A brief account of the firm, relating the taking over of the J.B.Addis' line of carving chisels (Plates CII & CIII, p.142 & 143), their successes with sheap shear manufcaturing and making edge tools is presented in Chapter III, p.129. The firm discontinued edge tools during World War I.

Another important joiner tool manufacturer operating during the last 30 years of the 19th Century at Sheffield was John Wilson. He is first noted in the 1868 Sheffield Directory at Wm. Marples & Sons Hibernia Works. It is quite possible that he formerly worked under Jabez Shaw, a contract worker at the Marples' Hibernia Works. With this arrangement the Marples' firm would furninsh the materials and provide the facilities for manufacturing, but Shaw would supply the labor and contract at an agreed price and time to produce a given number of tools, according to a sample or pattern from the Sheffield List. Thus the Marples' firm became merchants and factors rather than supervising the manufacturing. A substantial portion of the cutlery business was accomplished by this contract system - "for in these trades there were large numbers of small firms and independent out-

Telegraphic Address:—"KANGAROO, SHEFFIELD."

1851.
Prize Medal,
London,
Robt. Sorby & Sons.

1851.
Prize Medal,
London,
Marsden Bros. & Co.

1865.
Diplomas,
New Zealand.

1886.
Gold Medal,
Edinburgh.

1884.
Highest Award Silver Medal and Diploma, Edinburgh,
For Forestry Tools, &c.

ROBT SORBY.

ROBT SORBY
& SONS

"KANGAROO"

ROBT SORBY.

ROBT TREBOR
SHEFFIELD
YARRA-YARRA

1884.
Gold Medal,
Calcutta.

1887.
First Order of Merit
(Without Appeal),
Adelaide.

1888.
First Class Award,
Melbourne.

CORPORATE MARK.

TRADE OMEGA MARK

CORPORATE ODEN MARK.

THE MONIER-WILLIAMS SKATE

TRADE MARK

ROBERT SORBY & SONS, LIMITED,

... Comprising the Businesses of ...

ROBERT SORBY & SONS,
JNO. WILSON,
MARSDEN BROS. & CO.,
H. BURGON PATENT SHEAR CO.

KANGAROO WORKS,

Corner of Wellington Street and Trafalgar Street.

SHEFFIELD, ENGLAND.

ILLUSTRATED CATALOGUE

OF

EDGE TOOLS, ADZES and AXES, AUGERS, SAWS,

JOINERS', ENGINEERS,' PLASTERERS' AND IRON MOULDERS' TOOLS,

SCYTHES, HOOKS, and SHEEPSHEARS.

FORESTRY AND SEEDSMEN'S TOOLS.

Plate CXXX - Title Page 1907 Robert Sorby & Sons Trade Catalogue

workers and others who hired places in a larger manufacturer's shop and sold him the goods they made".[8] Becoming a contractor provided opportunity for financial advancement and the possibility of eventually owning an independent establishment. This apparently was the case with John Wilson who bought Jabez Shaw's business when the latter retired. Wilson continued to be a contract worker for Wm. Marples & Sons until 1879 when he mangaed his own business at Fitzwilliam St. In 1884 he moved to West St. building the Portland Works. In 1890 he advertised in the Sheffield Directory as: "manufacturer of trowel, brick, plastering and moulding tools, joiner, augers & screw boxes, bits, skates and saws. Portland Works, 222 West St.". In 1896 he bought out the Marsden Brothers incorporating their firm at the Portland Works, only to lose this and his own establishment to Robert Sorby & Sons in 1901. (See Table XII, p.195) John Wilson's 1871 advertisement in the Sheffield Directory, while a contractor at the Marples' Hibernia Works indicated that he had registered a design for a skate in addition to offering joiner tools and that he used steam power. (See Plate CXXXI)

MITCHELL, BROTHERS, & Co.s'

Prices of Joiners' Tools, &c.

FURNIVAL-STREET, SHEFFIELD.

Squares.

	3	4½	6	7½	9	12	15	18	24	30 in.	
Common Squares......	11s6d	15s	18s6d	23s	27s6d	37s	46s	58s	80s	108s	p.d.
Plated Squares........	16s	20s	21s	30s	37s	48s	64s	78s	108s	126s	—
Best Squares	21s	26s	30s	35s6d	41s	55s	72s	85s	110s	138s	—
Best Steel Plated do. ..	25s6d	30s	37s	44s	55s	69s	2½ in. Sash do. 24s				—
Bricklayers', brass blades	9s6d	11s6d	15s	19s6d	25s	London Pattern 34s6d					—
Coachmakers', all Steel	32s	39s	46s	53s	62s						—
Steel Drawing Squares	16s	18s6d	21s	27s6d	35s	41s	50s	62s			—

	6	9	10	12	15	18	21	24	30	36 inches.	
Best Square with Level		78s		90s	106s	120s	138s	154s	173s		p. doz.
Mahogany, or hard wood	14s	17s		21s	25s	31s	38s	44s	51s	62s	—
Hard wood T Drawing Squares with Bevil		23s		26s	29s	35s6d	41s	45s	52s	63s	—
Ditto with steel blade		34s6d		39s	46s	53s	60s	66s	76s	87s	—
Best do., brass plated		46s		53s	60s	67s	74s	80s	96s	115s	—

	4	5	6	7	8	10	12	14	16 inches.	
Mitre Square............					24s	30s	35s	41s	48s	p. dozen.
Steel Triangle Square with Mitre, Level, and Plumb.	18s6	23s	27s6	34s6	41s	55s	69s			—

Bevils.

	6	7½	8	9	10	12	15 inches.	
Common T Bevils..........		15s		17s6d		21s	27s6d	p.doz.
Plated do..............		22s		25s		29s	38s	——
Best do..............		27s6d		32s		42s	60s	——
Common Angle Bevils......	14s	17s		21s		27s6d		——
Plated do.	19s6d	23s		27s6d		35s		——
Best do.	27s6d	32s		38s		48s		——
Coachmakers' do., all Steel ..				36s		46s	50s	——
Bricklayers' do., brass blades	11s6d		15s6d		19s6d	24s		——
London Pattern	34s							
Com. Boat-builders' do., 2 brass blades			15s		17s	21s		——
Plated do. do..........			21s		23s	28s		——
Best do. do..........			31s		35s	41s		——

Com. brass spider-leg Bevil 46s Best, 57s Com. iron do., 40s Best, 52s ——

	¼	⅜	7-16	½	⅝	¾ in.
Flat shank Taper Bits..........				2s	2s6d	3s3d
Flat Nose and Shell Bits........	2s	2s6d	2s9d	3s3d	4s	4s6d
Chair Bits....................		2s2d		2s9d	3s6d	4s6d

Shipwrights' Gimblets, 9 inches long.. 7s6d 10 in... 9s 10s6d 13s6d

		1	1⅛	1¼	1⅜	1½	1⅝	1¾	1⅞	2 in.
Flat Centre Bits ½ to 1 in.	3s	3s6d	3s9d	4s	4s9d	5s6d	6s3d	7s	8s	9s
Plug Bits......................			12s	13s	14s					

Bell Bits, ⅛ to ¼ inch, 4s—5-16 to ⅜, 5s

	¾	⅞	1	1⅛	1¼	1½	1¾	2	2¼	2½	2¾	3	3½	4 in.
Tap Borers	4s9d	5s6d	6s6d	8s6d	10s6d	16s	22s							
Tap Bung..........								30s	84s	38s	42s	45s		

	⅛	3-16	¼	5-16	⅜	7-16	½	⅝	¾ inch.
Nose and Pin Bits....	3s6d	3s6d	3s8d	3s10d	4s	4s3d	4s9d		
Centre Bits..........	3s4d	3s4d	3s5d	3s5d	3s6d	3s7d	3s8d	3s9d	3s10d

Centre Bits....	⅞	1	1⅛	1¼	1⅜	1½	1⅝	1¾	1⅞	2 in.
	4s	4s3d	4s9d	5s	5s6d	6s	6s6d	7s9d	9s	10s

	s.	D.		s.	D.
Cooper's Croze Irons to 1¾ in. ..	6	6	Best Sash Bits	4	6
Best do. to 1½, 10s 1¾, 12s 2 in. p.d	14	0	Ditto Table	4	6
Do. Hawk neb, 1½ 16s 1¾ 18s 2 in.	20	0	Ditto Taper to ⅝	4	6
Best Hawk Nebbed Irons	55	0	Best Sexagon Rinder..........	4	3
Spool Bits to 5-16ths in. 3s6d ⅜	4	0	Ditto square do............	3	9
Wheelwright's to ¼ inch........	4	0	Ditto half-round do.	3	9
Ditto do, to 5-16ths. 4s6d. to ⅜,	5	0	Ditto Pin and Nose do. ⅛ to ⅜ in.	3	9
Cooper's Dowling Bits, ⅛ to ¼ in.	1	2	Ditto Turnscrews	3	9
Ditto Nosed	1	2	Ditto Centre Bits sorted ⅛ to 1 in	3	9
Ditto Spoon Nose	1	2	Ditto iron Countersinks	3	9
Brush Bit Gauges	6	0	Ditto single notched do.	3	9
Best Gimblet Bits, ⅛ to 5-16ths..	4	0	Ditto Rose do.	4	3
Best Bed Bits, to ½ in	5	0	Ditto Flooring Dowling Bits ..	3	6

*** If bright, 9d. If coloured, 2s. per dozen more. Bits made to any pattern.
Coopers' Vices, 5s 6s 7s 9s Coopers' Croze Stocks, 40s—Best do, 46s

Pads and Bits........	12	18	24	30	36	42	48 Bits
	6s11d	9s1½d	11s4d	13s6½d	15s9d	17s11½d	20s2d

Plate CXXXIIIa. - Mitchell, Brothers, & Co.'s 1828 Price List of Joiners' Tools
[Central Library, Sheffield, Archives, Marsh File 122]

Gauges.

	s.	d.		s.	d.
Marking Gauges, common	7	6	Leather Cutters' Beech Gauges	18	0
Ditto Plated	10	6	Ditto, Plated	23	0
Ditto, Head faced with brass ..	18	6	Ditto, Head faced with brass....	32	0
Ditto, Head and Stem do......	27	6	Ditto, Head faced with steel....	38	0
Ditto, Head faced with Steel ..	21	0	New Leather Cutters' Gauges ..	69	0
Ditto, Head and Stem do......	31	0	Mortice Gauges, split stem	11	0
Cutting Gauges, Beech........	10	6	Improved Mortice, split stem ..	18	0
Ditto, hard Wood, or Plated ..	14	0	Ditto, with brass slide	32	0
Ditto, best Brass Hooped	20	0	Ditto, best	39	0
Ditto, Head faced with brass ..	28	0	Best Mortice Gauge, plated	46	0
Ditto, Head and Stem do	35	0	Ditto. Head faced with brass....	52	0
Ditto, Head faced with steel....	27	6	Ditto, Head faced with steel. ..	55	0
Ditto, Head and Stem do	38	0	Ditto, with screw Slide	62	0
Improved do	48	0	Ditto, Head faced with brass....	76	0
New Cutting Gauge	25	0	Ditto, Head faced with steel....	80	6
Ditto, Head faced with brass ..	36	0	Improved do.	82	6
Ditto, Head and Stem do......	46	0	Side Gauges, single slide	10	6
Ditto, Head faced with steel....	36	0	Ditto, double slide............	14	0
Ditto, Head and Stem do......	49	0	Marking Gauges for Pencils....	7	0

Spokeshaves.

	2½	3	3½	4	5	6 inches.
Common Spokeshaves..............	13s6d	15s6d	16s6d	18s	20s	23s p. doz.
Plated do..................	17s	20s	22s6d	24s	29s	35s ——
Ivory Plated Spokeshaves..........	22s	25s6d	27s6d	29s		——
Common ditto with Screw Iron......	31s	35s	38s	41s	46s	53s ——
Plated ditto do..............	35s	39s	44s	48s	55s	64s ——
Plated Screw Ivory Spokeshaves	40s	44s	46s	53s		——
Double edged common do..........	23s	29s	31s6d	35s	42s	48s ——
Ditto Plated do..........	32s	39s	43s6d	46s	53s	64s ——
Common ditto, Screw Iron..........	41s	50s	55s	62s	73s	85s ——
Plated ditto, Screw Iron..........	50s	60s	66s	74s	87s	101s ——
Box Spokeshaves..................	17s6d	21s	23s	21s6d	30s	35s ——
Ditto, Screw Irons	35s	39s	44s	48s	55s	64s ——
Double edged Box Spokeshaves......	32s	39s	43s	46s	53s3d	58s ——
Ditto Screw Irons	51s	60s	66s	73s	87s	101s ——
Coopers' Spokeshaves..............		18s	20s	22s	25s	30s ——
Ditto inside do..................		20s	22s	24s	29s	——
Cast steel Spokeshave Irons	4s6d	5s	5s5d	6s	6s9d	7s6d ——
Coopers' Spokeshave Irons		6s6d	7s	7s6d	8s	8s6d ——

Direction Shaves, 24s — Second size, 28s. Cucumber Slicers, 20s ——

Coopers' Spokeshaves if boxed in, 3s extra.

Braces, Bits, Hand Pads, &c.

Best Braces, 9s6d each.	12	15	18	24	30	36	48 Bits.
Best Ditto, with Black Bits........	13s	15s	17s	19s	21s	23s	25s ——
Ditto with plated Bits	17s	19s	21s	23s	25s	27s	29s ——

If with bright Bits, 9d. If with brown Bits, 2s per dozen extra.

With Sash Dowling Bits, 10d per set extra.

Brace Pads, 33s—Best Do., 36s. Brace Heads, 20s—Best, 24s per dozen.

Common Coopers' Braces, 3s. Best Ditto, 4s. each.

	6	9	12	18	24 Tools.
Hand Pads..........	5s9d	7s6d	9s	11s	13s each.
Best do	8s	9s6d	11s6d	15s	18s6d —
Hollow Hand Pads ..	4s	4s9d	5s6d	7s	8s3d —
Best do............	5s9d	6s9d	7s6d	9s6d	11s —

Ditto Hand Pad Tools, 2s8d per dozen. Ditto with brass Pad, 3s6d per dozen.

Sportsmen's do. complete, 11s per dozen.

Brush Bits...............	1 to 10	11	12	13 Gauge.
	2s	2s3d	2s9d	3s6d

Cock Bits, ⅜ in. 11s9d. Centre Cock Bits, ⅜ in. 10s ½ in. 12s.

JANUARY, 1828.

Saw Pads, Levels, Scrapers, &c.

Small Saw Pads, 16s—Large, 19s6d—Best, 24s per dozen.

Small Drawing Knives, 8s6d—2nd size, 11s6d—3d. size, 15s6d—4th. size, 19s6d.

	s.	d.		s.	d.
Chair-makers' Saw Pads, per doz.	36	0	Best Flooring Awl Blades per doz.	0	11
Common Spirit Levels,.. ——..	28	0	Ditto handled............ ——..	2	8
Plated do............ ——..	35	0	Brad and Nail Punches.. ——..	1	6
Plated do. both sides .. ——..	46	0	Sportsmen's Turnscrews ——..	10	6
Brass Stocked do. ...... ——..	101	0	Best do.......... ——..	16	0
Unshouldered Brad Awls ——..	0	3	Best double Sportsmen's ——..	25	0
Best shouldered do...... ——..	0	6	Ditto with picker ...... ——..	28	0
Scribing Tool.......... ——..	4	6	Black Gun Vice........ ——..	13	0
Marking Awl.......... ——..	2	6	Bright do.......... ——..	17	0
Bolt Chisels .... ——..	5	2	Best Saw Sets, handled .. ——..	11	6
Common Scrapers ...... ——..	7	0	Drill Boxes............ ——..	18	6
Spring do.......... ——..	14	0	Ditto do. and Drills...... ——..	27	6
Scraper Steels.......... ——..	4	6	Drill Stocks.......... ——..	62	0
Best handled Brad Awls, ——..	1	3	Ditto and Drills .... ——..	76	0

Scraper Blades, from 3 to 6 inches, 1d. per inch.

Turnscrews, &c.

Turnscrews,	No. 1	2	3	4	5	6	7	8	9	10	11	12
	In. 3	4	5	6	7	8	9	10	11	12	13	14
Black, per dozen ..	4s6d	5s6d	6s6d	7s6d	9s	10s6d	12s	14s	16s	18s	20s	22s
Bright, do.	5s	6s	7s	8s	10s	12s	14s	16s	18s	20s	22s	24s
Black London, do ..	6s6d	8s	9s6d	11s6d	13s	15s	17s	19s	22s	25s	30s	35s
Bright do. do..	7s6d	9s	10s6d	12s	14s	16s	18s	20s	23s	26s	32s	36s

	s.	d.		s.	d.
Cast Steel Chisels, handled	6	6	Carpenters' Mallets, per dozen	18	6
Ditto Gouges, do.........	7	0	Joiners' do.	16	0
Beech Chalk Line Roller	2	0	Gentlemen's do.	10	0
Box do.... do..........	5	0	Hand Crows and Chisels	9	0

10 inches Saws and Frames, 38s—12 in., 42s—14 in., 48s—16 in., 58s—18 in., 69s

Screw Boxes for cutting Wood Screws.

¼	⅜	½	⅝	¾	⅞	1	1⅛	1¼	1⅜	1½	1⅝	1¾ inch.
3s6d				3s9d	4s	4s3d	5s	5s6d	6s	7s	7s6d	8s each.

2	2¼	2½	2¾	3	3¼	3½	4	4½	5 inches.
10s	13s	16s	20s	26s	36s	42s	54s	66s	80s each.

Spring Steel Drill Bows from 6 to 24 inches, 4d. per inch.

Wood Breast Plates, ... 9s Steel do.............. 14s

Wheelers' Rounders	⅝	¾	⅞	1	1⅛	1¼	1⅜	1½ inch.
	2s9d	3s2d	3s6d	3s10d	4s2d	4s6d	4s10d	5s3d each.

Skates.

	Without Straps. 1	2	3	4	With Straps. 1	2	3	4 size.
Common Skates	2s4d	2s8d	3s1d	3s4d	4s1d	4s8d	5s6d	5s10d
Common Fluted do......	2s10d	3s2d	3s6d	3s10d	4s10d	5s2d	6s	6s4d
Middle sort do......	3s6d	3s10d	4s4d	4s10d	5s6d	5s10d	7s	7s4d
Middle sort Fluted do....	4s	4s4d	5s	5s1d	6s	6s4d	7s6d	7s10d
Best hollow Ground do ..	6s6d	7s6d	8s6d	10s	9s6d	10s6d	12s	13s
Best hollow do., polished	10s6d	and Fluted		12s		13s	14s6d	
Inlaid do., 12s—Polished,	15s—Fluted, 15s					15s6d	17s	17s
Inlaid Skates, cased slots, 18s	Fluted, 19s6d					20s6d	22s	

Tool Chests.

Gentlemen's Oak Tool Chests, fitted up complete. each.	No 1	2	3	4	5	6	7	8	9	10	11	12	13
	21s	26s	33s	42s	53s	63s	73s	84s	105s	126s	157s	189s	210s

Garden Chests, No. 14, 30s—15, 40s each.

Carpenters' or Joiners' do.	No. 1	2	3	4	5	6	7	8	9	10	11	12
	£5.15	£8.13	£11.10	£13.16	£17.6	£23.	£.28.15	£34.10	£40.5	£46.	£51.15	£57.10

J. WILLIAMSON, PRINTER, 13, NORFOLK-ROW; SHEFFIELD.

Plate CXXXIIIa. - Mitchell, Brothers, & Co.'s 1828 Price List of Joiners' Tools
[Continued from page 198; Central Library, Sheffield, Archives, Marsh File 122]

The earliest known Sheffield price list of joiners' tools is by Mitchell, Brothers, & Co., dated 1828. (see Plate CXXXIIIa & b) At that date they were listed only as Edge Tool Makers in the Sheffield Directory, but apparently marketed joiner tools in exchange for their manufactured goods. They phased out of business by 1837.

The next known listing of joiner tools, and earliest for prices of planes, was by J.Wilks, Dealer in Tools, at Broad Street, dated 1829, is shown as Plate CXXXIV.[9] (title page & index only). Note the non-tool items (Skates & Busks [flat spring steel for ladies' corsets] sold by tool dealers at that date. Wilks retired in 1854 (See Plate CXXXV).

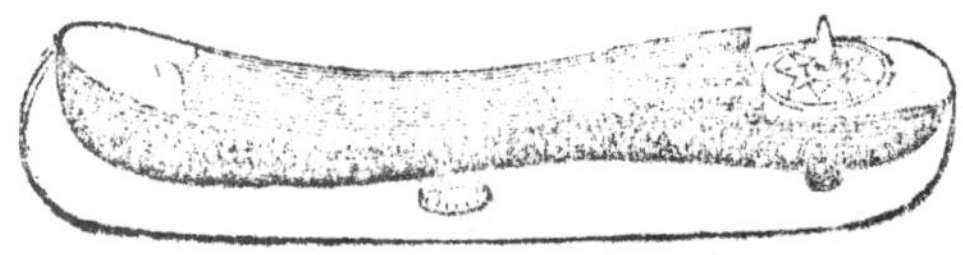

JOHN WILSON,

(Late JABEZ SHAW,)

WESTFIELD TERRACE, SHEFFIELD,

MANUFACTURER OF

JOINERS' TOOLS & SKATES,

By Steam Power for Home and Exportation, on the Shortest Notice and on the most reasonable Terms.

INVENTOR & MAKER OF A REGISTERED & PATENT SKATE.

266

Plate CXXXI. - John Wilson's 1871 Sheffield Directory Advertisement
[Note: Late JABEZ SHAW; address at Wm. Marples & Sons Hibernia Works; Skate Inventor]

ATTENTION IS PARTICULARLY REQUESTED TO THE FOLLOWING

CAUTION.

JOHN WILSON,

SYCAMORE STREET, SHEFFIELD,

THE ORIGINAL MANUFACTURER OF

SHOE, BUTCHERS', WEAVERS',

BREAD, COOKS', PALETTE, PUTTY, GLAZIERS',

AND

FARRIERS' KNIVES, BUTCHERS' STEELS,

&c. &c.,

Deems it right to apprize his Friends, that *not only does his Mark continue to be PIRATED both in this country and on the Continent*, but that certain unprincipled persons, who are deterred from striking his Mark only from a fear of being visited with Legal Proceedings, have been led by their cupidity to resort to *another mode of imposition*, and IN ORDER TO EVADE THE LAW, are vending articles marked—

WILSON, only; or, WILSON CAST STEEL
or, WILSON WARRANTED or, GEO. WILSON

In each case *omitting the Corporation Mark.*

OBSERVE!

All Knives, Steels, &c. of JOHN WILSON'S manufacture, are stamped with his Corporation Mark—"*Four Peppercorns and a Diamond,*" thus, ⁘◆ in one of the following forms:—

I. WILSON ⁘◆ — ⁘◆ I. WILSON CAST STEEL
⁘◆ I. WILSON WARRANTED — ⁘◆ I. WILSON SHEAR STEEL

Notice is hereby Given,

That all persons whomsoever, who shall counterfeit or imitate the Marks used by the said JOHN WILSON, or who shall sell any Shoe, Bread, Butchers', Cooks', or any other *KNIVES*, or any Butchers' *STEELS*, stamped with any such *counterfeit or imitated marks*, will, immediately after they shall be discovered, be proceeded against in one of the Courts of Law or Equity; and the said JOHN WILSON hereby invites all persons, to whom any such fraudulent practices as the above mentioned are, or may be known, to give him *Information*, assuring them that they shall be LIBERALLY REWARDED.

Plate CXXXII - Another John Wilson's 1849 Advertisement in Sheffield Directory

This John Wilson was a knife cutler, who worked at Sycamore Street in Sheffield during the mid-19th Century. It is not known whether he was any relation to the John Wilson, joiner tool maker noted in Plate CXXXI.

Note that his advertisement calls to attention difficulties experienced with his: *"Mark continue to be Pirated both in this country and on the Continent"*.

J. WILKS,

Broad Street, Park, Sheffield,

MANUFACTURER AND

DEALER IN TOOLS

OF EVERY DESCRIPTION,

FOR

JOINERS, CARPENTERS, CABINETMAKERS, COACHMAKERS, | WHEELERS, COOPERS, CHAIRMAKERS, TURNERS, | CARVERS, SHIPWRIGHTS, MASONS, and BRICKLAYERS.

ALSO,

SKATES, PLANES, SAWS, & EDGE TOOLS,

AND

Gentlemen's, Carpenters', Turners', Gardeners',

AND OTHER

TOOL CHESTS,

OF ALL SORTS.

TURNING ENGINES, &c. &c. COMPLETE.

Ridge's Braces, Bits, &c. at Ridge's Prices.

SHEFFIELD:
PRINTED BY H. A. BACON, INDEPENDENT OFFICE.
1829.

[Note Wholesale Discounts in Right Column]

INDEX.

Plate CXXXIV - Title Page and Index to J.Wilks' 1829 Tool Catalogue
[Central Library, Sheffield, Archives, Marsh File #122]
(Note the penned discounts)

Previous discussion concerning edge tools referred to the Pawson & Brailsford 1870 SHEFFIELD LIST and the arrangement whereby any prominent manufacturer being a Member of the Cutlers' Company could have a trade catalogue prepared from the same printing cuts. (see page 153) The same was true of joiner tool manufacturers. The earliest surviving trade catalogue known to have been printed from the 1870 Sheffield List printing cuts was for William Marples & Sons. The title page of this catalogue is shown as Plate CXXXVI. p.193. Four pages from this 1870 Wm. Marples & Sons' Catalogue are shown: Squares, p.11 (Plate CXXXVII); Spirit Levels, p.12,(Plate CXXXIX); Marking Gauges, p.13,(Plate CXLI) and Spokeshaves, p.14,(Plate CXLIII). The corresponding 1870 Sheffield List Plates are shown opposite to these: Squares, D3,(Plate CXXXVIII); Spirit Levels, D5,(Plate CXL); Marking Gauges, D6,(Plate CXLII) and Spokeshaves, D9,(Plate CXLIV). While the arrangements of the plates differ, it will be noted that the individual items were printed from the same cuts. Many of these cuts were used through the remainder of the 19th Century in making new lists, trade catalogues and advertisements. Similar Sheffield Illustrated Lists were published by Pawson & Brailsford again in 1876 and 1889.[10] Plates CXLV and CXLVI respectively show pages from the Ward & Payne, c1900 and Thos. Ibbotson,1903 Trade Catalogues, which were both printed by Payson & Brailsford.

JOSEPH WILKS,

WHITESMITH AND IRONMONGER,

42, FURNIVAL ROAD, CORN EHCHANGE,

SHEFFIELD,

DESIRES to return to the Public his heartfelt thanks for the extensive patronage accorded to him during the long period of Forty Years; and begs to inform them that he has retired from the above business in favour of his Children, MARY ANNE, JOSEPH, and EDWARD JOHN WILKS, who have for some years assisted him in his Business, and will henceforth carry on the same on their own account, under the Firm of " WILKS BROTHERS & Co," for whom he earnestly requests a continuance of those favours which he has himself so long enjoyed.

WILKS BROTHERS & Co.,

(Successors to Joseph Wilks,)

BEG to assure the Public that no exertions or pains shall be wanting on their part to conduct the above Business in all its Branches, in the same satisfactory way which earned for their Father so extensive a patronage, and they respectfully request for themselves a continuance of the like favours.

Sheffield, December 15th, 1854.

WILKS BROTHERS & CO.

(SUCCESSORS TO JOSEPH WILKS,)

IRONMONGERS, WHITESMITHS,

And Manufacturers of Wrought Iron Gates, Palisades, Tomb Railing, Hurdles, Fancy Park Fencing, Hot-houses, Book Safes and Deed Chests, Mangles, Double and Single Weighing Machines, Sack Barrows, Improved Horse and Hand Chaff Cutters, Turnip Cutters, Bean and Oat Mills, Garden Rakes and Reals, Cast Steel Daisy Rakes, Paring Spades and Machine Knives, Hay and Manure Forks, Dock Grubs, Blocks, Joiners' Cramps, all kinds of Bright Coffin Furniture, Brass and Steel Hames, Saw Tangs, Buckles and Bows, Tillers and Boxes, Wind Vanes, Block and Coil Chains, Backbands, Traces, &c.

☞ Agents for PRICE'S Patent Fire-resisting and Thief-proof SAFES.

40 & 42, FURNIVAL ROAD, CORN EXCHANGE,

SHEFFIELD.

Plate CXXXV - Joseph Wilks Announces His Retirement - December 15, 1854
[1856 Sheffield Directory Advertisement]

In the first printing of the 1870 Wm. Marples & Sons' Catalogue the following statements are noted, respectively concerning files and planes:

> *"In consequence of considerable advance, both in Wages and Materials, it was unanimously resolved at a General Meeting of the File Manufacturers, held at Cutlers' Hall, Sheffield, on the 30th Jan., 1854 that the following list of prices be adopted by the Trade on the 1st Feb., 1854."*

> *"Price List of Planes, Adopted by the Trade, August 24, 1854."*

On July 1, 1872 Wm. Marples & Sons issued a letter noting increases of prices of edge tools. (See Plate CXLVII) The prices of Planes remained the same but the prices of files were increased according to a "General Meeting of the File Manufacturers at Cutlers' Hall on May 16, 1872".

WILLIAM MARPLES & SONS,

HIBERNIA WORKS,

SHEFFIELD.

July 1st, 1872.

Dear Sir,

Be good enough to cancel all prices on pages 1, 2, 3, and 4, also prices of Nos. 273 to 276 inclusive on page 6, and also prices of Nos. 1 to 46 inclusive on page 39 of our Illustrated List, substituting for them the prices on this sheet.

Yours faithfully,

WILLIAM MARPLES & SONS.

In consequence of considerable advances. both in Wages and Materials, it was unanimously resolved, at a General Meeting of Edge Tool Manufacturers, held at the Cutlers' Hall, Sheffield, on the 25th June, 1872, that the following List of Prices be adopted by the Trade from the 1st July, 1872.

LIGHT EDGE TOOLS. (Pages 1, 2, 3)

Plate CXLVII - Notice of Edge Tool Price Increases July 1, 1872.

Similarly the prices of joiner tools were raised July 1, 1873. William Marples & Sons sent out a Price List of Joiners' Tools stating:

> "Prices in this List supersede Prices of corresponding articles on Pages 10 to 17 inclusive of our Illustrated Price List - *In consequence of considerable advances both in Wages and Materials, it was unanimously resolved at a General Meeting of Joiners' Tool Manufacturers, held at Cutlers' Hall, Sheffield, on 25th June, 1873, that the following List of Prices be adopted by the Trades from 1st of July, 1873.*"

It is believed that these prices remained in effect for all Sheffield tool manufacturing firms for the remainder of the 19th Century, but were raised slightly soon after 1900. It seems incredible that these firms made and marketed about the same products without free competition, but apparently that is the way it was from 1860 - 1900. Discounts from the list were varied according to the class of the purchaser in amount of orders and credit rating. There appears to have been considerable indebtedness among the various manufacturers and if word was circulated that one was selling at lower than trade prices or extending discounts resulting in lower prices, pressure was applied to settle debts or comply with the trade prices, thus forcing the price cutter in line to maintain agreed fixed prices.

WILLIAM MARPLES & SONS, SHEFFIELD.

11

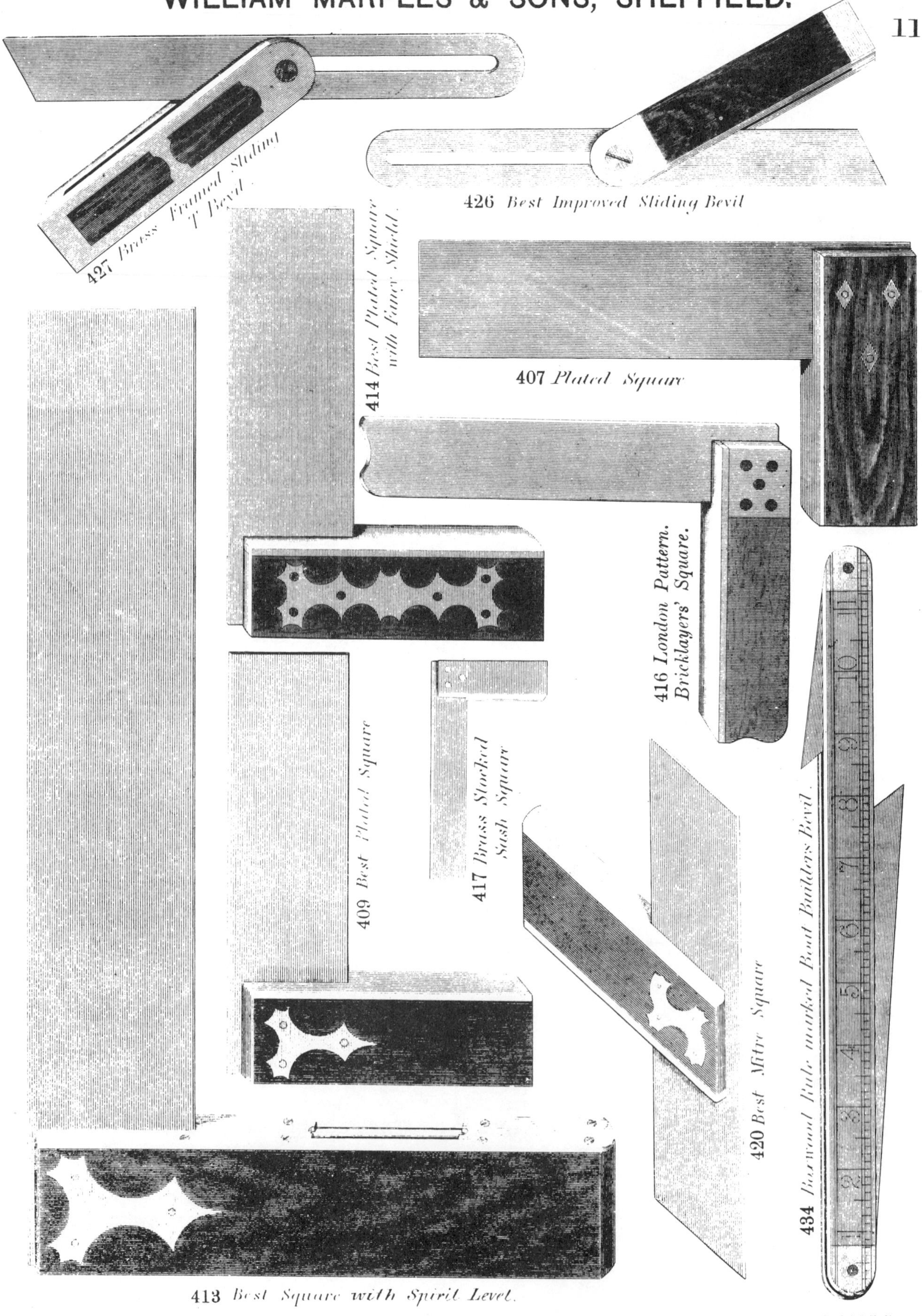

Plate CXXXVII - Squares, page 11, 1870 Wm. Marples & Sons' Catalogue

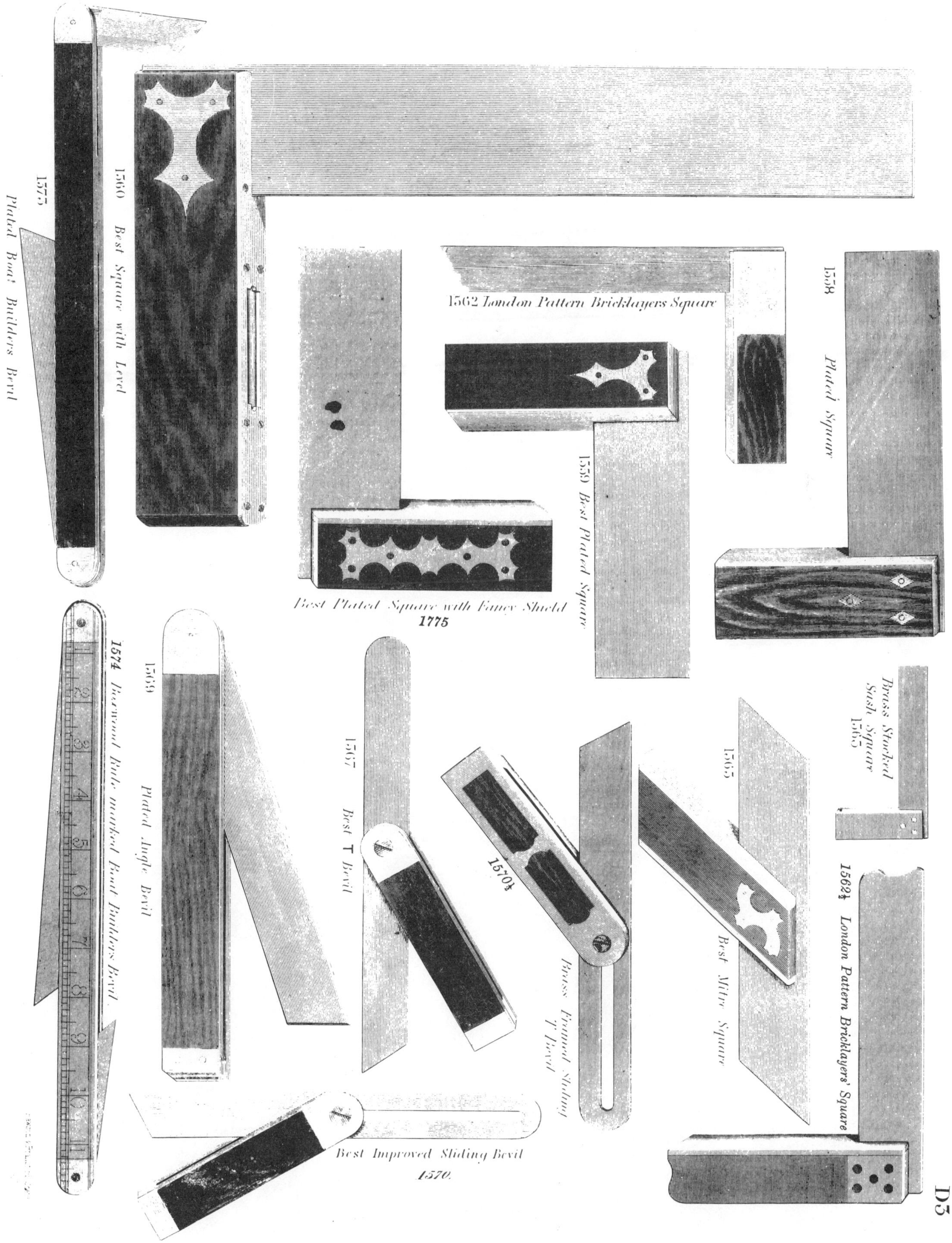

Plate CXXXVIII - Squares & Bevils, Plate D3, 1870 Sheffield List

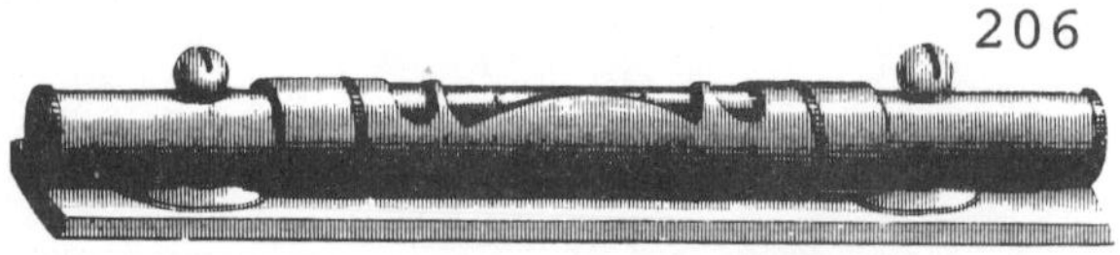

445 *Engineers' Brass Tube Level, with Guard.*

460 *Best Proved Spirit Level Tube.*

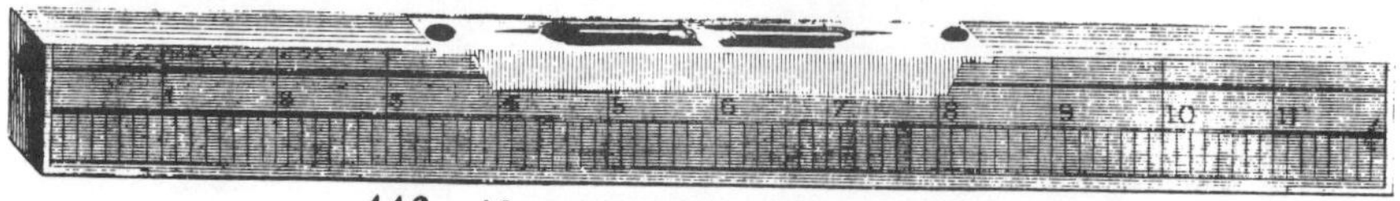

446 *Masons Rule Spirit Level*

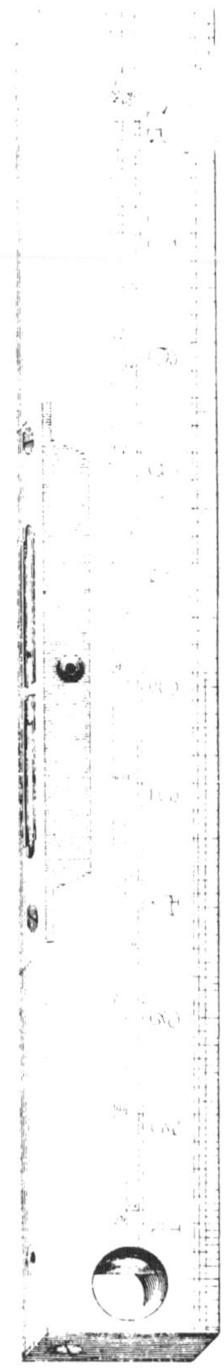

447 *Rule and Plumb Spirit Level.*

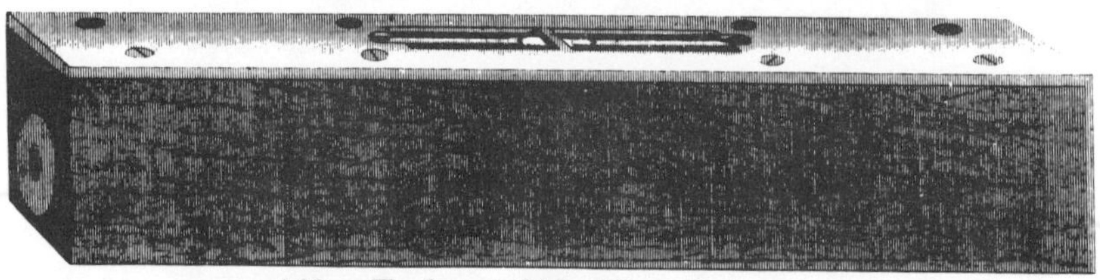

442 *End and Side View Spirit Level.*

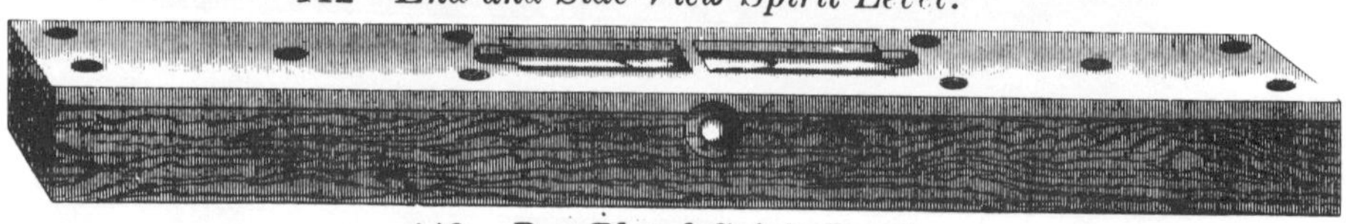

440 *Best Plated Spirit Level.*

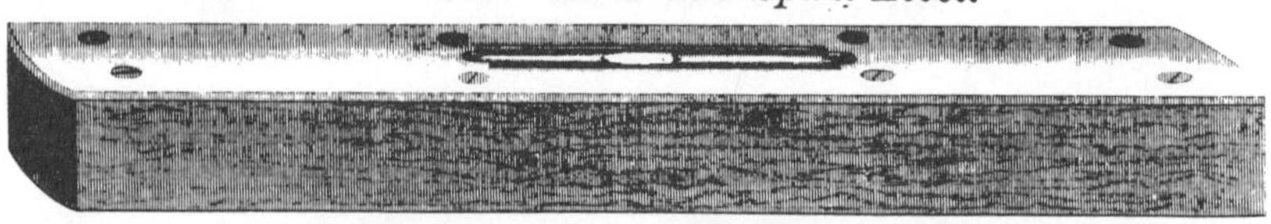

439 *Light Mahogany Spirit Level.*

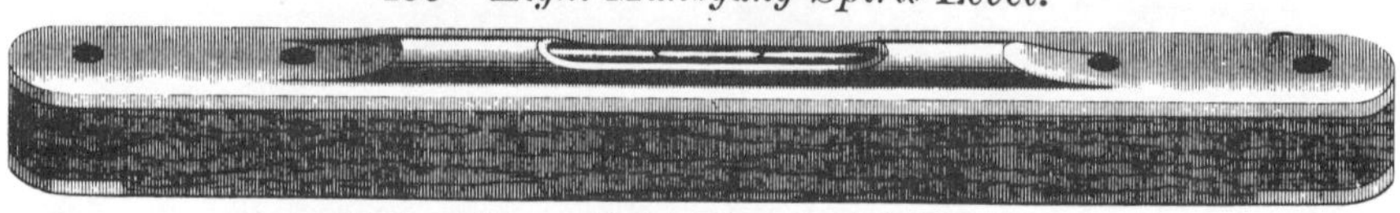

453 *Ebony Spirit Level with Raised Tube*

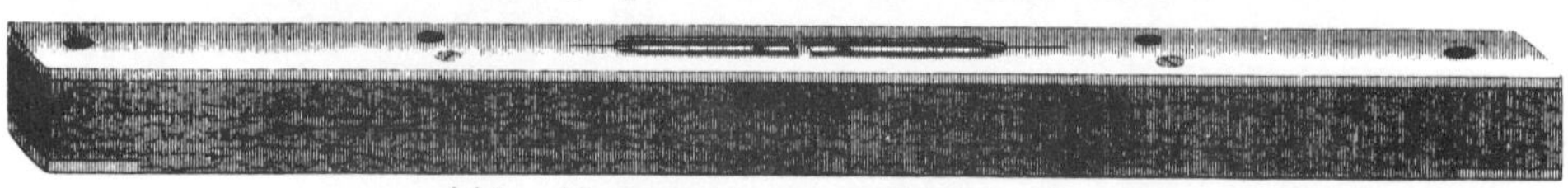

449 *Narrow Rosewood Spirit Level*

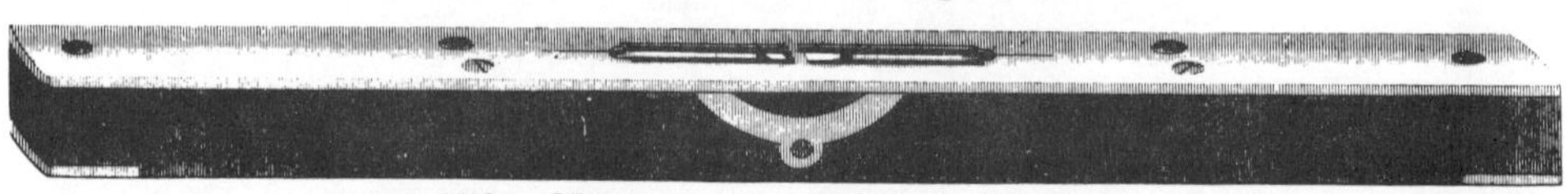

450 *Narrow Ebony Spirit Level.*

451 *Rosewood Scotch Spirit Level.*

452 *Ebony Scotch Spirit Level.*

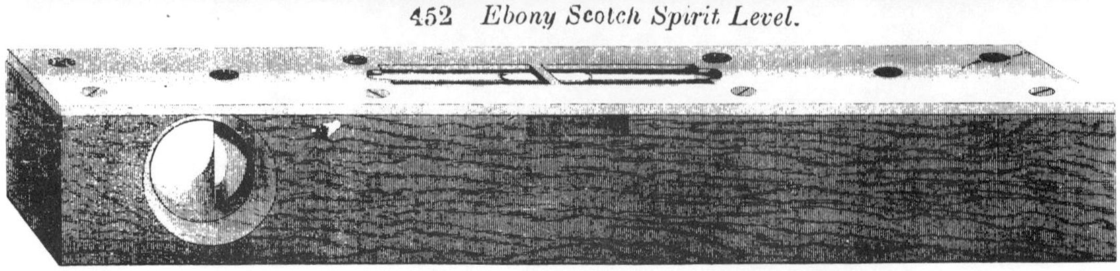

443 *Horizontal and Plumb Spirit Level.*

Plate CXXXIX - Spirit Levels, page 12, 1870 Wm. Marples & Sons' Catalogue

D 5

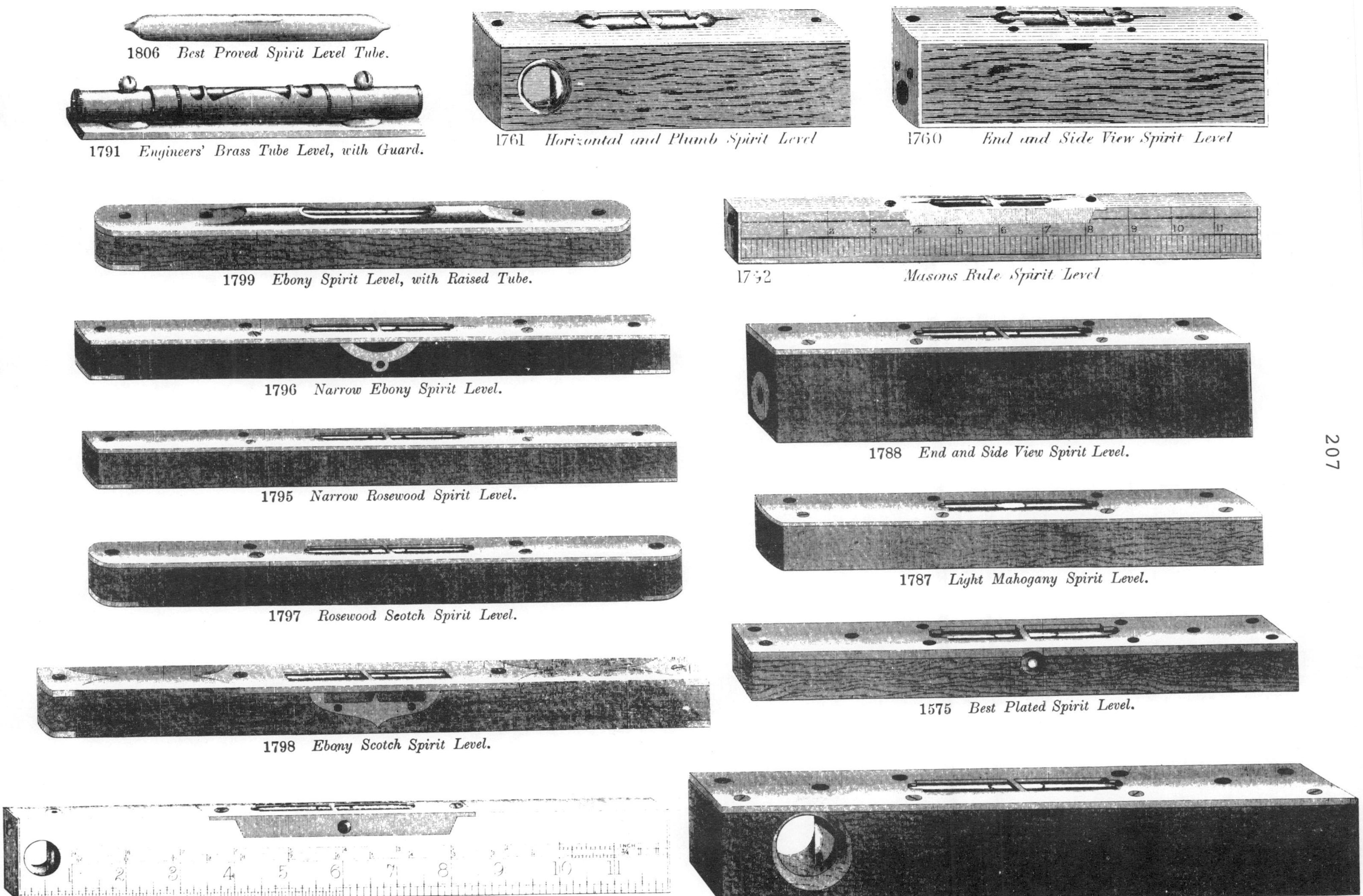

1806 *Best Proved Spirit Level Tube.*

1791 *Engineers' Brass Tube Level, with Guard.*

1761 *Horizontal and Plumb Spirit Level*

1760 *End and Side View Spirit Level*

1799 *Ebony Spirit Level, with Raised Tube.*

1792 *Masons Rule Spirit Level*

1796 *Narrow Ebony Spirit Level.*

1788 *End and Side View Spirit Level.*

1795 *Narrow Rosewood Spirit Level.*

1787 *Light Mahogany Spirit Level.*

1797 *Rosewood Scotch Spirit Level.*

1575 *Best Plated Spirit Level.*

1798 *Ebony Scotch Spirit Level.*

1793 *Rule and Plumb Level.*

1789 *Horizontal and Plumb Spirit Level.*

Pawson & Brailsford Lithographers Sheffield

Plate CXL - Spirit Levels, Plate D5, 1870 Sheffield Illustrated List

WILLIAM MARPLES & SONS, SHEFFIELD.

461
Beechwood Marking Gauge

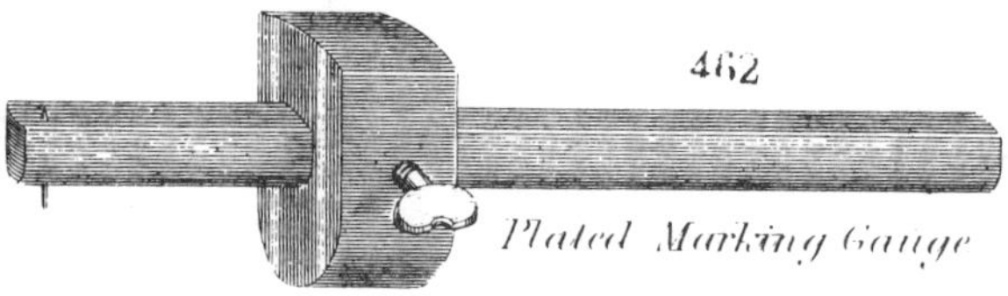
462
Plated Marking Gauge

469
Hardwood Brass Hooped Cutting Gauge

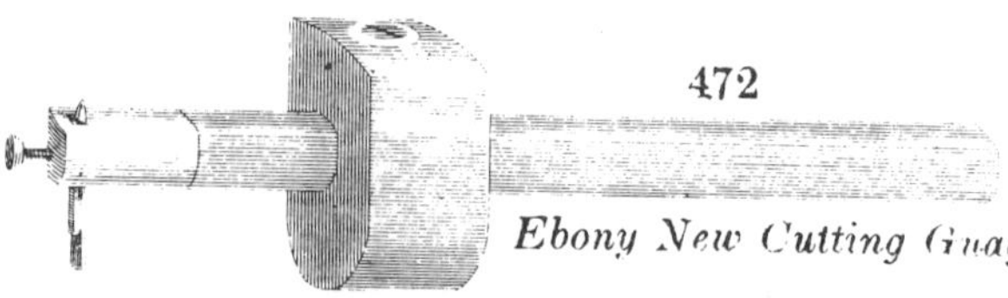
472
Ebony New Cutting Guage

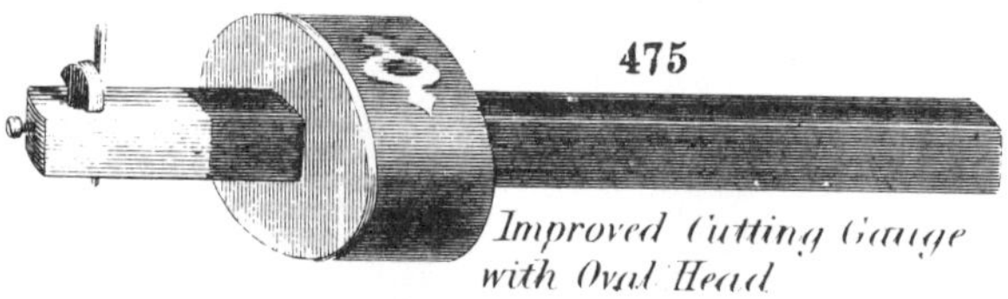
475
Improved Cutting Gauge with Oval Head

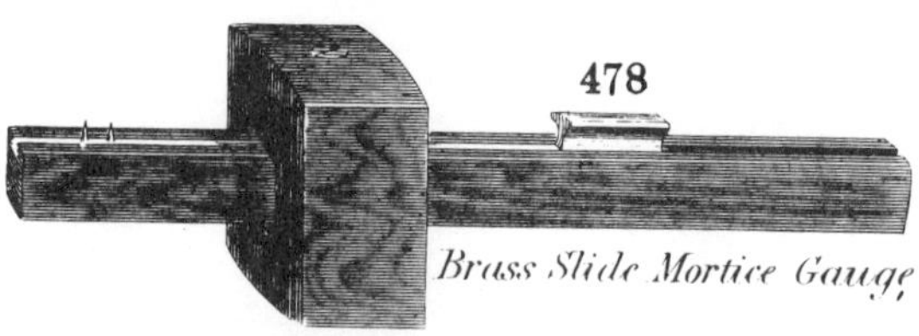
478
Brass Slide Mortice Gauge

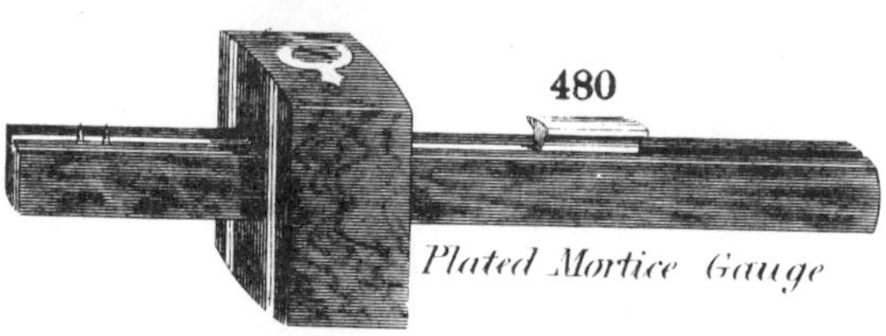
480
Plated Mortice Gauge

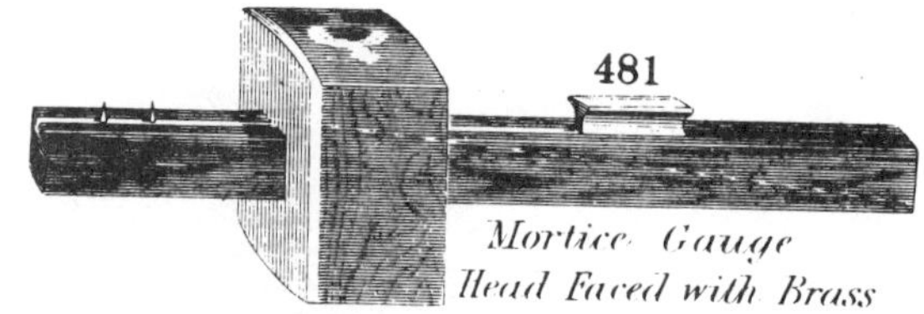
481
Mortice Gauge Head Faced with Brass

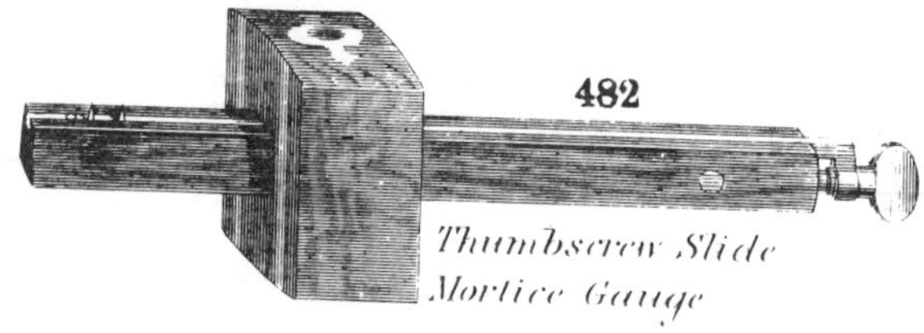
482
Thumbscrew Slide Mortice Gauge

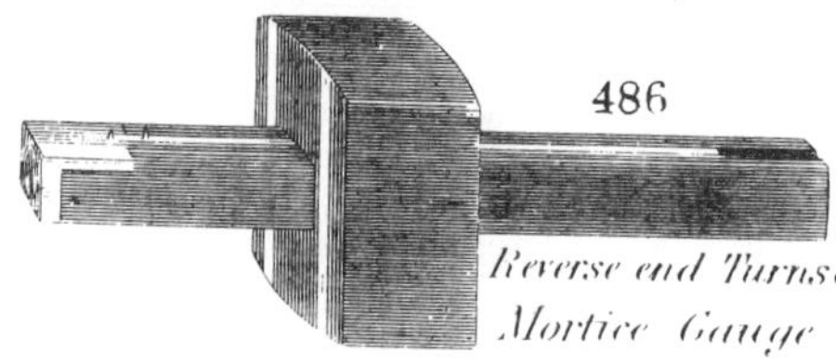
486
Reverse end Turnscrew Slide Mortice Gauge

489
Improved Brass Stem Mortice Gauge with Oval Head

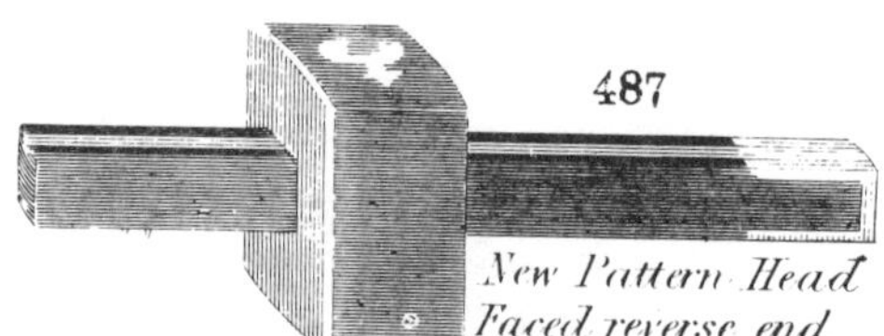
487
New Pattern Head Faced reverse end Mortice Gauge

488
Improved Mortice Guage with Oval Head

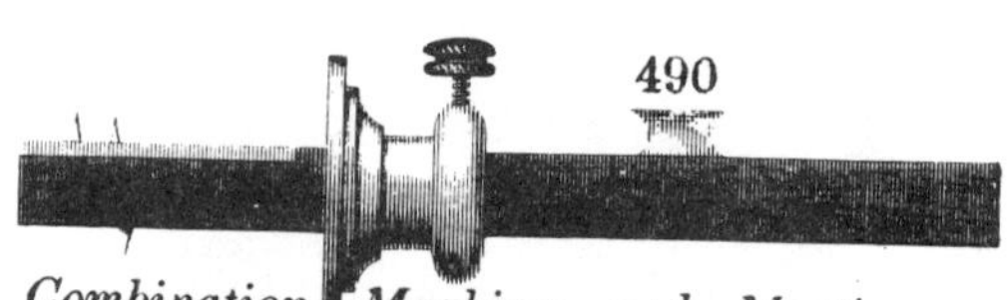
490
Combination Marking and Mortice Guage with Brass head.

Plate CXLI - Marking & Mortice Gauges, 1870 Wm.Marples & Sons Catalogue, page 13

D6

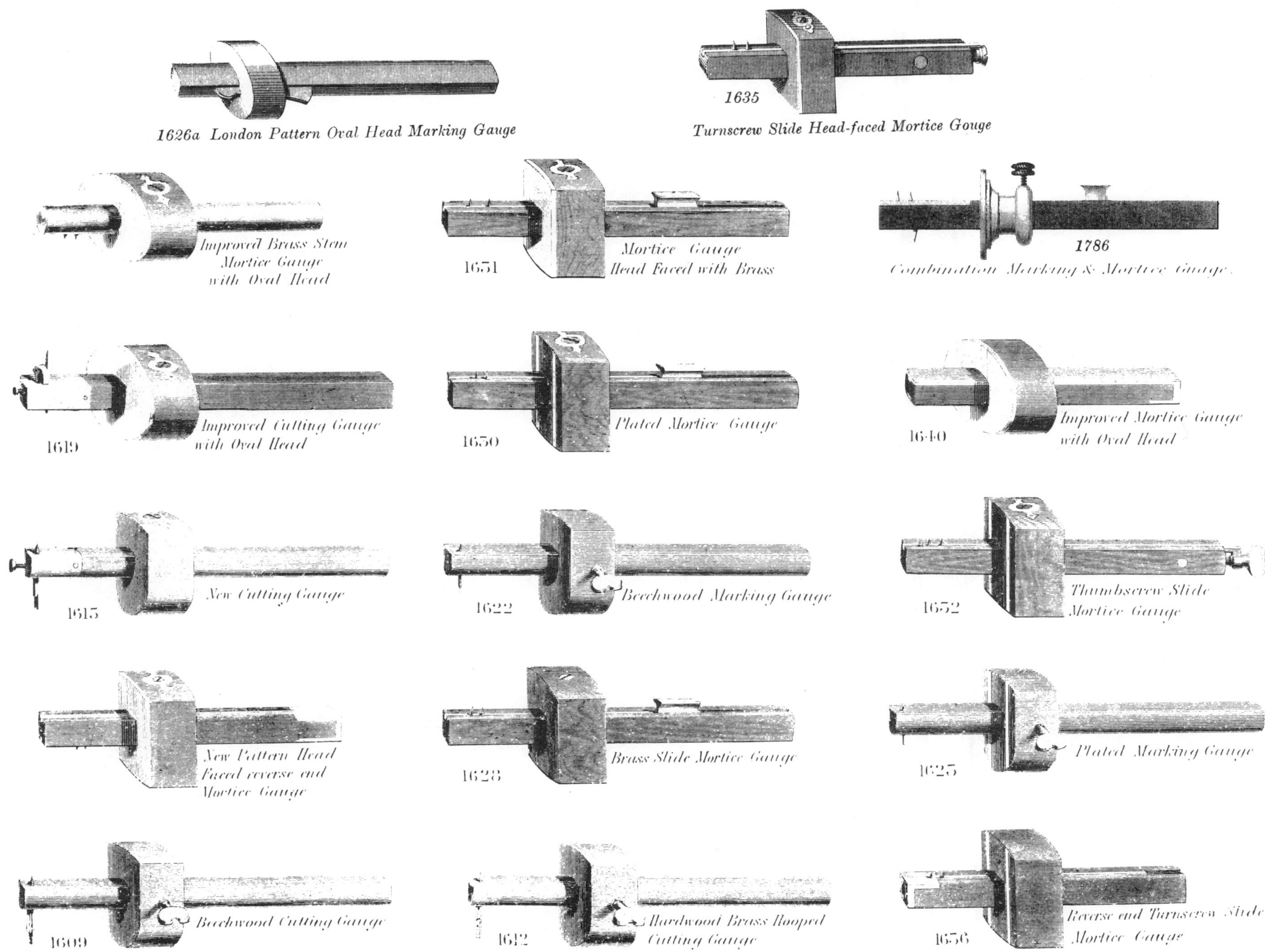

Plate CXLII - Marking & Mortice Gauages, Plate D6, 1870 Sheffield Illustrated List

WILLIAM MARPLES & SONS, SHEFFIELD.

466

Beechwood Cutting Gauge

526

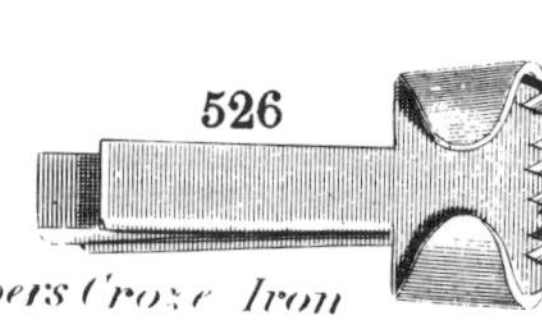

Coopers Croze Iron

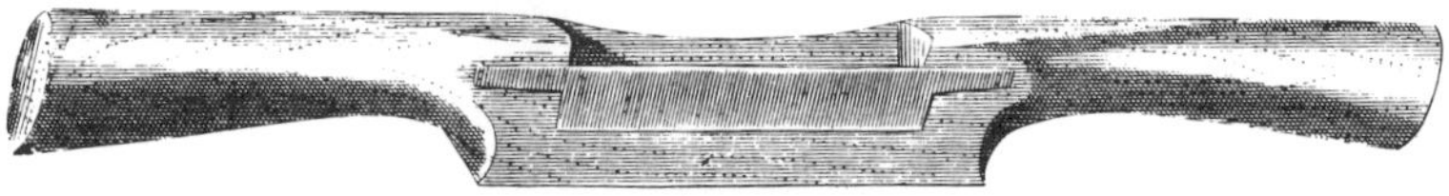

501 *Beechwood Spokeshave*

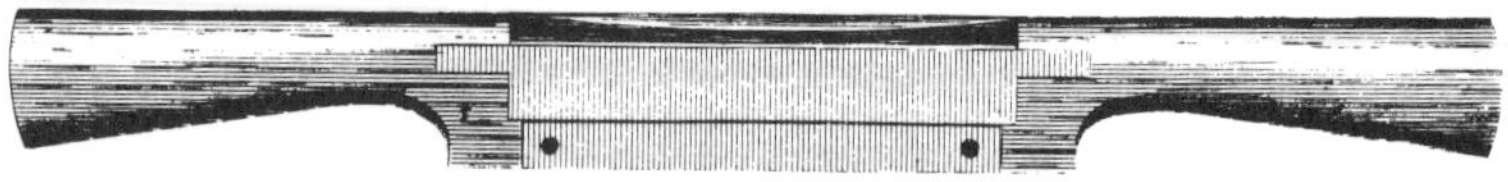

502 ***Beechwood Plated Spokeshave.***

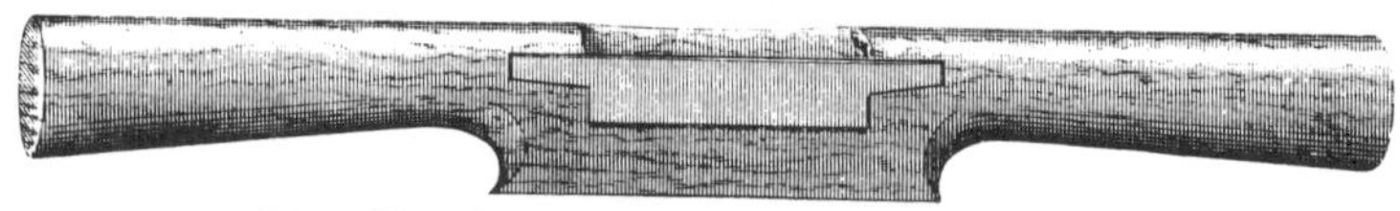

506 ***Beech London Pattern Coachmakers' Spokeshave.***

504

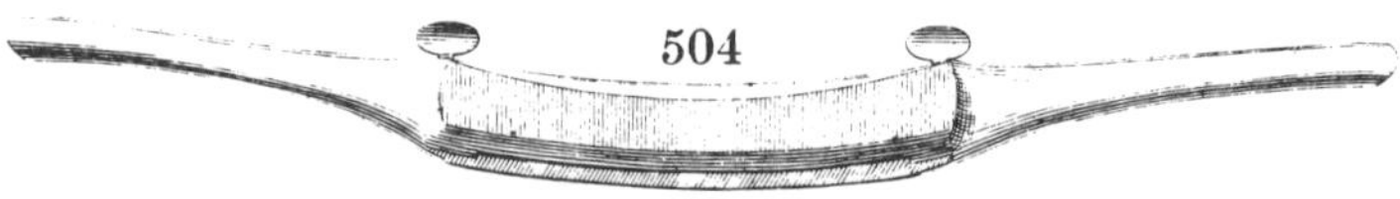

Beechwood Plated Screw Iron Spokeshave

507 *Boxwood Spokeshave*

522

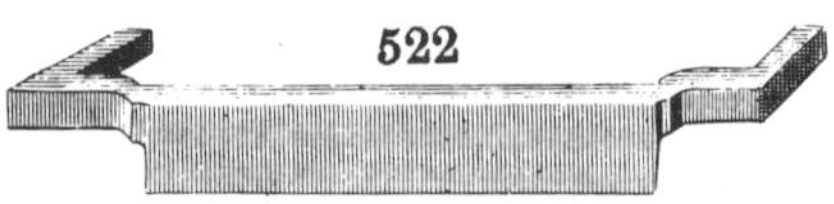

Coopers Spokeshave Iron

518

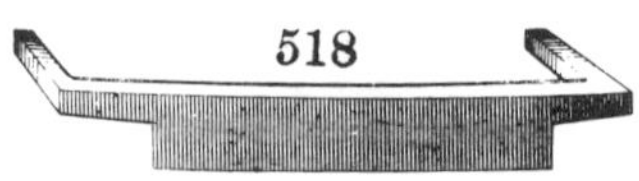

C.S. Spokeshave Iron.

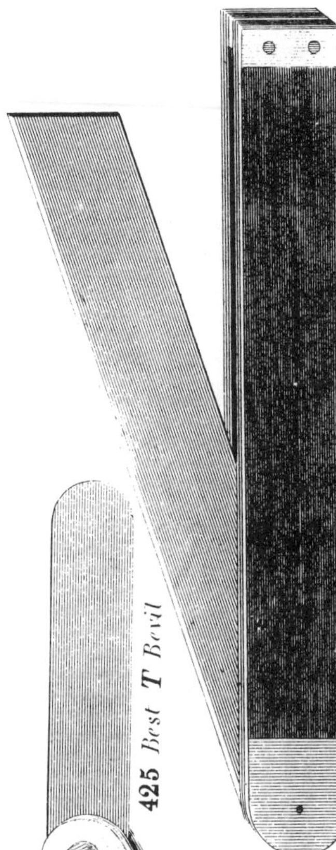

428 *Plated Angle Bevil*

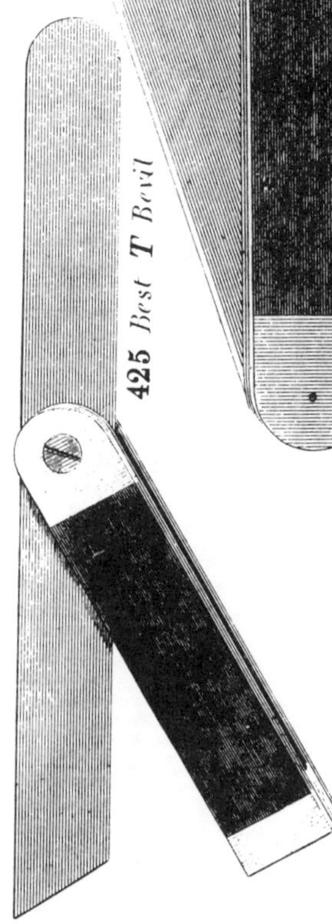

425 *Best* **T** *Bevil*

523

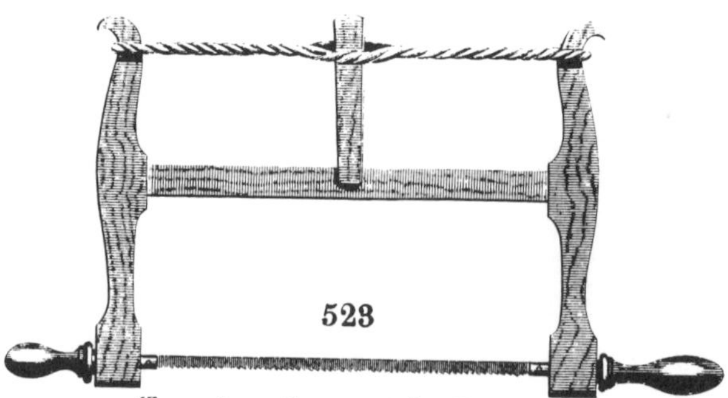

Turning Saw with Frame

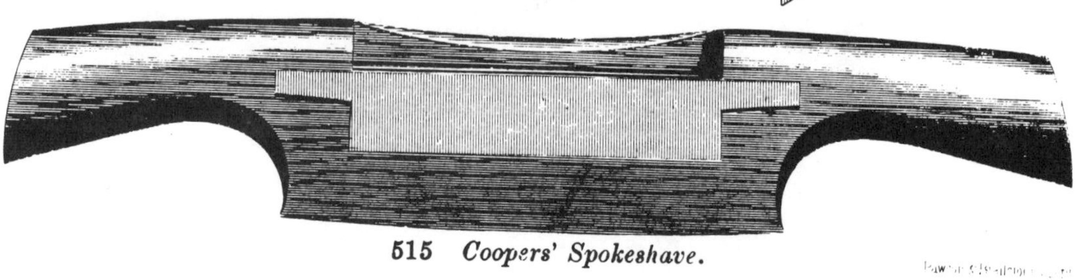

515 ***Coopers' Spokeshave.***

Plate CXLIII - Spokeshaves, Bevils, &c., 1870 Wm.Marples & Sons' Catalogue, p.14

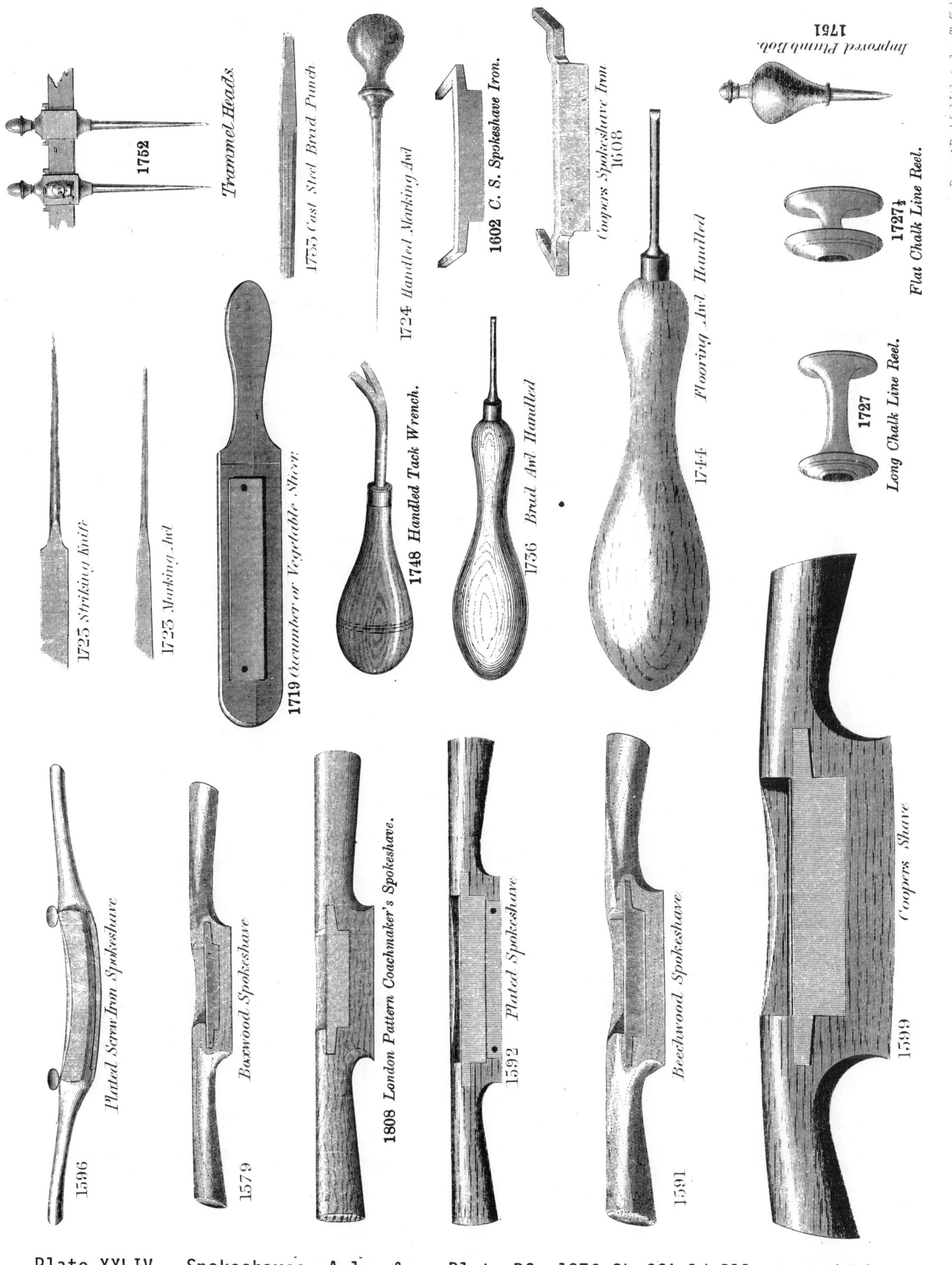

Plate XXLIV - Spokeshaves, Awls, &c., Plate D9, 1870 Sheffield Illustrated List

WARD & PAYNE, CORPORATE MARK. W P SHEFFIELD.

1093. *Rosewood Brass Slide Mortise Gauge.*

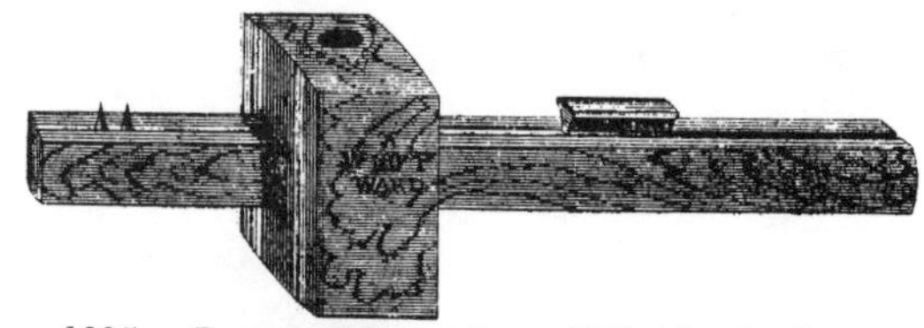

1095. *Rosewood Plated Brass Slide Mortise Gauge.*

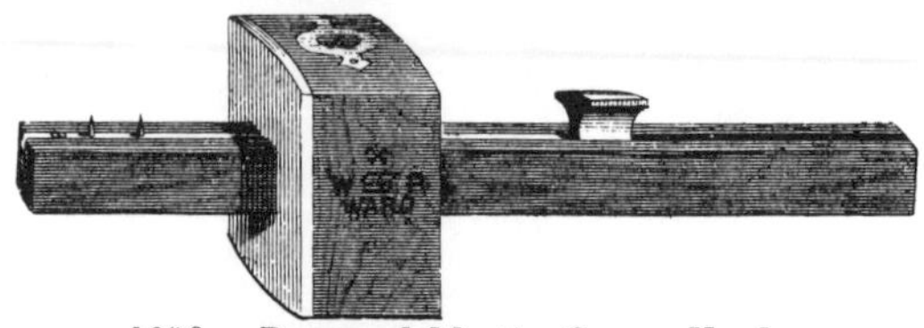

1096. *Rosewood Mortise Gauge, Head faced with Brass.*

1097. *Rosewood Thumbscrew Slide Mortise Gauge.*

1099. *Rosewood Thumbscrew Head and Slide Mortise Gauge.*

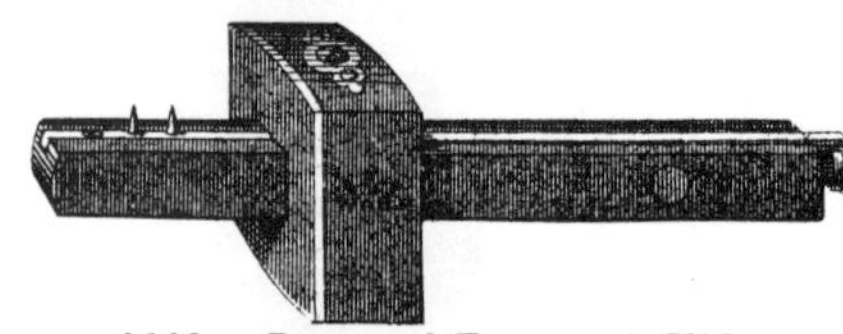

1101 *Rosewood Turnscrew Slide Head-faced Mortise Gauge.*

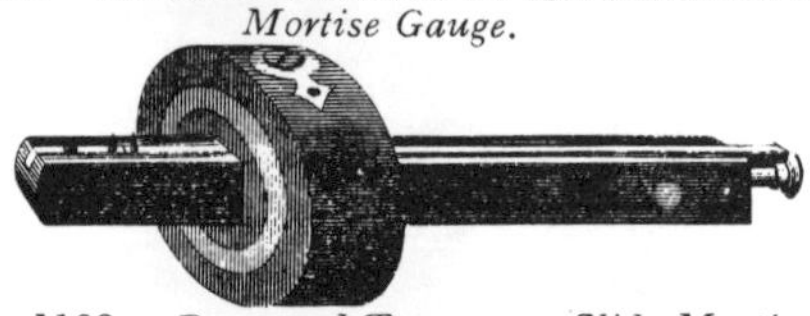

1103. *Rosewood Turnscrew Slide Mortise Gauge, Oval Ring Plated.*

1114. *Ebony Brass Stem Mortise Gauge, with Oval Head.*

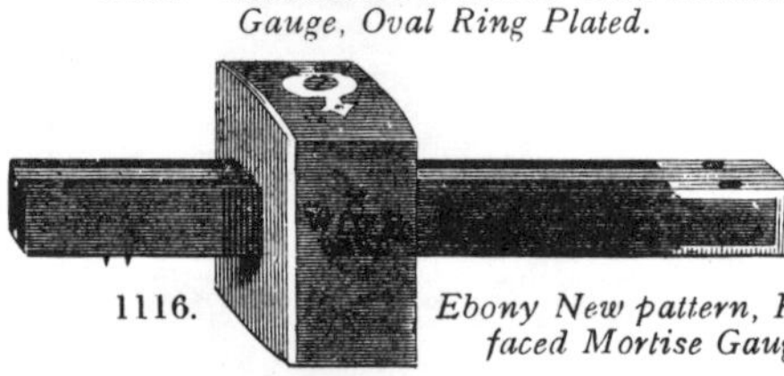

1116. *Ebony New pattern, Head-faced Mortise Gauge.*

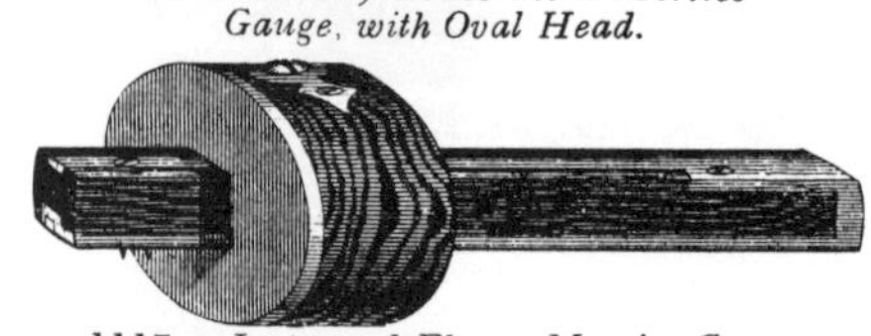

1117. *Improved Ebony Mortise Gauge, with Oval Head.*

1118. *Improved Ebony Oval Head Mortise Gauge.*

1119. *Improved Ebony Oval Head 3-pin Mortise Gauge.*

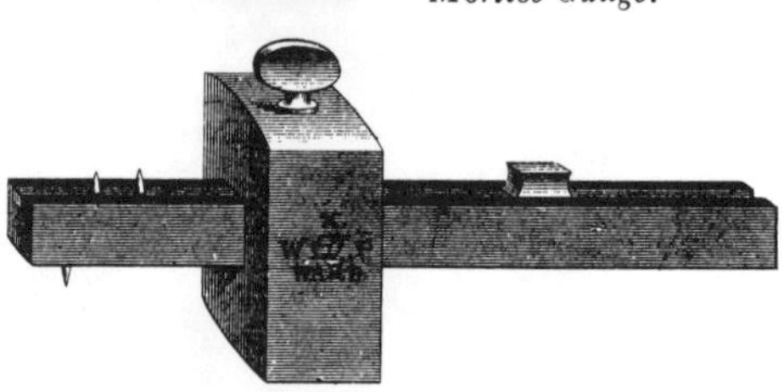

1121. *Canadian pattern Combination Gauge.*

1123. *Rosewood Combination Gauge, Thumbscrew end.*

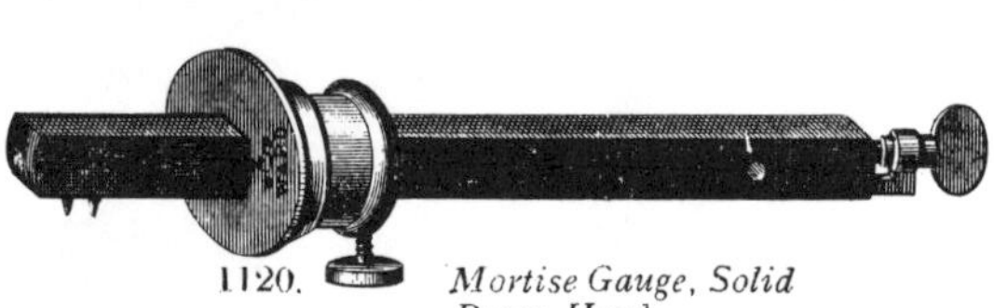

1120. *Mortise Gauge, Solid Brass Head.*

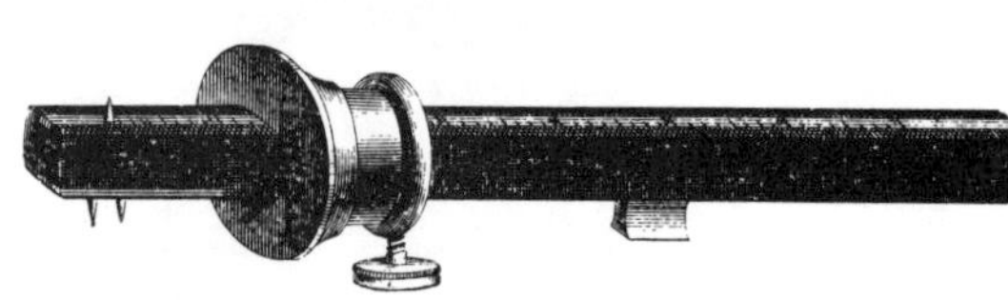

1124. *Ebony Combined Marking and Mortise Gauge.*

Plate CXLV - Marking & Mortice Gauages, page 85, Ward & Payne Catalogue, c.1900

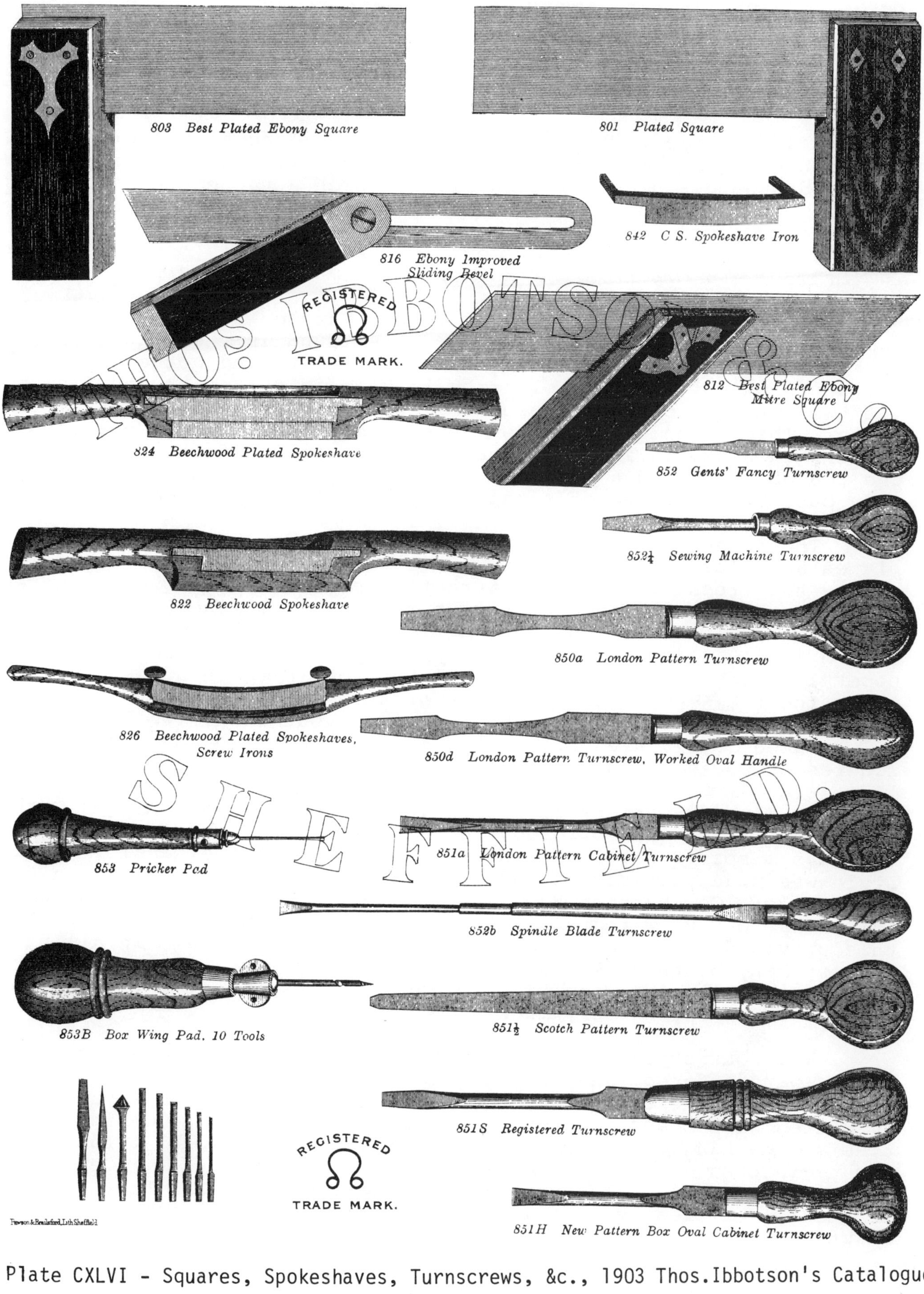

Plate CXLVI - Squares, Spokeshaves, Turnscrews, &c., 1903 Thos.Ibbotson's Catalogue

Examintation of artifacts is by far the best manner to appreciate both their utility and beauty. Several photographs of various joiners' tools follow (Figs. 13 - 30). It is suggested that these designs be compared with those previously illustrated appearing on the Sheffield Illustrated List and the 1870 Wm. Marples & Sons' Catalogue. (Plates CXXXVII - CXLIV)

A wide variety of plated squares are evidenced from Figs. 13 & 14. Function was combined with art. The brass plates set in routings on the wooden handle held the iron rivets which attached this to the steel blade. Numerous configurations of plating were offered. (See Fig. 14) For more complete information concerning the use and construction of these and other joiner tools, consult R.A. Salaman's *Dictionary of Tools used in the woodworking and allied trades, c1700-1970* [London, 1975].

The equipment formerly used at Wm. Marples & Sons to perform the routing operation on a wood handle for setting in the plates is shown in Fig. 15. The router was set in a brest plate and actuated by a bow. The depth of cut was controlled by the tip height of the cutting points. The two cutting points followed the contour of a hardened steel template. This technique had been developed in the cutlery business for setting in a sheild into the pearl or bone scale of the handle.[11] The router is called a parcee or parcer. For squares the usual plates were cast brass, filed on a taper to set into the routed depression.

The rosewood, mahogany or ebony handles with the polished brass plates and blued tempered steel blade provided pleasing contrasting colors that resulted in very attractice products. The best squares were made with brass face plates on both edges of the handles, as seen by the mitre square at the bottom right of Fig. 13 and the bottom left square in Fig. 14. The square in the upper right of Fig. 13 and the upper left of Fig. 14 have brass plates set in the handles about 2/3 along their lengths. The function of this is to provide a support for the handle to rest along an edge during layout work. Two bevels, formerly called bevil, are illustrated in Fig. 18. Two photo-etched squares by David Flather and also Henry Brown & Sons are seen in Figs. 15 & 16.

A wide variety of spirit levels, sometimes combined with plumbs and rules were offered, see Fig. 19. The square at the upper left has a spirit level within the inner edge of the handle. This style was made as early as 1837, as a similar design appeared in the 1837 Cutler & Co. Pattern Book. The level at the extreme right has a rule that may be withdrawn, held below the top plate. When pushed in, this provided a protecting cover for the glass bubble. In all probability many of these levels and squares were made in Birmingham by the firms: John Rabone & Sons, Edw. Preston & Sons and I. & D. Smallwood. Rules were not extensively made at Sheffield.

Numerous styles and designs of marking, mortise[12] and cutting gauges were offered, several of which are illustrated in Figs. 20 & 21. These were made with square, round or oval heads, which were frequently plated with brass to provide for wear resistance. Combination mortise and marking gauges were also manufactured. Fig. 22 illustrates two registered designs for mortice gauges by Fenton & Marsden and William Marples , also see Plates CXLVIII and CXLIX. The nature and protection offered by a Registered Design will be discussed in the Chapter concerning braces.

Fig. 13. - 3½" to 20½" Squares and Two Mitre Squares

Fig. 14. - Various Designs of Plated Squares

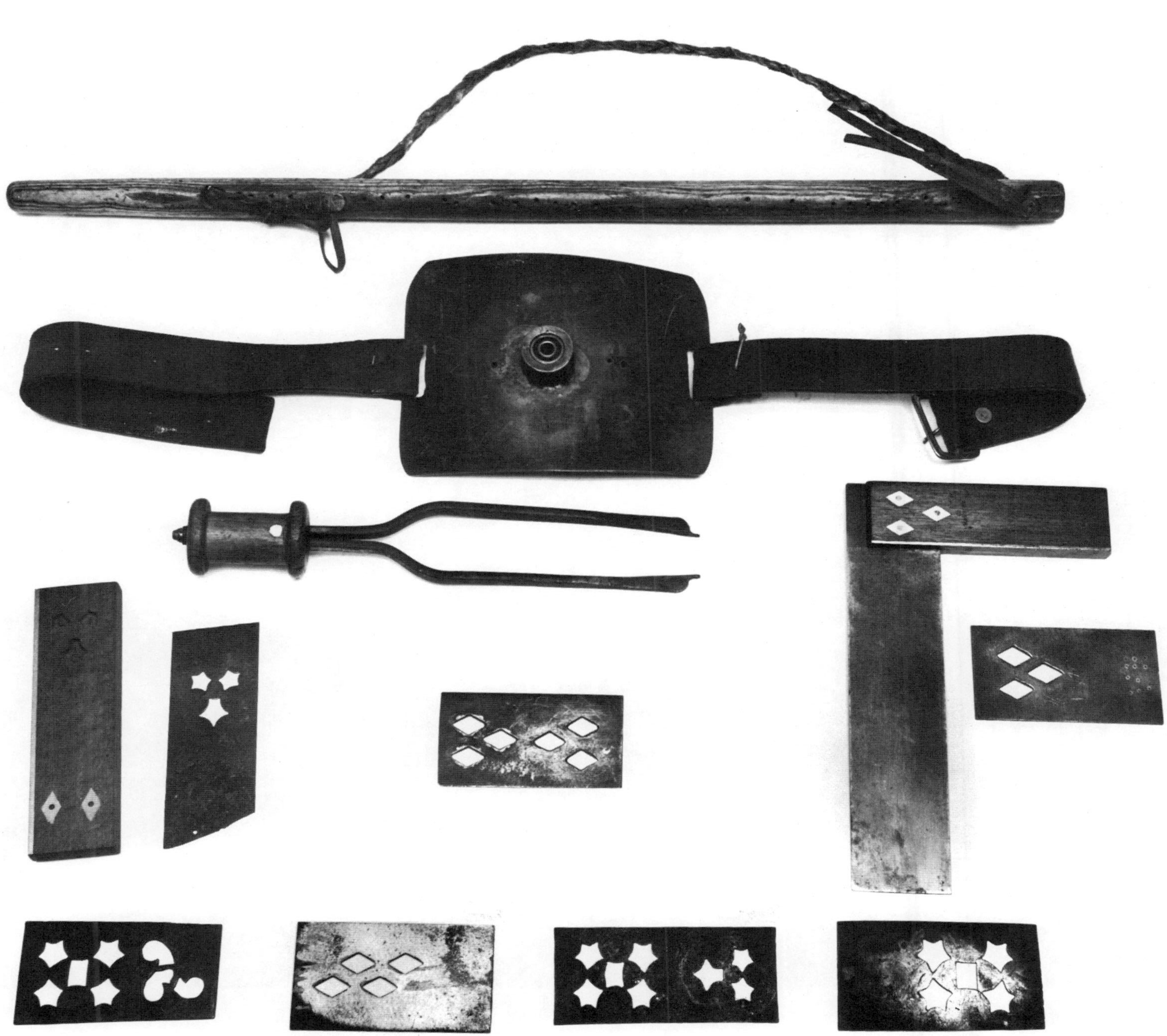

Fig. 15. - Bow, Brest Plate, Parcee, Templates formerly used by Wm. Marples & Sons for Square Plates
[Collection of Donald Wing]

Fig. 16. - Photo-etching on Blades of Squares, see Fig. 17
Top by Henry Brown & Sons; bottom by David Flather

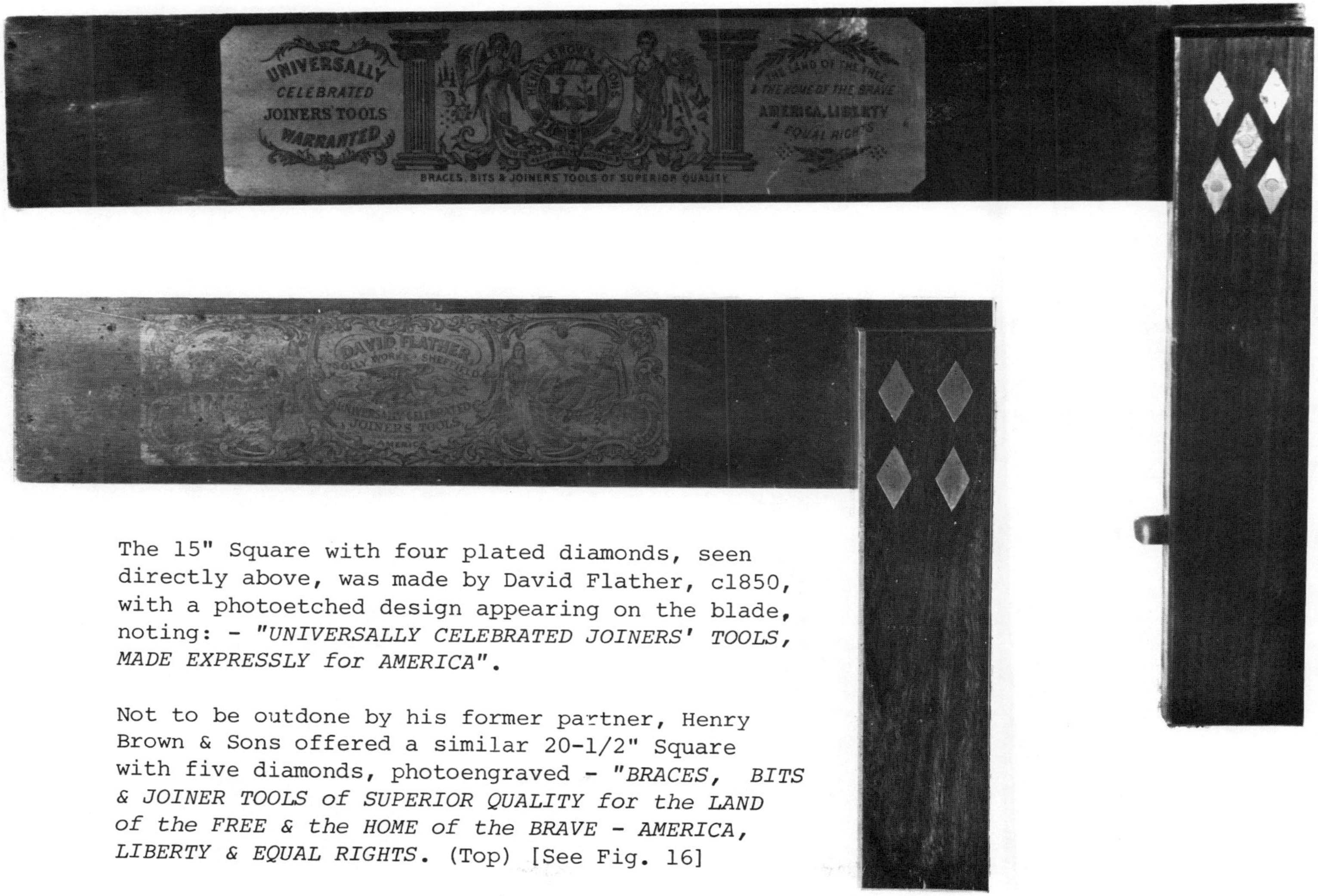

The 15" Square with four plated diamonds, seen directly above, was made by David Flather, c1850, with a photoetched design appearing on the blade, noting: - *"UNIVERSALLY CELEBRATED JOINERS' TOOLS, MADE EXPRESSLY for AMERICA"*.

Not to be outdone by his former partner, Henry Brown & Sons offered a similar 20-1/2" Square with five diamonds, photoengraved - *"BRACES, BITS & JOINER TOOLS of SUPERIOR QUALITY for the LAND of the FREE & the HOME of the BRAVE - AMERICA, LIBERTY & EQUAL RIGHTS*. (Top) [See Fig. 16]

Fig. 17. Photo-etched Squares. Top by Henry Brown & Sons; bottom by David Flather Details of Etching shown in Fig. 16. [Note brass plate support on handle, top square]

Fig. 18 - 20" and 6" Sliding Bevels in Rosewood and Brass Caps

Fig. 19. - Spirit Levels, Rule Levels and Plumb Level

Fig. 20. - Marking and Mortise Gauges

Fig. 21. - Marking and Mortise Gauges

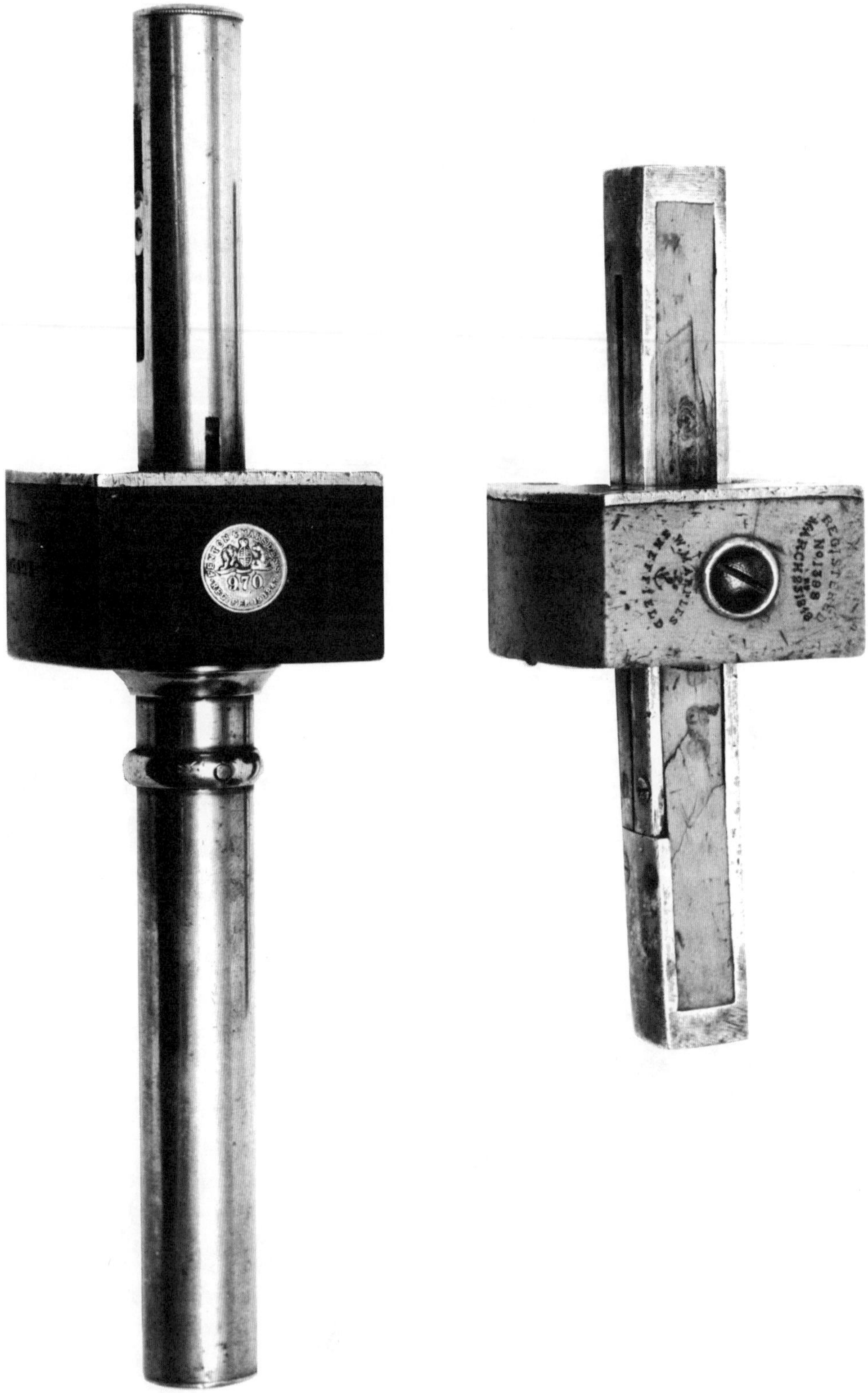

Fig. 22. - Registered Designs of Mortice Gauges: No. 970 Fenton & Marden, (left); No. 1398 Wm. Marples & Sons, (right)
[Fenton & Marsden Gauge Collection of James Cooley]

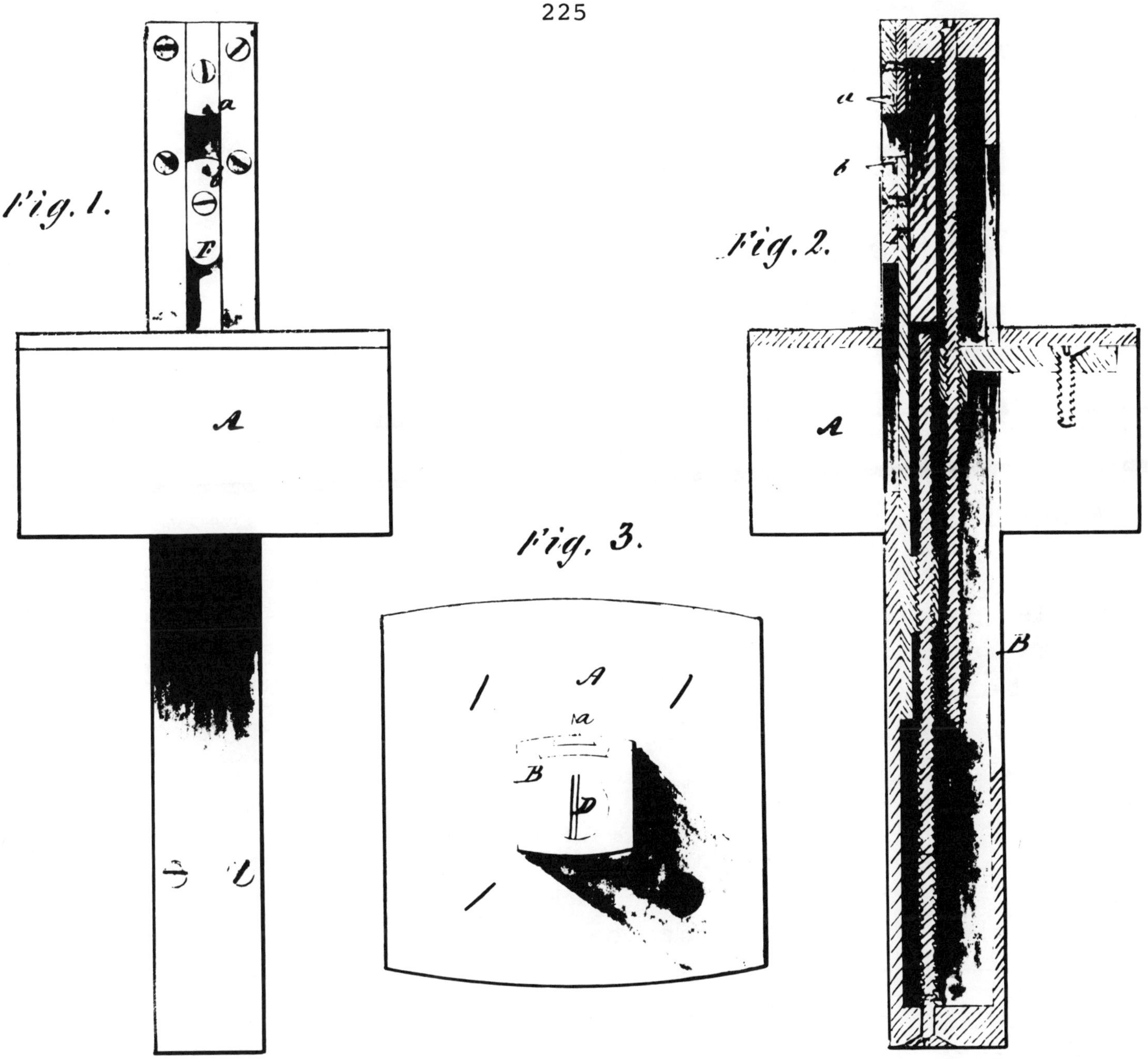

Plate CXLVIII - William Marples' Registered Design No.1398 for Mortice Gauge

Description - Figure 1 is a plan, Figure 2 a longitudinal section and Figure 3 an end view of this gauge. A is the head, B the stem which is formed partly of wood and partly of metal, the wooden part C is hollowed out for the reception of two screws D and E, the former of which (D) acts upon the head A so that it may be screwed further up or down upon the stem B and thereby regulate the distance of the mortice from the edge of the wood; the other screw (E) acts upon the sliding piece F and enables the gauge points a,b to be set to the size of the morticing chisel; the heads of the screws D and E are flush with the ends of the stem and can only be turned by a screw driver, thus there is no liability of the gauge being altered by accident. The parts of this design which are not new in so far as the shape and configuration thereof are all the parts in excepting those marked A B & D.

Class BT45/7 William Marples of Spring Lane, Sheffield March 23, 1848

FENTON AND MARSDEN'S
REGISTERED MORTICE GAUGE, No. 970,

MANUFACTURED BY

MARSDEN BROTHERS AND SILVERWOOD,

Edge and Joiners' Tool Manufacturers,

SHEFFIELD.

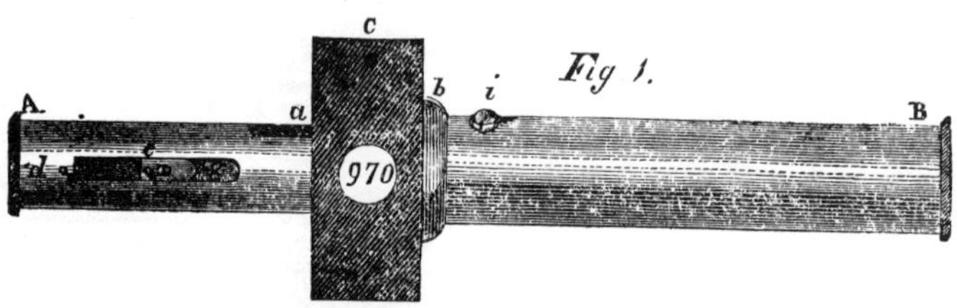

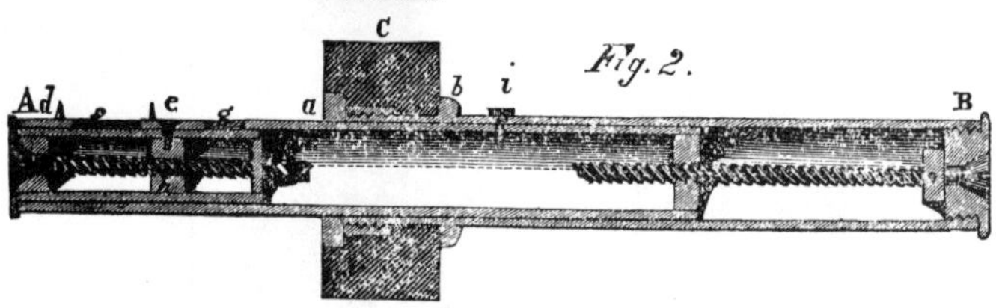

Fig. 1—*A a*, *B b*, is the Stem, which instead of being a square piece of wood, as heretofore, is formed of two hollow Brass Tubes, the one, *A*, sliding into the other, B. The Head, C, of Ebony, or other hard wood, as usual, furnished internally with a Brass Nut, by which it is screwed tight up to the Shoulder, *b*, on the Tube, B; *d* iss a Marking Point fixed to the Tube *A*; *E* is a second Marking Point, which traverses in the Slot *f*, *g*, and is attached by a small Screw to an internal Nut, *h*, shewn in Section Fig. 2. By turning the Screw at B the Tube *A* is drawn into, or forced out of, the Tube B, and the Marking Point, *d*, thereby brought nearer to, or farther from the Head, *c*. A similar Screw at the end, *A*, gives a backward or forward motion to the Marking Point *e*, and so by the movement of these two Screws the Marking points can be set to any required distance with great facility, and when set they cannot be accidently shifted in using. The arrangements for effecting this object are plainly shewn in Section Fig. 2. The Tubes are prevented from turning in each other by a Screw, *i*, in the Tube B, which traverses a Slot in the inner Tube, *A*.

The Proprietors and Manufacturers of the above Mortice Gauge feel confident that the preference will be claimed by such Workmen as value CORRECTNESS with ECONOMY OF TIME AND LABOUR. The advantage it possesses over all others, is that which is so essential in this article—when once fixed in the required position

IT CANNOT BE ALTERED BY ACCIDENT,

which the ordinary Gauges are liable to, thereby causing serious inconvenience, which, by this invaluable invention, is altogether avoided, and

CORRECT WORKMANSHIP INSURED.

Plate CXLIX - Fenton & Marsden's Registered Design No.980 for a Mortice Gauge
[Registered March 4,1847, Class BT45/5] Advertisement from 1849 Sheffield Directory

Another graceful tool usually made from beech, but also offered in boxwood and sometimes in rosewood, was the spokeshave. These were produced by bow sawing blanks and shaping with a draw knife. The mouth was shaped using a back saw and chisel. Frequently a strip of brass or bone was inserted and screwed across the mouth forming a plated spoke-shave with improved wear resistance. The iron was set in, having been formed with tapered tails. A more sophisticated form employed brass wing nuts, applied to threaded tails of the iron. (see lower left and lower center, Fig. 23) A modification of this, employing screws tapped into a straight iron, was obtained in a registered design No.12169 by William Marples & Sons, see lower right Fig. 23. Other joiners' tools illustrated in Fig. 23 include a rosewood T-square with adjustable bevel (left), mahogany scratch stock (top), rosewood panel gauge (right), stair rail shaves in boxwood (adjacent to the T-square and panel gauge) and a nest of spokeshaves in the center. Additional spokeshaves are illustrated in Figs. 24 & 25. Fig. 25 also includes some turnscrews (screw drivers) and two push drills, adjacent to the spokeshaves at the right and left ends.

Fig. 23. - Spokeshaves and Miscellaneous Joiners' Tools [Left - Rosewood T-Square with Adjustable Bevel; Left - Rosewood Panel Gauge; Top - Mahogany Scratch Stock; Right & Left Adjacent to the T-Square and Panel Gauge - Stair Rail Shaves; Center and Bottom - Miscellaneous Spokeshaves.

See Text Page 226

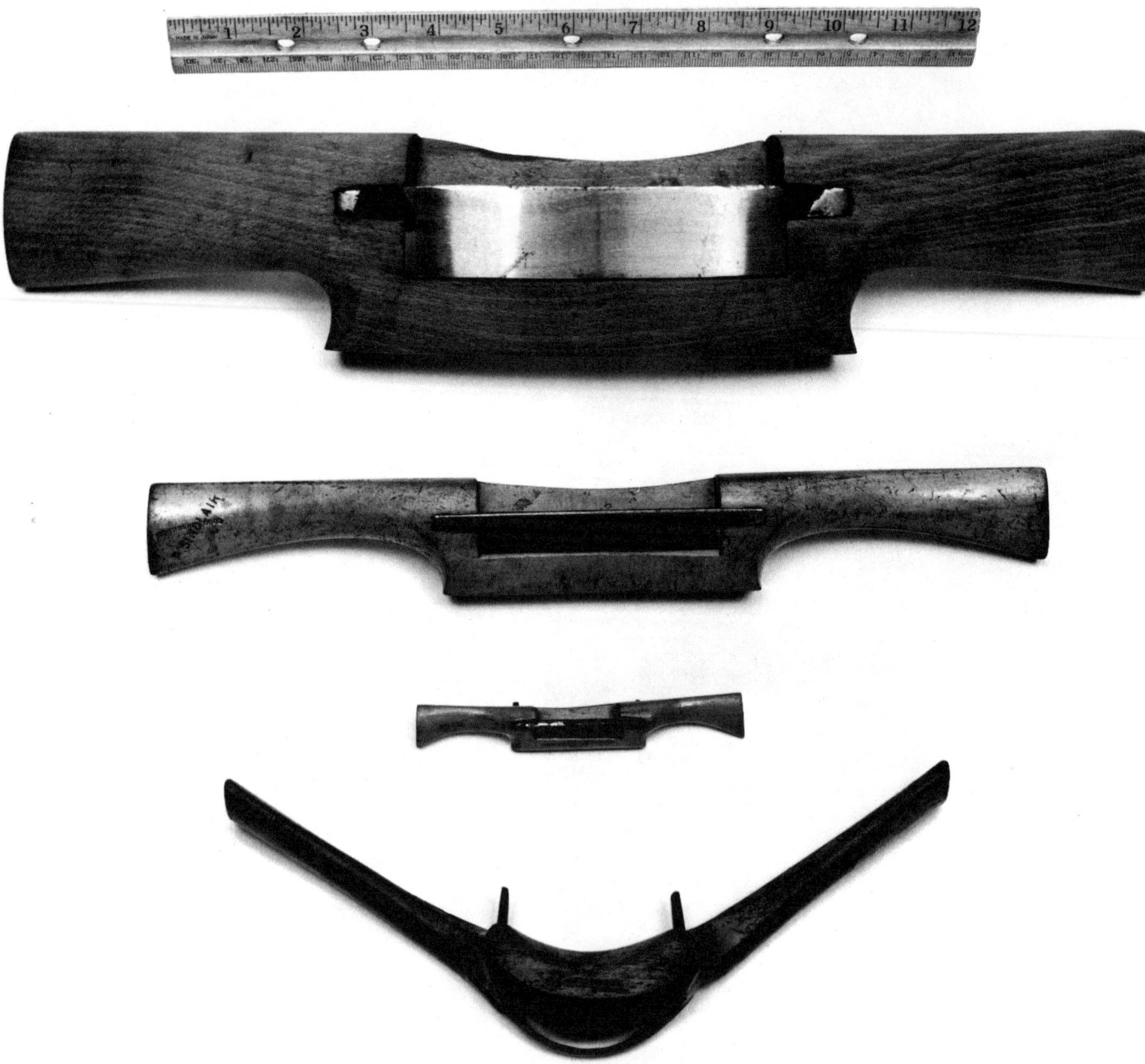

Fig. 24 - Spokeshaves [From Top to Bottom - 16" Coopers' Spokeshave in Beech; 11-1/2" Conventional joiner's Spokeshave in Boxwood; 3-1/2" Cabinetmaker's Spokeshave in Boxwood; Cabinet Maker's Curved Sopkeshave in Boxwood]

Some miscellaneous joiners' tools, including three back or mitre saws are shown in Fig. 26. A wider variety of back saws, stiffened with brass, which contributed to their attractive appearance as well as being functional, are illustrated in Fig. 28. The unique bow saw shown in this illustration was patented by William Cotter[14] of Bloomsbury (London), September 10,1872 (Letters Patent No. 1756).* His claims were achievement of deeper cuts by elimination of the usual stretcher bar, a spindle locking arrangement for fixing the saw blade, tension rods, and a screw ferrule arrangement for regulating the tension of the saw blade. This saw was made by Schlesinger Co. of London. A conventional bow saw along with several saw sets, a veneer saw and two hand pad saws are shown in Fig. 29. A 9" mitre saw with a unique mahogany case, made from a single piece of wood, is shown in Fig. 27.

*[A copy of the Cotter Patent Drawing is shown in Plate CLI, page 236]

Fig. 25. - Turnscrews (Screwdrivers), Push Drills and Spokeshaves

Fig. 26. - Miscellaneous Joiners' Tools. Center - Three Mitre Saws; Top (Right & Left) Ebony Filled Brass Plated, Sliding Bevels made by J. Howarth & Sons; Right Center - Ebony Filled, Brass Plated Mitre Square; Bottom Right - Brass Adjustable Bevel (probably for a mason)

Fig. 27. - Unique Case, Made from Solid Mahogany, and 9" Mitre Saw
Top - Saw in Case; Middle - - Case with Saw Withdrawn; Bottom - Saw

Several of the saw sets illustrated in Fig. 29 appear on page 15 of the Wm. Marples & Sons 1870 Catalogue (Plate CL). The saw set in Fig. 29 in the upper right part of the bow saw has a turn screw tip; similar to Wm. Marples #539; the one at the bottom center has a slide guard, similar to Wm. Marples #541. The one within the bow saw at the lower left is referred to in the Wm. Marples Catalogue as "Patent Plyer Saw Set", #545. This appears to be a copy from the engraving in the Richard Timmins & Sons Pattern Book (Birmingham, c1845), in which this is referred to as "American or Patent Sawset" The American Patent is not known and is assumed among those destroyed in the 1836 fire at the Patent Office in Washington, D.C. In all probability the saw set #545 offered by Wm. Marples & Sons was made in Birmingham, as was likely the case of the one in Fig. 29. The saw set to the right of this in Fig.29 also has a turn screw tip.

Fig. 28. - Wm. Cotter's Patented Bow Saw and Several Mitre Saws. (See Text, page 228)
[See Copy of Cotter's Patent Drawing No. 1750, June 11,1872,Plate CLI, page 236)

Fig. 29 - Conventional Bow Saw, Various Sawsets, Veneer Saw and Two Hand Pad Saws.

WILLIAM MARPLES & SONS, SHEFFIELD.

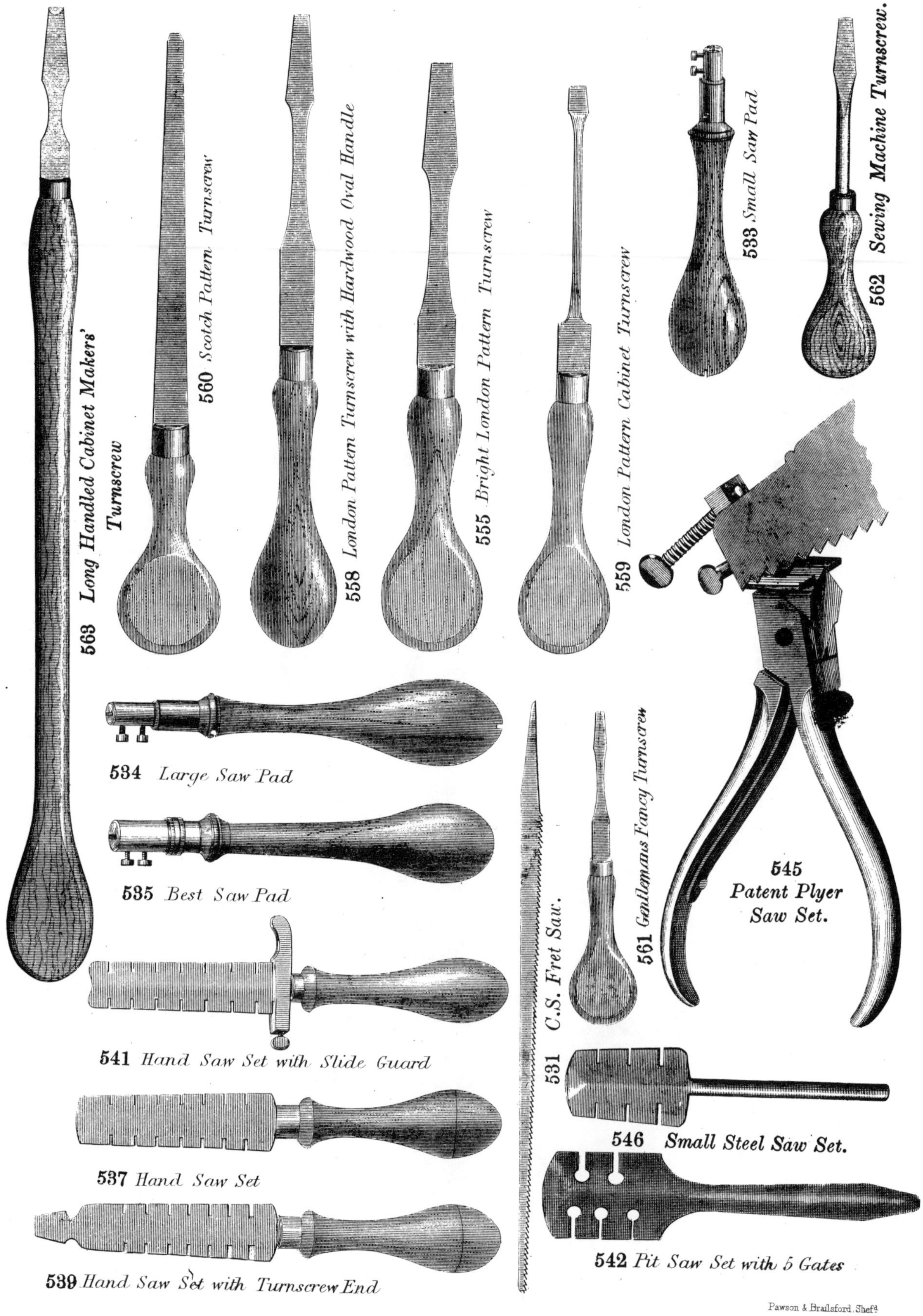

Pawson & Brailsford, Shefd.

Plate CL - Saw Sets and Turnscrews [Srewdrivers], 1870 Wm.Marples & Sons' Catalogue

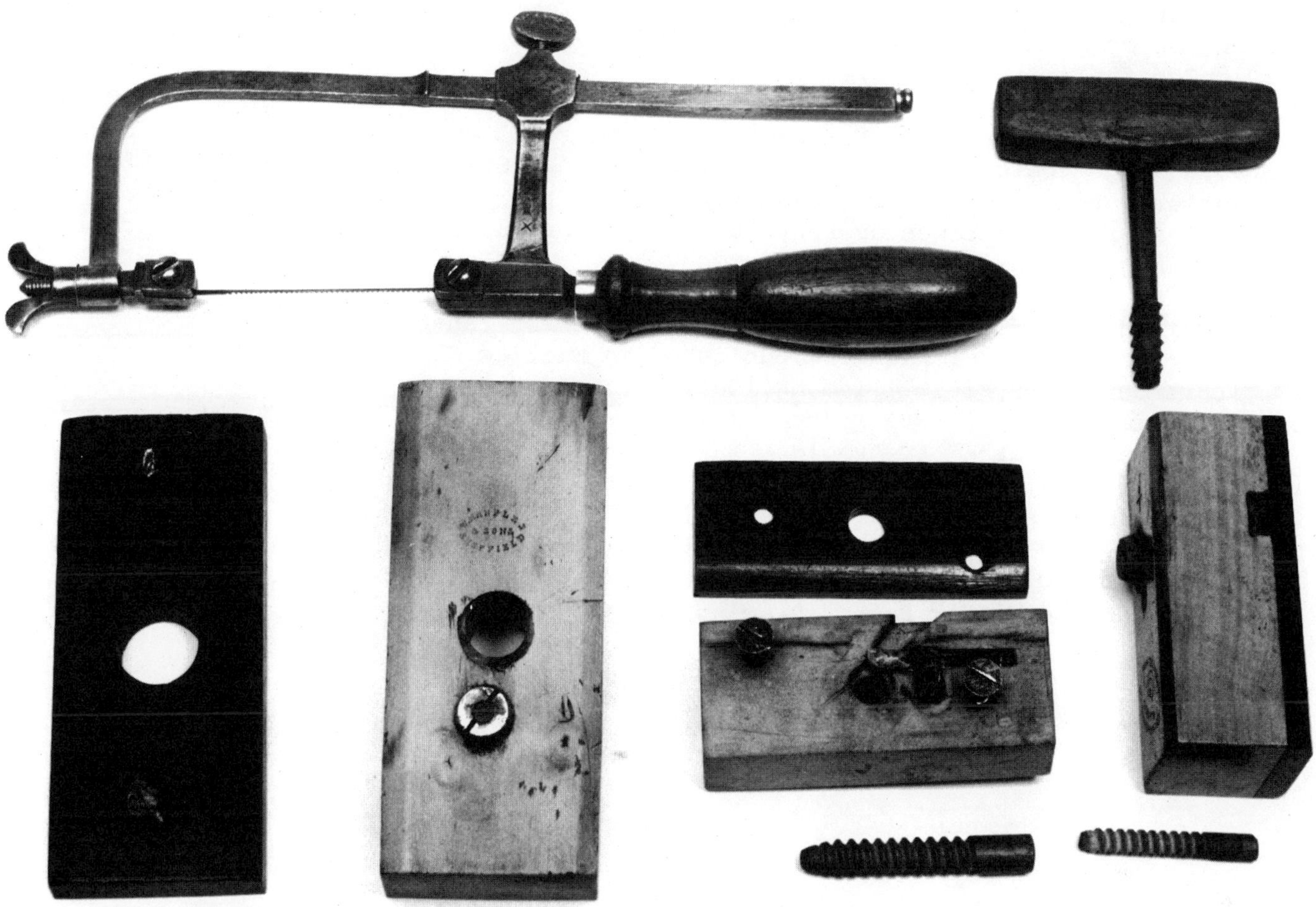

Fig. 30 - Jeweler's Saw and Screw Boxes and Tap by Wm. Marples & Sons

Three screw boxes are shown at the bottom of Fig. 30. The one at the extreme right is intact, while the other two (left and center) have been taken apart by removing the two screws holding the two sections together. Note that heavier section holding the cutting knife is made of box wood, while the entering section is made of rosewood. This makes an attractive design and assures the user of readily knowing which end to enter the dowel to start a thread.

1. *Tools for the Trades & Crafts, op.cit.*,p.17
2. *ibid,* The engraving at the bottom of the photograph is Plate 8 (p.56 *TT&C*) believed to have originally been engraved around 1810. See discussion page 13.
3. *Century of Progress Yorkshire Industry and Commerce* [London,1893], p.120
4. It was my pleasure to examine these while visiting Robert Sorby & Sons. Ltd. in Sheffield, April 1980.
5. *The Company of Cutlers in Hallamshire in the County of York*,[Sheffield,1978],p.29
6. Spear and Jackson claim tenure to 1760 and still operate at Sheffield, but only added edge tools to their saws in 1845. Robert Sorby claim tenure to 1624 by cutlery, but Robert Sorby began making edge tools in 1828. Although Robert Sorby & Sons, Ltd. was taken over by Hattersly & Davidson in 1923, the firm still is managed as and independent company.
7. *Century of Progress Yorkshire Industry and Commerce, op. cit.*, p.112
8. M.Walton, *Sheffield - Its Story and Achievement*,[Sheffield,1948],p.195
9. Wilk's prices of planes were published in *William Chapple 1876 Price List of Planes*, [this publisher,Fitzwilliam,NH,1980], p.15-20
10. A *facsimile* of the *1889 Edition Illustrated Trade List of Prices of Sheffield Goods* was published by Arnold & Walker, London, December 1974
11. C.A.Turner, *A Sheffield Herritage. An anthology of photographs and words of the cutlery craftsmen.* [Sheffield, 1978], p.31
12. The preferred Sheffield spelling during the 19th Century was"MORTICE"
13. *Tools for the Trades & Crafts, op. cit.*, p.108,Plate 52, Item#670A.
14. A Copy of the Cotter Patent Drawings,Figs.1-5 are shown below.(Plate CLI)

FIG. 1.

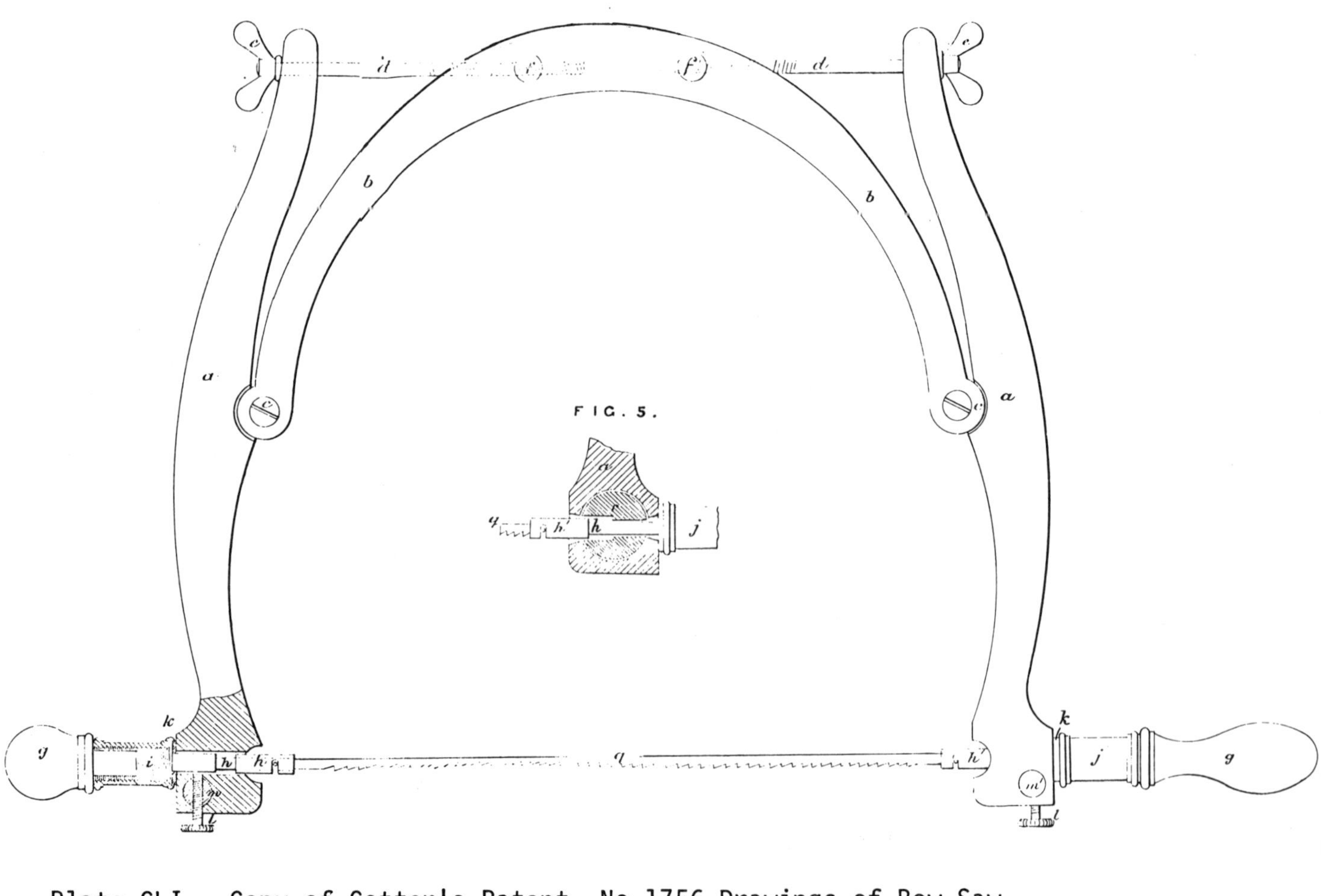

Plate CLI - Copy of Cotter's Patent No.1756 Drawings of Bow Saw
[Courtesy of Her Majesty's Patent Office, Orpington, Kent, England]

APPENDIX III

CHECK LIST OF 19th Century Sheffield Joiner Tool Makers [See Notes at End of List]

ABBEY, Charles [B]	Division St.		-1828-1830-
ADSETTS, William	Burgess St.	*Pool Works*	-1854-1857
	South St.		-1859-
ALLCARD, Clarence Wm.	Union Lane		-1865-1884-
ALLCARD, William	Norfolk Lane	[forger]	-1863-
	Pond St.		-1891-
ALLEN, J. [B]	Rockingham St.		-1833-1837-
ARNOLD, Charles	Love St.		-1859-
	Eldon St.		-1862-1865-
	Bridge St.		-1879-1884-
ASH, William [B,E]	Norfolk Lane		-1825-1841-
ATKINSON, BROTHERS [B,E,S]	Chester Lane		-1868-1879-
ATKINSON, Charles [B,E]	West Street		-1854-
	Trippet Lane		-1856-1857-
	Chester Lane		-1859-1865-
	" "		-1884-1900-
ATKINSON, Eustace Carey	Monuement St.		-1884-
	Division St.		-1887-
	Devonshire Lane		-1891-
	Rockingham St.		-1893-1900-
AXE, George	Pepper Alley		-1825-1837-
AXE, John	Pepper Alley		-1845-1863-
AXE, George & Thomas	Pepper Alley		-1841-
BAKER, John & Sonns [E]	Monument St.	*Monument Works*	-1876-1881-
	Harmer Lane		-1884-1900-
BARNES, Frederick & Co. [S]	Solly St.	*Corn Hill Works*	-1865-1900-
BARBER, Ralph	Bailey Lane		-1825-1828-
	Brook St.		-1845-1860-
BARLOW & HOBSON [B]	Highfield		-1857-1863-
	London Road		-1863-1872-
BATTIE, John & Son [E]	Bernard St.	*Pacific Works*	-1884-1891-
BAXBY, John [B]	Carver St.	*Dilligence Works*	-1860-1872-
	Broad Lane		-1876-1887-
BAXBY, William [B]	St. Thomas St.	*Enterprise Works*	-1876-1881-
BEE, James [B]	Townhead St.		-1814-1828
BIRDS, George	Backfield		-1854-1862-
	Cambridge St.		-1863-1872-

BIRTLES, William [B]	St. John's Road		-1884-
	Nunnery St.		-1891-
	Lord St.		-1893-1896-
BLACKTIN, Henry	Broad Lane		-1871-1894-
BLOOMER, John	Church St.		-1830-
	Rockingham St.		-1833-1845-
BLOOMER & PHILIPPS [B]	Brook St.		-1849-
	Bailey Lane		-1852-
BLYDE, Edwin & Co.	Lambert St.	*Charleston Works*	-1884-1896-
BRADSHAW, William	Eyre Lane		-1881-1891-
	Hermitage St.		-1893-1900-
BROOKES BROTHERS [S]	Eldon St.		-1856-1865-
BROOKES, William & Son [E,S]	Rockingham St.		-1845-
" " & Sons [E,S]	Carlisle St.	*Howard Works*	-1852-1865
	Attercliffe		-1871-1884-
	Woodbury St.		-1891-1900-
BROWN & FLATHER [B]	Fargate	*King's Arms*	-1828-1830-
	Hollis Croft		-1833-
	Solly St.	*Solly Works*	-1837-1849-
BROWN, Hnery & Sons [B,E,S]	Rockingham St.	*Western Works*	-1852-1887-
	Sorby St.		-1889-1900-
BRUFTON, John Thomas [B]	Fitzwilliam St.		-1881-1887-
BRUMBY & MIDDLETON [E,S]	Lambert St.		-1893-1896-
BURGESS, Henry	Monmouth St.		-1881-
	Broomhall St.		-1884-1896-
BURKINSHAW, William & Son [B]	Union Lane	*Burgess Works*	-1876-1881-
	Eyre Lane		-1884-1888-
BURKINSHAW, William	Eyre Street		-1893-1900-
BURYS & Co. [E,S]	Penistone Road		-1863-1900-
BUTCHER, William [E]	Eyre Street		-1821-1825-
BUTCHER, Wm. & Samuel [E,S]	Furnival St.		-1857-1899-
BUTCHER, BROWN & BUTCHER [E,S]	Eyre Lane		-1828-
CAM, James [E,S]	Howard St.		-1833-
CAM, Joseph	Charles St.		-1849-
	Eyre Lane		-1852-
CAM, Joseph & Sons [E]	Union St.		-1856-1861
	Trafalgar St.		-1863-1865-
	Button Lane	*Lawrence Works*	-1868-1900-

CAM, John	Arundel Lane Eyre Lane		-1837- -1841-1852-
CAMMEL, Charles & Co. [E]	Savile St.	*Cyclops Works*	-1854-
CANDOW, Walter	Tudor St.		-1845-
CAWTHORPE, George	Egerton St.		-1876-1881-
CHAPMAN, James [B]	Woodside Lane	*Industry Tool Works*	-1868-1900-
CHUMBLEY & EDLEY	Porter St.		-1825-
CHUMBLEY, EDLEY & HAWKSWORTH [B]			-1830-
CLEGG & BARTON	Sylvester Lane		-1884-1891-
COLQUHOUN & CADMAN [B.E]	Arundel St.	*Douglas Works*	-1881-1900-
COOPER, Culthbert	Hermitage St.	*Nelson Works*	-1879-1881-
COTTAM, William & Co. " " & Sons [B]	Thomas St. New Edward St. Shoreham St.		-1868- -1871-1872- -1887-1891-
CRADDOCK, Joseph	Gell St. Hedford St.	*Ebor Works*	-1891- -1893-1900-
CURTIS, James & Co.	Wicker Lane		-1849-
CURTIS & HOUSLEY	Wicker Lane Trafalgar St.	*Lever Works*	-1852- -1854-
DALTON, P. & CO.	Norfolk Lane		-1849-1852-
DARWIN, Samuel & Sons	Shude Hill		-1825-1828-
DAVENPORT, James [S]	Napier St. Headford St. Napier St.	*Clifton Works*	-1879- -1884-1891- -1895-1899-
DeGARRS, Henry	Egerton St. Harwood St.	*Exchange Works*	-1876-1879- -1887-1891-
DeGARRS & ALLEN	Harwood St.		-1893-1900-
DIXON, James	Norfolk St.		-1814-
DRABBLE, John	Duke St. West Hill Lane Carver Lane		-1841-1845- -1861- -1862-1871-
EATON, W.H. & T.		*Park Works*	-1837-
EADON, Moses [E.S] " " & Sons [E,S]	Norfolk Lane Savile St.	*President Steel Works*	-1841-1852- -1856-1865-
EDLY, William [B]	Porter St. Eldon St. Trafalgar St.		-1837- -1845- -1849-1852-

ELLIS, George	Carver St.		-1854-
	Ecclesall Road		-1856-
	Carver St.		-1857-1861-
" " & Sons [B]	Coalpit Lane		-1862-1865-
	Holly Lane		-1868-1872-
	Rockingham St.		-1876-1879-
	Edgerton St.	*Exchange Works*	-1884-1891-
ELLIS, George	Edgerton St.		-1895-1899-
ELLIS, John Henry	Jepson St.	*Jepson's Wheel*	-1893-1894-
	St. Mary's Road		-1895-1900-
ELLIS, Samuel	Brown St.	*Ellis's Wheel*	-1876-1896-
FARR, John	Young St.		-1841-
	New Road		-1845-
	Eccleshall Rd.		-1849-1860-
FELLOWS, William	Campo Lane		-1845-
	Chester Lane		-1849-1852-
	Thomas St.		-1857-1872-
FENTON & MARSDEN [B.E]	Rockingham St.		-1841-
	Bridge St.		-1845-
FLATHER, David [B]	Solly St.	*Solly Works*	-1852-1854-
" " & Sons [B,E]	" "		-1856-1900-
FLETCHER, John [E,B]	New St. Lane		-1833-
	High St.		-1837-
	Broad St.		-1845-
FURNISS, Chas.	Rockingham Lane		-1849-1860-
	Union Lane	*Wentworth Works*	-1861-
	Burgess St.		-1868-1872-
FURNISS, Thomas	Coalpit Lane		-1837-
GILLOT, William	Bard St.		-1884-1900-
GLAURET, L. & C. [S]	Love St.		-1891-
	Furnival St.	*Wallace Works*	-1893-1900-
GREAVES, Isaac [E]	Broad St. Lane	*Park Works*	-1865-1881-
GREEN, William [B]	Westbar Green		-1852-
	Cleveland Place		-1856-
	Portland St.		-1859-
GREENWOOD, Frederick [B]	Eldon St.		-1837-
GRAY, Charles & Sons	Watrey St.	*Philip's Works*	-1865-1889-
GROVES, R. & Sons [E,S]	Snow Lane		-1865-
	" " & Allen St.	*Beehive Works*	-1871-1884-
	Henry St.		-1891-
GUION, Hnery & Co. [B]	Eyre Lane		

HALE BROTHERS [E]	Allen St.	*Moorfield Works*	-1884-1891-
HALL, George	Granville St.		-1862-1871-
HANDLEY, Arthur	Meadow St.		-1893-1899-
HANDLEY, John	Fitzwilliam St.		-1893-1894-
HARDMAN, Arthur	Orange St. Love St.		-1881- -1891-
HARRIS, Frederick [B]	Trafalgar St. South St.	*Wells Works*	-1865- -1868-1884-
HARRISON, Frederick Charles	Trafalgar St.	*Trafalgar Works*	-1868-1872-
HARRISON, John & Sons [E]	Dronfield	*Unstone Mills*	-1891-1900-
HARRISON, William	Water St. Fargate		-1868- -1871-1872-
HARRISON BROTHERS	Furnival St. Union Lane		-1854- -1856-1857-
HAWKES, Samuel	Broomhill St. Bath St.		-1849- -1852-
HAWKSWORTH, Henry	Earl St.		-1833-1837-
HAYWOOD, CLARKE & CO. [B]	Rockingham St.		-1849-
HEARNSHAW BROTHERS [E]	Sorby St.	*Sorby Street Works*	-1879-1900-
HEATHCOTT & LINLEY	Fragate		-1841-
HERBERT, Joel	Harvest Lane		-1868-1872-
HERRING, John & Sons [E.S]	Broad St. Lane	*Navigation Works*	-1863-1900-
HEWITT, John & Co.	Cambridge St.	*Cambridge Steel Works*	-1876-1900-
HILL, Anthony [E]	Hoyle St.		-1825-
HOGAN, Dennis F.	Rockingham St.		-1891-1896-
HOLDEN, Nathan	Hollis Croft Mitchell St. Barber St.		-1868-1876- -1879-1884- -1887-
HORTON, David F. [B]	Woodland View		-1876-1881-
HORTON, George [B]	Thomas St.		-1849-1863-
HOWARTH, James [B,E,S]	Fitzwilliam Rd. Bath St.	*Broomspring Works*	-1856-1857 -1860-1862-
" " & Sons [B,E,S]	Bath St.		-1863-1913-

HURST, Jno & Son [B]	Allen St.		-1849-1852-
	Smithfield		-1854-1872-
	Vale Rd.		-1876-
HUNTER, Michael & Son [E]	Andrew St.		-1860-1863-
IBBOTSON, Henry	West St.		-1852-
	Edward St.		-1856-
IBBOTSON, Thomas [E,S]	Charles St.		-1821-1823-
" " & Co. [E,S]	" "	*Paternoster Row*	-1825-1905-
IBBOTSON & ROEBUCK			-1821-1825-
IBBOTSON & SAVILLE	Bridge St.		-1828-1830
IBBOTSON & WILSON	Eyre Lane		-1841-
JACKSON, Henry J.	Fitzwilliam St.		-1876-1879-
	Gell St.	*Ebor Works*	-1881-
JACKSON, William & Co. [B,E.S]	Sheaf Island	*Sheaf Island Works*	-1856-1857-
	Pond Hill		-1859-1861-
	Green Lane		-1887-1900-
JENKINSON, Amos & William	Eldon St.		-1854-1860-
JENKINSON, Amos [B]	Rockingham St.		-1862-1892-
" " & Son [B]	" "		-1893-1900-
JOWETT, John [E]	Countess Rd.		-1891-1900-
KIRK, Peter & Son	Norfolk Lane		-1845-
KIRK, Peter [E]	St, Mary's Place		-1862-1876-
LAW, Philip [E]	Carver St.		-1841-1856-
LINLEY, Geprge	Coalpit Lane		-1825-1828-
LOCKWOOD BROTHERS [E,S]	Arundel St.		-1884-
	Spital Hill	*Spital Hill Works*	-1859-1900-
MANSFIELD, Richard [B]	South St.		-1856-1863-
MARPLES, Charles	Button Lane		-1876-1879-
" " & Son	Carver St.		-1881-1891-
MARPLES, Isaac	Charlotte St.		-1841-
	Broomhall St.		-1852-1856-
	Monmouth St.		-1857-1860-
	Broomhall St.		-1861-1876-
MARPLES, Isacc & Albert Brothers	Broomspring Lane		-1879-1900-
MARPLES, Joseph, jr. [P]	Cambridge St.	*York Works*	-1868-1884-
MARPLES, Joseph & Son [B]	Carver St.		-1849-1872-

MARPLES, Joseph & William	Carver St.		-1876-1884-
MARPLES, Robert [B]	Sharrow Head		-1833-
	Hermitage St.	Hermitage Works	-1837-1860-
" " & Son [B,E]	" "		-1861-1931-
MARPLES, Wm. [sr.]	Sheffield Moor		-1814-
	Rockingham Lane		-1821-1828-
	Solly St.		-1833-1841-
MARPLES, Wm. [jr.] [B]	Broomhall St.		-1833-
	Broomspring Lane		-1837-1854-
	Westfield Terrace	Hibernia Works	-1856-1860-
" " & Sons [B,E,S]	" "		-1861-1972-
MARPLES, Wm.	Coalpit Lane		-1852-
" & MANSFIELD	Arundel		-1854-
" , Wm	Arundel		-1856-
MARSDEN BROTHERS & Co. [B.E]	Bridge St.	Bridge Street Works	-1854-1859-
MARSDEN BROTHERS & SILVERWOOD [B,E]	Bridge St.		-1849-1852-
MARSHALL & JOWETT	Arundel Lane		-1849-
MAW, Kirgarth	Sheffield Moor		-1814-1815-
MAW & Co. [E]	Rockingham St.		-1821-
MAW & STALEY [B.E]	Rockingham St.		-1825-1828-
MEMMOTT, Wm. [S]	Sylvester Lane		-1852-1856-
MITCHELL, Robert [S]	Carver St.		-1856-1859-
MITCHELL, W.B & Co. [S]		Suffolk Works	-1837-
MORRISON, John Carver [B]	Lee Croft		-1857-1900-
MORRISON & PARKER [B]	Rockingham Lane		-1845-1852-
MORT, Henry & Co. [E,S]	Castle Hill		-1854-
	Carver St.		-1859-
	Solly St.		-1863-1872-
MOULSON BROTHERS [B.E,S]	Tudor St.		-1828-
	Division St.		-1830-1891-
	Green Lane		-1892-1894-
MUNKS, John	St. Mary's Road		-1887-
	Napier St.		-1891-1899-
MURFIN BROTHERS	May St.		-1895-1899-
OAKES, William [E]	Spitalfield		-1857-
OAKES & SMITH	Spitalfield		-1852-1856-
OUTRAM, William [E]	Spitalfield		-1856-

OUTRAM, William & Son [E,S]		*Delhi Works*	-1857-1891-
PADLEY, John & Co.	Chester St.		
PARKIN, Charles [B,E]	West St. Lane		-1862-1863-
	Carver St.	*Carver Street Works*	-1865-1872-
	Rockingham St.		-1876-1881-
PARKIN, Charles & Son[B]	Charlotte St.		-1884-1910-
PARKIN, Henry [E]	Clarence St.		-1860-
	Carver St.		-1861-
PARKIN, POTTS & DENTON[E]	Arundel St.		-1823-
PARKIN & MARSHALL [E,S]	Furnival St.	*Telegraph Works*	-1865-1872-
PARKER & THOMPSON [B]	Rockingham St.		-1854-1861-
PASLEY, Richard	Rockingham St.		-1879-1884-
PASLEY, Henry [B]	Edward St.	*SHEFFIELD GENERAL LETTER WORKS*	-1884-1900-
PEARSON, F.G. [S]	Furnace Hill	*Hope Works*	-1865-1900
PENESTONE, WHEATCROFT & SINGLETON	Arundel St.		-1814-1828
PILKINGTON, PEDIGOR & STORR [B]	Chester Lane	*Chester Works*	-1854-
	Headford St.		-1856-
PILKINGTON, Charles & Thomas [B]	Headford St.		-1857-
	Fitzwilliam St.		-1859-1863-
PILKINGTON, Thomas Peacok [B]	Fitzwilliam St.		-1865-1872-
POPPLE, George W.	Division St.		-1879-1884-
	Norfolk		-1891-1900-
RATLEDGE G. & Sons	Hermitage St.		-1884-1887-
REANEY, William [E]	Bernard St.		-1863-
REANEY, Wm. & Sons [E]	Bernard St.		-1865-1871-
REANEY & WOOD [E,S]	Bernard St.	*BERNARD WORKS*	-1841-1861-
RIDGE, Alfred [B]	Ecclesfield	*Exchange Works*	-1854-1879-
	West St.		-1881-
RIDGE, Alfred & Sons [B]	Rockingham St.		-1884-1900-
RIDGE, William [B]	Carver Lane		-1856-
	Norfolk St.		-1857-1859-
	Carver Lane		-1860-1868
RIDGE, Mrs. Sarah Ann	Carver St.		-1871-1879-

RIDGE, William	Ecclesfield		-1881-1900-
RIMMINGTON, R.	Roberts St.		-1865-
	Brightmore St.	*Brightmore Works*	-1868-1872-
RIXHAM, Charles Hnery [B]	Cambridge St.		-1871-1884-
	Townhead St.		-1891-
ROBERTS & ASH	Eyre St.		-1814-1815-
ROCK, Thomas	Suffolk St.		-1852-
	Copper St.		-1854-
" " & Son [B]	" "		-1856-1865-
ROFFEY, Samuel	West St.		-1895-1900-
ROSS, David	Trippet Lane		-1868-1872-
" " & Sons	Court		-1891-
ROSS, Peter [B]	Townhead St.		-1856-1859-
" " & Sons [B]	" "		-1860-1876-
ROUND, William [B]	Bailey Lane		-1814-1815-
	Allen St.		-1821-
	Norfolk Lane		-1825-1830-
ROUND, G.	Allen St.		-1845-1852-
ROYLE, William	Button Lane		-1833-
	Angel Row		-1837-
	Norfolk Lane		-1852-1863-
SAVERSON, George	Sheffield Moor		-1814-1815-
SAUNDERSON, William	Peacroft		-1825-1828-
SAVILLE, Charles [E]	Rockingham Lane		-1833-
	South St.		-1837-1849-
	Eyre St.		-1852-
SCHOFIELD, George [B,E]	Norfolk St.		-1825-
	Sheldon Row		-1845-1857-
SCHOFIELD,George & John [B]	Norfolk Lane		-1833-1841-
SCHOFIELD,Isaac	Rockingham St.		-1865-
	Broomhill		-1876-
SCHNETZLER, COLLIG & Co.	Arundel		-1895-1899-
SHAW, Jabez	Brown St.		-1857-
	Westfield Terrace	*Hibernia Works*	-1857-
	Bowden St.		-1859-
	Westfield Terrace	*Hibernia Works*	-1860-1863-
SHAW, John	Wicker		-1814-1815-
SHELDON, Thomas & Son [E,S]	Trafalgar St.		-1856-1860-

SLACK, Robert Shaw [E]	Regent Terrace		-1865-
	Broomspring Lane	*Perserverance Works*	-1868-1900-
SLATER, John	Hollis Croft		-1825-1830-
SLATER, Joseph [B]	Rockingham St.		-1833-1837-
" . J.T. & J.	" "		-1841-
SMITH, S. & Sons [B]	Bowden St.		-1891-1899-
SMITH, Samuel	Bowden St.		-1887-
SMITH & SCHOFIELD [B]	Fitzwilliam St.		-1881-
	Bowden St.		-1884-
SORBY, John & Sons [E,S]	Spital Hill [Lockwood Bros,after 1861]		-1821-1900-
SORBY, Robert & Son [E,S]	Carver St.		-1849-1980-
STALEY, Thos. & Wm. [B]	Rockingham St.		-1830-
STALEY, William [B]	Rockingham St.		-1833-1837-
STANLEY, Charles	Carver St.		-1860-1863-
STANSFIELD & SHIPSTONE	Carver St.		-1881-1887-
STEELE, Charles	Trafalgar St.	*Trafalgar Wheel*	-1881-1884-
	Carver		-1887-
	Deveonshire		-1891-1900-
STONES, Frederick [B]	Commercial Place		-1821-1828-
	Arundel Lane	*Albion Work*	-1833-1841-
	Orchard Place		-1845-1849-
SWIFT, Michael	Coalpit Lane		-1828-1857-
SWINSCOW & MANUEL [B,E]	Fitzwilliam St,		-1841-
	Coalpit Lane		-1845-1852-
	Backfield		-1854-
	Broomhall St.		-1856-
TAYLOR, Henry [E,S]	Fitzwilliam Rd.		-1849-1859-
TAYLOR, Joseph	Solly St.	*Cambridge Works*	-1868-1872-
THOMPSON, Francis	Infirmary Rd.	*Gatefield Works*	-1865-
	Lower Allen St.		-1868-
	Allen St.		-1871-1872-
THOMPSON, Henry [B,E]	Broomhall St.		-1876-1891-
TURNER, NAYLOR & Co.	Castle Hill		-1868-1872-
" " & MARPLES	John St.	*Northern Tool Works*	-1876-1891-
" " & Co.	" "	" " "	-1893-1963-
TURNER, Thomas [E]	Queen St.	*Orion Works*	-1845-1879-
" , Mrs. Sarah	" "		-1881-1884-
" , Thomas	Leopold St.		-1887-1899-

TYE, William	Green Square		-1833-1841-
	Charles Lane		-1852-
	Brown St.		-1856-1860-
" " & Sons	" "		-1862-1872-
WALES, Thomas [B,E]	Pond St.		-1861-1891-
	Denby St.		-1895-1896-
	John St.		-1898-1900-
WARBURTON BROTHERS [E]	Furnival St.		-1845-1849-
	Eyre Lane		-1852-
WARBURTON, Charles [E]	Rockingham St.		-1856-1857-
WARD, David [E]	Trippet Lane		-1821-
" " & Son [E]	Portobello St.		-1825-1833-
" " & Co. [E]	" "	*Ward's Works*	-1841-
WARD & PAYNE [B,E,S]	West St.		-1845-1965-
WILSON, John [E]	Westfield Terrace		-1868-1872-
	Fitzwilliam St.		-1879-
	West St.	*Portland Works*	-1884-1896-
JOHN WILSON, MARSDEN BROTHERS	West St.	*Ortland Works*	-1897-1901-
WILSON, Samuel [B]	Bow St.		-1845-
	Trippet Lane		-1852-1856-
WILSON, HAWKSWORTH, ELLISON & Co. [S]	Arundel Lane		-1856-1857-
	Carlisle St.		-1859-1861-
	Effingham Road		-1863-1868-
	Carlisle St.		-1872-1879-
WOOD, D. & Co. [E]	Effingham Rd.		-1863-1865-
WOOD, John	Sheldon St.		-1891-1899-
WOOD, Joseph	Howard St.		-1854-
	Cumberland St.		-1856-
WOODCOCK, James [E]	Trafalgar St.		-1868-1872-
" " & Sons	Young St,		-1893-1900-
WORSTENHOLME, John	Smithfield		-1825-
	Moorfield		-1833-
WRIGHT, Benjamin	Briggs St.		-1887-
	Bridge St.		-1891-
WRIGHT, James [B,S]	Highfield		-1833-1837-
WRIGHT, J. & Sons	Trippet Lane		-1821-
	Highfield		-1825-
	George St.		-1830-
WIRGHT & Son	Highfield		-1814-1816

WRIGHT, Samuel [E]	New George St.		-1837-
WRIGHT, Thomas	Highfield		-1841-1856-
WRIGHT, William	Arundel St.		-1893-1900-
YOULE, John [E,S]	North Church St.	*Orioj Works*	-1884-1891-

Notes: B= made or sold Braces. E = Made or sold Edge Tools
S= made or sold Saws. P = Made Planes

* *

FITZWILLIAM STREET, SHEFFIELD.

HENRY TAYLOR,

MANUFACTURER OF

EDGE TOOLS,

ENGRAVERS', CARVERS', DIE-SINKERS', SILVER-SMITHS', AND ALL KINDS OF TURNERS' TOOLS, &c.

ALSO, MANUFACTURER OF HAND-DRAWN CAST STEEL.

Miscellaneous Advertisement

Left: Henry Taylor, former partner of James Howarth, advertising in 1851 Staffordshire Directpry

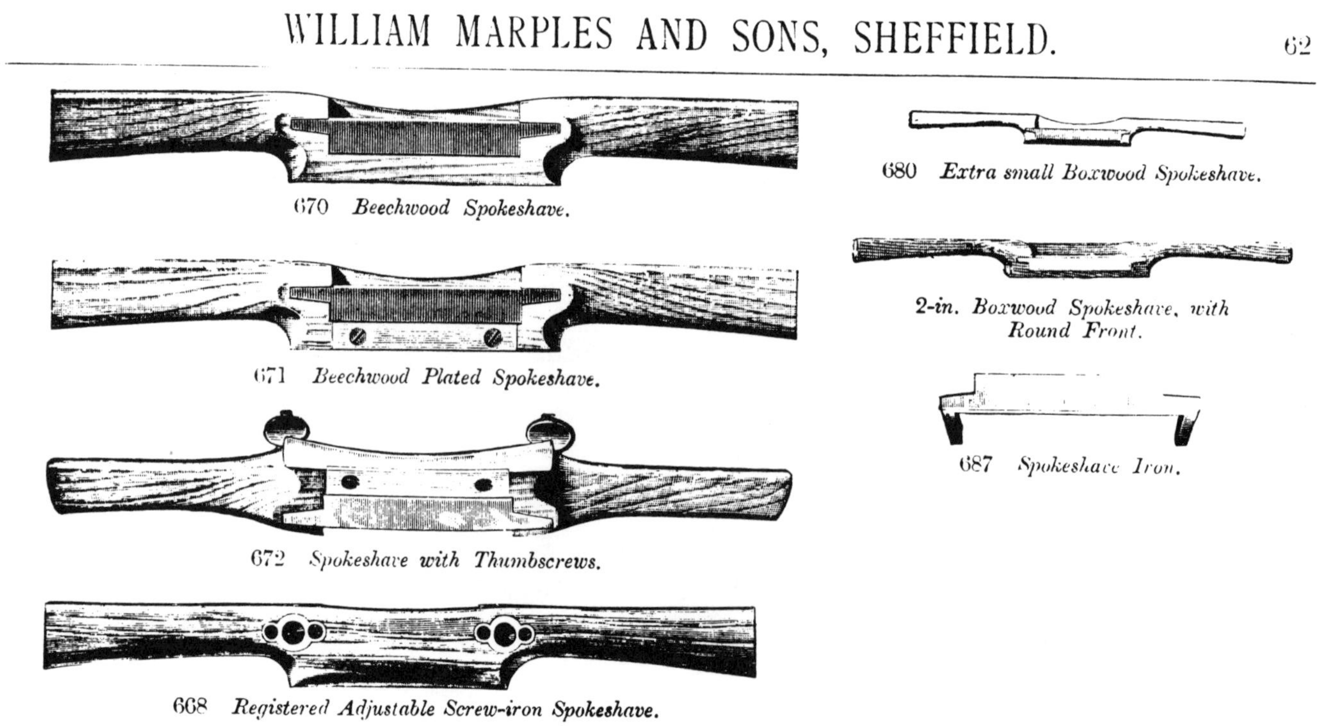

Different Types of Spokeshaves Illustrated in 1891 Wm.Marples & Sons Catalogue
[See Fig. 23, p.227 as all of these are illustrated therein]

Chapter V

Planes

The intent of this section is to comment on the nature of the plane business during the latter half of the 19th Century at Sheffield, tabulate some price lists, and illustrate and discuss some unusual English and Scottish planes.

The trade of planemaking does not appear to have been as important an activity during the 19th Century at Sheffield as it was in London, Birmingham, Bristol and Manchester. A check list of actual planemakers appears in Table XIII. The most important planemaking firm was that of Thomas Turner & Co., 1841 - 1912. After Thomas Turner's death in 1880, his wife Sarah ran the firm, until her son became of age. Many of the Sheffield firms dealing in edge and joiner tools supplemented their product line with planes, which they purchased from Turner or either other Sheffield planemakers or firms at other locations. (Table XIV)

The most important contribution to planemaking came from the Sheffield edge tool makers in manufacturing high quality cast steel tipped plane irons. Cursory examination of the Check List of Plane-Iron Makers prepared by W.L. Goodman reveals that the majority of firms making these during the entire 19th Century were located in Sheffield.[1] Noted among the Check List of Edgetool Makers (Appendix II, pages 170 - 182) are 44 firms that offered these plane irons. From personal examination of thousands of planes in United States the most common imprints observed on such irons are: William Ash & Co., James Cam, Wm. Greaves & Co., Aaron Hildick, James Howarth & Sons, Thomas Ibbotson & Co., Philip Law, Moulson Brothers, Samuel Newbould, I. & H. Sorby, W.Butcher and Ward & Payne. Also frequent are: Barton Brothers,[2] Isaac Greaves, R.Groves & Sons, R.B.Mitchell, Mottram & Co., Robert Sorby, and Wm. Weldon. Aaron Hildick appears to be the most prolific of moulding plane iron makers. His advertisement for "Plow Bits, Rabbet, Moulding and all kinds of Plane Irons, etc." is shown on Plate CXII, page 151.

Plate CLII - James Marsh & Co.

Advertisement in 1862 Sheffield Directory

This was the only year that this firm was listed.

JAMES MARSH & CO.,

ORION WORKS, NORTH STREET, SHEFFIELD,

WHOLESALE MANUFACTURERS OF

PLANES AND SKATES,

EDGE TOOLS, JOINERS' TOOLS, AUGERS, &c.

Sole Proprietors and Makers of the

PATENT SCREWLESS SKATE,

Unequalled for Cheapness and Durability.

Table XIII-Check List of Plane Makers Working in Sheffield during 19th Century

Name or Firm	Address	Dates
BARSTOW, Thos.	Barker Pool, Fargate	-1854-1860-
BARSTOW, W. & Son	126 Barker's Place	-1845-1852-
BURTON, George	15 Westbar	-1849-
BURTON, William	Shale Moor	1814-1815
	24 Bailey St.	-1823-1829-
	16 Pinfold St.	-1837-
	88 Campo Lane	-1841-
BURTON, William & Son	15 Westbar	-1845-
HASLER, William	Bow St.	-1825-
	2 Holly St.	-1828-
	100 Fargate	-1829-1837-
	14 - 16 Fargate	-1837-1852-
	16-1/2 Fargate	-1854-
	16 Fargate	-1856-1865-
HEATHCOTT, Charles	47 Rockingham St.	-1829-1823-
	83 Fargate	-1837-
	50 Fargate	-1845-
HEATHCOTT & LINLEY	5 Exchange Place	-1839-
	46 Fargate	-1841-1845-
MARPLES, Joseph	Cambridge St., York Works	-1871-1884-
TURNER, Thomas	16 Bower	-1841-
	35 - 37 Queen St.	-1845-1872-
	77 - 79 Queen St.	-1876-1879-
TURNER, Mrs. Sarah	77 - 79 Queen St.	-1881-1883-
TURNER, Thos. & Co.	77 Queen	-1884-
	57 Leopold	-1889-1912-

[Above dates and Addresses from Sheffield Directories]

Imprints of all of the above on planes have been reported, except for Thomas Barstow, George Burton, and William Burton & Son.[3] The latter two names apparently have not been previously reported. The most common imprints of the above makers appear to be planes by Joseph Marples and Thomas Turner. In all probablity the majority of planes made by all above had imprints of the dealers noted in Table XIV.

Table XIV - Sheffield Firms Dealing In Planes during 19th Century

BARBER, Ralph	-1833-1860-
BROWN, Henry & Sons	-1852-1900-
* CAMMEL, Charles % Co.	-1854-
DIXON, James	-1787-1824-
FENTON & MARSDEN	-1841-1845-
* GRAY, C. & Sons	-1861-1865-
HAWKSWORTH & EDLEY	-1833-
HAYWOOD, Joseph & Co.	-1865-
HOGG, James & John	-1833-1841-
* HOWARTH, James & Sons	-1863-1913-
IBBOTSON, Thomas & Co.	-1860-1905-
* JINKINSON, Amos & Sons	-1893-1900-
KIRK, Peter	-1862-1876-
KIRK, Peter & Son	-1845-
LARGE, William	-1860-
LONG, H.G. & Co.	-1849-1861-
MARPLES, Joseph, jr.	-1871-1884-
* MARPLES Robert & Sons	-1861-1900-
MARPLES, William & Sons	-1861-1900-
MARSDEN BROTHERS	-1854-1896-
* MARSH BROTHERS	-1852-1900-
* MARSH, James & Co.	-1862-
* MARSHES & SHEPHERD	-1841-1849-
OUTRAM, William & Son	-1857-1891-
PARKIN, Charles & Son	-1884-1881-
PEARSPN, F.G. & Co.	-1884-1910-
PENESTON & WHEATCROFT	-1811-1823-
POPPLE, George	-1879-1900-
RIDGE, Alfred	-1854-1881-
RIDGE, Alfred & Sons	-1884-1900-
SHEPPERD, J.	-1861-
SKELTON, C.T. & CO.	-1868-1900-
Smith, Joseph & Sons	-1824-1833-
SORBY, Robert	-1828-1839-
SORBY, Robert & Sons	-1841-1980-
SPEAR & JACKSON	-1845-1900-
STEEL, George	-1852-1856-
TURNER, Thomas & Co.	-1841-1912-
* TURNER, J.M. & Co.	-1865-
TURNER, NAYLOR & Co.	-1868-1872-
*WARD & PAYNE	-1845-1916-
* WILKS, Joseph	-1829-1861-
WILSON, HAWKSWORTH & Co.	-1856-1879-

Two others not listed in the Directories as selling planes but who issued trade Catalogues noting planes were:

*Greaves, Isaac	-1825-1900
*Johnson, Christopher	-1876-1896-

THOMAS TURNER,
PLANE AND JOINERS' TOOL
Manufacturer,
35. 37 AND 39, QUEEN STREET, SHEFFIELD.

All kinds of TOOLS, for Joiners, Cabinet Makers, Coopers, Carpenters, Coach Makers, Wheelwrights, Millwrights, Chair Makers, Brush Makers, &c.

GENTLEMEN'S CHESTS OF TOOLS COMPLETE,
FROM £1. TO £20.
JOINERS' AND CARPENTERS' Ditto £10. to £40.
SAWS, FILES AND EDGE TOOLS OF EVERY DESCRIPTION.
Planes Made to any Pattern or Drawing, or Altered and Repaired.

Plate CLIII - Advertisement by Thomas Turner 1854 Sheffield Directory
[Probably the most prolific plane maker in Sheffield during 19th Century]
(Collection of the Author)

* *These are new listings of Plane Dealers since the 1978 2nd Edition W.L.Goodman, BRITISH PLANEMAKERS from 1700.*

It is quite possible that many of the firms listed in Table XIV hired contract workers or outside workers to make planes in their shops. It is also possible that some of the makers listed in Table XIII might have been among such contractors. Peter Kirk was an example of a small time master ("little mester") working 1845-1860 in Sheffield. His 1854 advertisement in the Sheffield Directory indicated that he manufactured augers, edge tools, planes, joiner's tools, saws, files, gimlets, brace bits, &c. (See Plate CLIV) It is suggested that he likely made augers, gimlets and brace bits and traded with other "little mesters" and firms for the other products.

The firm of William Marples & Sons provides an excellent example of a large concern marketing a complete line and large volumes of planes made under contract. During the last quarter of the 19th Century arrangements were made with John Moseley & Son, London planemakers, to be their agents and obtain their supplies of planes for resale from this source. By 1891 William Marples & Sons had purchased the Moseley firm. After the death of the London manager in 1904, the equipment was removed to Sheffield and set up in the former old factory of Robert Sorby & Son. "Many of the Moseley workmen removed to Sheffield and continued in the Company's employment, and as no men were available of this trade in Sheffield, additional planemakers were engaged from Glasgow, York and Birmingham".[4] The following quotations are transcribed from various Wm. Marples & Sons Trade Catalogues: -

January 1, 1891 - *"William Marples and Sons are owners of the celebrated TRADE MARK - "MOSELEY"/ & SON and will quote for Tools of that Brand on application."* [This ofcourse was the same brand as offered with imprint of Wm.Marples & Sons]

April 1, 1909 - "Our London Plane Business has been removed to Sheffield, where we have built a commodious Plane Factory, equipped with Modern Machinery, and a complete Saw Milling Plant for breaking up Timber from the Log. We are now the only firm in England manufacturing Beechwood Planes with their Irons, in every detail and in every stage, from the Raw Material to the Finished Article. [This is evidence of the declining state of the wooden plane industry in England at that date]

December 1828 - Centenary Edition - "Our Plane Factory is equipped with all Modern Machinery and Saw Mill Plant. We are the only Firm in England manufacturing Beechwood Planes with their iron in every detail and in every stage, from the Raw Material to the Finished Article. "WILLIAM MARPLES & SONS"PLANES. *We can also supply Planes in the celebrated 'John Moseley & Sons' make. Price List on Application."* [here again the Marples' and the Moseley' were the same with different imprints]

October 1959 - "Over 125 years of Plane making, coupled with modern methods of manufacture, render it possible for us to offer this high-class range of Beechwood Planes at popular prices. We are the only firm in this country manufacturing Beechwood Planes with Irons, in every detail and in every stage from Raw Material to Finished Article. *SHAMROCK BRAND."*[5]

January 1, 1883 - "N.B. - We are sole Wholesale Agents for the above celebrated London Planes [Moseley & Son].

The largest plane manufacturer in Yorkshire during the 19th Century appears not to have been at Sheffield, but York. This was Robert Varvill & Sons, Ebor Works at Bridge Street, along the Ouse River. The history of these celebrated works is reported to have dated back to 1793.[6] Numerous Varvills worked at plane making during the 19th Century.[7] This particular establishment was founded by Robert Varvill and his brother, John in 1820. This later became R.Varvill & Sons, when controlled by John and Michael Varvill, sons of Robert. In 1890 over 80 hands were reported to have been employed at this large factory of seven storeys in height and frontage of one hundred feet.[8] Four travellers, salesmen with horse carts, circulated through the county peddling their joiners' coachbuilders', smiths', engineers' tools and ironmoners' supplies. In all probability the plane irons used at this establishment were from Sheffield edgetool makers. Pobbibly some Varvill planes without imprints were taken in trade at Sheffield for plane irons.

Plate CLIV - Advertisement of Peter Kirk in 1854 Sheffield Directory
(Collection of Author)

Dated price lists provide valuable information and insights concerning the types of tools being offered and their relative costs. The earliest known surviving price list of planes at Sheffield was by J.Wilks, ironmonger and dealer in tools, dated 1829.[9] (Table XV) The next known dated price list is that of Marsh Brothers, 1849. (See Plate LXXXXVI, p.118, showing Title Page). A slightly earlier undated one of Marshes & Shepherd,c.1846-1848, but having the same prices as the 1849 Marsh Brothers' List appears as Table XVI. The next two are respectively from the 1870 and 1876 Sheffield Illustrated Price Lists, (Tables XVII a&b; Table XVIII). As a consequence of increases in prices in edge tool forging and grinding, plane prices were advanced in 1824, 1836, 1846, 1864 and 1872. The list of plane prices appearing with the James Cam or Marshes & Shepherd price list published with the *Joseph Smith Explanation or Key to the Various Manufactories of Sheffield* (EAIA, 1975) is believed to have been issued 1838-1846. This price list with the others in Tables XV - XVIII provides plane prices at Sheffield from 1829 through the remainder of the 19th Century.[10]

PRICES OF PLANES.

No.	Article	Single s.	Single d.	Double s.	Double d.
694	Smooth Plane	2	2	3	6
695	Do. 2¼ Iron	2	6	3	10
696	Do. 2⅜ do	2	8	4	0
697	Do. 2½ do	2	10	4	2
698	Jack Plane	2	9	4	6
699	2½ Iron, 17 inches long	3	2	4	9
700	Pannel Plane	3	6	5	2
701	Do. do. and Slipt	4	2	5	10
702	Trying Plane	4	6	6	2
703	Do. 24 inches long	4	10	6	6
704	Long Plane	5	2	6	10
705	Jointer, 28 inches	5	9	7	6
706	Do. 30 inches	6	0	8	0
707	Strait Block, 2¼	3	0	4	6
708	Mitre do. with stop and Iron reversed	9	0		
709	Iron do. steel faced	25	0		
710	Compass Smoothing	3	0	4	0
711	Toothing Plane	3	0		
712	Toothing Plane, 2 Irons	4	3		
713	Fork Shaft Plane	3	6	6	0
714	Raising Smoothing, with Fence and Screw in Stop	5	6	7	6
715	Do. full Grove and Do	6	0	8	0
716	Do. Jack, with Smooth & Screw in Stop	7	9	10	0

A

No.	Article	s.	d.
717	Astragal	2	6
718	Do. ¾ and ⅞	2	8
719	Do. 1 Inch	2	10
720	Do. Box Moulding	3	3
721	Do. Astragal and Hollow Sash	3	2
722	Do. with Quirk Boxed	3	8
723	Do. and Hollow Boxed	3	0
724	Do. Dovetail Boxes	5	9
725	Templets for do. per pair	1	6

B

No.	Article	s.	d.
726	Bead Plane to ⅜ inch	2	10
727	Do. do. and slipt	3	2
728	Do. Thick Boxed, above ½ inch	3	6
729	Do. Shouldered, Boxed and Slipt	4	0
730	Do. Dovetail, Boxed and Slipt	5	0
731	Do. Stick Torus and Bead	3	10
732	Do. ⅞ do. do	4	0
733	Do. 1 Inch do. do	4	3
734	Cock Beads, Boxed	2	10
735	Doubled do. not Boxed	3	2

C

No.	Article	s.	d.
736	Cove and Bead	3	6
737	Do. do. ¾ and ⅞	3	9
738	Do. do. 1 inch	4	0
739	Do. do. double Squares	3	10
740	Do. do. ¾ and ⅞ do	4	2
741	Do. do. 1 inch	4	6
742	Chair Foot Plane	3	2
743	Cornice Plane per inch size O G moulding	2	6

D

No.	Article	s.	d.
744	Dovetail Plane for Keying dado	5	0
745	Do. Shouldered Boxed and Edge	5	9
746	Do. Wood Stop	5	0

F

No.	Article	s.	d.
747	Filister side	2	6
748	Do. For Drawer bottom	2	6
749	Cock Bead Filister	4	8
750	Filister Brass Slip Stop and Boxed	6	6
751	Moving Filister Wood Stop	5	0
752	Do. Do. Screw in Slide	5	6
753	Do. Do. Brass Slip Stop	6	3
754	Do. Do. Boxed	6	9
755	Do. Slip Stop Shouldered Box and Tooth	7	3
756	Do. Do. with Wedged Tooth	8	0
757	Do. Do. Improved Brass Screw-up Stop	9	0
758	Do. Do. Dovetail Boxed	10	6
759	Sash Filister, Brass Slip Stop	8	0
760	Do. Do. Plough Screw Stop	10	6
761	Do. Do. Shouldered Boxed	11	6
762	Do. Do. Ferruled Stem	14	0
763	Do. Do. Extra Work on Stem and Fenced	16	0
764	Do. Do. Wedged Tooth and Improved Stop	17	0
765	Do. Do. Dovetail Boxed	19	0
766	Fluting Plane	2	10
767	Do. Do. with Handle	3	2

G

No.	Article	s.	d.
768	Groving Planes for Slit Deal	6	0
769	Dado Do. with Wood Stop	4	9
770	Do. Do. Brass Screw Stop	7	9
771	Do. Do. for Drawer Bottom Boxed	3	0

H

No.	Article	s.	d.
772	Hollows and Rounds per pair	4	8
773	Hand Rail Plane Moving Fence	6	0

N

No.	Article	s.	d.
774	Necking Plane to ⅜	3	6
775	Do. Do. Boxed	4	0
776	Do. Do. to work part of a Cove	3	6

O

No.	Article	s.	d.
777	Ogees	2	6
778	Do. Slipt	3	2
779	Ogees ¾ and ⅞	3	6
780	Do. 1 Inch	3	10
781	Do. 1⅛ Do	4	2
782	Do. 1¼ Do	4	6
783	Quirk Ogees	3	6
784	Do. Slipt	3	10
785	Do. ¾ and ⅞	4	3
786	Do. 1 Inch	4	8
787	Do. 1⅛ Do	5	0
788	Do. 1¼ Do	5	6
789	Quirk Ogees and Bead Double Boxed	5	0
790	Back Ogees to ¾	3	3
791	Do. Do. Double Square	4	0
792	Do. Do. Bead at Foot to ¾	5	0
793	Do. Square at Top	6	0
794	Do. Do. ¾ and ⅞	6	4
795	Back Ogees and Beads 1 Inch	6	8
796	Base Ogees	3	0
797	Ovolo Sash to ¾	2	9
798	Do. With Box Moulding	3	6
799	Quirk Ovolos	3	8
800	Do. With Bead at Foot to ¾	5	4
801	Do. Ovolos Do. Do. to ⅞	5	8
802	Do. Do. Do. Do. to 1 Inch	6	0
803	Old Woman Tooth	1	9

P

No.	Article	s.	d.
804	Plow with Wood Stop and 6 Irons	15	6
805	Do. Do. Screw Do. Best Iron	18	0
806	Do. Ferruled Stem at one end	19	0
807	Do. Do. Do. both ends	20	0
808	Do. Extra Work on Stem and Fence	22	0
809	Do. with Sliding Fence	23	0
810	Do. with German Wood, Screw Stem	26	0
811	Do. with Iron Screws to Stem	28	0
812	Do. Small Circular	24	0
813	Do. Large Strait and Circular, with 12 Irons complete	36	0
814	Picture Frame Plane	5	0
814A	Do. Do. Italian	5	6

Q

No.	Article	s.	d.
814B	Quarter Rounds	2	6
814C	Do. Do. with Fence	3	2
814D	Rabbet Planes Square	2	6
814E	Do. Do. Skew	2	9
814F	Do. Do. 1¼ Do	3	0
814G	Do. Do. 1½ Do	3	3
814H	Do. Do. 2 Inches	3	6
814I	Side Rabbet, each	2	8
814J	Reed Plane, 2 Reeds	4	8
814K	Do. 2 Irons	6	6
814L	Do. 3 Reeds	5	0
814M	Do. 4 Do	5	9
814N	Do. 5 Do	6	6

S

No.	Article	s.	d.
814O	Snipe Bill Shouldered and Boxed	3	3
814P	Do. Slide Do	3	9
814Q	Side Rounds	2	6

T

No.	Article	s.	d.
814R	Table Plane, per pair	4	8
814S	Tambour Plane	3	6

York Planes at the regular York List of Prices.

Table XV - J.Wilks' 1829 Price List of Planes

In studying these lists it is apparent that English terminology or nomenclature is some instances differed from American. This is not surprising, as different practices did exist. However, the wooden plane industry in America was greatly influenced by its English herritage at least until 1850. Many of the same types and sizes of planes were almost identical. There were three fields of significant differences: sash work, grooving planes, and metal planes.

In making sash for windows in England two separate planes were used: a sash fillister (filletser, in America) for cutting the rebate (rabbet, in America) and a moulding plane for shaping the interior section. For holding the glass and cutting the rabbet two types of sash fillisters were used: one which *threw the shavings on the bench* and one which *threw the shavings off the bench*. These have been described by Nicholson.[11] Several illustrations of sash fillisters follow in this Chapter, one of which appears as item #2135 on the 1876 Sheffield Illustrated List. (Plate CLVII). 'On the bench' and 'off the bench' fillisters are noted as items #2132 and #2138 on the 1870 Sheffield Price List (Table XVIIa) with several extra options. The sash fillister like the plow plane had an adjustable fence, usually of the sliding type. The arms attached to the fence protruded through slots cut in the body of the plane and were locked with tapered wedges. The better grades were furnished with brass ferrules on the ends of the side arms, which prevented splitting when tapped with a mallet. Boxwood was applied to the edge of the sole for wear resistance. This increased the cost in proportional to the degree of boxing: i.e., corner; shoulder; shoulder dovetail or double-dovetail. R.A. Salaman has described these forms and methods of boxing the fillister.[12] Several different forms of boxing are shown in Fig. 31A . A side router (pricker) was usually furnished with sash fillisters to prevent tearing at the edge during planing. Indeed the English sash fillister is a plane of great beauty.

After making the rabbet cuts on both sides of the stock (stuff), this was placed in a sash board and each side planed using a selected moulding plane of an ovalo, Gothic or O.G. desired shape. In America the rabbet and the sash were planed simultaneously, called *"stick and rabbet"*. This practice is believed to have been developed in America late in the 18th Century, when some enterprising joiner modified a plane iron used for making a tongue so that a board could be planed to form a rabbet on one side and the sash contour on the other. While the earliest of these planes had single irons, it soon became apparent that two irons for each cut performed better and permitted easier sharpening of the irons. Thereafter double iron planes were offered, sometimes as separate planes attached together by adjusting screws to regulate the width between sash and rabbet.[13] It is now apparent that this technique was used to a limited degree in England at least by 1849. The Marsh Brothers 1849 Catalogue notes such a plane - *#963 "to Stick the Moulding and Rabbet the Glass"*. (Plate CLV) Note the Moulding Plane #978 adjacent to this to the right: *"2 Irons, It is easier to keep in order and Works better than Planes with 1 Iron"*. Such planes were apparently used earlier in England for the *James Cam or Marshes & Shepherd List* (c.1838-1845) notes *"Rustic Sash Plane to 'stick & Rabbit"*.

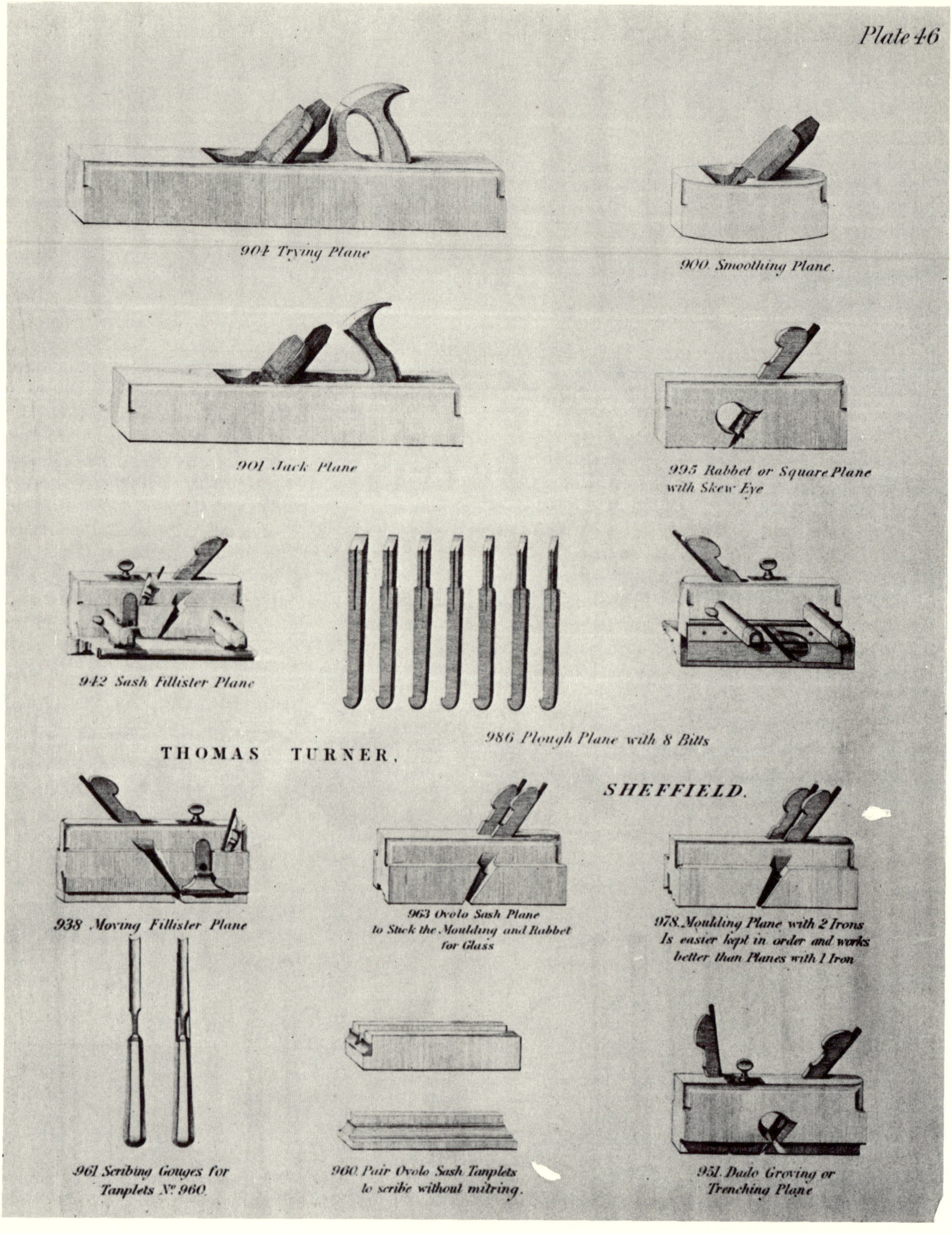

Plate CLV - Illustrations Accompanying the 1849 Marsh Brothers Trade Catalogue
[Note Planes Offered were made by THOMAS TURNER]

LIST OF PRICES OF PLANES.

Table XVI - Page from Marshes & Shepherd Trade Catalogue, c1838-1846

Plate	No.		German Steel Single Irons. s.	d.	Cast Steel Single Irons. s.	d.	German Steel Double Irons. s.	d.	German Steel Double Irons, Solid Nuts. s.	d.	Cast Steel Double Irons, Solid Nuts. s.	d.	Cast Steel Double Irons, Top Irons all C Steel hardened, and Brass Nuts. s.	d.
46	900	Smoothing Plane, 2 in. Iron	2	2	2	5	3	0	3	3	3	6	3	10 each
		Do. do. 2⅛ in. do.	2	4	2	7	3	2	3	5	3	9	4	1
		Do. do. 2¼ in. do.	2	6	2	9	3	4	3	7	4	0	4	4
46	901	Jack Planes, 2 in. Iron	2	10	3	1	4	0	4	3	4	6	4	10
		Do. do. 2⅛ in. do.	3	0	3	3	4	3	4	6	4	9	5	1
		Do. do. 2¼ in. do.	3	3	3	6	4	6	4	9	5	0	5	4
—	902	Panel Plane	3	6	3	9	5	3	5	6	5	9	6	1
—	903	Do. do. slipped	--	--	--	--	5	6	5	9	6	6	6	10
46	904	Trying Plane, 22 in. long	4	6	4	9	5	9	6	0	6	3	6	7
		Do. do. 24 in. do.	4	10	5	1	6	0	6	3	6	6	6	10
		Do. do. 26 in. do.	5	2	5	5	6	3	6	6	6	9	7	1
—	905	Jointer do. 30 in. do.	5	3	5	7	6	9	7	0	7	6	7	10
—	906	Compass do.	2	6	2	9	3	6	3	9	4	0	4	4
—	907	Forkstaff do.	3	0	3	3	5	0	5	3	5	6	5	10
—	908	Smoothing Raising Plane	4	6	4	9	6	6	6	9	7	0	7	4
—	909	Jack do. do.	6	0	6	3	8	0	8	3	8	6	8	10
—	910	Smooth Fascia, 2¼ in. Iron	4	6	4	10	5	8	5	11	6	3	6	7
—	911	Do. do. with Fence	5	2	5	6	6	4	6	6	6	10	7	2
—	912	Badger Plane, 2¼ in. Iron	4	6	4	10	6	4	6	7	6	10	7	2

Pl	No.		s.	d.
—	913	Toothing Plane	3	0
—	914	Do. do. 2 Irons	4	0
—	915	Spout Plane	4	6
—	916	Gentlemen's Smoothing Plane	1	9
—	917	Do. Jack do.	2	4
—	918	Mitre Plane	4	0
—	919	Do do. Box-wood Plug in front	6	6

A

Pl	No.		s.	d.
—	920	Astragals, to ¾-inch	2	2
		Every ⅛ extra, 2d.		
—	921	Astragal and hollow sash & templets, boxed, per pair	7	6
—	922	Astragal and hollow sash, 2 mouths	6	0
—	923	Do. with templets, extra	1	6
—	924	Lamb's tongue sash and templets, per pair	7	0

B

Pl	No.		s.	d.
—	925	Beads, boxed, to ¾-inch	2	6
—	926	Do. slipped, to ¾-inch	3	0
		Every ⅛ extra, 2d.		
—	927	Bead, slipped for torus	3	4
		Every ⅛ extra, 3d.		
—	928	Beads, Cock, boxed	2	6
—	929	Beads, Double Cock	4	0
—	930	Chair foot	2	6
—	931	Beads, double boxed	2	10
		Every ⅛ extra above ¾, 3d.		

C

Pl	No.		s.	d.
—	932	Cove and bead, to 1 inch	3	0
—	933	Cornice, per inch	3	0

F

Pl	No.		s.	d.
—	934	Fluting Plane	2	6
—	935	Filister, moving	4	3
—	936	Do. do. brass slip stop and tooth	5	6
—	937	Do. do. brass screw stop and forked tooth	9	6
46	938	Do. do. with improved fence	10	6
—	939	Filister, Sash, wood stop	5	6
—	940	Do. do. box screw stop	8	0
—	941	Do. do. and capped stems	11	6
46	942	Do. do. shoulder boxed forked tooth and improved stop	16	0
—	943	Filister, for Cock bead	3	6
—	944	Filisters, moving or sash, dovetail boxed extra	2	0

G

Pl	No.		s.	d.
—	945	Grooving Planes for slit deal, to ¾-inch in size per pair	5	6
—	946	Do. ⅞ and 1 inch	5	10
—	947	Grooving Planes Improved, to ¾-in.	5	9
—	948	Do. do. ⅞ and 1 inch	6	1
		Every ⅛ extra per pair. 4d.		
—	949	Grooving, long, with handles, 1¼ and 1½-inch	9	6
		For every ⅛-in. extra per pair, 6d.		
—	950	Grooving, dado or carcase, wood stop	4	0
46	951	Do. do. do. screw stop	6	0
—	952	Gothic sash and templets	7	0
—	953	Grooving for drawer bottom	3	0

H

Pl	No.		s.	d.
—	954	Hollows and rounds, per pair	4	0
—	955	Hand-rail Planes according to pattern and No. to set		
—	956	Table Planes, per pair	4	0
—	957	Guages for do.	0	6

N

Pl	No.		s.	d.
—	958	Necking, to 1-inch	3	0
—	959	Nosing, to 1-inch	3	2
		Every ⅛ 2d. extra.		

O

Pl	No.		s.	d.
46	960	Oveloe Sash templets, to scribe with gouges, without mitring, ℘ pair	3	3
46	961	Gouges for ditto, per pair	1	6
—	962	Oveloe sash and templets, per pair	6	0
46	963	Do. do. with two mouths	5	0
—	964	Do. with templets, extra	1	6
—	965	Oveloes, common or ogees, to ¾-in.	2	2
		Every ⅛ extra, 2d.		
—	966	Oveloes and ogees, quirk, to 1-in.	3	0
		Every ⅛ extra, 4d.		
—	967	Ogees, back, to ¾-inch	2	6
—	968	Do. do. and double square	3	0
—	969	Do. do. and beaded boxed to 1-in	3	6
—	970	Ogees, quirked and slipped, to 1-inch	3	4
—	971	Do. do. and cock bead, to 1-in.	3	4
—	972	Ogee, Grecian, to 1-inch	3	6
—	973	Oveloe, do. to 1-inch	3	6
—	974	Do. do. double boxed	3	10
		Every ⅛ extra 4d. and upwards 6d. per ⅛ extra.		
—	975	Oveloe, quirk and bead, to 1-in.	3	6
		Every ⅛ extra, 4d.		
—	976	Old woman's tooth	1	9
—	977	Ogee, quirk or Grecian, with raking square	4	6
46	978	Do. do. improved, with two irons	6	0

P

Pl	No.		s.	d.
—	979	Plough Plane, single plate, one iron	5	0
—	980	Do. do. double do. do.	6	6
—	981	Do. do. three irons, wood stop	8	0
—	982	Do. do. six irons, do.	10	0
—	983	Do. do. screw stop, 8 irons..	14	6
—	984	Do. do. do. and capped stems	16	0
—	985	Do. with bright irons, do.	17	0
46	986	Ploughs, best, with eight irons, and extra work on fence	19	0
—	987	Improved stop do.	20	0
—	988	Screw stems do.	22	0
—	989	Ploughs to work straight or circular	36	0

R

Pl	No.		s.	d.
—	990	Reed Planes, fenced to work more reeds	5	0
—	991	Reed Planes, single reed	3	4
		Do. do. two reeds	3	6
		Do. do. three reeds	4	6
		Do. do. four reeds	6	0

S

Pl	No.		s.	d.
—	992	Side rabbet, per pair	4	6
—	993	Side rounds, per pair	4	6
—	994	Square rabbet, to 1¼-inch	2	2
		For every ⅛ inch 2d. extra.		
46	995	Skew rabbet, to 1¼ inch	2	4
—	996	Skew or square, boxed edges	4	8
—	997	Snipe bills, per pair	5	0
—	998	Do. shoulder boxed, per pair	6	6
—	999	Side snipe, per pair	7	0
—	1000	Spring Plane	5	0
—	1001	Coopers' crozes, to 16 inch	4	6
—	1002	Sun Planes 5s. 6d. to	7	0

All Quirked Moulding Planes, with box-wood, dovetailed at bottom, extra .... 5d.

Do. do. do. two irons, extra .... 1s. 6d.

N.B.—EVERY DESCRIPTION OF PLANES MADE TO DRAWINGS ON THE SHORTEST NOTICE.

JOINERS' TOOLS.

No. of Article.	PLANES.	Single Iron.	Double Iron	C.S. Brass Nutted Irons
2001	Smoothing Planes, 2 inches	2/4	3/2	3/9
2002	do do 2 1/8 do	2/6	3/4	3/11
2003	do do 2 1/4 do	2/8	3/6	4/2
2004	do do 2 3/8 do	..	4/	4/6
2005	do do 2 1/2 do	..	4/6	5/
2006	Jack Plane to 17 in. 2 do	3/	4/2	4/8
2007	do do 2 1/8 do	3/2	4/5	4/11
2008	do do 2 1/4 do	3/5	4/8	5/2
2009	Pannel Planes	..	5/6	6/
2010	Slipped Pannel Planes	..	6/6	7/
2011	Trying Planes, 22 inches	5/	6/	6/6
2012	do do 24 do	5/3	6/3	6/9
2013	Long Trying Planes, 26 inches	5/6	6/6	7/
2014	Jointer Trying Planes, 28 do	6/	7/	7/6
2015	do do do 30 do	6/3	7/3	7/9
2016	Coopers' Jointers	18/	24/	
2019	Compass Smooth, 2 inches	2/9	3/9	4/3
2020	do Box Stop	..	6/	6/6
2021	do Screw Stop	..	7/6	8/
2022	Toothing, 2 inches	3/3		
2023	Forkstaff or Hollow	4/	5/6	6/
2024	Gentlemen's Smoothing	2/	3/	3/6
2025	Gentlemen's Jack	2/6	3/9	4/3
2026	Straight Block, 2 1/4 inches			
2027	Smooth Raising	4/9	6/9	7/6
2028	do Badger Mouth	..	7/6	8/
2029	Jack Raising	6/6	8/6	9/
2030	Jack Badger Mouth	..	9/	9/6
2031	Badger Plane	..	..	8/
2032	do do 2 fences	..	..	12/6
2033	Plated Smoothing, 2 inches	6/9	7/9	8/6
2033½	Plated Smoothing, with moveable iron front, to 2 inches	7/6	8/6	9/
2034	Spout Planes	5/		
2035	Stower Planes to Round, 3 inch	8/		
2036	Facia Planes, 2 1/4 inches	6/		
2037	do with fence	6/9		
2038	Mitre Planes, 2 1/4 inch	3/6	4/6	5/
2039	do with skew eye	5/	6/	6/6
2040	do do with box front	6/6	7/6	8/
2041	Jack Grooving Planes to 1 inch, pair....	10/		
2044	do do with iron tenon....	10/		

Moving Fences, 2/ extra.
Brass Nuts to common Double Irons, 3d. extra.
Planes, with hardened Top Irons, 4d. extra.
Do. with solid nutted Top Irons, 3d. extra.

No. of Article.			
	A		
2045	Astragals, 1/4 to 5/8 inch	2/2	each
2046	do 3/4 inch	2/2	,,
2047	do 7/8 inch	2/5	,,
2048	do 1 inch	2/8	,,
2049	do with boxed mouldings, 1/4 inch ..	3/2	,,
2050	Astragal and hollow sash	3/0	,,
2051	do quirk boxed	3/9	,,
2052	do quirk hollow	4/0	,,
2053	do quirk dovetail	5/9	,,
2054	do templets	1/6	pair
	B		
2064	Beads unslipt, 1-16 to 5/8 inch inclusive	2/6	each
2065	do do 3/4 inch	2/6	,,
2066	do do 7/8 inch	2/9	,,
2067	do do 1 inch	3/0	,,
2068	do slipt, to 1/2 inch	3/0	,,
2069	do do shouldered boxed	3/9	,,
2070	do do dovetail boxed..............	5/0	,,
2071	Beads to stick torus, to 5/8 inch	3/9	,,
2072	do do 3/4 inch	4/0	,,
2073	do do 7/8 inch	4/3	,,
2074	do do 1 inch	4/6	,,
2075	do do dovetail boxed to 5/8 inch..	5/10	,,
2076	do do do 3/4 inch..	6/0	,,
2077	do do do 7/8 inch..	6/2	,,
2078	do do do 1 inch..	6/6	,,
2079	Set of 9 Beads, slipt to 1/2 inch	26/	set
2080	do do 3 torus	31/	,,
2081	Set of 9 double boxed Beads	27/6	,,
2082	do do do 3 torus	35/	,,
2083	do dovetail boxed Beads	44/	,,
2084	do do do 3 torus	48/	,,
2085	Bead square and cove to 5/8 inch	7/	,,
2086	do do 3/4 inch	7/6	,,
2087	do do 7/8 inch	8/	,,
2088	do do 1 inch	9/	,,
	C		
2090	Cock Beads	2/6	each
2091	do do double	4/	,,
2092	Cove and Bead to 1 inch	3/	,,
2098	Cove and double square, to 1 in.	3/6	,,
2104	Chair foot Plane	3/2	,,
2105	Chair splat Plane	4/8	,,
2106	Curled shaving Plane....................	3/2	,,
2107	Cylinder boxed	3/8	,,

No. of Article.			
	D		
2115	Door Plane	5/9	each
2116	do. for keying Dado, two brass stops....	11/6	,,
2117	Dado grooving, wood stop................	4/6	,,
2118	do brass stop....................	6/6	,,
	F		
2125	Fillister, moving wood stop	4/9	each
2126	do do screw at side	5/6	,,
2127	do boxed edge and tooth	6/	,,
2128	do Shoulder boxed slip stop............	7/4	,,
2129	Fillister, shoulder brass screw at top	8/4	,,
2130	do do plough stop	9/6	,,
2131	do do dovetail boxed	13/6	,,
2132	do Sash plough stop to throw the shaving on the bench	10/9	,,
2133	do do shoulder boxed	11/9	,,
2134	do do ferruled one end	13/	,,
2135	do do do both ends	14/	,,
2136	do do do do extra work..	15/6	,,
2137	If to throw shaving off the bench, extra..	—/6	,,
2138	Fillister sash, with tooth, to throw the shaving off the bench..........	17/	,,
2139	do do dovetail boxed	18/6	,,
2140	do do do iron screw stems	27/	,,
2141	do side	2/6	,,
2142	do drawer	2/8	,,
2143	do cock bead brass slip stop and boxed..	6/6	,,
2144	Double Fillister Plane		,,
2145	Fillister sash, with self-regulating screw and solid handle		
2146	Fluting Planes	2/6	,,
2147	do cut out behind..............	3/2	,,
	G		
2160	Gothic sash and templets	7/6	pair.
2161	Grooving Planes, bevel for bookcase shelf..	8/6	each
2162	do dado wood stop	4/8	,,
2163	do do plough stop	7/	,,
2164	do drawer bottom boxed	3/	,,
2165	do slit deal, 3/8 to 3/4 inch	5/6	pair.
2166	do do 7/8 in	5/10	,,
2167	do do 1 in	5/10	,,
2168	do do 1 1/4 in..................	6/6	,,
2169	do do moving fence with 3 pair of irons.	12/6	each
2170	do large as Jack Plane, with 1 pair of irons.	15/6	,,

JOINERS' TOOLS.

No. of Article.			
	H		
2180	Hollows and Rounds to No. 18	4/	pair.
2182	do do Joiners', 72/ per set of 18 pairs.		
2183	do Cabinet, 56/ per set of 14 pairs.		
2184	do Joiners' with snipes bills, 108/10 set of 18 pairs		
2185	do Cabinet do 89/6 set of 14 pairs		
2186	Hand rail Planes	5/	each
2187	do moving fence	5/10	,,
2188	do best 18/		per set of 3
2189	do do circular 27/6		,,
2190	do Router's dovetail boxed 39/		,,
	L		
2195	Lamb's tongue sash and templets, to $\frac{5}{8}$ in... Every $\frac{1}{8}$ inch, 6d. per pair extra.	7/6	pair.
	N		
2200	Necking Planes, to 1 in..................	3/	each
2204	Nosing Planes, to 1 in.................. Every $\frac{1}{8}$ inch, 3d. extra.	3/2	,,
	O		
2204	Ovalo, with two mouths.................. With templets, 1/6 extra.	5/6	each
2205	Ovalo sash and templets to $\frac{5}{8}$ in.	6/6	pair
2206	do 2 in. bar templets	7/6	,,
2207	do with improved brass templets to $\frac{5}{8}$ in.	12/	,,
2208	do do 2 inch bar to do ..	13/	,,
2209	Ovalos, common................ to $\frac{3}{4}$ in.	2/2	each
2211	do do .... $\frac{7}{8}$ in.	2/5	,,
2212	do do .... 1 in.	2/8	,,
2214	do Quirk .... 1 in.	3/	,,
2217	do do and square .... 1 in.	3/6	,,
2221	do do and Astragal ...... 1 in.	4/	,,
2226	do do and Bead........ 1 in.	3/6	,,
2230	do Grecian and Bead........ 1 in.	4/	,,
2233	do boxed, small	3/6	,,
2234	do do large	3/10	,,
2236	OG, common .... to $\frac{3}{4}$ in.	2/2	,,
2237	do do .... $\frac{7}{8}$ in.	2/5	,,
2238	do do .... 1 in.	2/8	,,
2239	do slipped .... $\frac{3}{4}$ in.	2/10	,,
2252	do back .... $\frac{3}{4}$ in.	2/6	,,
2253	do Quirk, with double square	3/	,,

No. of Article.				
	O			
2254	OG and Astragal	to 1 in.	4/	each
2260	do Quirk and Astragal	1 in.	4/	,,
2267	do do and Bead	1 in.	3/6	,,
2272	do do and Quirk Bead	1 in.	4/	,,
2278	do Grecian	$\frac{5}{8}$ in.	3/6	,,
2279	do do	$\frac{3}{4}$ in.	3/6	,,
2280	do do	$\frac{7}{8}$ in.	3/6	,,
2282	do do new pattern	1 in.	3/6	,,
2287	do do do and Bead..	1 in.	4/	,,
2292	do do do & Quirk Bead	1 in.	4/6	,,
2302	do Grecian and Hollow	1 in.	4/	,,
2307	do Quirk Cove and Square......	1 in.	4/	,,
2312	do and Bead, with Astragal on top	1 in.	5/	,,
2318	do & Quirk Bead, with Astragal on top to	1 in.	5/6	,,
	P			
2320	Plough Plane, 1 iron		5/	each
2321	do do double plate		6/6	,,
2322	do do wood stop................		8/	,,
2323	Ploughs, wood stop, 6 irons		10/	,,
2324	do screw stop, 8 irons		14/6	,,
2326	do do ferruled two ends		16/	,,
2327	do do bright irons		17/	,,
2328	do do iron screw stems......		30/	,,
2329	Ploughs, with improved stops		20/	,,
2330	do with gauged irons, brass behind irons..		2/	extra
2331	do with ivory gauge in stems.........		6/	,,
2332	Ploughs, with circular and straight work, with 12 irons, 2 stops, and 2 pair plates, ivory gauge stems, and iron gauge		52/	each
2333	Best Ploughs, with extra work on fence....		19/	,,
2334	do improved stop, rivetted steel plate		22/	,,
2335	Beech Screw Stems, with improved stops ..		22/	,,
2336	Box do do do ..		25/	,,
	R			
2340	Rabbet Planes, square, to $1\frac{1}{4}$ inch		2/4	each
2341	do do skew $1\frac{1}{4}$ inch		2/4	,,
2342	do do do $1\frac{1}{2}$ inch		2/8	,,
2343	do do do $1\frac{3}{4}$ inch		3/	,,
2344	do do do 2 inch		3/4	,,
2345	do boxed edges $1\frac{1}{4}$ inch		4/8	,,
2346	do do reversed skew..............extra		/5	,,
2347	do do with tooth .... do		/6	,,
2348	Rabbet side		2/3	,,

No. of Article.			
	R		
2349	Reed Planes, 2 reeds....................	3/6	each
2350	do do 3 do	4/6	,,
2351	do do 4 do	6/	,,
2352	do do 5 do	7/6	,,
2353	do do 6 do	9/	,,
2354	do do to follow	4/6	,,
2355	do do with fence	5/	,,
2356	do do single reed	2/3	,,
2357	Round side, with eye holes	2/3	,,
2358	do quarter	2/6	,,
2359	do do with fence	3/3	,,
2360	Rabbet Plane Iron	12/	,,
2361	do do with two irons..........	18/	,,
	S		
2365	Snipes Bills shouldered boxed	3/6	each
2366	Side Snipes	3/9	,,
2367	Shutter Planes	5/8	pair
	T		
2370	Table Planes	4/	pair
2371	Gauge for do	/6	each

COACHMAKER'S PLANES, &c.

2380	Smoothing Planes, to $1\frac{3}{4}$ inch	3/9	each
2381	Compass Planes do	4/3	,,
2382	do do ......per set of 6......	25/6	,,
2383	Concave do to $1\frac{3}{4}$ inch	4/6	,,
2384	Jack Planes, with double toat	5/2	,,
2385	Rabbet Smoothing Planes.............	4/6	,,
2390	T Rabbet Planes	3/6	,,
2391	T Compass Planes	4/	,,
2392	Spoke Planes, plated	7/6	,,
2393	Side Light Planes	9/9	,,
2394	Jarvis do brass face	13/	,,
2395	do do iron face	12/	,,
2396	Shaft Ovalos..........................	7/	,,
2397	Tonguing Planes	5/6	,,
2398	Boxing Ploughs	8/	,,
2399	Pistol Routers	14/	pair.
2400	Fence Grooving Planes	9/6	,,
2401	do do plated	12/	,,
2402	Boxing Routers	4/	,,

N.B.—Planes made to any kind of Drawing or Pattern.

Some Illustrations of Planes from the 1870 Illustrated Sheffield List

Plate CLVI

2361

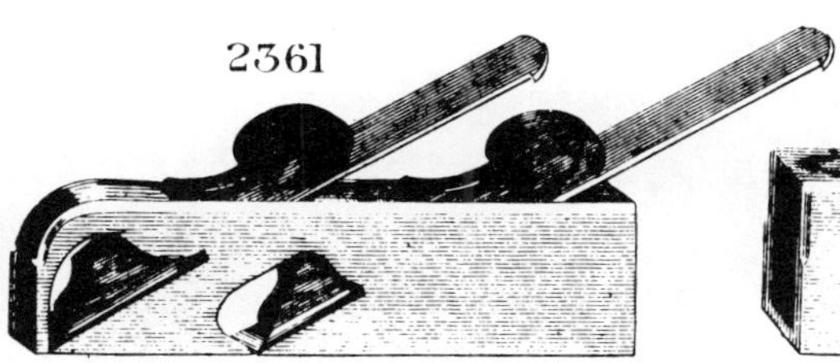

Iron Rabbet Plane with two Irons.

2560

Iron Rabbet Plane.

2204

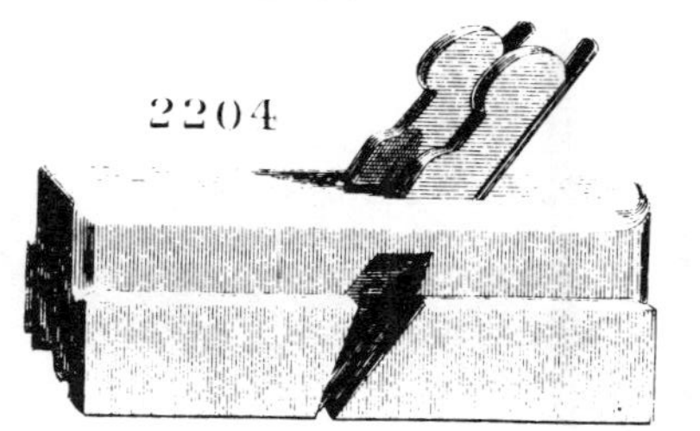

Ovalo Sash Plane.

2145

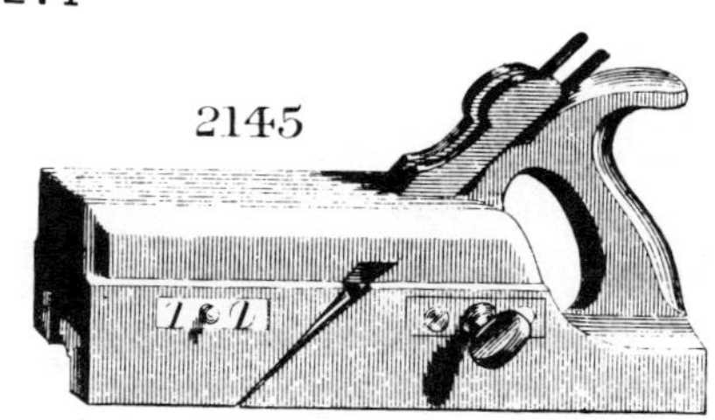

Sash Plane with Self Regulating Screws and Solid Handle.

JOINERS' TOOLS.

PLANES.

No.	Description	Single	Double	Cast Steel Brass Nutted Irons
2001	Smoothing, to 2 inch	2/6	3/3	3/9
2002	Do. 2⅛ „	2/8	3/6	4/
2003	Do. 2¼ „	2/10	3/9	4/3
2004	Do. 2⅜ „		4/3	4/9
2005	Do. 2½ „		4/9	5/3
2005A	Do. with iron fronts, 2¼ inch			9/
2005B	Do. with solid iron face, 2¼ inch			9/
2006	Jack, 14 to 17 inch to 2 inch	3/6	4/3	4/9
2007	Do. do. 2⅛ „	3/8	4/6	5/
2008	Do. do. 2¼ „	3/10	4/9	5/3
2008A	Do. do. 2⅜ „		5/6	6/
2008B	Do. do. 2½ „		5/9	6/6
2009	Panel		5/9	6/6
2010	Do. slipped		6/9	7/6
2031	Badger, 2¼ inch			8/6
2011	Trying, to 22 inch	5/6	6/3	6/9
2012	Do. 24 „	5/9	6/6	7/
2013	Do. 26 „	6/3	7/	7/6
2014	Do. 28 „	6/9	7/6	8/
2015	Do. 30 „	7/3	8/	8/6
2019	Compass Smoothing, to 2 inch	2/10	3/10	4/3
2019A	Do. do. 2⅛ „			4/6
2019B	Do. do. 2¼ „			4/9
2020	Do. do. with Brass Screw Stop to 2 inch			7/
2020A	Do. do. 2⅛ „			7/3
2020B	Do. do. 2¼ „			7/6
2022	Toothing, coarse or fine, to 2 inch	3/3		
2022A	Do. two irons	4/3		
2023	Fork Stail, to 2 inch	4/	5/6	6/
2017	Mast Plane		7/	to 7/6
2018	Toy Smoothing	1/		
2024	Gentlemen's Smoothing	1/8	2/3	
2025	Do. Jack	2/10		
2016	Coopers' Jointers to order			

A

No.	Description	Price
2045	Astragal, to ⅝ inch	2/4

Every ⅛ to 1 inch, 3d. extra; above, 4d. extra.

B

No.	Description	Price
2064	Beads, boxed to ⅝ inch	2/6
2065	Do. slipped to ⅝ inch	3/

Every ⅛ inch, 3d. extra.

No.	Description	Price
2071	To answer for Beads and Torus, to ⅝ inch	3/6

Every ⅛ inch, 4d extra.

No.	Description	Price
2081	Beads, double boxed, to ⅝ inch	3/

Every ⅛ inch, 3d. extra.

No.	Description	Price
2082	Beads, 10 to the set, to 1 inch	27/6
2083	Do. do. slipped do.	32/6

If any in the set are Torus, extra price is charged.

C

No.	Description	Price
2090	Cock Beads, boxed	2/6
2091	Do. double	4/3
2092	Cove and Bead to 1 inch	3/3

F

No.	Description	Price
2125	Moving Fillister, brass slip stop and tooth	6/6
2126	Do. brass screw stop and forked tooth	10/
2127	Do. with improved fence	10/9
2128	Do. do. do. dovetail boxed	14/6
2132	Sash Fillister, wood stop	6/6
2133	Do. box screw stop	9/6
2134	Do. and capped stems	12/
2135	Do. left-handed shoulder boxed, forked tooth and improved stop	18/
2136	Do. do. do. dovetail boxed	21/
2144	Double Fillister Plane	
2146	Fluting Plane	3/

G

No.	Description	Price
2161	Grooving Planes for slit deal, to ¾ inch .. per pair	5/6
2162	Do. do. ⅞ inch, 5/10; 1 inch „	6/

Every ⅛ inch, 6d. per pair extra.

No.	Description	Price
2163	Long Grooving, with handles, 1¼ and 1½ inch „	10/

Every ⅛ inch, 1/ per pair extra.

No.	Description	Price
2117	Dado or Carcase Grooving, wood stop	4/6
2118	Do. do. brass screw stop	6/6
2160	Gothic Sash and Templets	8/
2164	Grooving for drawer bottom	3/
2169	Moving Grooving, with 3 pair irons	13/

H

No.	Description	Price
2180	Hollows and Rounds, per pair, to No. 15	4/4
2181	Do. do. skew mouth, per pair, do.	4/10
2182	Half set Hollows and Rounds, 9 pairs	40/
2183	Full set Hollows and Rounds, 18 pairs	80/
2184	Half set Skew Mouth Hollows and Rounds, 9 pairs	44/6
2185	Full set Skew Mouth Hollows and Rounds, 18 pairs	89/

Gauges for ditto, 6d. each.
Hollows and Rounds above No. 15, 4d. per pair extra.

L

No.	Description	Price
2195	Lamb Tongue Sash and Templets, per pair	8/
2196	Do. do. do. do. ¾ inch, per pair	8/6

N

No.	Description	Price
2200	Necking, to 1 inch	3/2
2201	Nosing, to 1 inch	3/4

Every ⅛ inch, 4d. extra.

O

No.	Description	Price
2209	Common Ovaloes or Ogees, to ⅝ inch	2/4

Every ⅛ inch, 3d. extra to 1 inch; above, 4d. extra.

No.	Description	Price
2214	Quirk Ovaloes and Ogees, to 1 inch	3/6

Every ⅛ inch, 4d. extra.

No.	Description	Price
2252	Back Ogees, to ¾ inch	3/
2253	Do. and double square	3/6
2271	Do. and beaded boxed, to 1 inch	4/
2272	Quirk Ogees, slipped, to 1 inch	3/10
2273	Do. and cock beaded, to 1 inch	4/
2278	Grecian Ogees, to 1 inch	4/
2230	Do. Ovaloe, to 1 inch	4/
2231	Do. do. double boxed, extra	/4

Every ⅛ extra to 2 in., 4d. per ⅛; 2 in. and upwards, 6d. per ⅛.

No.	Description	Price
2226	Quirk Ovaloe and Beads, to 1 inch	4/

Every ⅛ inch, 4d. extra.

No.	Description	Price
2268	Old Woman's Tooth	2/9

P

No.	Description	Price
2321	Plough Plane, double plate, one iron	7/
2322	Do. do. wood stop, 3 irons	8/6
2323	Do. do. do. 6 irons	10/6
2324	Do. do. screw stop, 8 irons	15/
2326	Do. do. do. capped stems	16/
2327	Do. do. do. bright irons	17/6
2333	Best Plough, with extra work on fence	19/6
2334	Improved Stop do. do.	21/
2335	Beechwood Screw Stems do. do.	23/
2336	Boxwood do. do. do.	27/
2337	Ploughs to work straight or circular	40/

R

No.	Description	Price
2337	Reed Planes, fenced to work more reeds	5/
2338	Do. single reed	3/4
2349	Do. two reeds	3/6
2350	Do. three reeds	4/6
2351	Do. four reeds	6/
2348	Side Rabbets, per pair	5/6
2357	Side Rounds „	4/8
2340	Square Rabbets, to 1¼ inch	2/6
2341	Skew do. to 1¼ inch	2/6

1⅜ inch, 2/8; 1½ inch, 2/10; every ⅛ inch above. 3d. extra.
Boxed edges, skew or square, double price.

S

No.	Description	Price
2364	Snipe Bills, per pair	5/6
2365	Do. shoulder boxed, per pair	8/
2366	Side Snipe, per pair	7/6
2366A	Coopers' Crozes, to 16 inch	6/
2366B	Rustic Sash and Templets, per pair	6/6
2205	Ovaloe do. do. „	7/
2207	Do. Sash and brass-end Templets, per pair	11/

Ovaloe Sash for 2 in. stuff, 6d. per pair extra.

No.	Description	Price
2207A	Gouges for ditto, per pair	2/
2207B	Scribing Plane for ditto	3/8
2372	Astragal and Hollow Sash and Templets, „	8/
2373	Ovaloe Sash Plane, to stick and rabbet, with 2 irons	6/
2374	Do. do. with Templets	7/6
2375	Astragal and Hollow to stick and rabbet, with 2 irons	8/
2376	Do. do. with Templets	9/6
2377	Gothic, to stick and rabbet, with 2 irons	8/
2378	Do. with Templets	9/6

T

No.	Description	Price
2370	Table Planes, per pair	4/6
2370A	Trenching Planes, wood stop	4/6
2370B	Do. do. screw stop	6/6

SUNDRIES.

No.	Description	Price
2410	Improved Beech Bull-nose Plane, brass mounted	3/
2411	Boxwood Bull nose Planes	3/6
2412	Do. Chariot do.	4/3
2413	Iron Bull-nose Planes	6/
2414	Gun Metal do.	8/
2415	Iron Chariot Planes	8/
2416	Gun Metal do.	9/
2417	Beech Thumb Planes	1/9
2418	Boxwood do.	2/9
2419	Coachmakers' Jarvis Planes	12/6

Planes made to Drawing or Pattern.

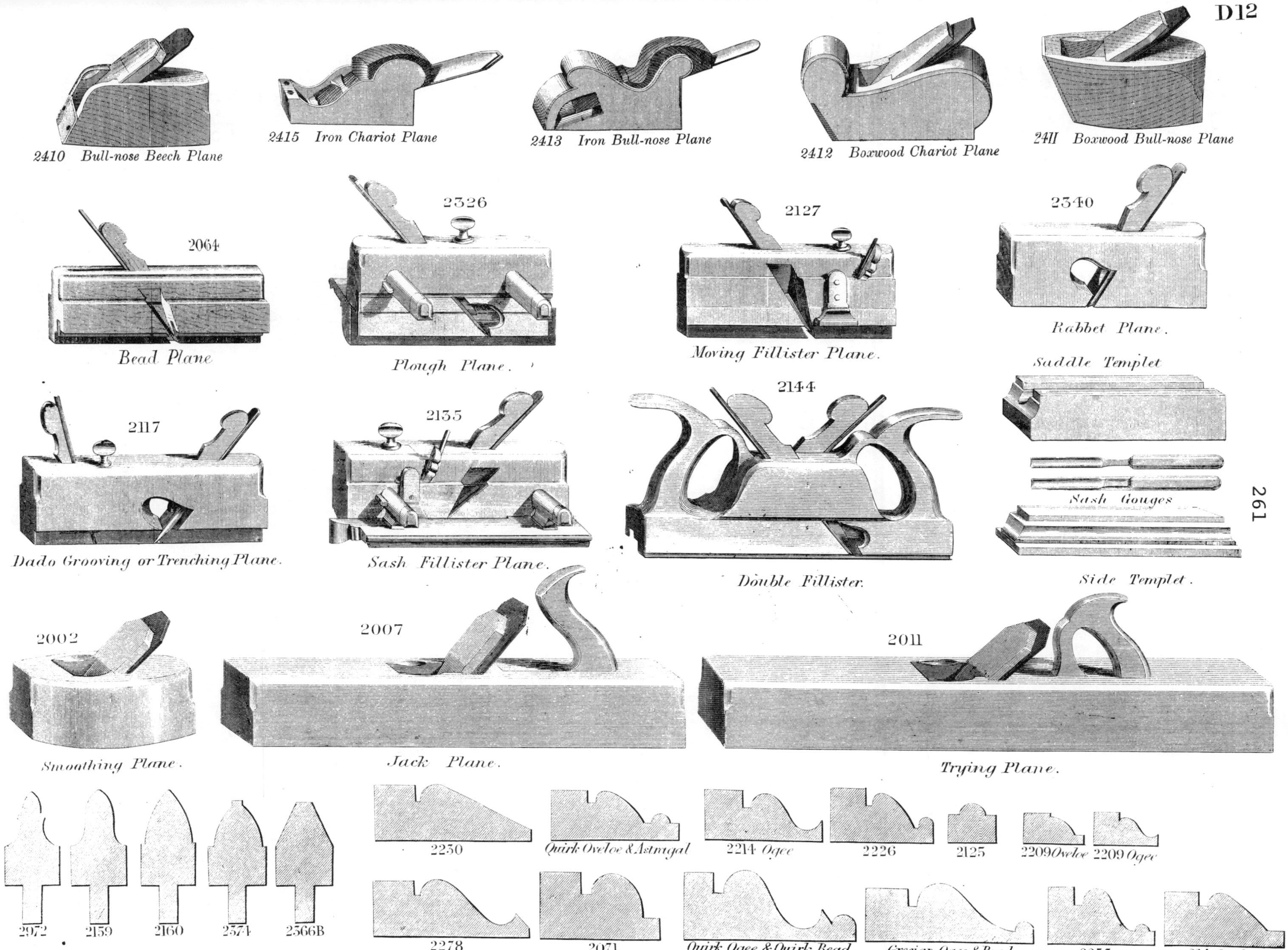

Plate CLVII - Planes Shown on 1876 Sheffield Illustrated Price List [Plate D12]

Grooving was accomplished by a plane with an iron skate for guiding the cutting iron, an adjustable fence for positioning the cut at proper distance from the edge of the stuff being grooved and a stop for regulating its depth. In England such a plane was called a'plough' and was generally made from quartered beech.[14] It was usually sold with a set of eight interchangeable cutting irons, varying in width from 1/8" to 9/16". The fence was attached to two arms, friction fitted and mounted through slots of the same cross-sectional shape, cut in the stock. Thus the fence could be adjusted by sliding the arms and locked securely by wedge in each slot. The iron skate was fitted in a groove, cut sligthly deeper than a rabbet in the plane stock sole, and riveted through the stock. The least expensive was furnished with a slotted, friction fitted wood stop. Extra features, increasing the price and quality of the plough, included: brass screw stops; brass ferrule tipped arms, boxwood splined fences and screw arm stems for adjusting the fence.

Plane hardware in the form of cast brass screw stops, slide stops, iron skate plates and rivets for attaching these through the stock of grooving planes and fillisters were readily available. Table XIX lists prices of such hardware from the 1829 J.Wilks' Catalogue.

Plough and Filister Stops, Plough and Groving Plates, &c.

575	Plough Stops.........................dis	26	0
576	Do. Plates..............................	20	0
577	Sash Filister Stops......................	25	0
577A	Improved Sash Filister Stops	26	6
577½	New Do. Do....................	25	0
578	Moving Filister Stops Screw at side	9	0
579	Do. Do. Screw at top	13	6
580	Do. Plough Fashion	17	0
580½	New Do............................	25	6
581	Dado Groving Stops	13	6
582	Groving Plane Plates......................	4	6
583	Double Iron Screws 6	6 per gro. net	
584	Do. Do. Broad Heads 9	0....Do....	
585	Long Plough Rivets 5	0....Do....	
586	Short....Do........................ 3	0....Do.. .	

Table XIX
Plough and Fillister Hardware

1829 J.Wilk's Catalogue

Plate CLIX illustrates such hardware from an unidentified Birmingham Pattern Book, printed on paper watermarked 1803, indicating these items were available early in the 19th Century. Undoubtedly considerable of this material was exported to the United States before such was domestically manufactured.

At what date the screw stem plough first appeared in England is not presently known. It may have been a development introduced from the Continent, possibly from Germany. However, Continental plows usually had the arms attached to the plane stock and the fence slid laterally along these, whereas the English and American types had the arms attached to the fence, adjusting through the plane stock. From the 1829 J.Wilks' Catalogue the plough designated "German wood, screw stems" indicated that such was then available. Plough planes with screw stems are also noted on the Marshes & Shepherd List and the 1870 Illustrated Sheffield List.

Recently a plow plane with screw arms, having a rather coarse thread pitch, by Thomas Napier of Philadelphia, Penn. has been found.[15] This must date before 1811 (date of Napier's death) and after the early 1780's after he emigrated from Edinburh, Scotland. He is also known to have made sliding arm plows at Philadelphia.[16] It is suggested that Napier may have learned the technique of making such screw arms while working in Edinburgh,cl770 and brought this knowledge to America.

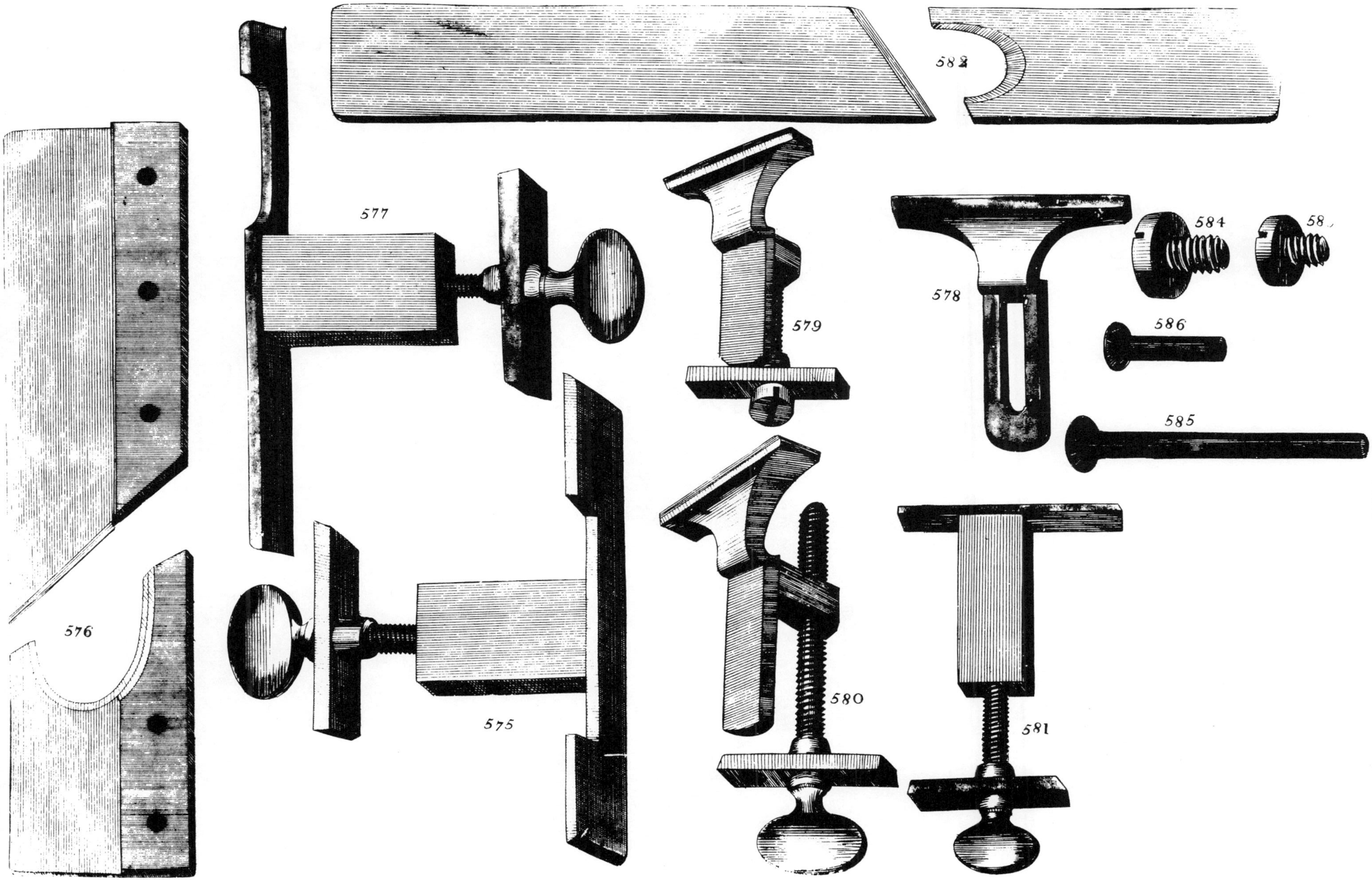

Plate CLIX - Plough and Fillister Hardware from an Unidentified Birmingham Pattern Book [paper watermarked 1803]

Fig. 31B Earliest Known American Screw Arm P ow Plane
Imprint: *T.NAPIER PHILAD^a^.*
c.1810

[Photo and Plane - Collection of Robert D. Graham,jr.]

It is not uncommon to find screw arm plow planes having imprints by makers from Scotland. Frequently these have closed handles. Some appear with imprints from the north country of England. Such a plow, along with a double-dovetailed boxed screw arm fillister, made by John Cockbain of Carlisle, Cumberland, c.1860,[17] is illustrated in Fig. 32. It is rare to find such planes made in the Mid-lands or southern England.[18]

In America the screw arm plow is believed to have first become popular about 1830 in the Philadelphia area.[19] It is quite possible that this may have been an influence of the work of Thomas Napier. After 1840 screw arm plows with handles were commonly offered by planemakers throughout United States. The better grades were in applewood, boxwood, rosewood and ebony, or combinations thereof, instead of the usual beech. However the slide arm plows continued to be optionally offered at least through 1872 as less expensive items. In 1872 the Greenfield Tool Co. offered a 4th grade slide arm plow at $4.00 as the least expensive(#497) of some 58 different choices of plow planes. The most expensive was #553, Handled ebony plow, ivory trimmed (tips of screw arms),silver trimmed (bands on screw nuts),boxed plate on fence at $22.80.

The 1828 J.Wilks' Price List (Table XV) notes item #813 as a *Plow Large Straight and Circular with 12 Irons*. Item No. 989 on the Marshes & Shepherd List (Table XVI) notes *Ploughs to work Straight or Circular*. These ploughs are more clearly spelled out on the 1870 Sheffield Illustrated List (Table XVIIb) Item#3233 - *Ploughs with Circular and Straight Work with 12 Irons and 2 Pair of Plates*. These were plows furnished with interchangeable fences and skates each set separately to preform either straight or circular work.[21] One of these rare planes, unfortunately missing the fence and skate attachments for performing circular work, is illustrated in Fig. 33. The skate may be readily readily removed as it is held to the plane body by screws set in brass plates on the other side of the stock. Then this may be replaced with a curved skate, held in an adapter frame. A curved fence attachment may then be fixed to the regular fence, screwed into the brass plates thereon. This particular plane was made by Joseph Marples at York Works, Sheffield, c.1874. The body and fence are cocobola wood with boxwood screw arms and plate, dovetailed into the fence.

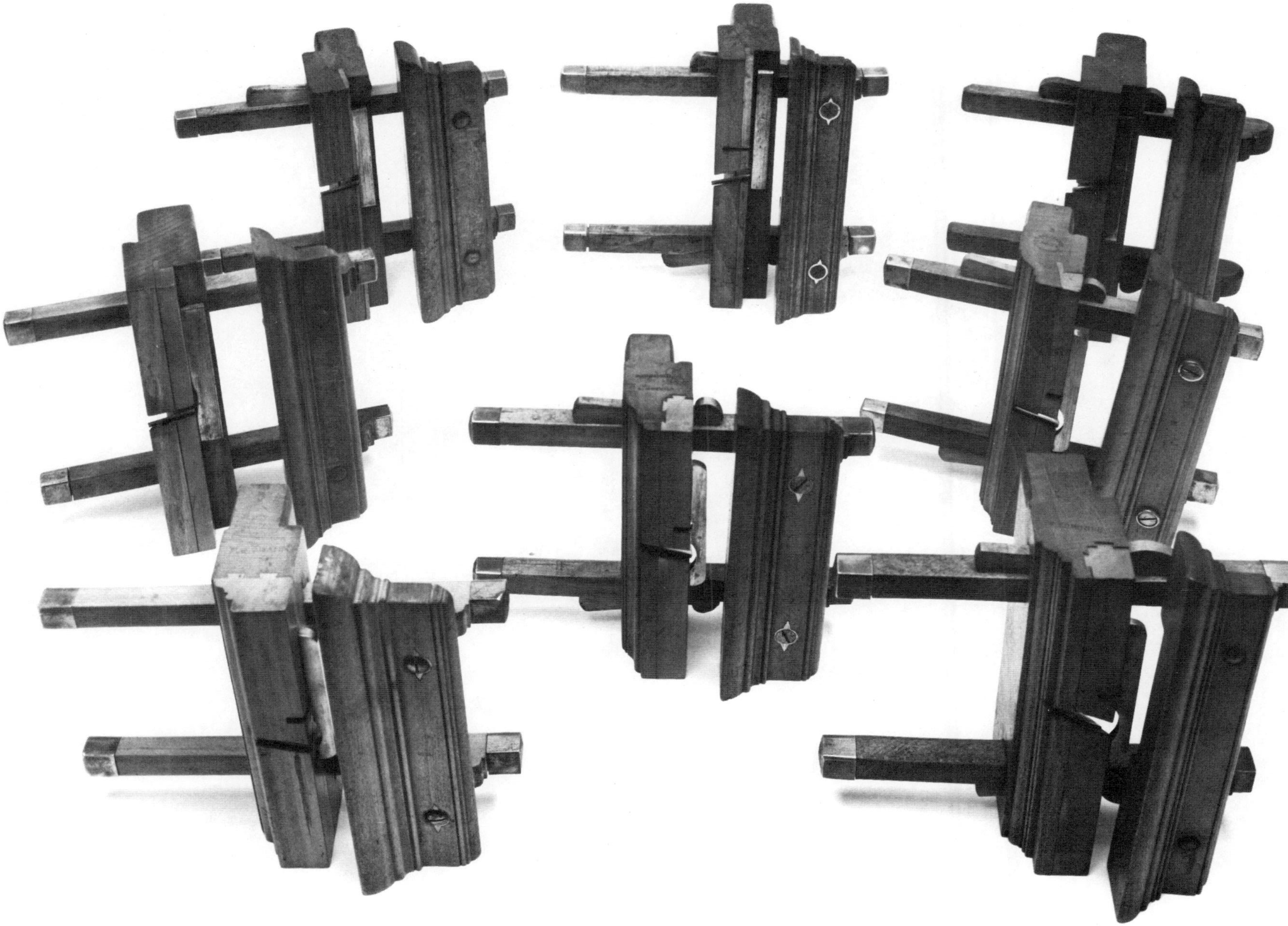

Fig. 31 A Various Methods of Boxing Fillisters [bottom & center; double dovetail; upper right no boxing; others - shoulder boxing]

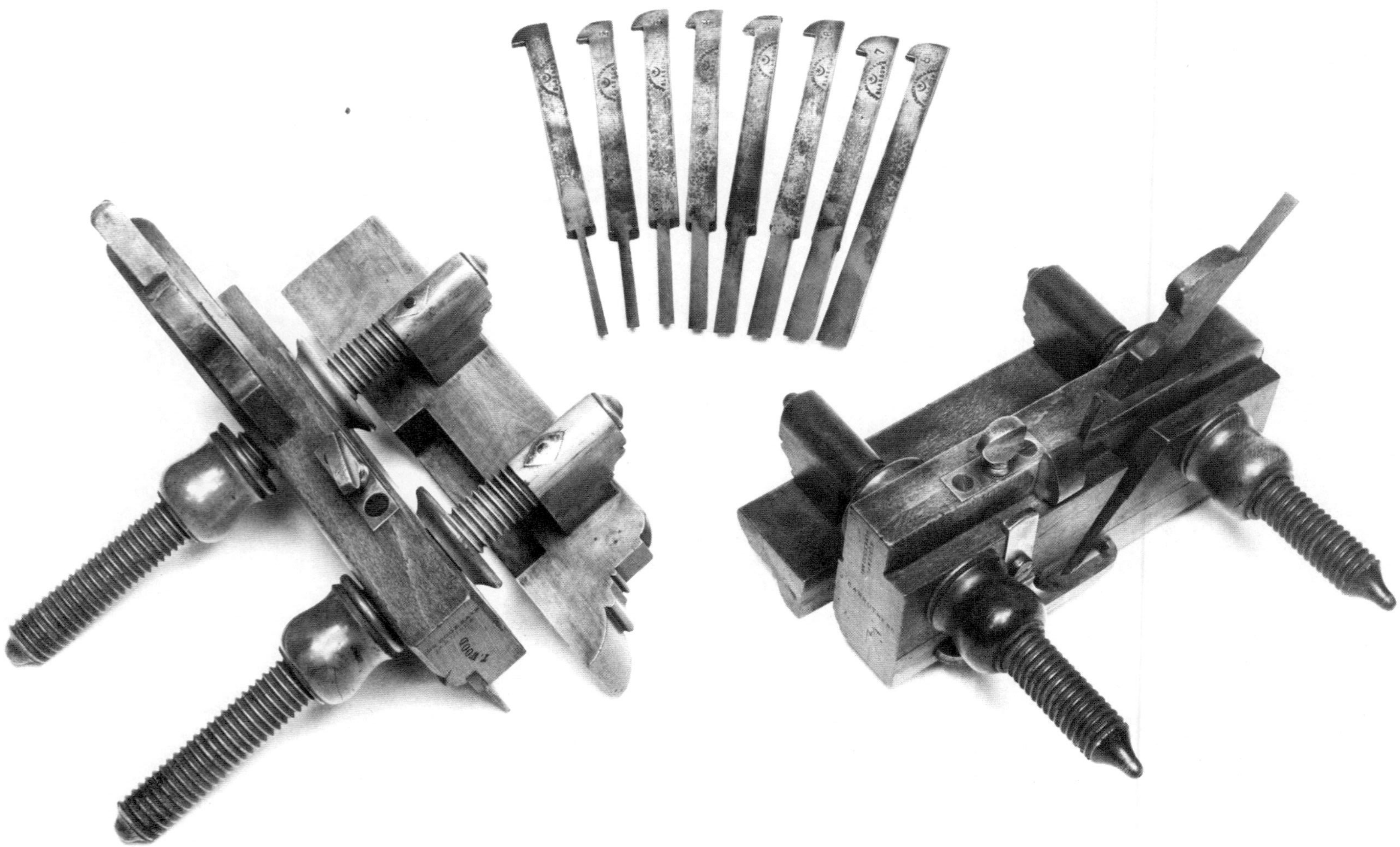

Fig. 32 - Handled Plough Plane with Screw Arms (left); Double Dovetailed Boxed Sash Fillister (right)
Planes made by John Cockbain, Carlisle, Cumberland (north England)
[Set of 8 Irons by Alex.Mathieson & Sons, Glasgow, Scotland]

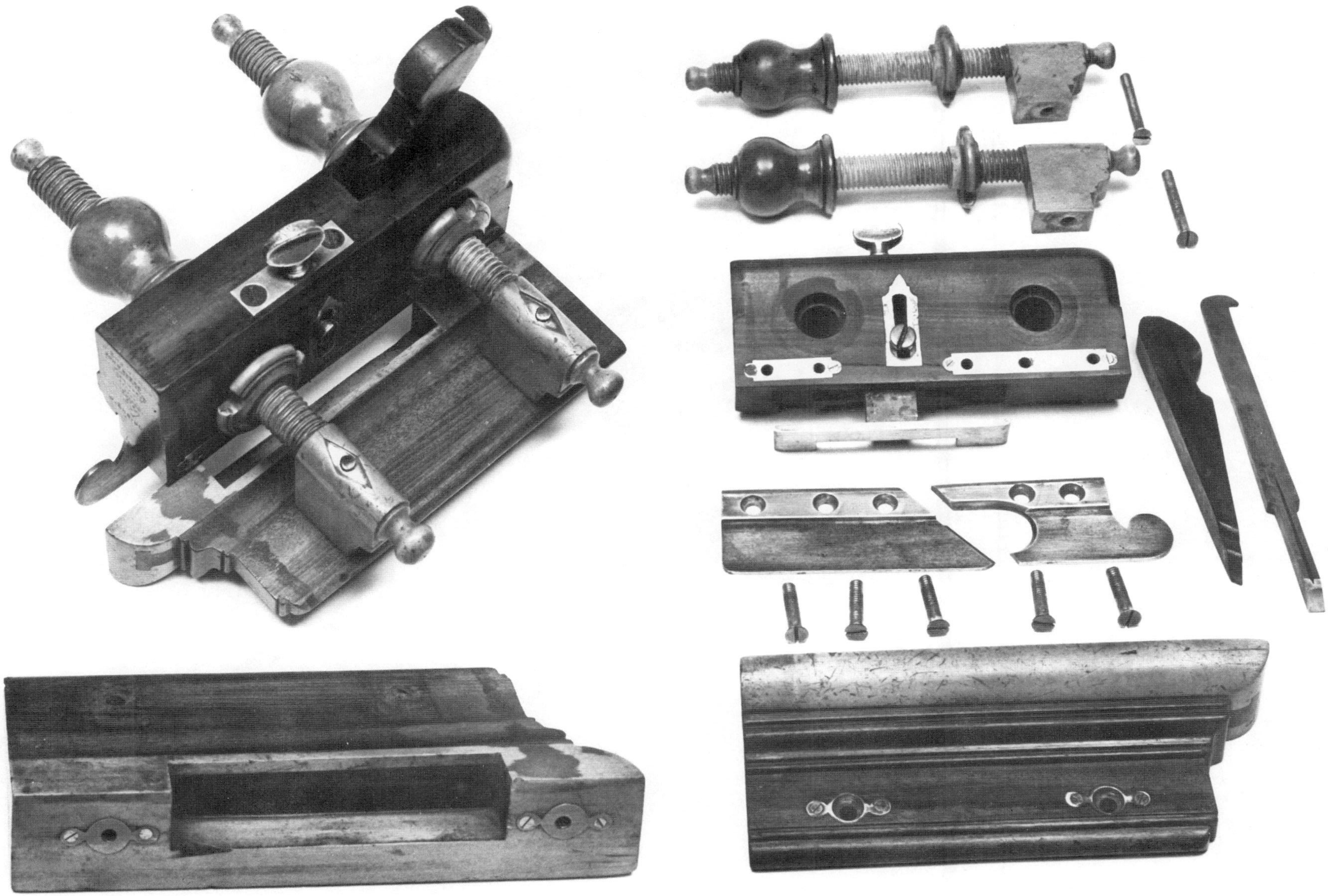

Fig. 33 - Combination Plough for Straight and Circular Work by Joseph Marples, York Works, Sheffield, c1874.
[Plane body and fence in cocobola wood; fence plated with dovetailed boxwood; circular attachment missing]

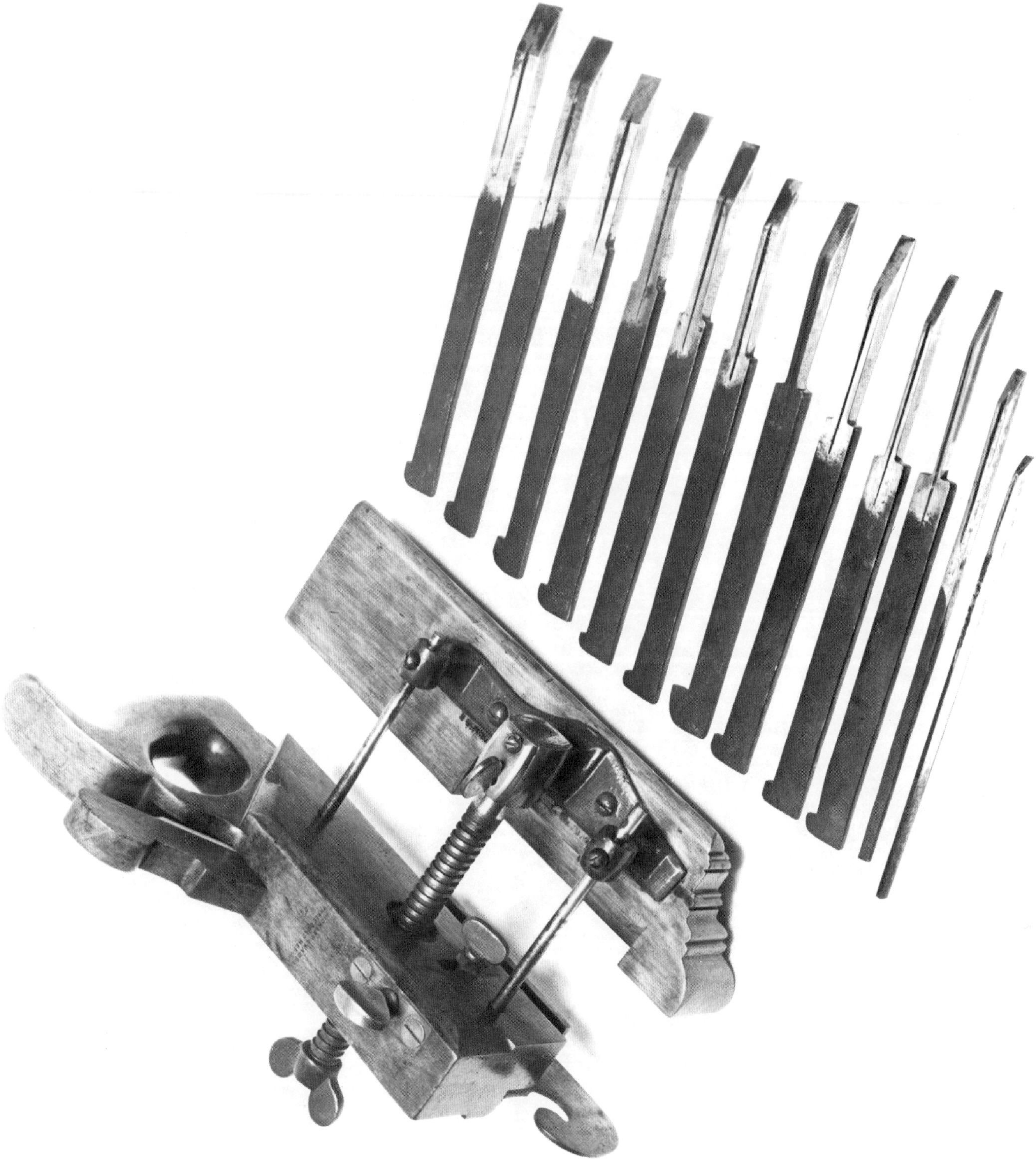

Fig. 34 - Screw Arm Adjustable Plough Plane by D. Kimberley of Birmingham with set of 12 Irons.

Fig. 35 - Solid Handle, Steel Bridle, Improved Angular Brass Stems (Ebony Filled) Plough Plane
by Alex. Mathieson & Sons, Glasgow, Scotland

Patent No. 2848 for Improvements in Plough and Sash Fillister Planes was granted to William Kimberley of Highgate Tool Works, near Birmingham. (applied for February 24, 1887) This consisted of a central transversing metal screw, attached to the fence and passing through an adjusting box in the body of the plane.[22] Turning the screw, by means of a wrench or wing handle, adjusted the distance of the fence. This arrangement is illustrated in Fig. 34. This plane was made by D.Kimberley & Sons, Birmingham. W.Kimberley was later granted Patent No. 16,689 (applied for September 1,1894) for a further improvement, consisting of a locking screw to fasten the regulating screw, once the adjustment had been set.

Alex. Mathieson & Sons, Glasgow, Scotland offered a line of many different plow planes.[23] Among the most unusual of these was their No.11. This had a solid handle, steel bridle, improved angular brass stems (ebony filled) and a brass lined fence. The bridle provided for adjustment of the fence position. This is illustrated in Fig. 35, and was their most expensive plow, listed in 1899 for 54/. The least expensive was a slide arm plow, brass ferruled with a wood stop at 8/8.

Many special purpose wooden planes were made in Great Britain, frequently using exotic woods. Included in such designs were chamfering, bull-nose and compass planes. Several of these are illustrated in Figs. 36 - 38. Iron bottoms and toe plates could be purchased from joiner tool houses to provide for wear resistant surfaces for smoothing planes. Some wide complex moulding planes are shown in Fig. 36.

It is apparent that many ideas for improvements in planes that were later fully developed in the United States had an earlier appearance in England. These included bench plane iron depthening adjustment, multipurpose iron combination planes, variable sole contour adjustment for compass planes and combination moulding and rabbeting sash plane.

Fenn's Patent Improved Smooth Plane, c1857, has been previously reported.[25] While no patent has been located covering this feature, this plane was listed in the 1857 *General Catalogue of Engineers' and Mechanics' Tools* of Joseph Fenn published at London in 1857. The method of adjusting the depth of the cutting iron is evident from examination of Fig. 39. This technique of depth control preceded Leonard Bailey's Patent in United States. A wooden plane with steel sole, adjustable for curvature, forming a compass plane is shown in Fig. 38. A set screw at each end controlled the position of the steel sole. This plane is imprinted with the name FENN on the front of the stock, and may have also been manufcatured at Joseph Fenn's London factory.

After studying the detailed specifications of nine pages and several drawings of James Silcock's Patent No. 10,033, granted January 31, 1844, it is evident that his design for a metal combination plane preceded that of either Phillips or Miller in America.[26] The only known, yet reported, surviving example of this Silcock's Patent Combination Plane is illustrated in Fig.40 , p.274. A drawing from the Patent shows the elevation and top views of this plane in Plate CLX, p.275.

The nine claims specified by James Silcock in this Patent are listed on page 276.

Fig. 36 - Wooden Planes. Top: Left and Right, Mahogany Smoothing Planes with Handles. [Note Cast Iron Toe Plate in Plane at Right]; Left center, Mahogany 'Old Woman's Tooth'Router; Right Center Bull nose plane. Bottom: Left; Complex Single Iron 3" width moulding plane with handle; Center; 3" Complex Moulding Plane without Handle. Right; 3" Complex Moulding Plane with Handle and 3 Irons. [The latter had the advantage of easier to sharpen and boxed for wear resistance]

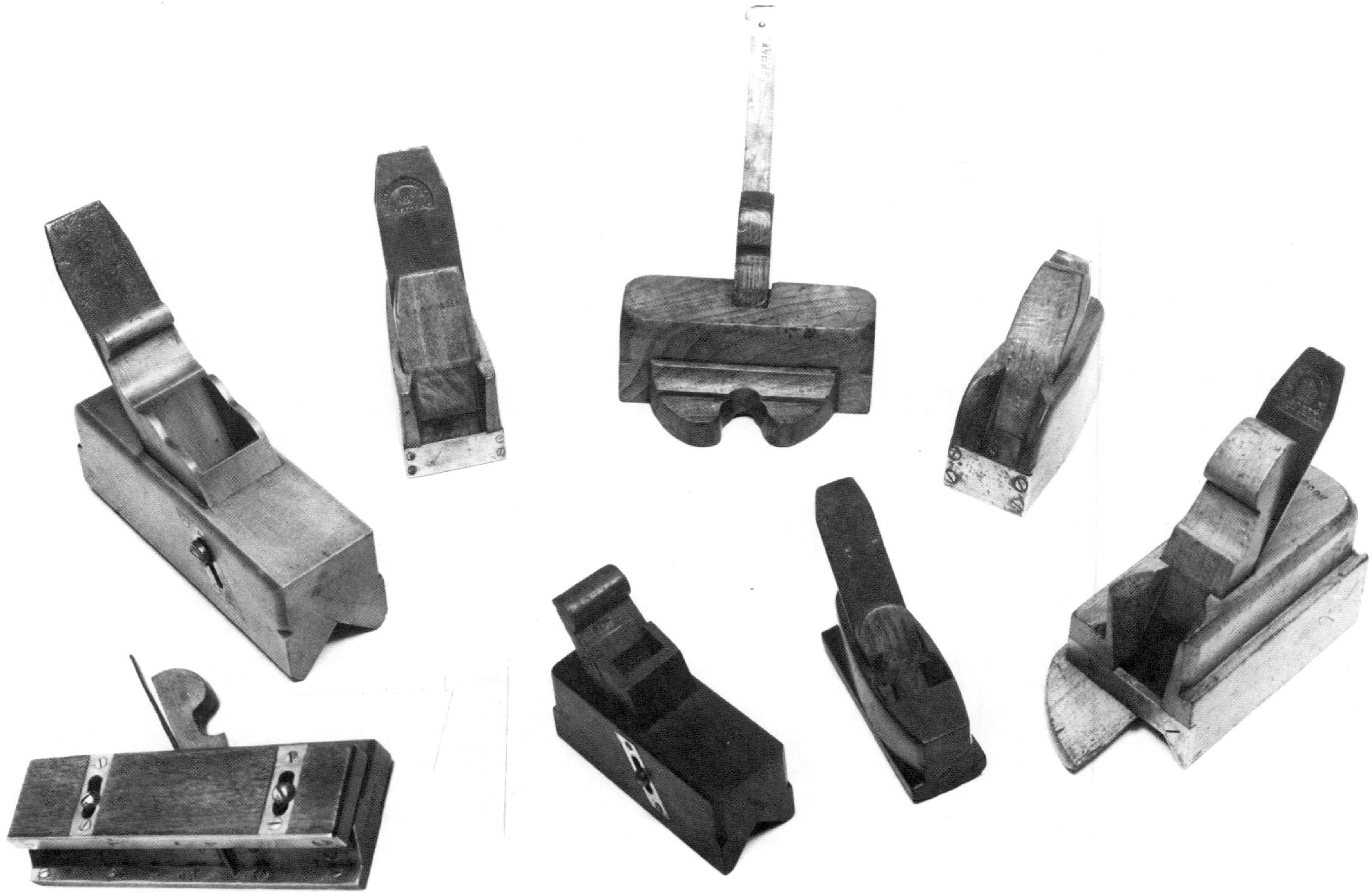

Fig. 37 - Wooden Planes. (Top Row - Left to Right) Adjustable Boxwood Chamfer Plane; Bull Nose Plane; Old Woman's Tooth Router; Bull Nose . (Bottom Row - Left to Right) Adjustable Bevel Plane; Rosewood Chamfer Plane; Small Smoothing Plane; Adjustable Beech Chamfer Plane.

Fig. 38 - Wooden Planes. (Top Row, Left to Right) Smooth Plane with Cast Iron Front Plate; Window Sill Drip Plate Plane; Smoothing Plane with Cast Iron Sole Plate. (Bottom Row, Left to Right) Compass Plane with Adjustable Boxwood Front Plate, Mahogany Adjustable Compass Plane with Steel Sole. [This can be set on each end by a screw in a slotted plate. This plane is stamped Fenn (possibly London). The manual adjustment probably predated Leonard Bailey's US Patent 113,003, March 28, 1871 for a Circular Plane). The plane on the right has an iron plate adjusted by a brass wing screw from the top.

Fig. 40 - James Silcock's Patent Combination Plane [Collection of Kendall H. Bassett] (see pages 270,275 & 276)

FIG. 5.

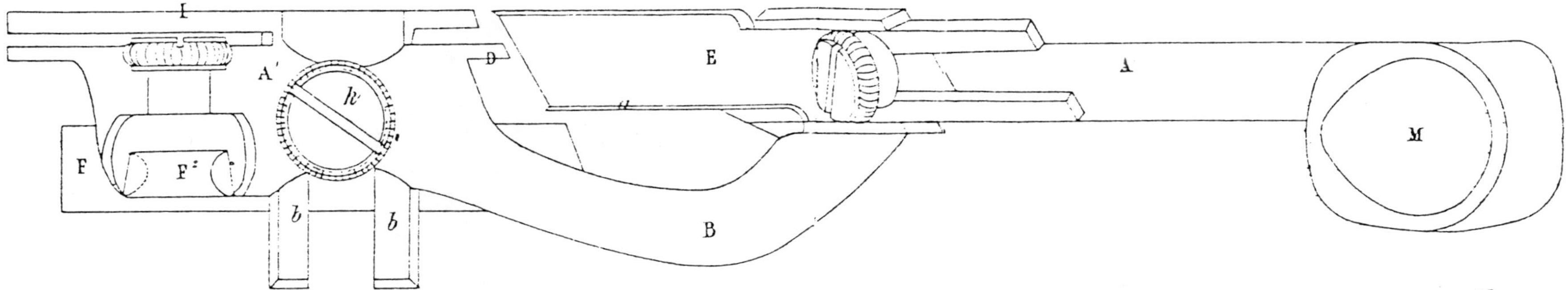

FIG. 6.

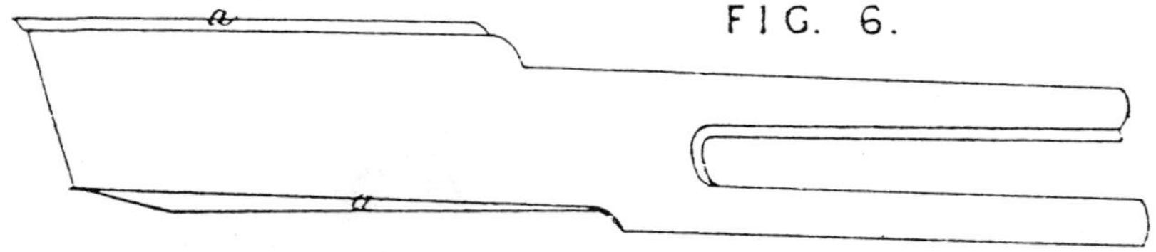

FIG. 4.

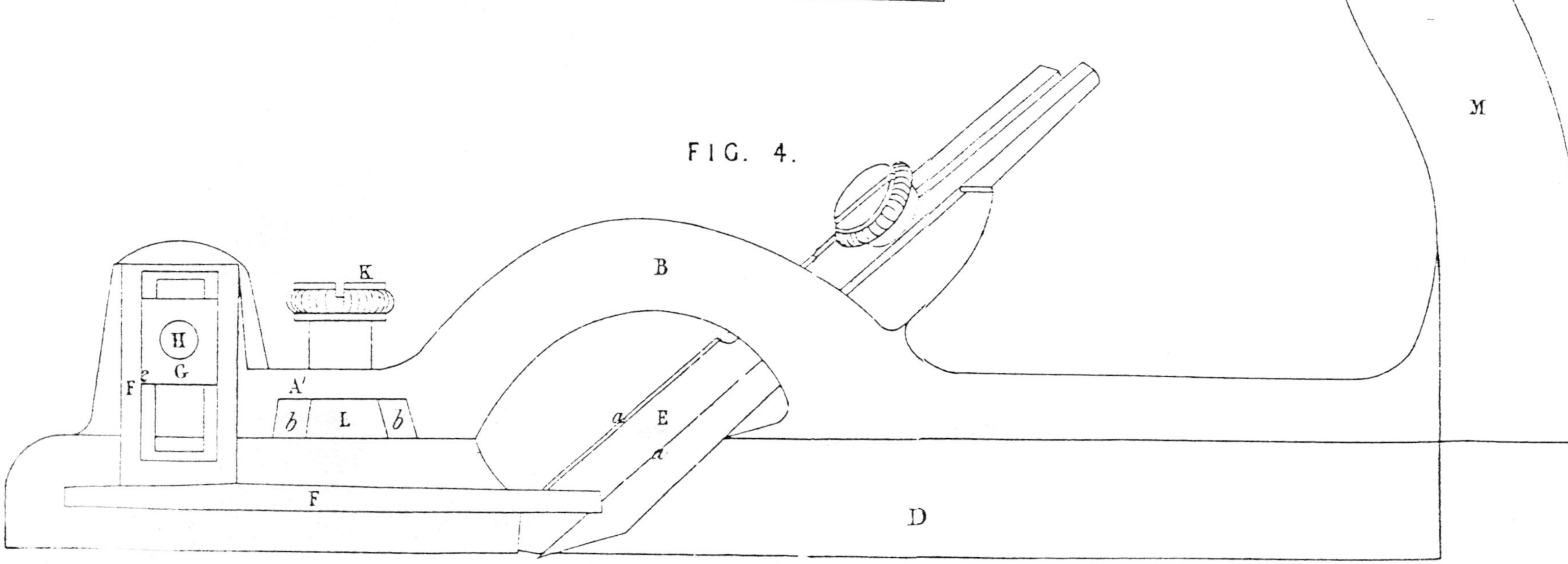

Plate CLX - Patent Drawing of James Silcock's Combination Plane #10,033, Jan. 31,1844, Elevation & Top Views

James Silcock made the following claims in his Letters Patent No. 10,033, dated January 31,1844 for his invention: "Certain Improvements in Planes" -

1. The making of double fillister planes whereby they can fillet boards of all sizes, from about three-eighths of an inch to about three inches, and can also be adapted to the several purposes of a filleting plane, a side fillister, a sash or back fillister, and a skewed rabbet plane.
2. The making fluting or grooving planes.
3. The making of dado grooving planes.
4. The making of trying planes of a number of pieces connected together.
5. The making of moulding or beading planes.
6. The securing the irons of planes to their beds or seats by means of screws and nuts instead of by wedges as usual.
7. The use of chases with travelling nuts and screws for varying the distance between the fences and cutting irons of planes.
8. The use of stops for regulating the height of fillets or depth of grooves.
9. The casting in one piece of those parts of planes which are metal. (usually cast iron coated with tin or zinc)

Fig. 39 - Fenn's Patent Improved Smoothing Plane [See Page 270]

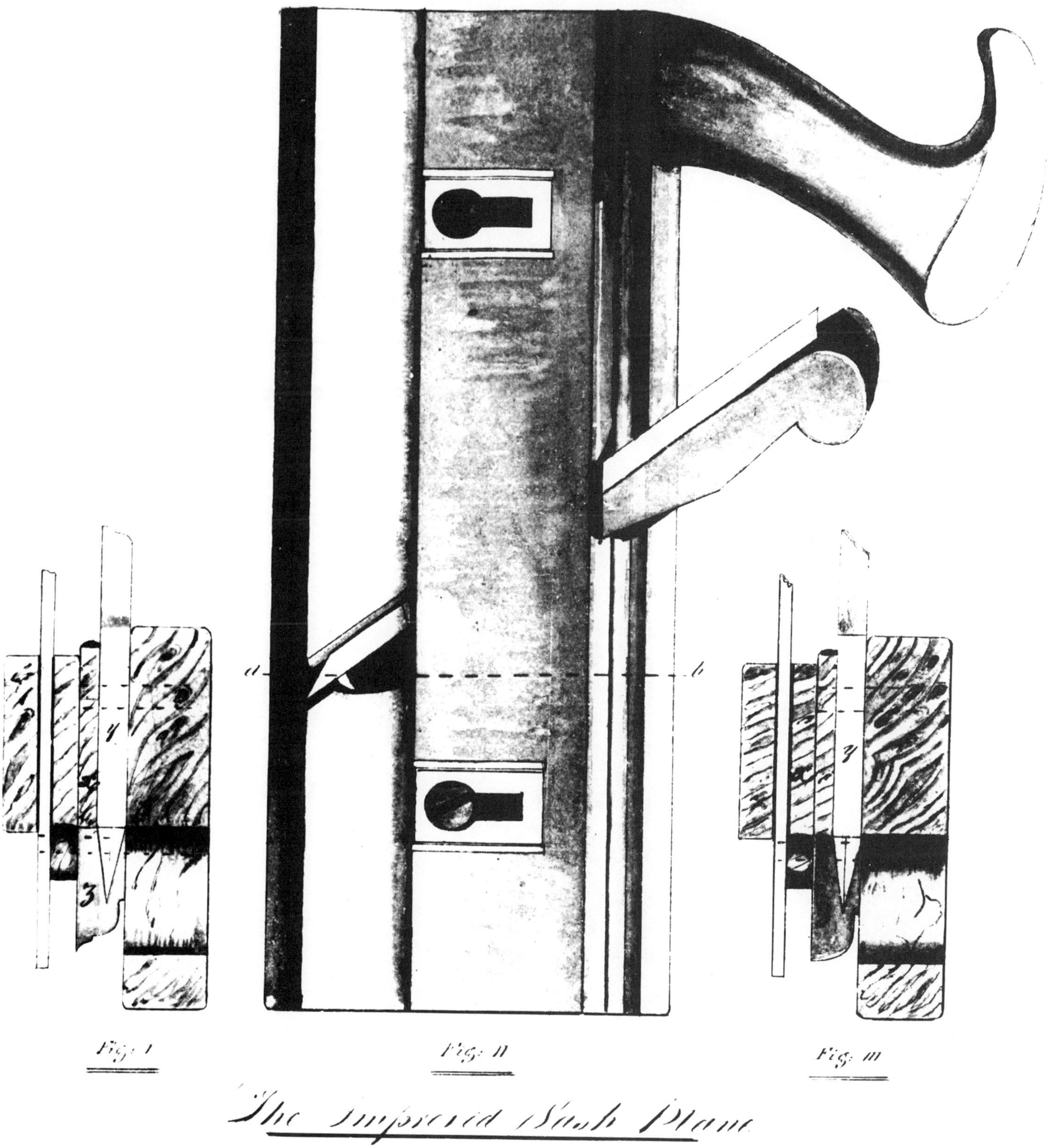

Plate CLXI - Drawing Accompanying George Sievers'Registered Design for Improved Sash Plane. No. 549 November 24, 1853.

George Sievers of Cannon Street, Bindley, Yorkshire was granted Registered Design No. 549, November 24, 1853 for an improved sash plane: "in the forming or working the moulding on the front edge of sash bar simultaneously with rebating the back edge to the same".

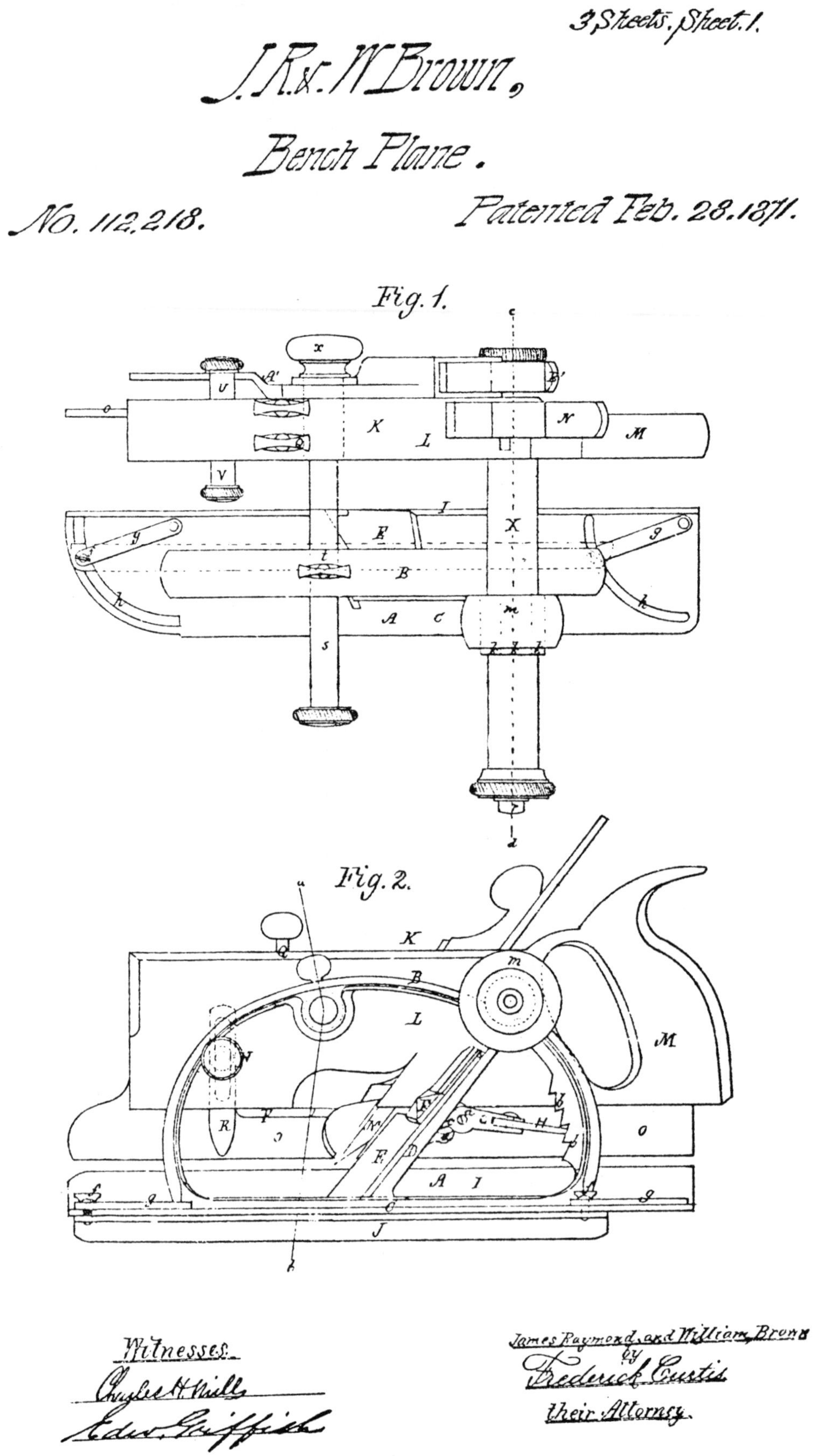

Plate CLXII - Patent Drawing from USA Patent No. 112,218 J.R. & W. Brown Improvement in Carpenters' Planes - February 28,1871.

Fig. 41 - J.R. and W. Brown Combination Plane - US Patent No. 112,218, Feb. 28,1871.

James Raymond Brown and William Brown of Boston, Massachusetts were granted US Letters Patent No. 112,218, Feb. 28,1871 for Improvement in Carpenters' Planes. In their application they stated they were: - "subjects of the Kingdom of Great Britain, at present residing at Boston, USA." Fig. 41 illustrates this plane, without fence, which can be noted from the Patent Drawing, shown as Plate CLXII. Their specifications stated: "We have embraced in the combination about to be explained devices for rebating lumber for the manufacture of doors and for various other purposes; also devices which constitute a matching-plane for operating upon and preparing boards of various dimensions,and, finally, devices for plowing a cross-groove, the latter constituting one side of the matching-plane, and the whole operating as hereinafter explained". They also claimed that considerable economy in space was gained which enabled a mechanic to obtain at a small comparative cost, and in a compact and efficient form, the substitutes for several classes of planes.

With these various innovative ideas it may seem strange that subsequent manufacture did not follow to any large extent in Great Britain, but succeeded later in United States. The probable reasons were lack of communications between inventors and manufacturers with limited capital for risking radical departures. Traditional English practices were indeed difficult to change. The owners of such tool manufacturing enterprises were more concerned with maintaining profitable returns from conventional items rather than uncertain investments in new products. In America it was different. Free enterprise with competition was the manner of "building a better and less expensive mouse trap". In the case of the combination plane by Silcock, it was probably too early to realize the potential of this idea. By 1871 the Browns probabily considered that United States offered better sales' markets for their design, and consequently patented this here instead of England. However it was too late. That year Stanley Rule & Level Co. added the Miller's Patent combination plane to their line of successful Bailey designed metal planes.[27]

The appearance of metal planes in Great Britain is usually associated with Stewart Spiers of Ayr, Scotland about 1840.[28] Spiers, an apprenticed cabinetmaker,is said to have started making such planes after picking up a rough casting in Glasgow and finishing it for resale. If such was the case, metal planes must have been in existence before this date. In any event Spiers made a line of smooth, panel, jointing, mitre, shoulder, rabbet, chariot, bull-nose and thumb planes that soon became very popular. Some of these were made with steel soles so finely dovetailed into iron castings that it is difficult to observe the joints. Several Spiers' planes are illustrated in Fig. 42.

Thomas Norris of London claimed to have been established in 1860, though he is first listed in a London Directory of 1873, working at 57 York Road, Lambeth. The firm was followed by T.Norris & Son. Their 1914 Trade Catalogue called to attention the Norris Patent No.11526/13 for lateral and vertical adjustments of the cutting iron.[29]

After 1860 numerous other makers of metal planes appeared. Among these were: H.Slater, Clerkenwell, London; W.Thackeray, Leeds; Charles Badger, London; Isaac Green, London; and John Holland, London. Numerous styles and sizes of chariot, shoulder and bull-nose planes, both in iron and gun metal,[30] and often filled with exotic woods have been produced. These have become exceeding popular and in high demand by present tool collectors,as well as craftsmen. (See Figs. 43 - 50) Many of these do not have imprints of makers. A large portion of such planes were made by individual unknown craftsmen, who purchased rough castings from joiner supply houses and finished their own planes. This accounts for the great diversity in shapes.

The only two metal planes noted on the 1870 Illustrated Sheffield List were the #2360 and #2361 wood filled iron Rabbet Planes, respectively single and double irons. (p.259) Chariot and bull nose planes, both in iron and gun metal, were added to the 1876 Illustrated Sheffield List. (p.260-261)

By 1880 William Marples & Sons were offering a complete line of imported Bailey planes from Stanley Rule & Level Co. in competiton with their line of Moseley wooden planes. Their 1891 Catalogue noted a line of English Pattern Iron Planes and American Pattern Iron Planes, which appear to be modifications of Bailey's designs. (Plate CLXIII)

ENGLISH PATTERNS—IRON PLANES.

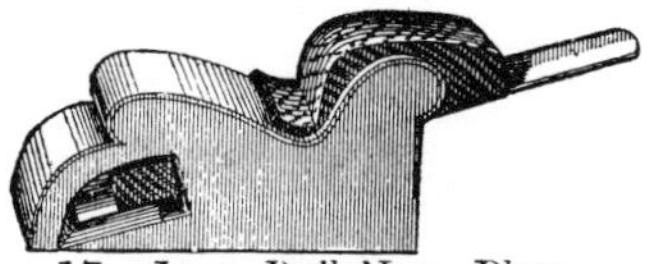
17 *Iron Bull-Nose Plane.*

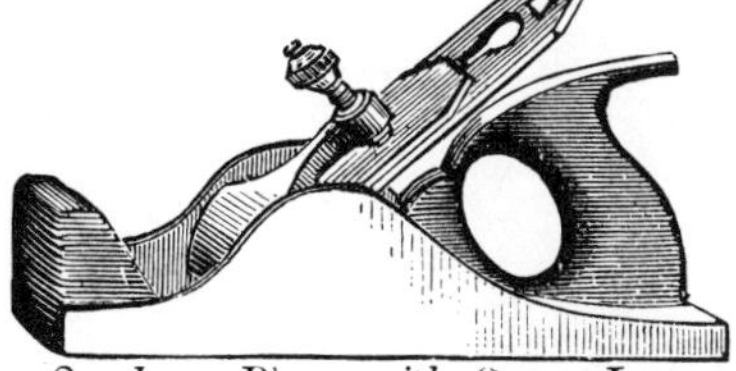
8 *Iron Plane with Screw Lever.*

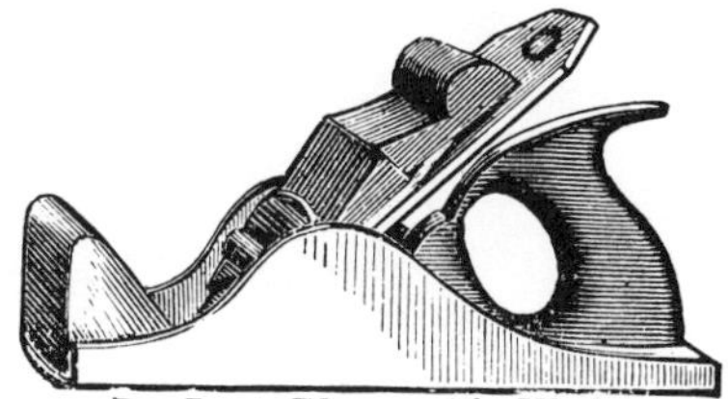
7 *Iron Plane with Wedge.*

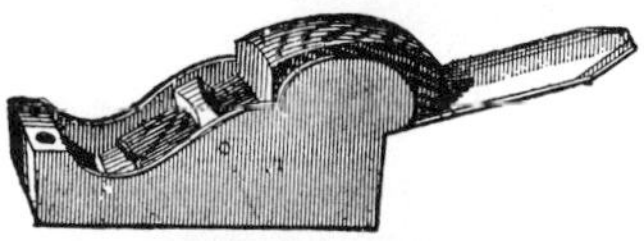
20 *Iron Chariot Plane.*

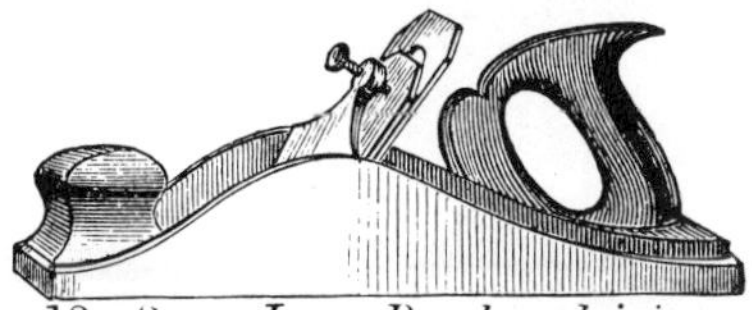
13 *Screw Lever Panel or Joining Plane.*

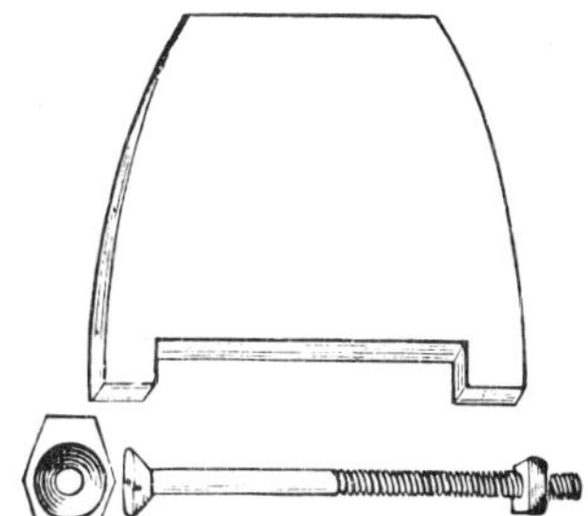
1089 A *Iron Fronts for Smooth Planes.*

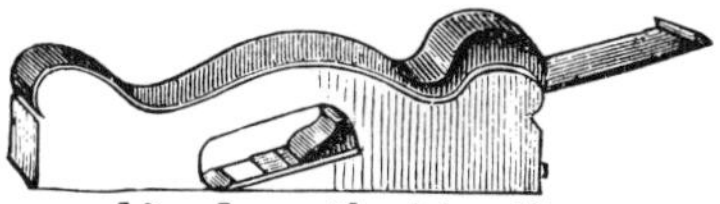
15 *Iron Shoulder Plane.*

4 *Iron Smooth Plane, with Screw Lever.*

1 *Iron Smooth Plane, with Wedge.*

AMERICAN PATTERNS—IRON PLANES.

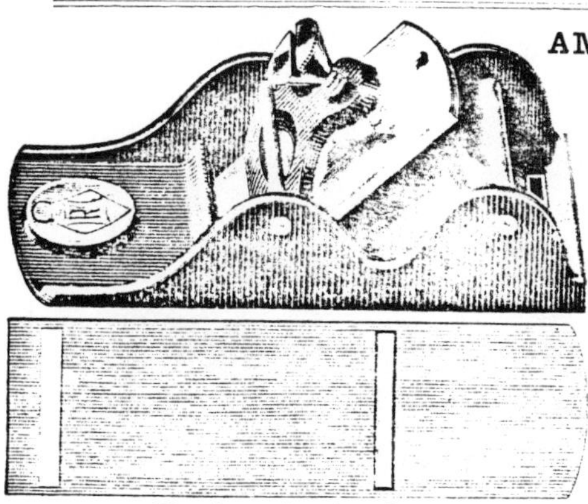
99 *Iron Duplex Plane.*

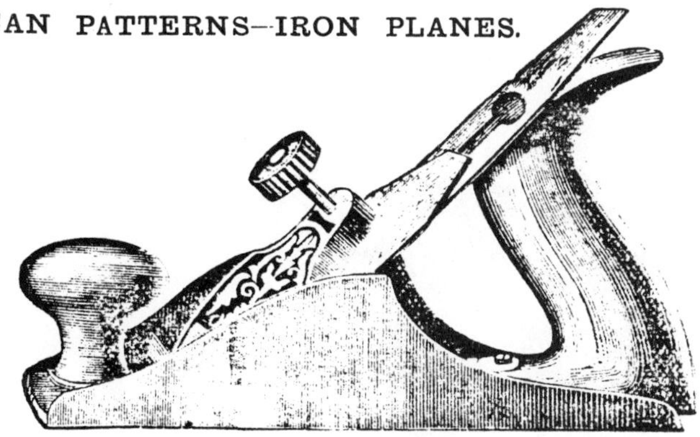
34 *Iron Smoothing Plane.*

173 C *Thin Plane Iron.*

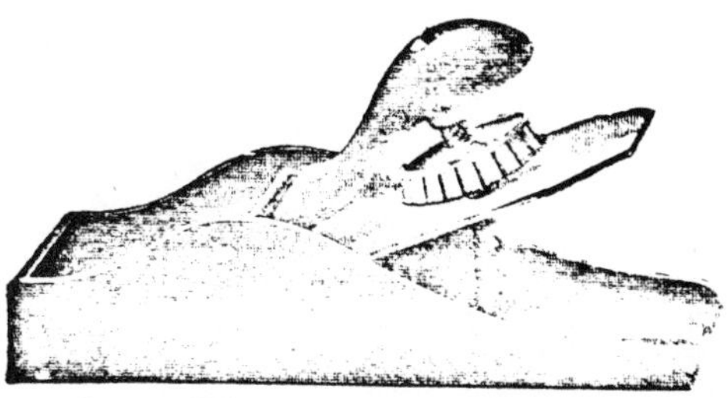
1089 *Hibernia Iron Chamfer Plane.*

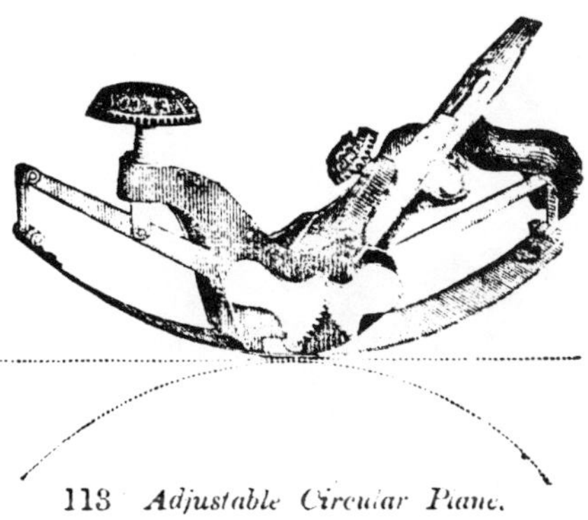
113 *Adjustable Circular Plane.*

100 *Iron Chamfer Plane.*

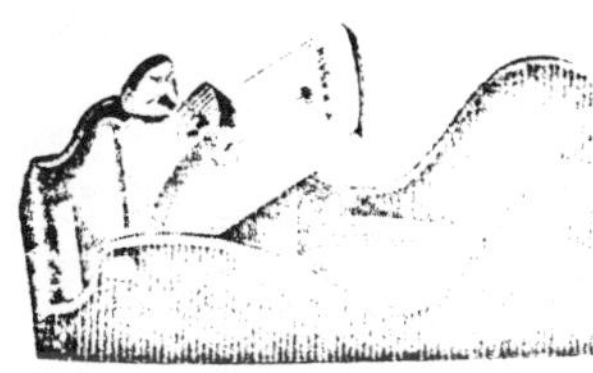
100½ *Iron Bull-Nose Plane.*

102½ *Iron Block Plane.*

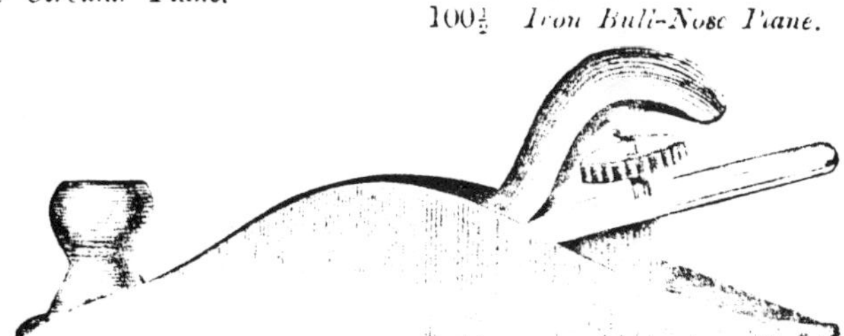
110½ *Iron Block Plane.*

Plate CLXIII - Iron Planes in William Marples & Sons 1891 Catalogue

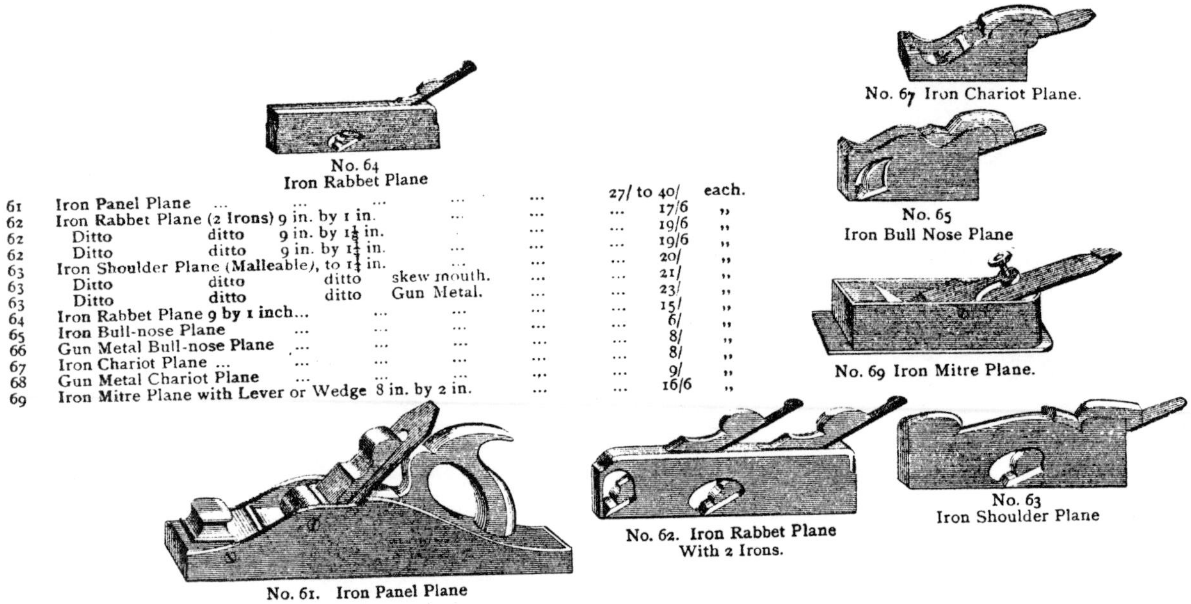

61	Iron Panel Plane	27/ to 40/	each.
62	Iron Rabbet Plane (2 Irons) 9 in. by 1 in.	17/6	,,
62	Ditto ditto 9 in. by 1⅛ in.	19/6	,,
62	Ditto ditto 9 in. by 1¼ in.	19/6	,,
63	Iron Shoulder Plane (Malleable), to 1¼ in.	20/	,,
63	Ditto ditto ditto skew mouth.	21/	,,
63	Ditto ditto ditto Gun Metal.	23/	,,
64	Iron Rabbet Plane 9 by 1 inch...	15/	,,
65	Iron Bull-nose Plane	6/	,,
66	Gun Metal Bull-nose Plane	8/	,,
67	Iron Chariot Plane	8/	,,
68	Gun Metal Chariot Plane	9/	,,
69	Iron Mitre Plane with Lever or Wedge 8 in. by 2 in.	16/6	,,

Plate CLXIV - Metal Planes Offered in A.Jinkinson & Sons Catalogue, c1890

The line of metal planes offered by Ward & Payne, Sheffield in their Catalogue, c1890, is illustrated in Plate CLXV.

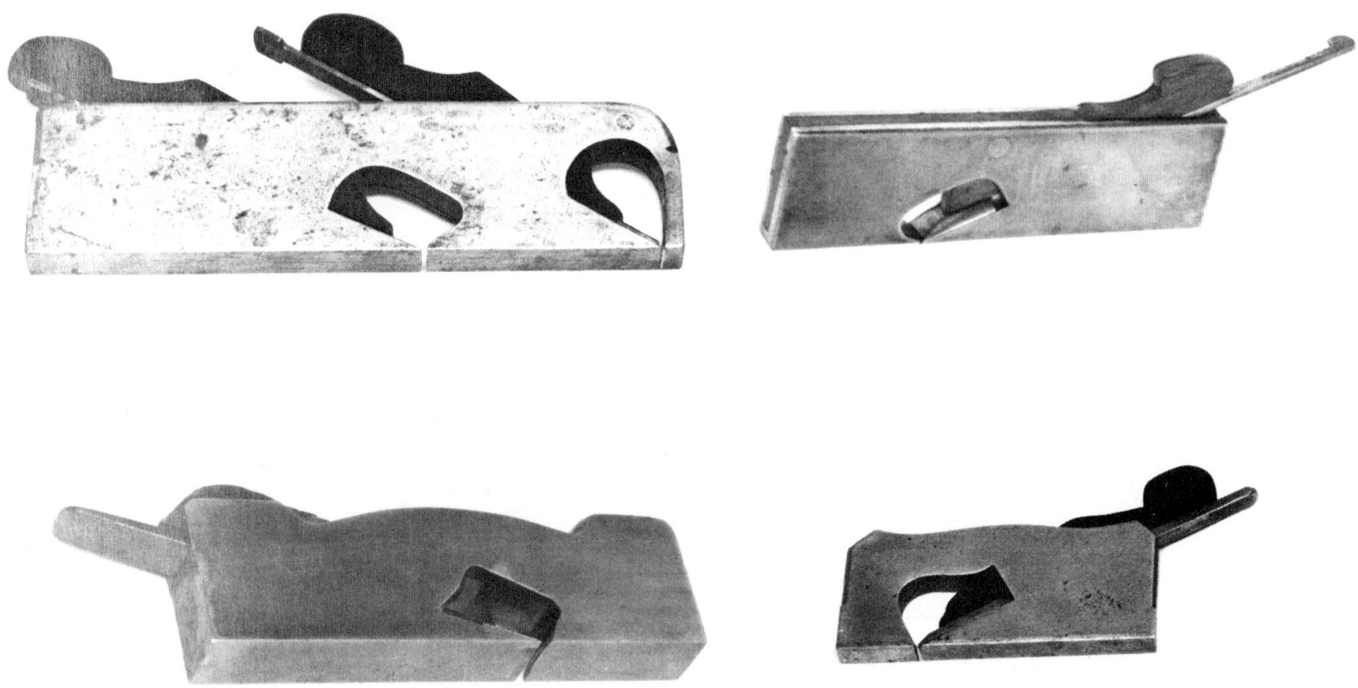

Fig. 43 - Miscellaneous English Pattern Metal Planes: Top - Left, Double Iron Rabbet Plane by Spiers, 9-1/4"x11/16"; Right, Iron Rabbet Plane by Norris, 6"x1/4"; Bottom - Brass Shoulder Planes - Left, 6"x1"; Right, 3-3/4"x3/8" [Makers Unknown]

WARD & PAYNE, CORPORATE MARK W P SHEFFIELD.

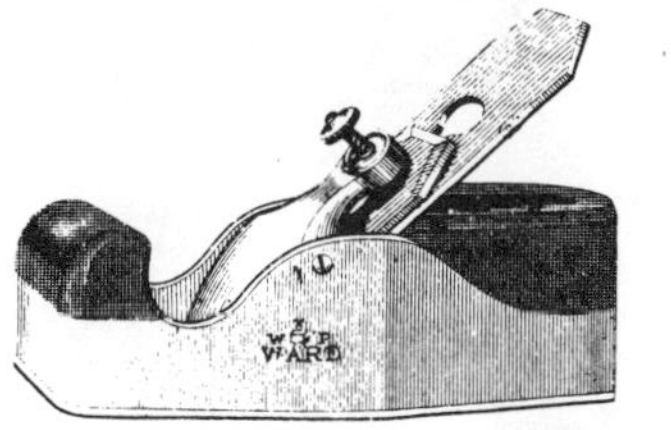

1560. *Iron Smooth Plane, with Lever.*

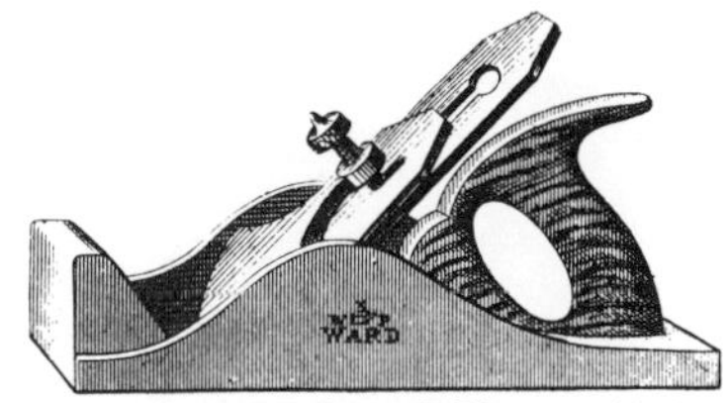

1564. *Handled Smooth Plane, with Lever.*

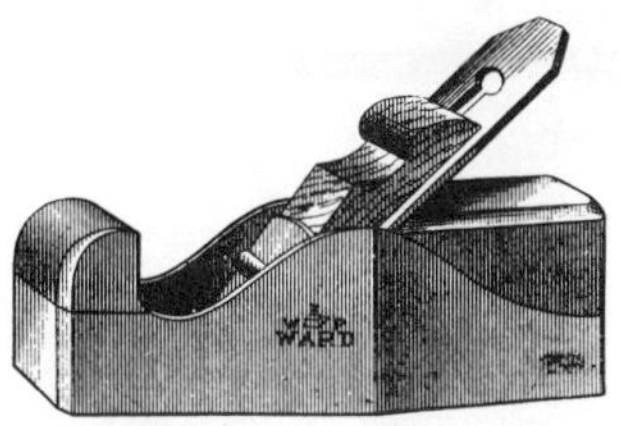

Iron Smooth Plane, with Wedge.

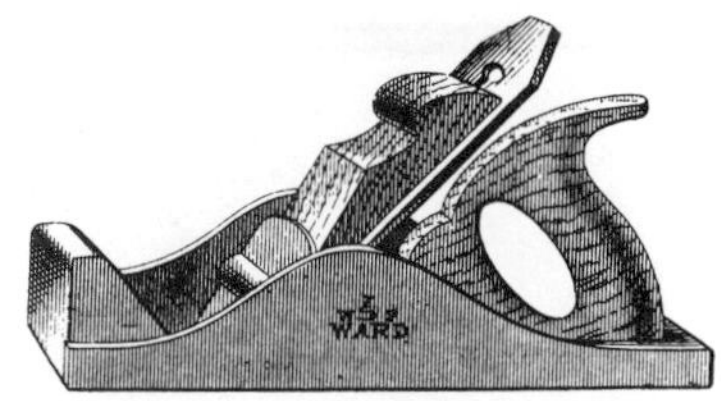

Handled Smooth Plane, with Wedge.

1568. *Iron Mitre Plane, with Lever.*

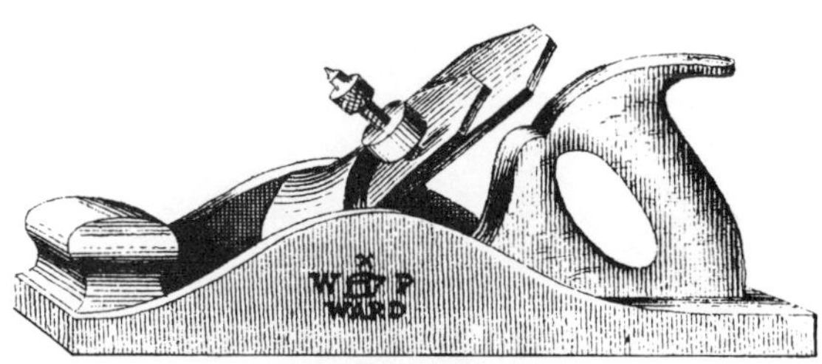

1570. *Iron Panel Plane.*

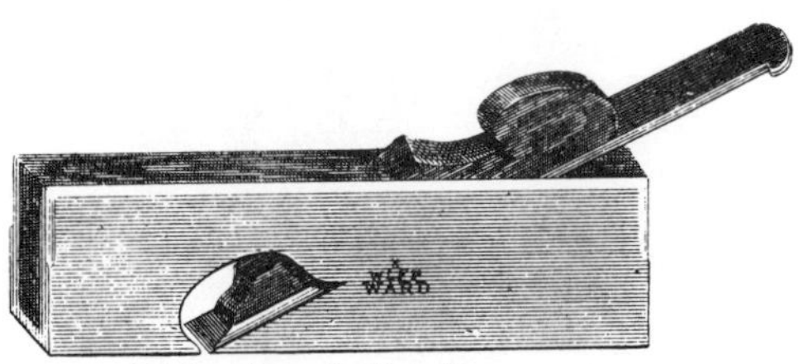

1572. *Iron Rabbet Plane*

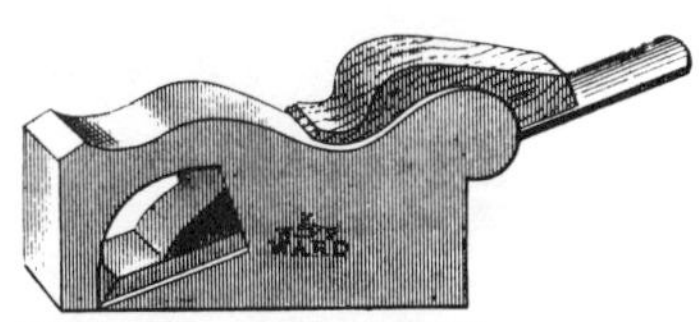

1573. *Iron Bull Nose Plane.*

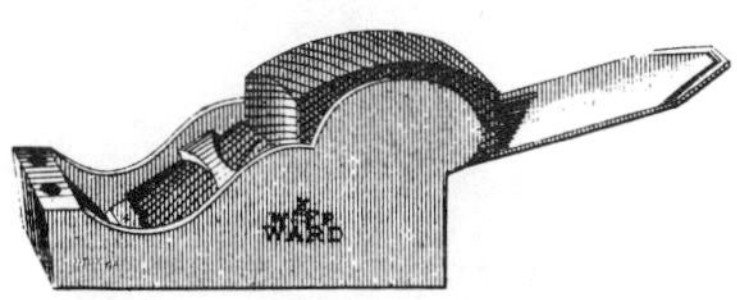

1576. *Iron Chariot Plane.*

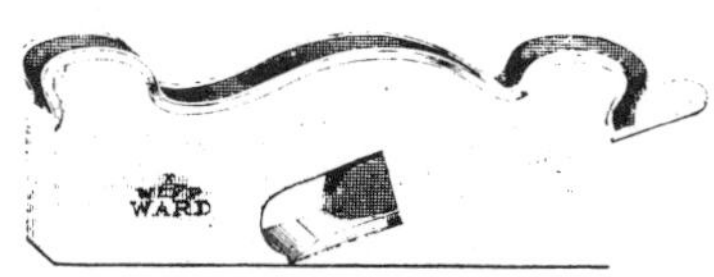

1579. *Iron Shoulder Plane*

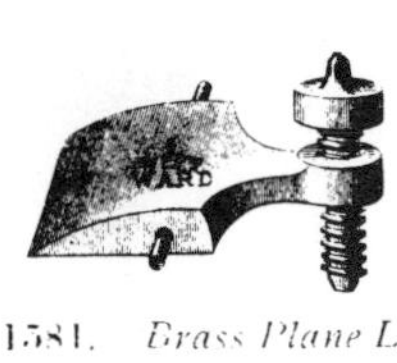

1581. *Brass Plane Lever.*

Plate CLXV - Metal Planes Offered by Ward & Payne, c1890

Fig. 42 - Metal Planes by Stewart Spiers of Ayr, Scotland. Front Row: Left, Dovetailed Steel Smoothing Plane with Lever, 6-3/4"x2-1/2"; Center, Dovetailed Steel Mitre Plane, 10-1/2"x2-3/4"; Right, Dovetailed Steel Smoothing Plane with Handle, Round Sides, 7-1/2"x2-5/8". Back Row - Left, Dovetailed Steel Joining Plane, 20-1/2"x2-3/4"; Center, Dovetailed Steel Panel Plane, 15-1/2" x2-3/4"; Right, Dovetailed Steel Smoothing Plane, Parallel Sides, 8"x2-3/8".

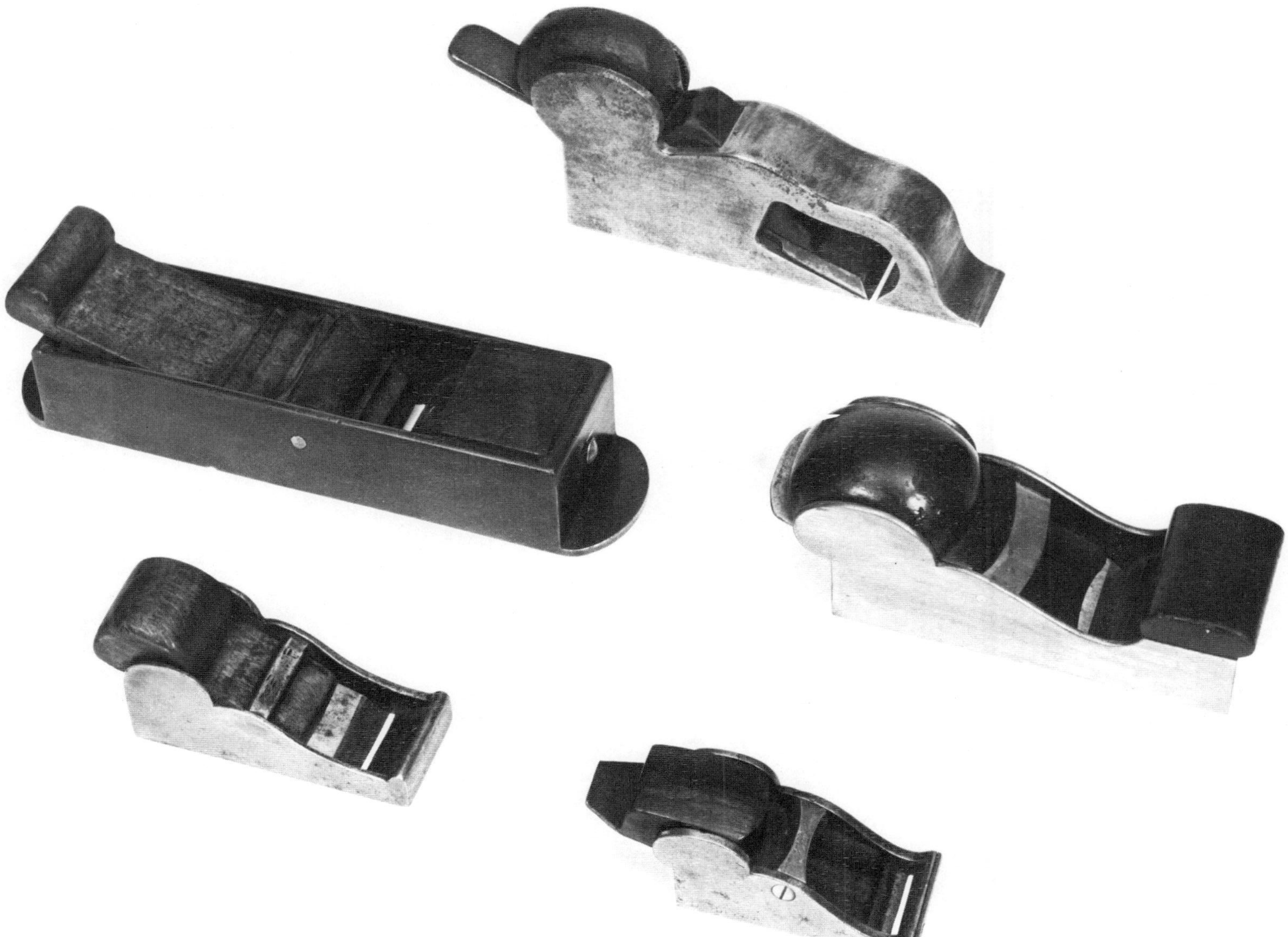

Fig. 44 - Miscellaneous English Pattern Iron Planes [Unknown Makers] - Top - Shoulder Plane, 7-1/8"x1-3/8"
Center, Left - Mitre Plane, 10-1/4"x2-3/8"; Right - Chariot Plane, 5-5/8"x2-3/16"; Front -
Bull Nose Chariot, Left 4"x1-5/8"; Right 3-1/8"x1-7/16"

Fig. 45 - Gun Metal Shoulder Planes - Top center, 7-3/4"x1-3/4" [Note: Protrusion at top for holding plane]; Middle left, 7-1/2"x1-1/2", Middle right, 8-1/8"x1-11/16"; Front left, 7-5/8"x1-9/16"; Front right, 8-1/8"x1-1/2".

Fig. 46 - Iron Shoulder Planes - Top left, 7-1/2"x1-3/8"; Top right, 8"x1-1/8"; Middle left, 7-3/4"x1-1/2"; Middle right, 8"x1-1/2"; Bottom center, 7-3/4"x1-1/2"

Fig. 47 - Miscellaneous Iron English Pattern Planes - Top left, Bull-Nose, 3-7/8"x1-3/16"; Top right, Rabbet Plane, 8"x3/4"; Center left, Bull-Nose, 4"x1'; Center right, small rabbet, 4-9/16"x7/16"; Front left, Bull Nose, 3-9/16"x1"; Front right, Double iron (right or left) Rabbet Trim, Wood Top, 4-3/4"x2-3/4"

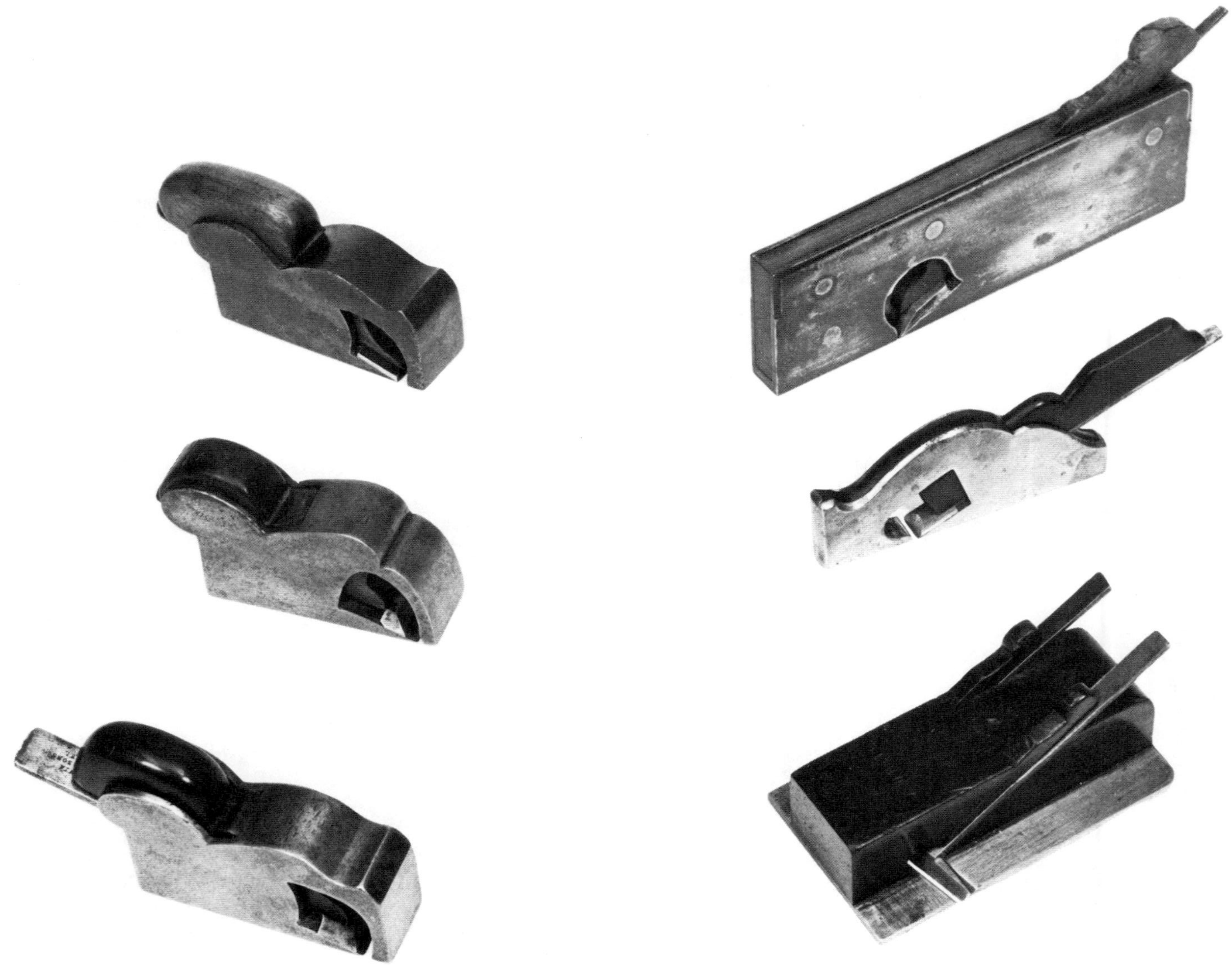

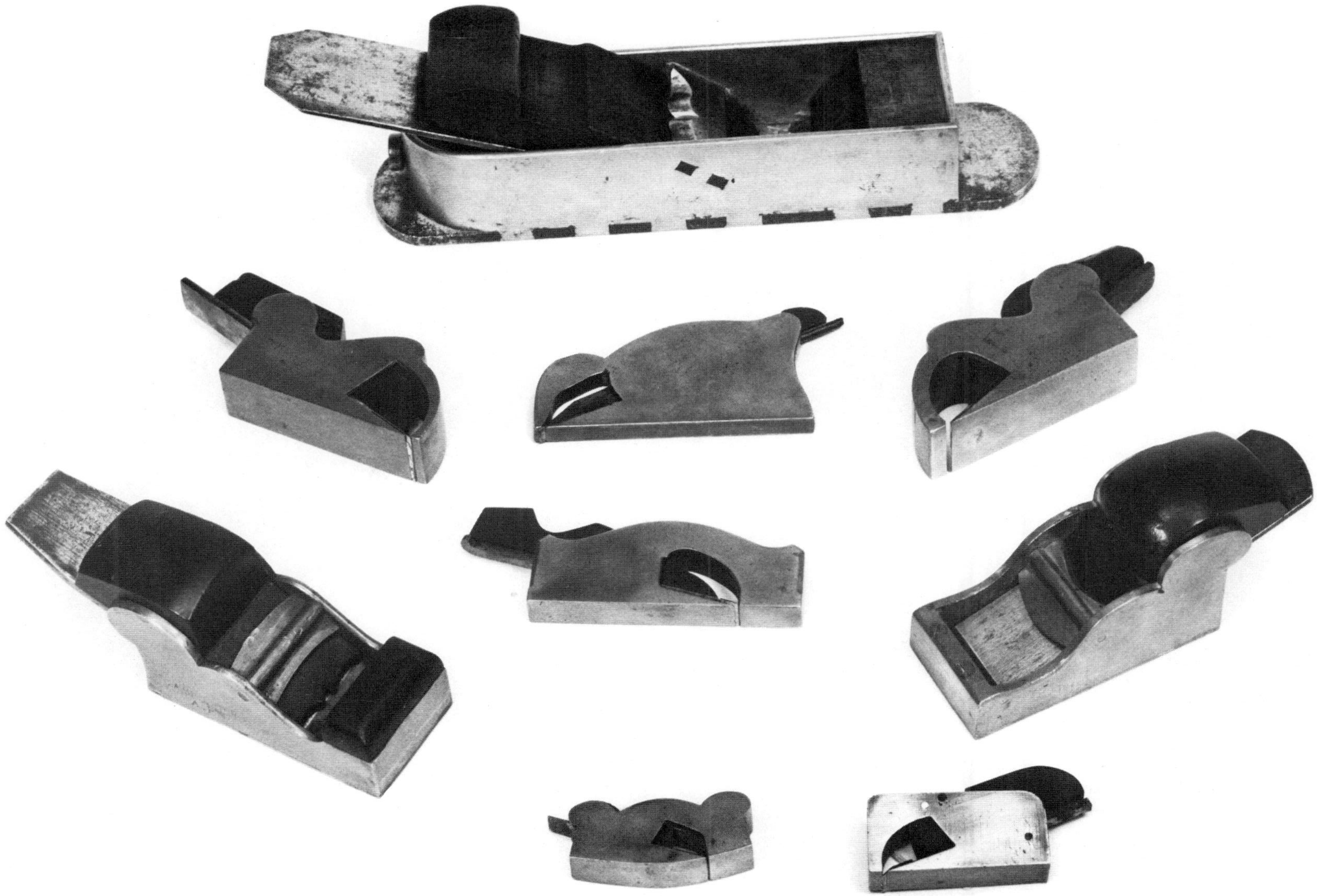

Fig. 48 - Miscellaneous English Pattern Gun Metal Planes - Top center, Dovetailed Mitre with 14in.Sole, 11-1/4"x2-1/2"; Third row - Left, Bull-Nose, 3-9/16"x15/16"; Center, Bull-Nose, 4-1/2"x3/8"; Right, Bull-Nose, 3-3/4"x1"; Second row - Left, Chariot, 4"x2-13/16"; Center, Small Rabbet, 3-7/8"x1/2"; Right, Chariot, 4"x2-3/4", Front center - small compass 2-1/2"x1/2"; front right - Small Rabbet, 2-1/2"x3/4".

Fig. 49 - Miscellaneous English Pattern Rabbet Planes - Top center, left iron, 7"x3-1/8", wood top; Left center, left iron, gun metal base, wood top, 8-1/4"x3"; Right center, Iron Side Rabbet, 5-3/4"x1/2"; Front right, right iron, 7"x3-3/4", wood top.

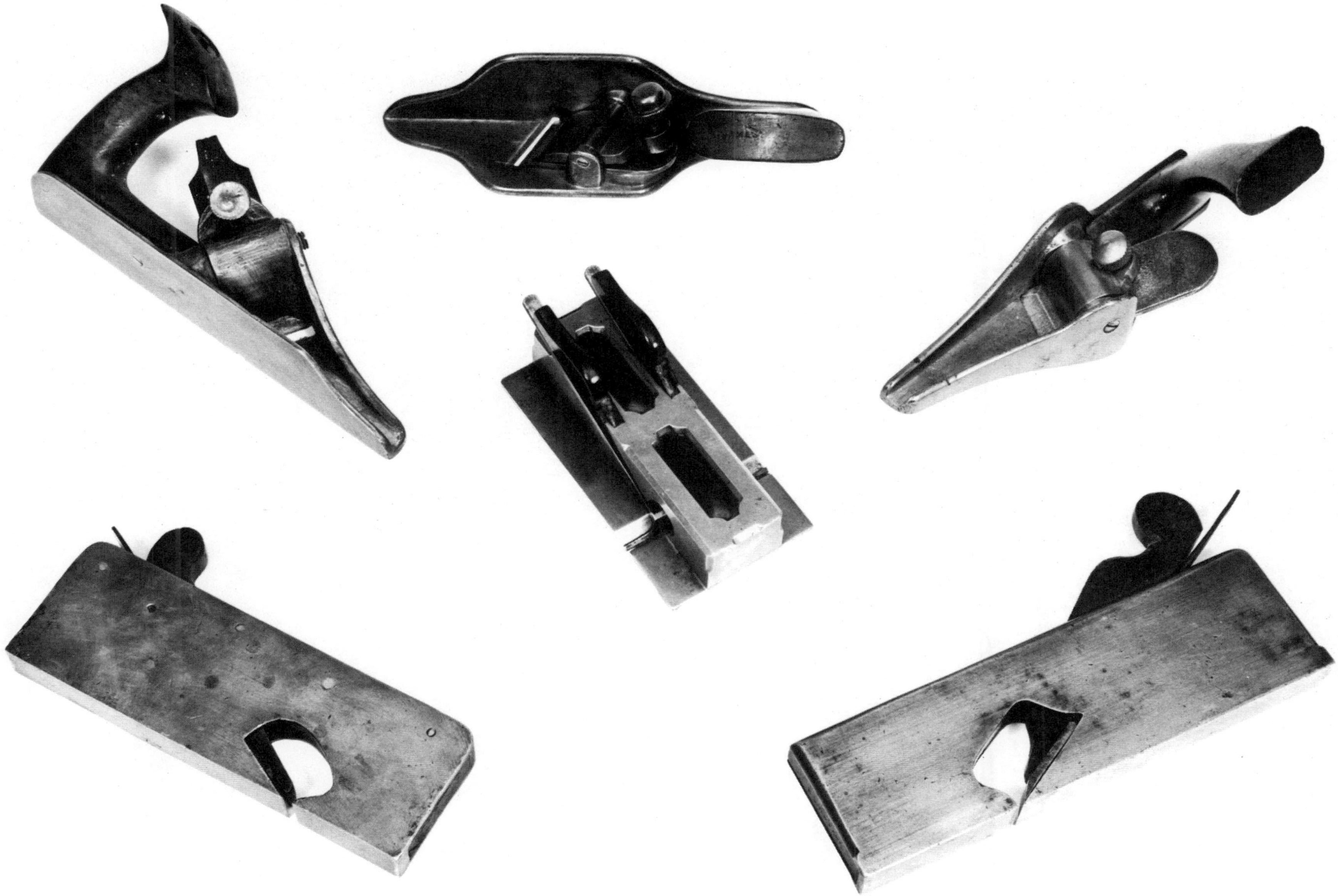

Fig. 50 - English Pattern Gun Metal Rabbet Planes - Left, Center and Right, Gun Metal Angular Shouldering Planes [George E.Bryant & Edward J.Witcombe, Letters Patent 4172, 16th Jan.1902] for Rabbet Trimming. 8-3/4"x2-3/4" Center, Double Iron (right & left) Rabbet Trimming Plane; Front, Left Rabbet Plane, 7"x3/4" Right, Rabbet Plane, 8"x3/4".

1. Wm.L. Goodman, *British Planemakers from 1700*, [2nd Edition,1978],pp.167-175.
2. Barton Brothers are not listed as Edge Tool Makers in the Sheffield Directories.
3. Wm.L. Goodman, *op.cit.*, Check List of Planemakers & Dealers, pp. 96-160.
4. *Tools for All Trades*,[Reprint of 1909 Wm.Marples & Sons Trade Catalogue by Mid West Tool Collectors Association, History of Marples Company,[1979],pviii
5. Wm. Marples & Sons continued making wooden planes in Sheffield until 1960. K.W.Hawley of Sheffield acquired many of the master planes.
6. *The Century of Progress Yorkshire Industry and Commerce*,[Vol.II,London,1873],p.218
7. Wm. L.Goodman, *op.cit.*, p.155
8. op.cit, [Reference #6], p.218
9. *Revised List 1876 William Chapple Plane Manufacturer*, [Ken Roberts Publishing Co, Fitzwilliam, NH , 1980], pp.15-20
10. James Howarth's Price list is after June 3,1846, Marsh File,Sheffield Central Librar
11. P.Nicholson, *Mechanic's Companion*, [Phila.,1849],Sect.21-23,pp.105-106];Reprinted *Wooden Planes in 19th Century America, pp.164-164*
12. R.A.Salaman, *Dictionary of Tools*,[London,1975], Fig.557-559,pp.377-378
13. *Wooden Planes in 19th Century America*, p.254-255
14. R.A.Salaman, *op.cit.*, p.346
15. Robert D.Graham, jr., correspondence, March 21, 1980
16. J.T.Stakes, "Thos. Napier Planemaker", *Chronicle of EAIA*, Vol30,No.1,Mar.1977,p.14-
17. W.L. Goodman, *op.cit.*, p.109, -1858-1862-
18. Greenslade of Bristol offered a plow plane with brass screw arms and nuts. c1865-1880
19. Isreal White of Philadelphia, Pa. USA patented a screw arm adjusting plow with slide arms, No. 12,1839. See Broadside Plate VI - *WP19thCA, op cit*, p.33
20. *Greenfield Tool Co. 1872 Trade Catalogue* [Reprint - Ken Roberts Publishing Co., Fitzwilliam, NH, 1978], pp.49-50
21. R.A.Salaman, *op.cit.*, p.346
22. E.N.Chapin & Solon Rust, US Patent 176,051, March 31,1868
23. *Alex.Mathieson Illustrated Price List of Woodworking Tools,1899*, [2nd edition, 1979, Ken Roberts Publishing Co., Fitzwilliam, NH], p.4
25. *WP19thCA, op. cit.*, pp200-201
26. Alan Beardmore,"Who Invented the Stanley Combination Plane?", *Woodworker*, [Hemel Hemstead,Herts,England], Sept.1976,pp.258-261
27. K.D.Roberts,*Stanley Rule & Level Co.'s Combination Planes*, [Fitzwilliam.NH,1975],p
28. K.D.Roberts,*Scotch and English METAL PLANES*, [Fitzwilliam,NH,1979], p.4
29. *ibid*, p.47
30. Gun Metal is an alloy containing 88% Copper; 10% Tin and 2% Zinc. It was used to make cannons.

* *

THE DUPLEX PLANE.

View shewing Plane for smoothing.

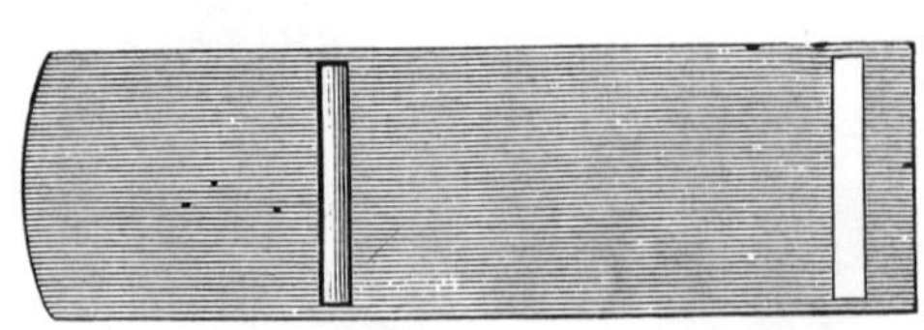

View shewing Face, &c.

The DUPLEX (Smooth or Chamfer) PLANE, No. 1 ... ... 13/- per dozen.

" " " " " " " 2 ... ... 22/- "

This Duplex Plane was offered by Wm. Marples & Sons Jan. 1,1883 and appears to have been and English Registered Design previous to the Stanley Rule & Level Co.'s No.130 Double End Block Plane [J.Campbell,US Patent Jan.30,1883] first offered in 1884 Catalog

Chapter VI

Braces and Bits

A brace for woodwork drilling may be defined as a crank tool having a pad or chuck at one end to hold a bit and a head or quill at the other end to support the rotational motion imparted by turning the handle of the stock. The diameter of the turning circle in rotating the handle is known as the'sweep'. The great advantage of using a brace compared to a gimlet or auger is the continuous motion imparted by the former. However, before the development of large auger bits, late in the 18th Century, such drilling was limited to small holes. A frequent application was pilot holes for follow up with auger drilling. A Boat Sway brace, having a relatively small sweep was used. (Fig. 51)

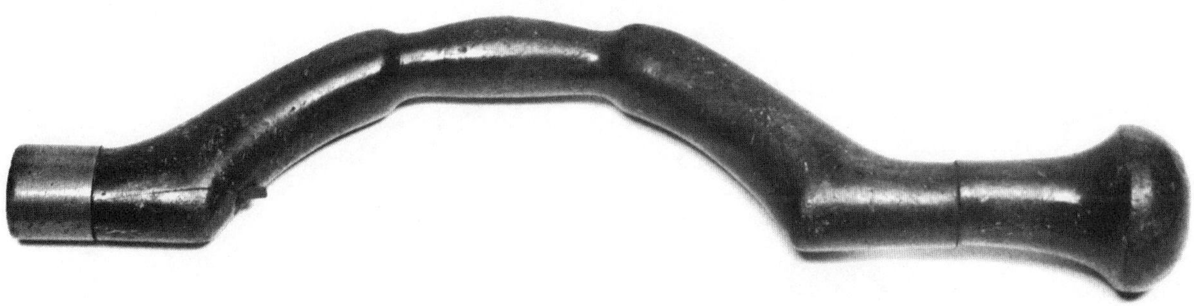

Fig. 51 - Shipwright's Boat Sway [12" long x 5" sweep]

Incidentally, contrary to some present beliefs, E.L'Hommedieu of Saybrook, Connecticut, USA was not the inventor of the spiral auger, but this was accomplished by Phineas Cook in England in 1770. (See Plate CXLVI)

Before considering the development of the brace in Great Britain during the 19th Century, it is appropriate to briefly discuss some historical background.[1] W.L. Goodman reported that the brace appeared "in Northern Europe as more or less a fully developed tool early in the 15th Century" and that the earliest surviving brace dates 1596.[2] In 1703 Joseph Moxon referred to such a brace as a *PIERCER*, noting - "*Its Office is so well known, that I need say little to it. Only you must take care to keep the Bitt straight to the hole you pierce less you deform the hole or break the Bitt. You ought to be provided with Bitts of several sizes, fitted into so many Padds*".[3] The inference is that interchangeability had been achieved by that date. Chairmaking and coopering trades offered exceptions to this practice where often bits were permenently fixed into the stock. (Fig. 52)

Fig. 52 - A Set of Chairmaker's Braces

Plate CXLVI

A Description and Explanation of Mr. Phineas Cooke's new constructed Spiral Auger.

Chap. IX, p.276, Fig I from *The Advancement of Arts, Manufacture and Commerce. Description of the Useful Machines Contained in the Repository of The Society for the Encouragement of Manufacture and Commerce.* by William Bailey, London, 1772.

A. The Head of the Auger, in which there is a round hole one inch and a quarter diameter to receive the drift or wooden handle which turns the auger.

B. The Shank of the Auger, is six inches and an half long, and five-eights of an inch square at its upper end, having its edges or angular points taken off.

C. An endless Screw, with a double worm or thread a quarter of an inch thick, and flat on their edges; these worms cut two spiral chips which pass through the two concave spiral channels of the auger, and are gradually discharged thereform without drawing out the auger till it has bored a hole three feet deep, or any other depth required.

D. The Point of the Auger, is a taper Screw with a double worm like gimblet, which pierces the wood much earlier and truer than common augers, and requires no picking with a gouge, which in the usual method is an unavoidable operation attended with a great deal of trouble and loss of time. It has been proved by repeated experiments, that this instrument does not want to be drawn out of the wood to discharge the chip, which is also the occasion of much labour and lots of time in boring with a common auger, as the accumulated chip in its cavity must be take out every four or five inches of wood bored therewith.

A trial of this new constructed auger was made in fir, oak, beach, and mahogany, in presence of the Committee of Mechanics, who was of the opinion that it effectively answered the purpose of discharging the chip during the time of boring; that it works more freely, is withdrawn more easily than common augers of the same bore, and promises to be a tool of great use in shipbuilding, &c. It was therefore resolved to recommend to give a bounty of thirty guineas to Mr. Cooke, for his invention, he leaving the instrument with the Society for use of the public, to which the Society agreed, May 1, 1770.

Acknowledgement of the above data is made to
The British Library - Science Reference Library
Holborn Branch, Chancery Lane, London

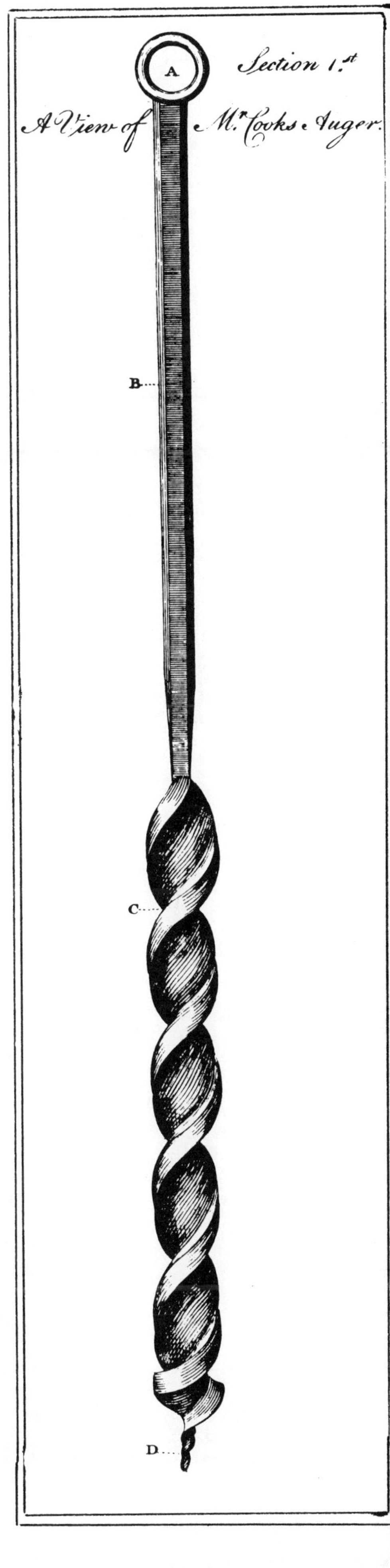

A late 18th Century English brace with an interchangeable pad (clothes pin type) is shown in Fig. 53. Another type of wooden pad had a round tailpiece cut across a diameter forming a shoulder to fit in a matching socket cut on the front end of the brace, making a fixed joint.[4] Other methods included tapered, truncated or square sockets on the front of the stock to accept similarly shaped tail pieced pads. Conically shapped tailpieces were also fitted into a stock, relying on friction to hold firmly while drilling. Tapping the tailpiece from the opposite protruding end permitted release. Two other later 19th Century methods of holding bits are shown in Fig. 55.[5] The brace at the top has a wing nut screwed into a reinforcing band, driven on the front end. The other brace has a forged, split iron spring chuck that could be either driven or screwed into the socket. This latter method was frequently used with cooper's braces. (See Fig. 54)

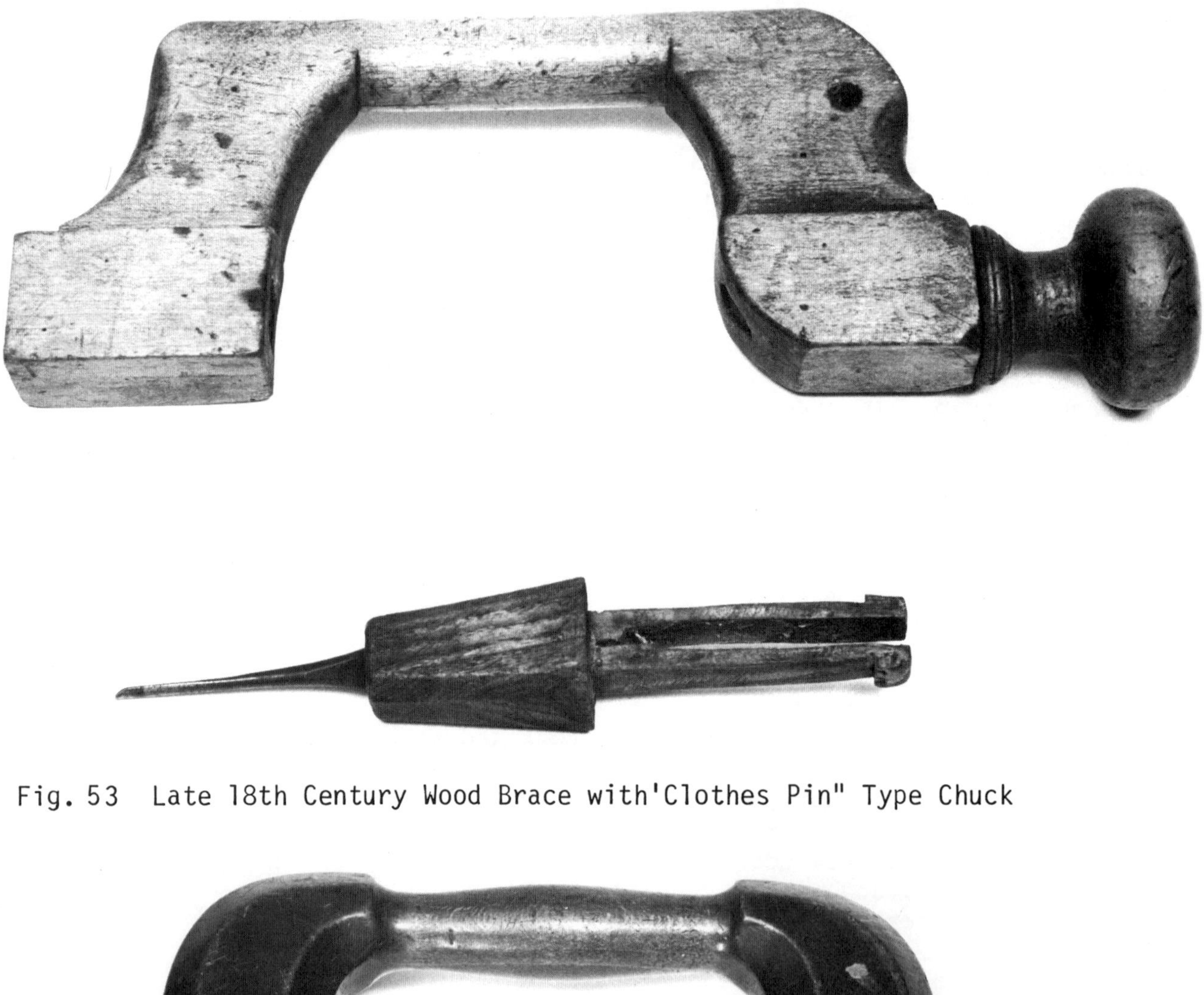

Fig. 53 Late 18th Century Wood Brace with'Clothes Pin" Type Chuck

Fig. 54 Cooper's Brace with Forged Iron Spring Chuck

Fig. 55 19th Century Wood Braces - Top - Wing Nut Screw Chuck
Center & Bottom - Forged Split Spring Chuck

A "Joiners' Spring Pad", constructed from metal was available at least by 1770. This was illustrated in Plate 35 of the *John Wyke Pattern Book, Liverpool*.[6] (See Plate CLXVII) This pad was fitted into the front of a wooden brace and held securely by a screw which fastened the two lugs through the brace. This became the most common method employed during the 19th Century. Such a spring catch, actuated by a button, was also used with hand pads. (see Fig.12, p.185) These nosepieces were made of both brass and iron, at least until the early part of the 19th Century, after which brass predominated.

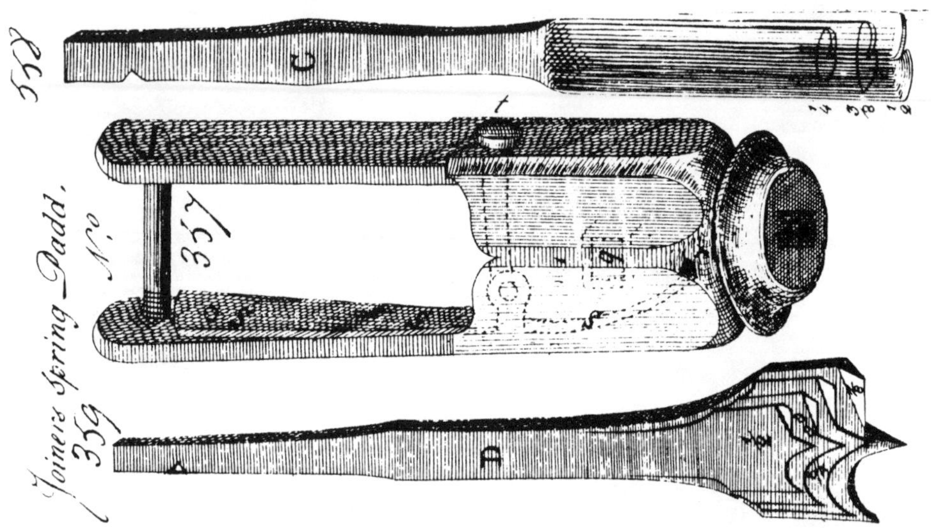

Plate CLXVII. Joiners' Spring Pad from *John Wyke Pattern Book, Liverpool*,c.1770
[The spring catch, R, made from hardened and tempered steel spring,S, attached to the inside of a lug of the nosepiece, was depressed by pushing the button,T]

A brace with such a brass chuck with the imprint of *SHEPLEY* [William], working at Soho, London (-1780-1799-)[7] as a planemaker is shown in Fig. 56 . The imprints of the makers or dealers of these early braces were placed on the stock near the head, rather than on the chuck which became the usual practice during the 19th Century. Similar braces with the imprint of *MUTTER* [George Mutter, 1766-1811][8] , another London planemaker, have been observed. In all probability both Shepley and Mutter had purchased such braces for resale,supplemented their plane lines. These were available in wholesale quantities at Birmingham or from tool agents in London. Incidentally, when viewing these early braces from the head towards the pad, the button was generally on the right side; whereas during the 19th Century, the button was generally placed on the left, except for those made in Scotland generally continued to have right side buttons.

This quick acting button spring chuck became in demand as a consequence of the large production of a great variey of steel bits manufactured principally at Birmingham. These were offered in sets of 12, increasing in multiples of 6 to 60. These were furnished with square tapered shanks, which were notched so they could be held securely in spring actuated pads, but easily released by pressing the button. The sets were made of drilling bits of the following diameter ranges: centre, ½" - 1"; shell, 1/16" -7/16"; nose, ¼" - ½"; sash, 3/16" - ½". Two types of turnscrew bits, plain & forked; three types of tapered rimmers - square, tapered and half-round; three types of countersinks, flat-pointed, rosehead and snail.[9] (See Table XX). Since Birmingham had become the leading center of bit production by 1800, it is not surprising that leadership in brace manufacture also prevailed there then.

Fig. 56 - Wooden Brace by William Shepley of London

[Note imprint on Stock above head]

[Left - Note Button on Right Side of Chuck]

Table XX - Sets of Cast Steel Brace Bits [*1928 Wm. Marples & Sons Catalogue*]

Number in Set	Shell Bits Number	Shell Bits From	Shell Bits To	Centre Bits Number	Centre Bits From	Centre Bits To	Turnscrew Plain	Turnscrew Fork	Countersinks Flat	Countersinks Rose	Countersinks Snail	Square	Rimmer Half Round
12	4	1/8	1/4	5	5/16	3/4	1			1		1	
18	4	1/8	1/4	8	5/16	1	1			1	1	1	
	(also 3/16 Sash Bit)												
24	5	3/32	1/4	11	5/16	1	1	1	1	1	1	1	
	(also 3/1b Sash Bit)												
30	6	3/32	5/16	14	1/4	1-1/4	1	1	1	1	1	1	
	(also: Sash Bits: 3/16, 1/4; Drill Bit: 1/4; Taper Bit)												
36	9	1/16	7/16	15	1/4	1-1/4	1	1	1	1	1	1	1
42	Set of 36 Bits and also: 5/16 Sash Bit; 1-3/8 Centre Bit; 3/8 Drill Bit; Swiss Bits: 1/8, 3/16, 1/4												
48	Set of 42 Bits and Also: 3/8 Sash Bit; 1-1/2 Centre Bit; 5/16 Drill Bit; Swiss Bits: 5/32,7/32, 9/32												
54	Set of 48 Bits and Also: 1-3/4 Centre Bit; 3/16, 7/16 Drill Bits Swiss Bits: 1/16, 3/32, 5/16												
60	Set of 54 Bits and Also: 1-5/8', 1-7/8, 2 Centre Bits; 1/2 Drill Bit Swiss Bits: 11/32, 3/8												

[All above sizes are diameters in inches or fractions thereof]

Many of these are illustrated on Plate D2 *1876 SHEFFIELD ILLUSTRATED LIST*
[See Plate CLXVIII - page 301]

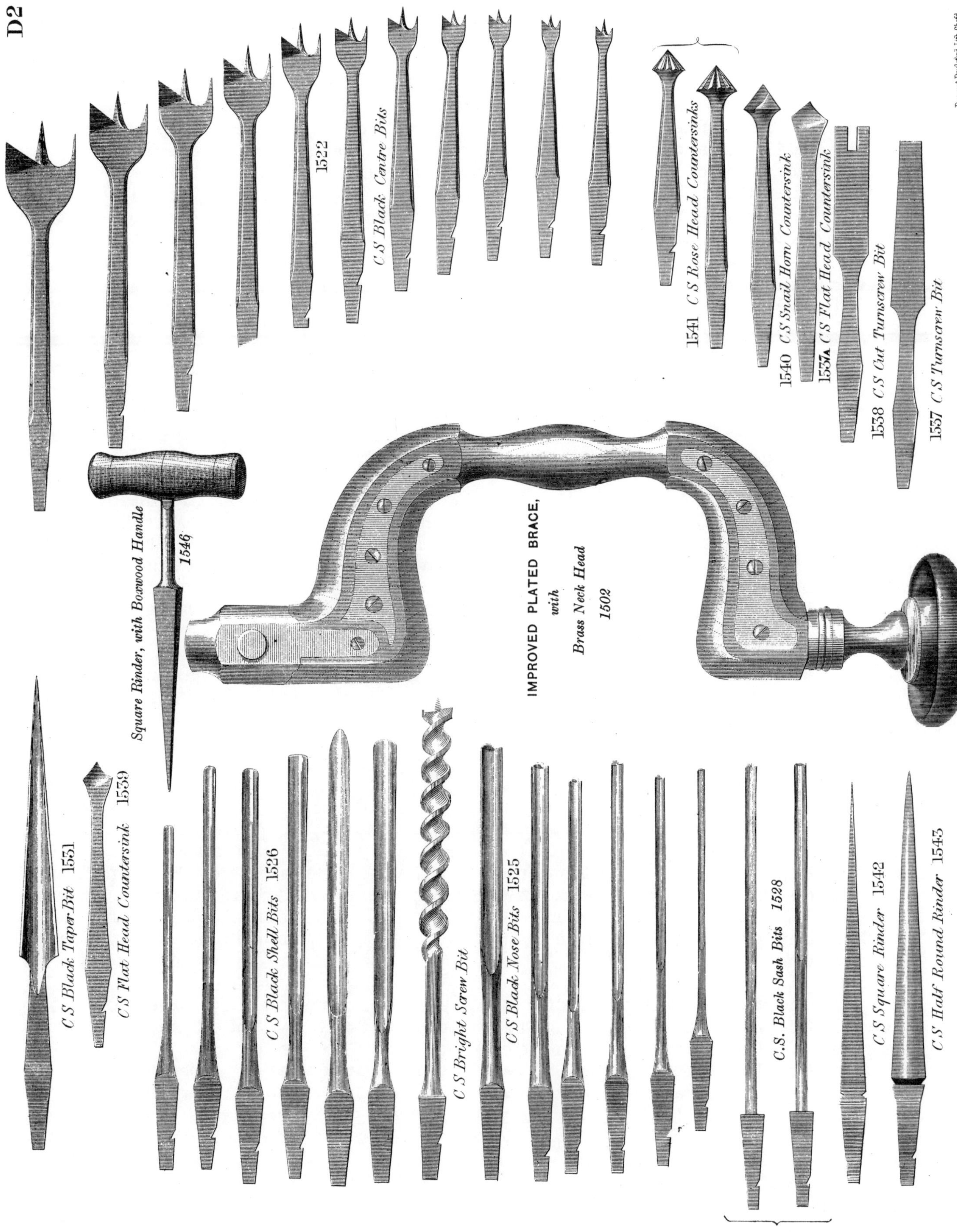

Plate CLXVIII - Bits and Improved Plated Brace - Plate D2, 1876 Sheffield List

Several Birmingham pattern books with copper plate engraved prints of such braces appeared early in the 19th Century. The earliest, though unidentified as to firm for whom it was produced, had watermarked paper dated 1798-1801.[10] It has therefore thought to be dated 1800/1804. This illustrated such a brace with a right-handed button brass pad and was referred to as a *"SCOTCH BRACE"*. A similar one, so also defined, from another unidentified Birmingham pattern book, printed on paper watermarked 1804, is shown, Plate CLXIX. A *Gentlemens' Iron Brace* is also seen on the same print. These were more expensive items for owners of estates, rather than intended for the working class. The 1810 Price List of W. & C. Wynn [earliest known surviving Birmingham dated price list] noted item #377 (Scotch Brace) at Eight Shillings six pence (8s6d or 8/6). This particular brace with a right-handed release button has been illustrated from their 1830 pattern book.[11] This print was undoubtedly made from the same copper plate engraving as referred to in the 1810 price list. From examination of these Birmingham Pattern Books it is evident that "Scotch Braces" with brass spring chucks had become an important production item there at the beginning of the 19th Century. The nomenclature of this as a *"Scotch Brace"* is believed to have been derived from a former usage of the word *scotch* meaning "an incision, cut, score or gash".[12] Formerly it had been suggested that this may have been derived from usage of this word meaning either "common" or "made in Scotland".[13]

A surviving price list of B.Freeth & Co. of Birmingham included *Best Black and Bright Bits* [Plates 4 - 9] and *Iron Brace* [Plate 17].[14] A pattern book has been located illustrating items corresponding to the same numbers on this price list, printed on paper bearing watermark date 1802. Another pattern book, printed from the same plates, had watermarked paper dated 1820. While no wooden braces were offered on this price list, Items No. 63 and 64 were respectively identified as *"brass and steel pads, 2s6d each"*. These were nosepieces to attach to such wooden braces, and are shown in Plate CLXX. Benjamin Freeth was noted as a Birmingham brace maker, 1818 - 1820. A brace with his imprint on the brass pad is shown in Fig. 57..

The Birmingham Commercial Directory of 1816-17 was the first to list Brace and Bit Makers as a specific trade. Seven persons were noted: Blackemore & Sons, Hurst St.; Timothy Bridgewater, Church St.; Joseph Glascott, Coleshill St.; Thomas Horton, Edgbaston St.; John Neeld, Bordesley St.; William White, Green St., and Wm. & Cornelius Wynn, Suffolk St. By 1828 the number had increased to ten and by 1833 to fourteen. Several trade cards of Birmingham "toy" or tool merchants illustrated such braces: Benj. Smith, 1821; W.Burton & Co., 1825; Thomas Ross, 1829; Richard Thomas, 1829.[15] It is indeed questionable as whether any of these merchants advertising such braces actually manufactured these.

As noted in the previous chapters concerning edge and joiner tools, after 1800 both of these industrial activities expanded rapidly at Sheffield. The first listing of Brace and Bit Makers as a trade appeared in the 1821 Directory. The five persons noted were: James Bee, Townhead St.; Maw & Staley, Rockingham St.; William Round, Allen St; Joseph Slater, Hollis Croft; and John Sorby, Spittal Hill. [Actually Maw & Staley were merchants & John Sorby and edgetool maker; both probably only traded in braces and were not makers] Three more were added in 1825: William Ash &Co., Norfolk Street; Wm. Saunderson, Peacroft St.; and John Wheatcroft, Broad Street. By 1830 the list has grown to seventeen, surpasing the number of firms then making these products as Birmingham. In 1852 twenty-six firms were listed at Sheffield making braces and bits.

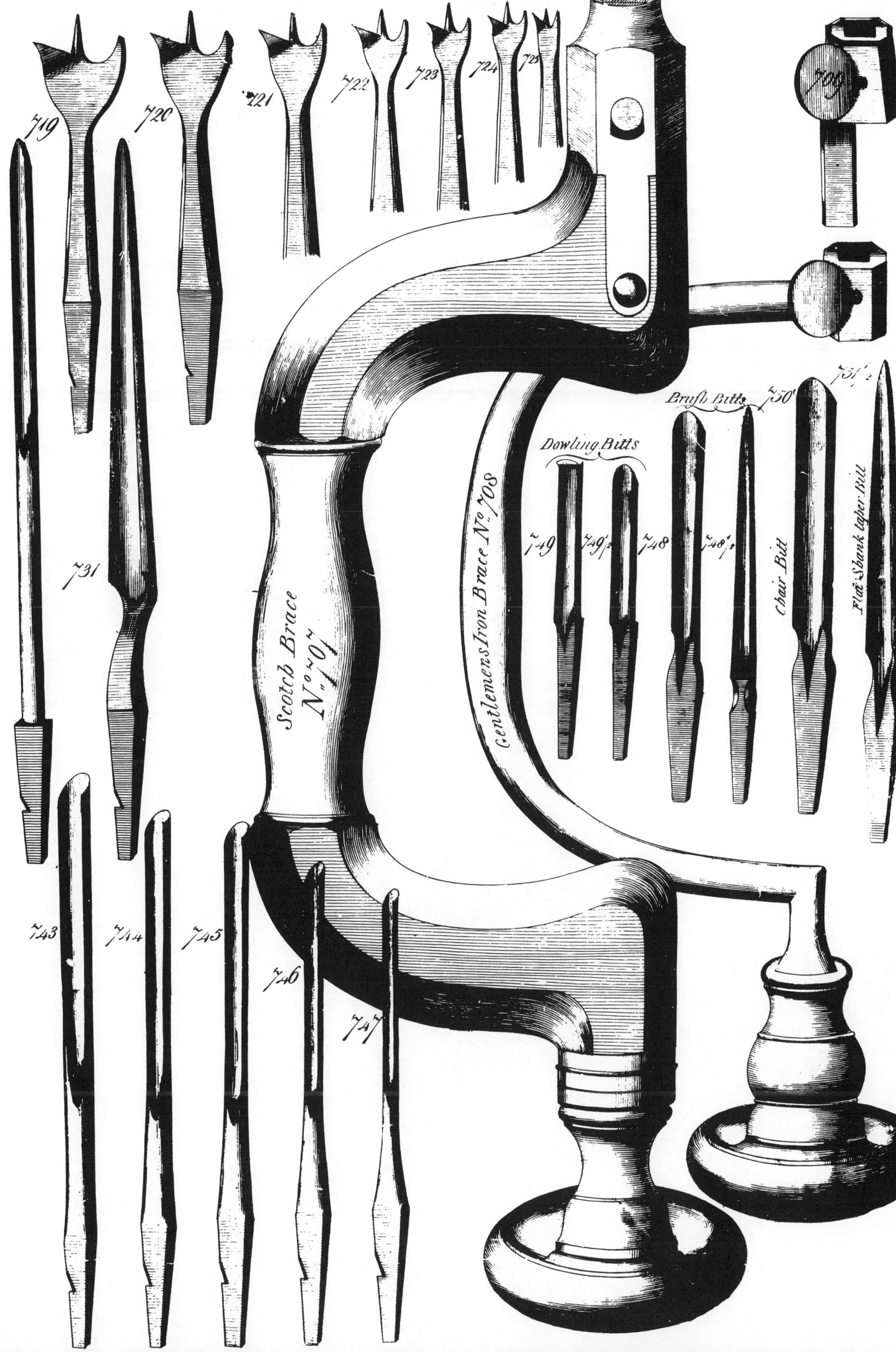

Plate CLXIX - Scotch Brace & Gentleman's Iron Brace and Bits, Unidentified Birmingham Pattern Book, cl802/1820

Fig. 57 - Wooden Brace with Brass Pad by Freeth, Birmingham
[Courtesy of "The Farmer's Museum, Inc. Cooperstown, NY]

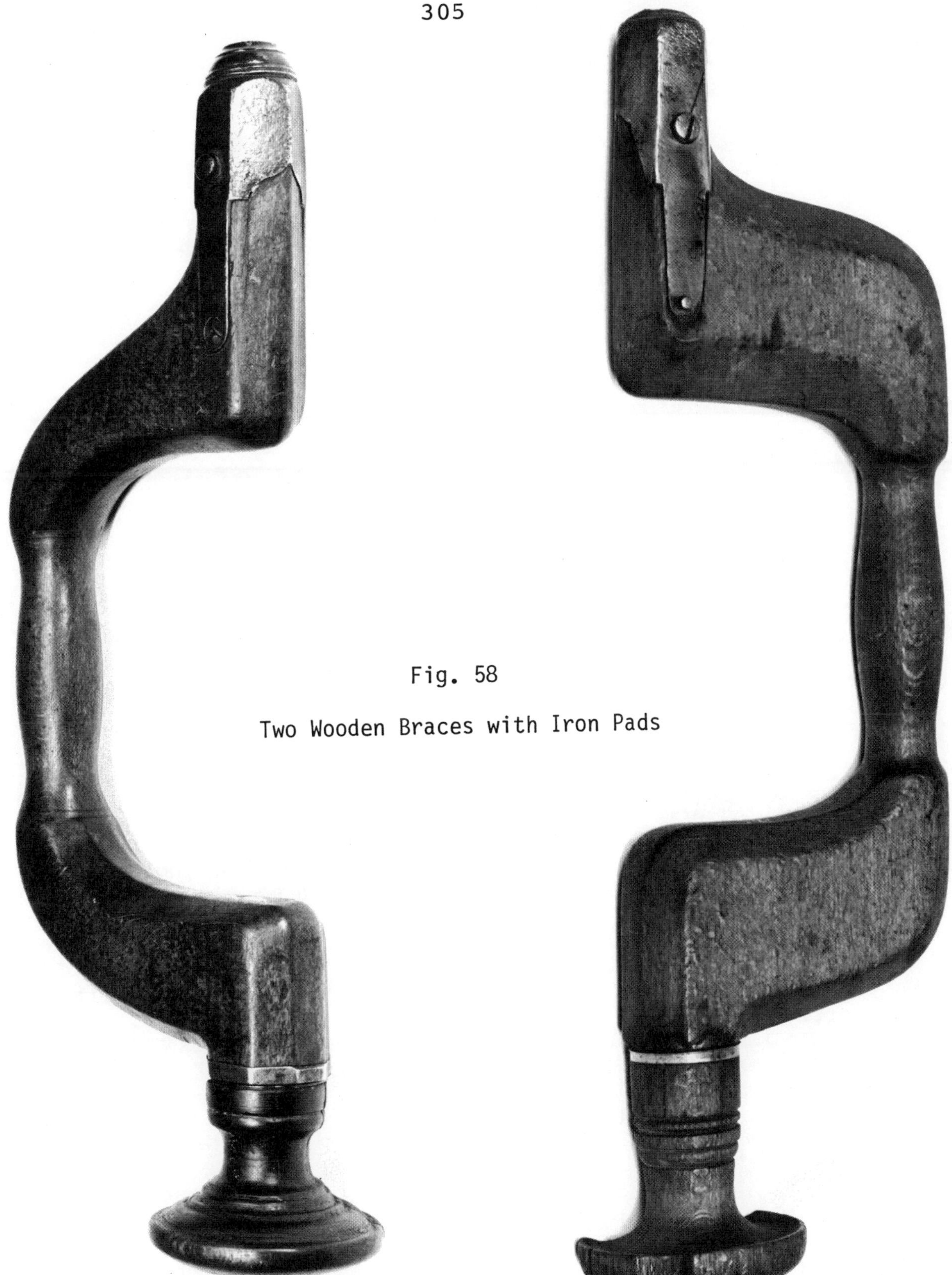

Fig. 58

Two Wooden Braces with Iron Pads

Two braces with iron pads, probably made in Birmingham during the late 18th Century, are illustrated above. Iron forging had developed into an art of high perfection at this industrial center. After the turn of the century such pads were generally available as brass castings which were less expensive and took precedence over iron for this application.

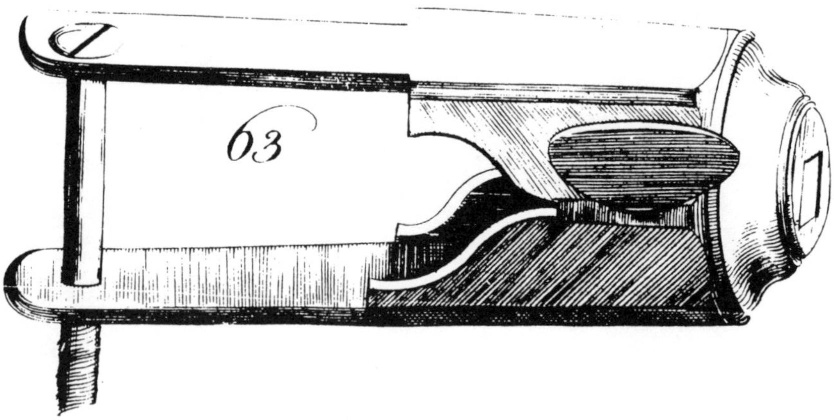

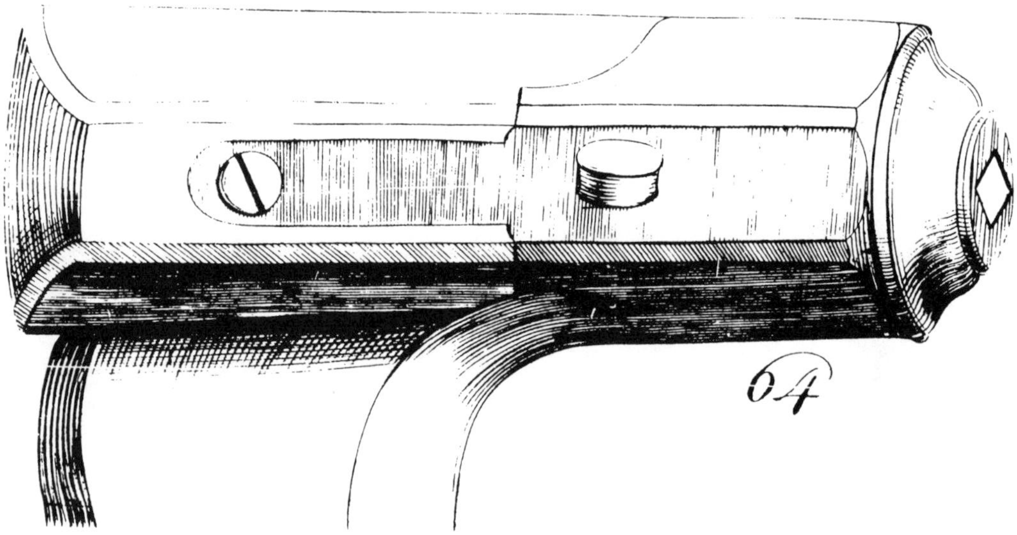

Plate CLXX . - Metal Nosepieces for Braces or Hand Pads [brass & steel pads]
[From Birmingham Patternbook; believed to be B.Freeth, c1802 or 1820]

The earliest known Sheffield price list of Joiners' Tools is by Mitchell, Brothers & Co., January 1828. (see Plate CXXXIIIa&b, p.198-199) This listed *BEST BRACES*, as the *SCOTCH* brace at Birmingham was to be hereafter known at Sheffield, at a price of 9s6d. The title page of the 1829 J,Wilks Price List was shown as Plate CXXIV. This noted "Ridge's Braces, Bits, &c. at Ridge's Prices"

BRACES, HAND PADS, &c.

281 Improved Plated Brace 13s

	12	18	24	30	36	42	48 bits
Black..	17s	19s	21s	23s	25s	27s	29s

If Bright 9d. p. doz. extra.—Do. Coloured Bits 1s. 4d. p. dz. extra

…9 Best Brace .... 9s.

	12	18	24	30	36	42	48 bits
Black........	13s	15s	17s	19s	21s	23s	25s each

If Bright 9d. p. doz. extra.—If with Collars 10d. p. set extra,
Coloured Bits 1s. 4d. extra.

…4½ Braces with Brass Necks, 2s. each extra,

…05 Brace Pads 33s

Best do.. .... 36s

307 Brace Heads 20s

Best do... .... 24s

Improved Brace Heads 36s

308 Brass Collars for Sash Bits 6s

309 Common Coopers' Braces .... 3s ea

Best do....................... 4s ea

		12	15	18	24	30	36
311	Gentln. Braces Black Bits	12s6d	13s7½d	14s9d	17s	19s3d	21s6d
311A	Do. Bright do..	13s3d	14s7d	15s10½d	18.6	21s1½d	23s9d

Plate CLXXI

Prices of Braces & Bits
J.Wilks 1829 Sheffield

Archives [Marsh File]
Central Library - Sheffield

J. WILKS,

Broad Street, Park, Sheffield,

MANUFACTURER AND

DEALER IN TOOLS

OF EVERY DESCRIPTION,

The 1829 J.Wilks' Price List is the earliest known dated reference to both *improved plated braces* and *brass necks*. This is considerably earlier than previously reported.[16] The *improved plated brace* was noted to sell at 13s, the *best brace* at 9s and 2s extra for the addition of a *brass neck*. These prices were to prevail at Sheffield through the 1876 Sheffield List.

The plated brace is believed to have been introduced at Sheffield in 1829 as a product of the firm of Brown & Flather. This concern consisting of the partnership of Henry Brown and David Flather was first listed in the 1828 Directory at King's Arms on Fargate Street. It is thought that they purchased the business of James Bee, who may have designed this style of brace before turning over his concern in 1828. Appearing in the 1814 Sheffield Directory, James Bee was the first listed bracemaker. An *Improved Plated Brace* with the imprint of *Brown & Flather* thought to have been made 1830-1840 is illustrated in Fig. 59 Both styles, *Improved Plated Brace* and *Best Brace* remained popular in England throughout the 19th Century. From the standpoint of quantities manufactured and numbers of different imprints of makers, the plated brace, made from the period of 1829 into the first quarter of the 20th Century, may be among the most historically significant tools produced at Sheffield. Not only was this product important to the domestic economy, but also influencial to export trade to America and the colonies. While a struggle for leadership in manufacturing resulted from 1845 - 1865 at Sheffield, considerable interchange of this product occurred among manufacturers and marketing firms, which has often made obscure the actual maker of such braces.

The probable method of making such a brace is indicated on the following sketch. A blank was sawed from a beech plank (A); chucked between centers and the handle turned on a lathe (B); The bow ends were shaped using a saw, draw knife and spoke shave and the head, separately turned in a lathe, and the brass pad assembled (C).

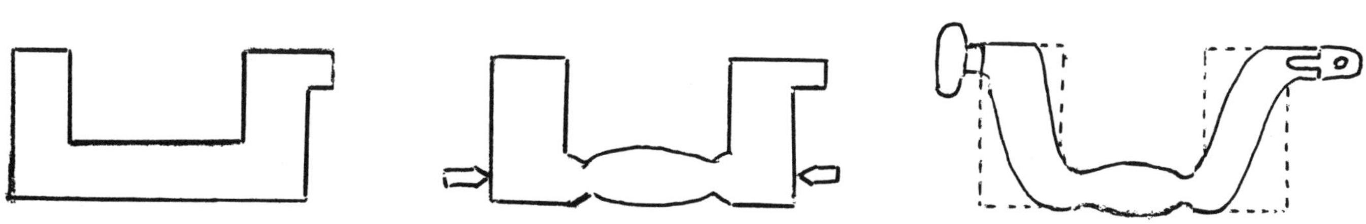

A B C

In making a plated brace each side of the stock was routed to accomodate a cast brass plate, approximately 3/32" thick. In all probability a jig for routing, similar to that used in preparing squares (See Fig.15,p.217) was employed. A set of four plates was cast, probably in a match plate. The inside surfaces were cast with pips, numbered from one to four, to denote position of a particular plate when assembled to the stock. An indication of the high degree of skill in such castings reveal the name of both the customer and brass foundry imprint on the inside surface. These brass castings were made in job foundries, which were numerous at Sheffield. The outside surface was hand filed on a jig, and the edge filed on a slight taper before fitted and glued into the stock. The plates were drilled and countersunk to accomodate four screws. The brass pad, cast with an iron liner and fitted with a spring button actuated catch, was attached to the stock. (See Fig.60)

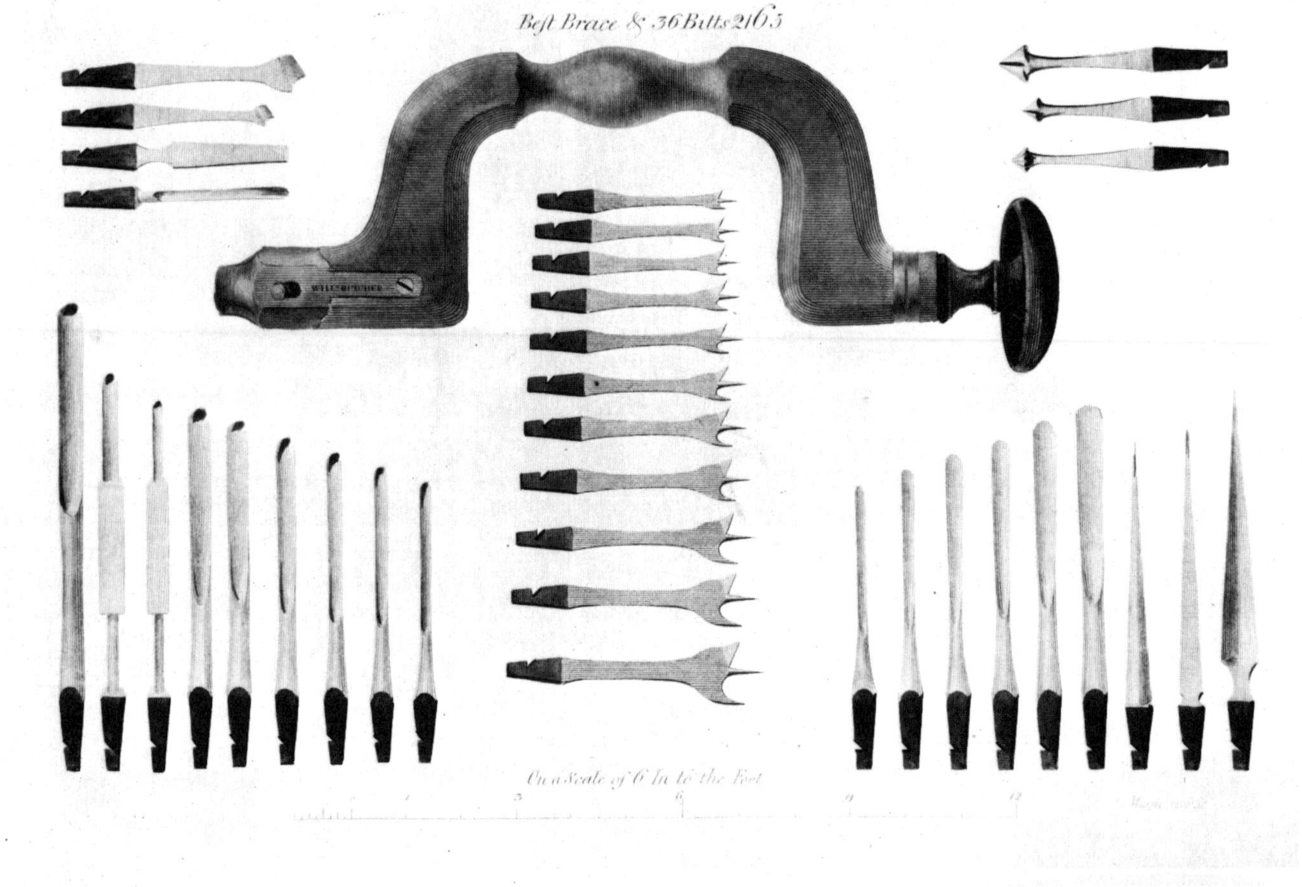

Plate CLXXII - Best Brace Imprinted Will$^{m.}$ Butcher with 36 Bitts
[from W.& S. Butcher Trade Catalogue - Courtesy, Henry Francis du Pont Winterthur Museum Libraries, item no. 80.453 TJ1200 B98*T.C.]

The above Plate CLXXII was photographed from the William & Samuel Butcher Catalogue, c1841 in the collection of the Winterthur Museum. The original engraving by Mequiner may have been made 1821-1825, at the time when William Butcher was in the edge tool business alone at Sheffield. In 1828 he was a partner in Butcher, Brown and Butcher, and after 1833 with his brother, Samuel in Wm. & Sm. Butcher. The paper on which the above photographed engraving was printed was water-marked 1841. However since the tools in this Catalogue were imprinted only with William Butcher and many of earlier style, it is my belief that the plates may have been engraved c1825 for an earlier printing of such a catalogue.

It is doubtful whether either Wm. Butcher or the later firm, Wm. & Sm. Butcher manufactured braces, but incorporated these into their catalogues to complement their line of edge tools. In all probability they traded their edge tools with some other joiner tool maker in Sheffield in order to offer a full line of tools in their catalogue. Publishing such a catalogue was a substantial expense at this date. An individual printing was probably limited to 200 copies or less.

Fig. 59 - Brown & Flather Braces, "improved plated"(left); "best" (right)

Top View Inside Brass Pad

Outside Surface Back Plate

Inside Surface Back Plate

Fig. 60a - Brass Pad and Back Plates

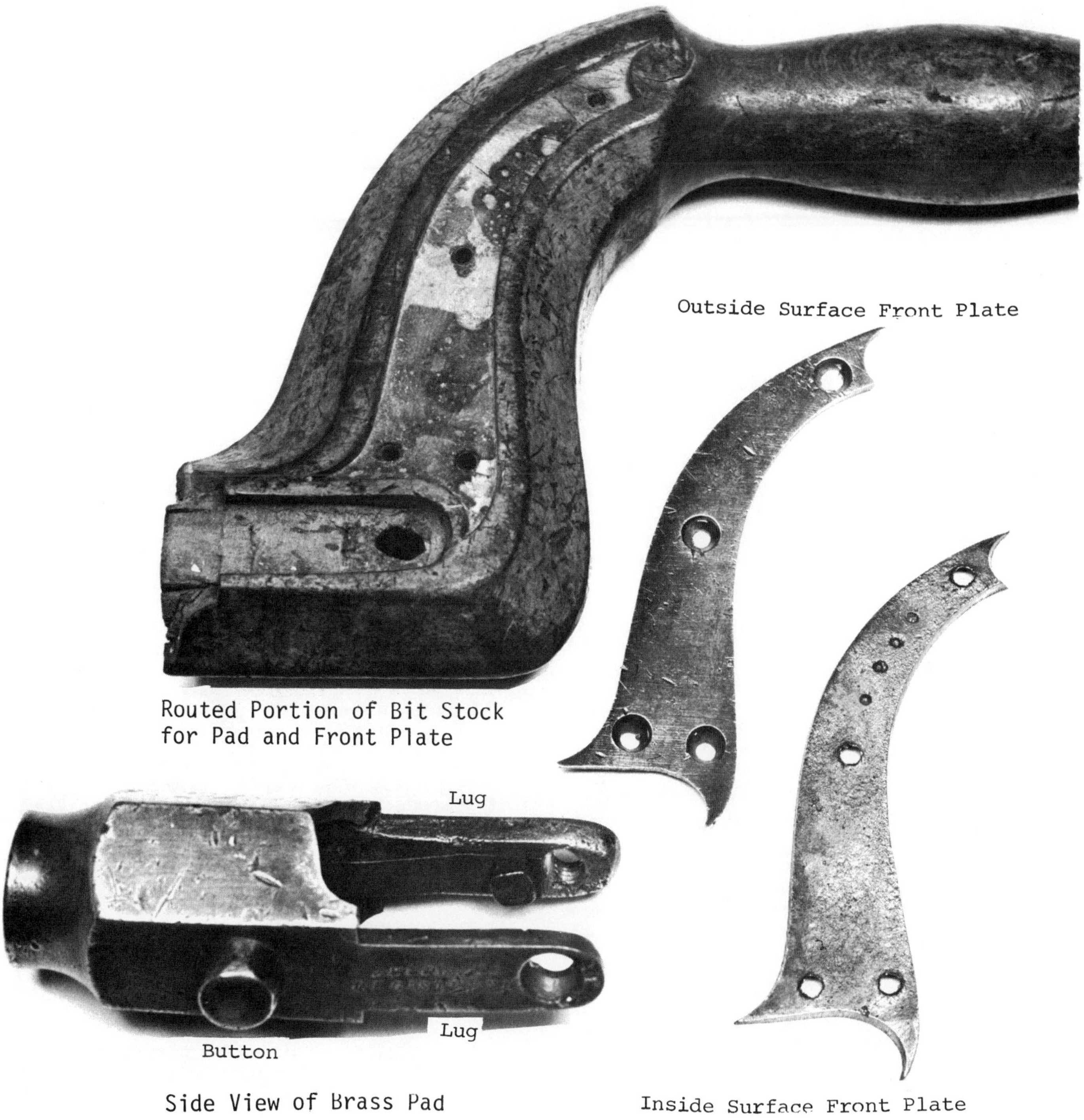

Fig. 60b - Brass Pad and Front Plates and Front Stock

The earliest known dated illustration of a Sheffield plated brace appeared in the 1841 Sheffield Directory by Brown & Flather. (Plate CLXXIII). That year they had moved from Hollis Court to Solly Street. Henry Brown secured Registered Design #306, November 2,1844 for a "Carpenters' improved Brace Head", claiming as new a socket piece with reverse screw with a metal piece screwed into the socket to hold the wooden cap secure. The details of this design are shown on Plate CLXXIV Brown apparently engaged the service of John Bottom to devise a patent concerning a brace for his use. John Bottom was first listed in the 1841 Sheffield Directory as "clock maker, machinist, &c." at Fitzwilliam St. In the 1845 Directory he was noted as "patentor of the anti-friction carpenter brace and roasting jackes, &c" at St. Philips Drive. His Letters Patent No.10.526 was granted February 20, 1844 for " Certain Improvements in Carpenters' Stocks and Braces" His claims were:

> First, the making the head of carpenters' braces to bear and turn on a point instead of a broad flat face, and securing the head to the neck, as before described.[Figs. 2 & 4]
> Second, the connecting the head by a joint [Fig. 7]
> Third, the mode of fixing the bit in the pad [Fig. 8, 10, 11]

Henry Brown purchased this Patent of John Bottom and thus became the Pantentee. Brown & Flather announced in the 1845 Sheffield Directory they were " SOLE MANUFACTURERS of *HENRY BROWN'S* PATENT ANTI-FRICTION BRACE", and cited the advantages. Actually it appears that they did not follow any of the above three claims in Bottom's specifications, except the most impressive title - 'ANTI- FRICTION BRACE'. They apparently modified the style of Brown's Registered Design, used an interior threaded brass neck with a double reverse thread on the holding screw and a threaded cap. They apparently capitalized in advertising this Patent Anti-Friction Brace, which may not actually have been covered by the Bottom patent.

BROWN AND FLATHER,
SOLLY WORKS, SOLLY-STREET, SHEFFIELD,
MANUFACTURERS OF
BRACES, CAST STEEL BITS,
JOINERS' TOOLS, SKATES,
SCREW BOXES, PLOW STOPS,
TOOL CHESTS,
AUGERS, GIMBLETS, &c., &c.,
OF SUPERIOR QUALITY, AND BEST WORKMANSHIP;
AND
SOLE MANUFACTURERS OF
HENRY BROWN'S
PATENT ANTI-FRICTION BRACE.

The Advantage of this Patent Brace above all others,
is its easy Working, and having no Friction as in other Braces. The Spindle and Socket Nut being of Hardened Cast Steel, and applicable to either Lignum Heads or Improved Brass Neck Heads. The Brass Neck has a Nut screwed into or upon the Socket piece, thereby preventing the Cap from working off. In the Patent principle the Lignum Head has a Brass Socket piece, screwed into the Neck part, which contains the Spindle and Nut.

Plate CLXXV 1845 Sheffield Directory

BROWN & FLATHER,
SOLLY WORKS, SOLLY ST., SHEFFIELD,
MANUFACTURERS OF
BRACES, CAST STEEL BITS,
JOINERS' TOOLS, SKATES,
SCREW BOXES, PLOW STOPS, TOOL CHESTS,
AUGERS, GIMBLETS, &c., &c.,
OF SUPERIOR QUALITY, AND BEST WORKMANSHIP;
AND SOLE MANUFACTURERS OF
HENRY BROWN'S PATENT ANTI-FRICTION BRACE.
Granted by Her Majesty's Royal *Letters Patent*.

THE ADVANTAGES OF THE PATENT ANTI-FRICTION BRACE, ABOVE ALL OTHERS,

is its EASY WORKING, having *no friction*, as in other Braces. The Spindle and Socket Nut being of hardened Steel. The Brass Neck has a Nut screwed upon, or into, the Brass Socket Piece, thereby firmly securing the Ebony Cap of the Head, and preventing its coming off, causing a firm steady working of the Brace.

BROWN & FLATHER ADVERTISEMENTS

Plate CLXXVI-1849 Sheffield Directory

Plate CLXXIII

Brown & Flather Advertisement
1841 Sheffield Directory

Earliest known dated illustration of Brass Plated Brace.

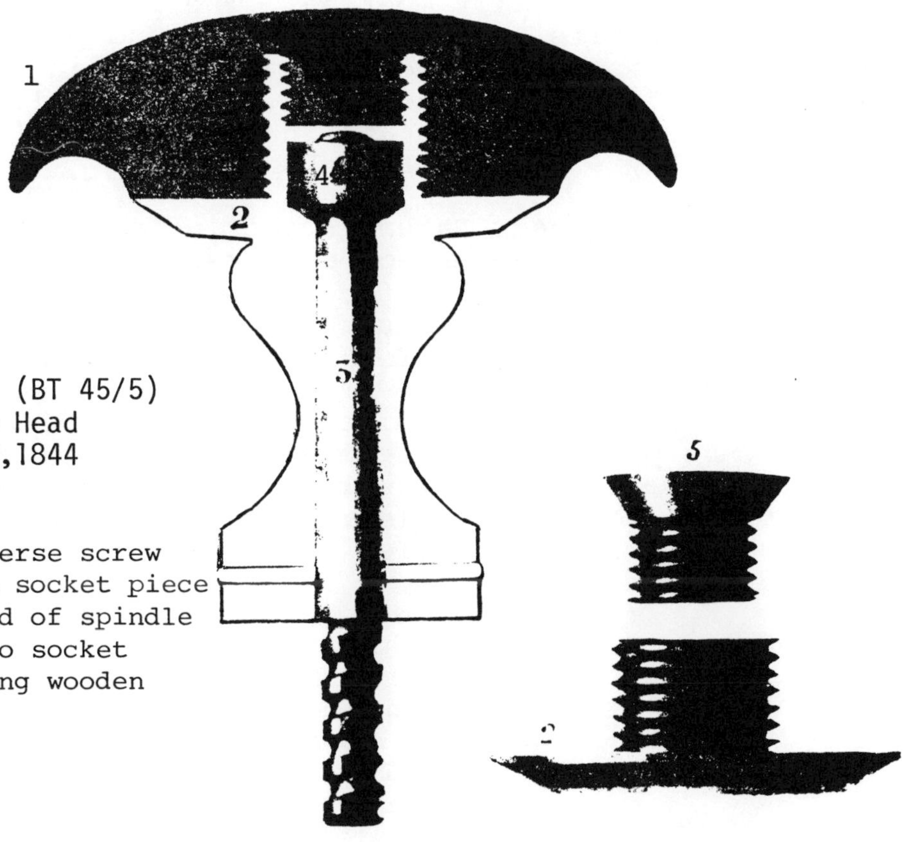

Plate CLXXIV
Registered Design No. 306 (BT 45/5)
Carpenter's Improved Brace Head
by Henry Brown, November 2,1844

1. A Wooden Cap
2. Socket Piece with a reverse screw
3. Spindle passing through socket piece
4. nut screwed upon the end of spindle
5. Metal Piece screwed into socket piece,2, thereby securing wooden cap, 1

The constructional details of what was claimed to be the PATENT ANTI-FRICTION BRACE (Henry Brown, Patentee imprinted on the brass chuck) were shown on Plate 50 of the 1849 Marsh Brothers Catalogue. (Plate CLXVII) The specifications with reference to that drawing are transcribed, as follows:

Patent ANTI-FRICTION BRACE

Fig. 1 A Wooden Cap
" 2 Socket Piece with a reverse screw
" 3 Steel Spindle hardened at the end passing through socket piece having a right and left handed thread or screw.
" 4 Two Nuts to screw upon the Spindle to secure it in its position.
Fig. 5 Brass Nut with a hardened Steel Screw both being screwed down Fig. 2 bringing the two Steel Screw Points together preventing friction at the Collar.
" 6 Brass Nut screwed into Fig. 2 securing the Wooden Cap

The prices of the Patent Anti-Friction Brace, along with those of the Best Plain Brace and the Improved Plated Brace, as manufactured in 1849, are noted on the Marsh Brothers Price List on Plate CLXXIX. The Patent Brace with brass neck demanded a premium of 4/6. Various extra charges are noted for Ebony Improved Brace Heads with Brass Necks or All Brass Heads. Ebony, Box or Rosewood Braces were priced 12/ more than the Improved Plated Brace, totaling 25/. The Improved Scotch Iron Brace cost the same as an Improved Plated Brace, 13/. The three styles of braces are shown on Plate 51 from the Marsh Brothers Catalogue. (p.318) Plate CLXXX. An improved Scotch Iron Brace (Patent) #711 with the brass head cap is shown in Fig. 61. The prices of bits are quoted in three colors: Black, Bright and Straw. Black refers to oxide finish after tempering without removing the scale; bright refers to polishing after tempering, removing the scale; and straw refers to color achieved directly from special tempering. The latter higher cost is due to the fact that the bits have to be cleaned and polished directly from hardening before tempering. The resulting loss in breakage of some bits during polishing reflects a higher cost. Incidentally, the first reference to a 'Scotch iron brace' appears in the 1830 Sheffield Directory offered by the firm Hanny & Kinder, Eyre Lane.

In having their Patent Brace included in the Marsh Brothers Catalouge, Brown and Flather not only received excellent publicity, but also this provided opportunity for export sales ahead of their competition. Marshes and Shepherd has established a sales' office in Philadelphia as early as 1817 and at New York in 1840.[17] This assured leadership in American sales for Brown & Flather. The 'golden years' for such export sales were 1840 - 1860. Listing in New York City Business Directories during this period indicate over 20 Sheffield firms offering imported steel, hardware, cutlery, saws, files and other tools. Additionally there were several tool stores (seven noted in 1849-50 NYC Directory). Among these was Charles S. Little, who advertised in the 1849-50 NYC Business Directory: -

General Furnishing of Hardware, Cutlery, Edge Tools, &c

Charles S. Little

Successors to Osborn & Little

[Charles Osborn being deceased September 14. 1846]

Importers and Dealers in

ENGLISH, GERMAN and AMERICAN HARDWARE & CUTLERY, &c.

No.33 Fulton Street, New York

The Old Stand Opposite the United States Hotel

(where business has been conducted for the last eighty years)

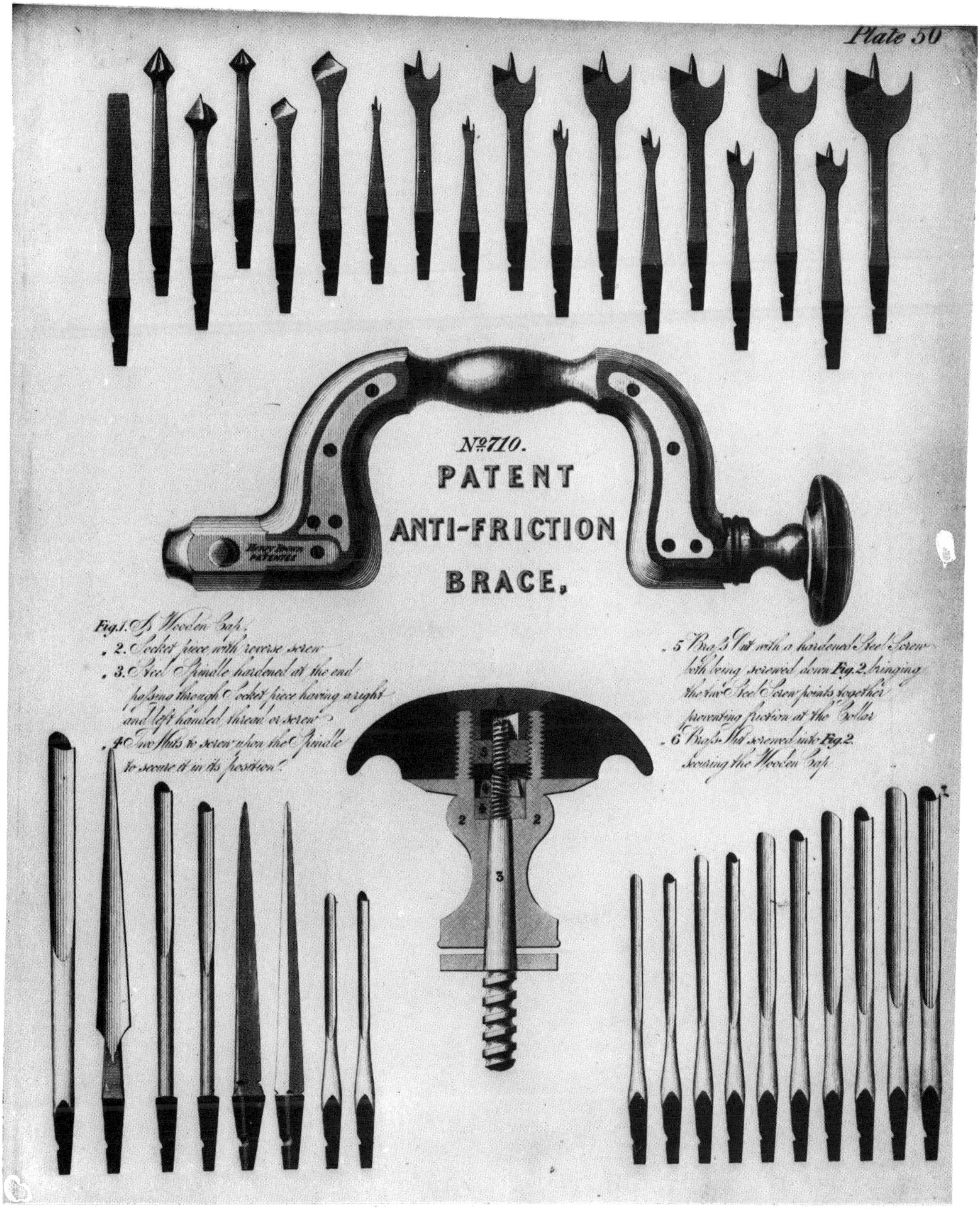

Plate CLXXVII - Patent Anti-Friction Brace Plate 50 from Marsh Brothers 1849 Catalogue
[Archives, Marsh File, Sheffield Central Library]

Plate CLXXVIII Drawings Accompanying Bottom's Patent 10,526

FIG. 2.

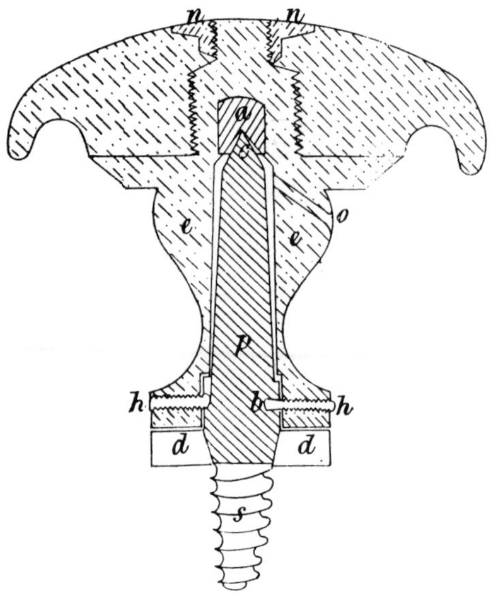

FIG. 4.

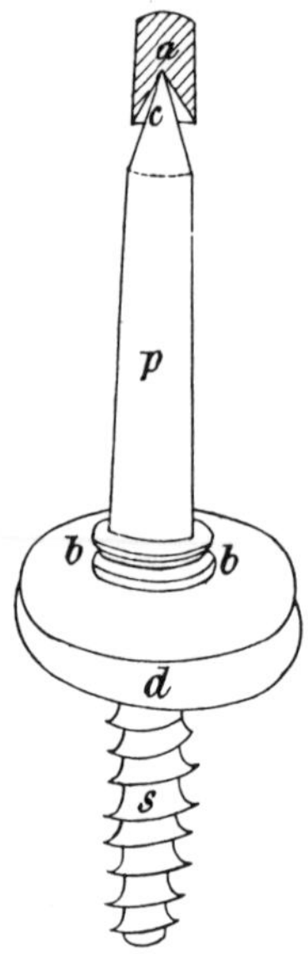

FIG. 7.

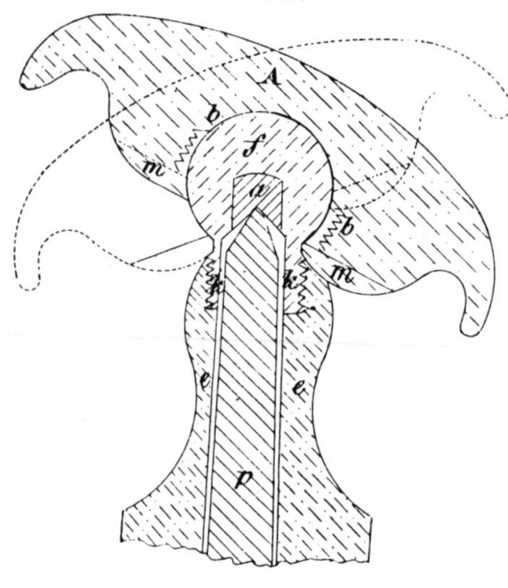

FIG. 10.

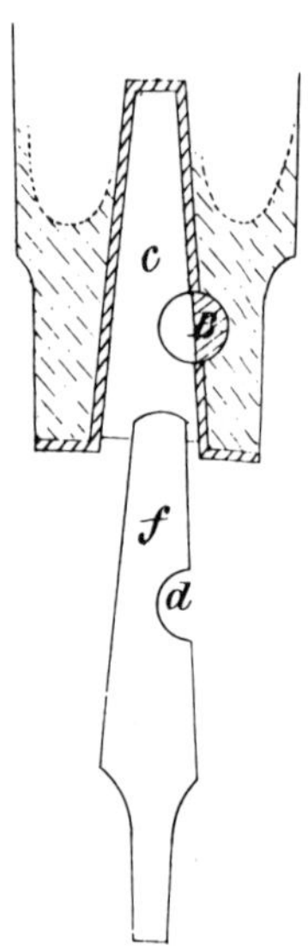

FIG. 11.

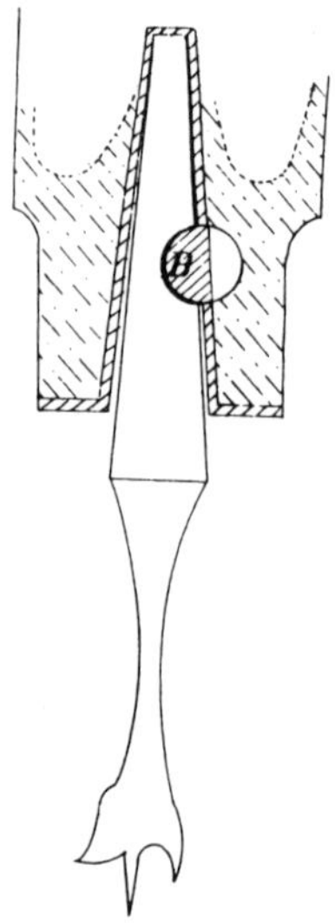

FIG. 8.

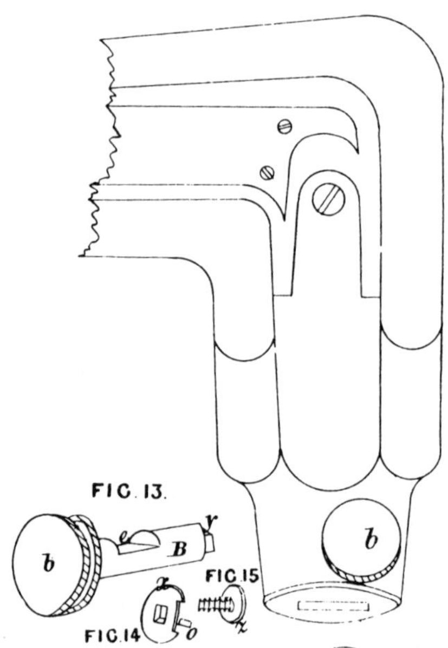

FIG. 13. FIG. 14. FIG. 15.

FIG. 12

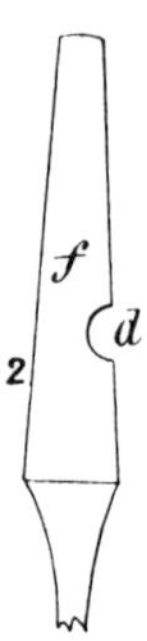

FIG. 3.

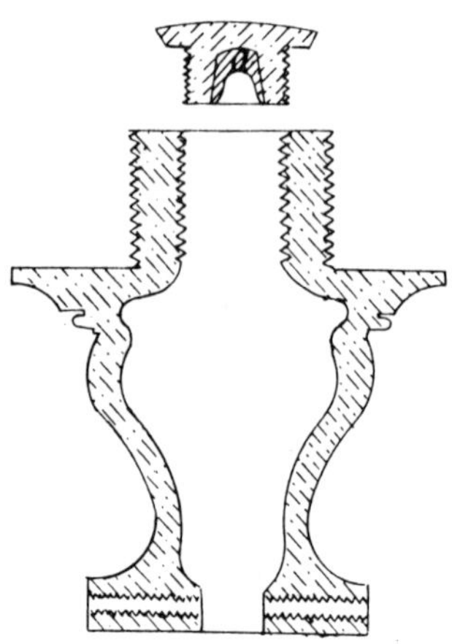

A.D. 1845. FEB 20. No. 10,526.
BOTTOM'S SPECIFICATION.

Plate CLXXIX

1849

LIST OF PRICES OF JOINERS' TOOLS

MANUFACTURED BY

MARSH BROTHERS & Co.,

POND WORKS, SHEFFIELD.

PLT.	No.			12	18	24	30	36	42	48	54	60 bits.
51	700	BEST PLAIN BRACE, polished, 9/ each.	If with C. S. Black Bits	13/8	16/	18/4	20/11	23/9	26/1	28/8	31/8	35/2 each.
			,, C. S. Bright Bits	14/7	17/4	20/2	23/2	26/6	29/3	32/4	35/9	39/9 ,,
			,, C. S. Straw Colored Bits	15/3	18/4	21/6	24/10	28/6	31/7	35/	38/9	43/1 ,,
51	701	Improved Plated Brace, polished 13/ each.	If with C. S. Black Bits	17/8	20/	22/4	24/11	27/9	30/1	32/8	35/8	39/2 ,,
			,, C. S. Bright Bits......	18/7	21/4	24/2	27/2	30/6	33/3	36/4	39/9	43/9 ,,
			,, C. S. Straw Colored Bits	19/3	22/4	25/6	28/10	32/6	35/7	39/	42/9	47/1 ,,

If with Improved Ebony Head and Brass Neck, or with all Brass Head, 2/ each extra

51 | 702 | Improved Scotch Iron Braces and Bits, same price as Improved Plated Braces and Bits, No. 701.

PLT.	No.			PLATE.	No.		
..	703	Best Lignum Vitæ Brace Heads	24/ doz.	..	706	Best Brace Pads	30/ per dozen
..	704	Improved Brace Heads, all Brass	48/ ,,	52	707	Brass Collars, for Sash Bits	1/ per pair.
..	705	Improved Brace Heads, Ebony, brass necks	48/ ,,	..	708	Common Coopers' Iron Braces....	3/ each.

PATENT ANTI-FRICTION BRACES.

PLT.	No.			12	18	24	30	36	42	48 bits.
..	709	Best Plain Brace, polished, with improved Brass Neck, 13/6 each.	If with C. S. Black Bits	18/2	20/6	22/10	25/5	28/3	30/7	33/2 each.
			,, C. S. Bright Bits......	19/1	21/10	24/8	27/8	31/	33/9	36/10 ,,
			,, C. S. Straw Colored Bits	19/9	22/10	26/	29/4	33/	36/1	39/6 ,,
50	710	Improved Plated Brace, polished with improved Brass Neck, 17/6 each.	If with C. S. Black Bits	22/2	24/6	26/10	29/5	32/3	34/7	37/2 ,,
			,, C. S. Bright Bits	23/1	25/10	28/8	31/8	35/	37/9	40/10 ,,
			,, C. S. Straw Colored Bits	23/9	26/10	30/	33/4	37/	40/1	43/6 ,,

.. | 711 | Patent Improved Iron Braces and Bits, same price as Patent Improved Plated Braces and Bits, No. 710.

.. | 712 | Ebony, Box, or Rosewood Plated Braces, 12/ each extra, to Plated Beechwood Braces, No. 701 or 710.

.. | 713 | Do Plain do 7/ do to Plain do No. 700 or 709.

CENTRE BITS, CAST STEEL.

PLT.	No.		To $\frac{3}{8}$	$\frac{7}{16}$	$\frac{1}{2}$	$\frac{5}{8}$	$\frac{3}{4}$	$\frac{7}{8}$	1	$1\frac{1}{8}$	$1\frac{1}{4}$	$1\frac{3}{8}$	$1\frac{1}{2}$	$1\frac{5}{8}$	$1\frac{3}{4}$	$1\frac{7}{8}$	2 inch.
52	734	Black	3/8	4/	4/	4/3	4/6	4/9	5/	6/	7/	8/	9/	10/	11/	12/	14/ doz.
		Bright	4/7	4/11	4/11	5/2	5/5	5/8	5/11	7/2	8/2	9/2	10/2	11/4	12/4	13/4	15/4 ,,
		Straw colour	5/3	5/7	5/7	5/10	6/1	6/4	6/7	7/10	8/10	10/	11/	12/4	13/4	14/4	16/4 ,,

PLT.	No.		Black.	Bright.	Straw Colour.
..	....	Cast Steel Centre Bits, assorted, $\frac{1}{8}$ to 1 inch..	4/2 ..	5/1 ..	5/9 per dozen.

PLT.	No.			$\frac{1}{8}$	$\frac{3}{16}$	$\frac{1}{4}$	$\frac{5}{16}$	$\frac{3}{8}$	$\frac{7}{16}$	$\frac{1}{2}$	$\frac{5}{8}$	$\frac{3}{4}$ inch.
52	735	Cast Steel Nose Bits	Black.......	3/6	3/9	4/	4/3	4/6	5/6	6/6	10/	14/ per doz.
52	736	Do Shell do	Bright	4/5	4/8	4/11	5/2	5/5	6/5	7/5	10/11	14/11 ,,
52	737	Do Spoon do	Straw Colour..	5/1	5/4	5/7	5/10	6/1	7/1	8/1	11/7	15/7 ,,

PLT.	No.		Black.	Bright.	Straw Colour.
..	....	Nose, Shell, or Spoon Bits, assorted to $\frac{3}{8}$ inch ..	4/2 ..	4/11 ..	5/7 per dozen.

PLT.	No.			$\frac{1}{8}$	$\frac{3}{16}$	$\frac{1}{4}$	$\frac{5}{16}$	$\frac{3}{8}$ inch.
52	738	Cast Steel Sash Bits	Black	4/6	4/6	4/6	5/	5/6 per dozen.
			Bright............	5/5	5/5	5/5	5/11	6/5 ,,
52	739	Do Table do	Straw Coloured	6/1	6/1	6/1	6/7	7/1 ,,
52	740	Do Gimblet Bits	Black	4/	4/	4/	5/	6/ ,,

PLT.	No.			$\frac{5}{8}$	$\frac{3}{4}$	$\frac{7}{8}$	1	$1\frac{1}{8}$	$1\frac{1}{4}$ inch.
52	741	Cast Steel Taper Bits	Black	5/6	7/6	9/6	11/6	..	.. per doz.
			Bright........	6/5	8/6	11/	13/	..	.. ,,
			Straw Colour ..	7/1	9/2	11/8	13/8	..	.. ,,
52	742	Do Centre Cock Bits, Black		15/	16/	17/	18/	..	.. ,,
52	743	Do do Plug do		12/	12/	13/	14/	15/	16/ ,,

52 | 744 | Do Bed Bits to 7-16, 6/; to ½ inch, 7/ per dozen.

PLT.	No.		Black.	Bright.	Straw Colour.
52	745	Do Brace Turnscrew, Plain	3/6	4/5	5/1 per dozen.
52	[illegible]	Do do [illegible]	3/6	4/5	5/1 ,,
52	[illegible]	Do Flat Head Countersink	4/	4/11	5/7 ,,
52	[illegible]	Do Snail Horn do	4/3	5/2	5/10 ,,
52	749	Do Rose Head do	4/6	5/5	6/1 ,,
52	[illegible]	Do Square Rim [illegible]	4/	4/11	5/7 ,,
[illegible]	[illegible]	Do Half-[illegible]	4/	4/11	5/7 ,,
[illegible]	752	Do Five, Six [illegible]	5/6	6/5	7/1 ,,
[illegible]	[illegible]	Do Bobbin [illegible]	5/6	6/5	7/1 ,,

BRACE BITS IN SETS, ASSORTED, UNFITTED.

No.		12	18	24	30	36	42	48	54	60 bits.
755	Black Bits, Cast Steel, assorted.	4/	6/	8/	10/3	12/9	14/9	17/	19/8	22/10 set.
	Bright do do do ..	4/11	7/4	9/10	12/6	15/6	17/11	20/8	23/9	27/5 ,,
	Straw Colour do do ..	5/7	8/4	11/2	14/2	17/6	20/3	23/4	26/9	30/9 ,,

No.		$\frac{1}{8}$	$\frac{3}{16}$	$\frac{1}{4}$	$\frac{5}{16}$	$\frac{3}{8}$	$\frac{7}{16}$	$\frac{1}{2}$	$\frac{5}{8}$	$\frac{3}{4}$	$\frac{7}{8}$	1 inch.
756	Chair Makers' Black Bits	..	..	3/	3/	3/	3/	3/	3/9	4/6	6/	7/6 per dozen.
	Do Bright do	..	..	3/9	3/9	3/9	3/9	3/9	4/6	5/3	6/9	8/3 ,,
757	Brush Makers' Black do	3/	3/	3/6	4/	4/6	..	..	..	..	..	.. ,,
758	Coopers' Black Dowling Bits . ..	1/9	1/9	1/9	2/8	3/	..	..	..	..	..	.. ,,

PLATE 51

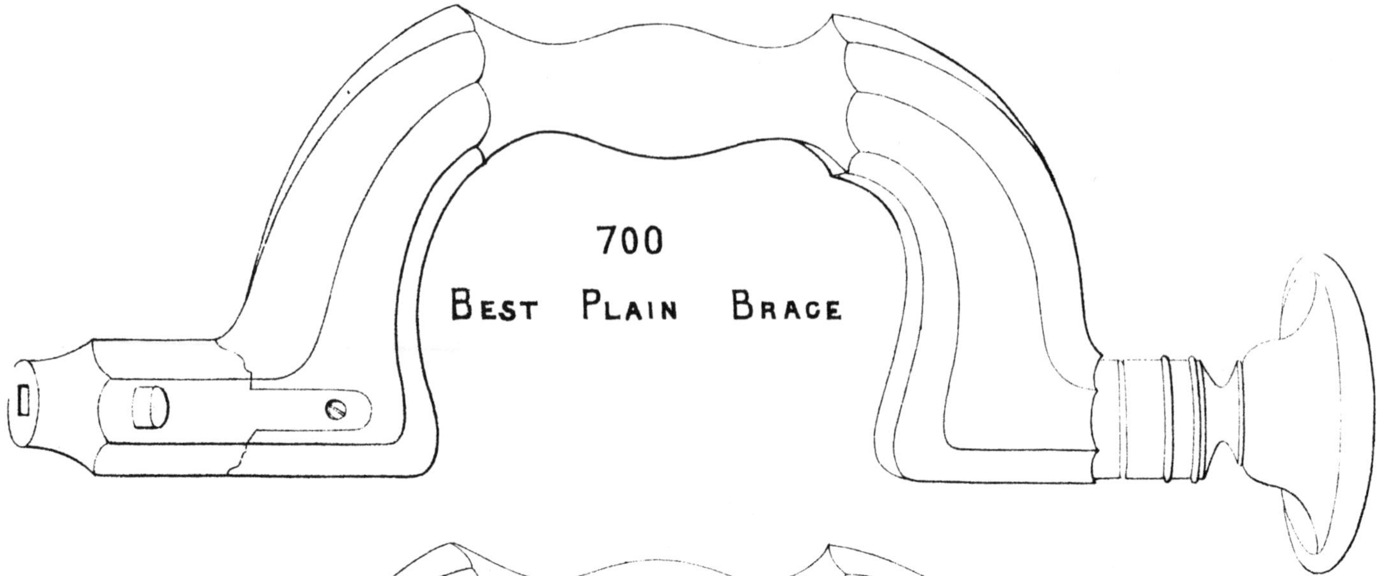

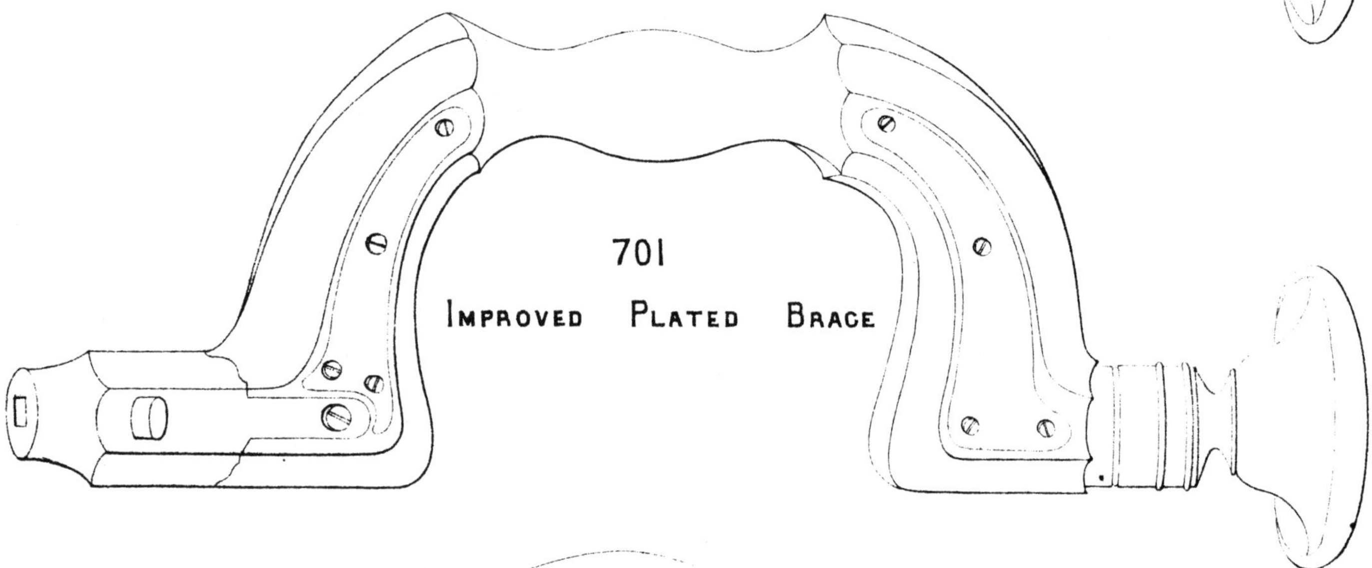

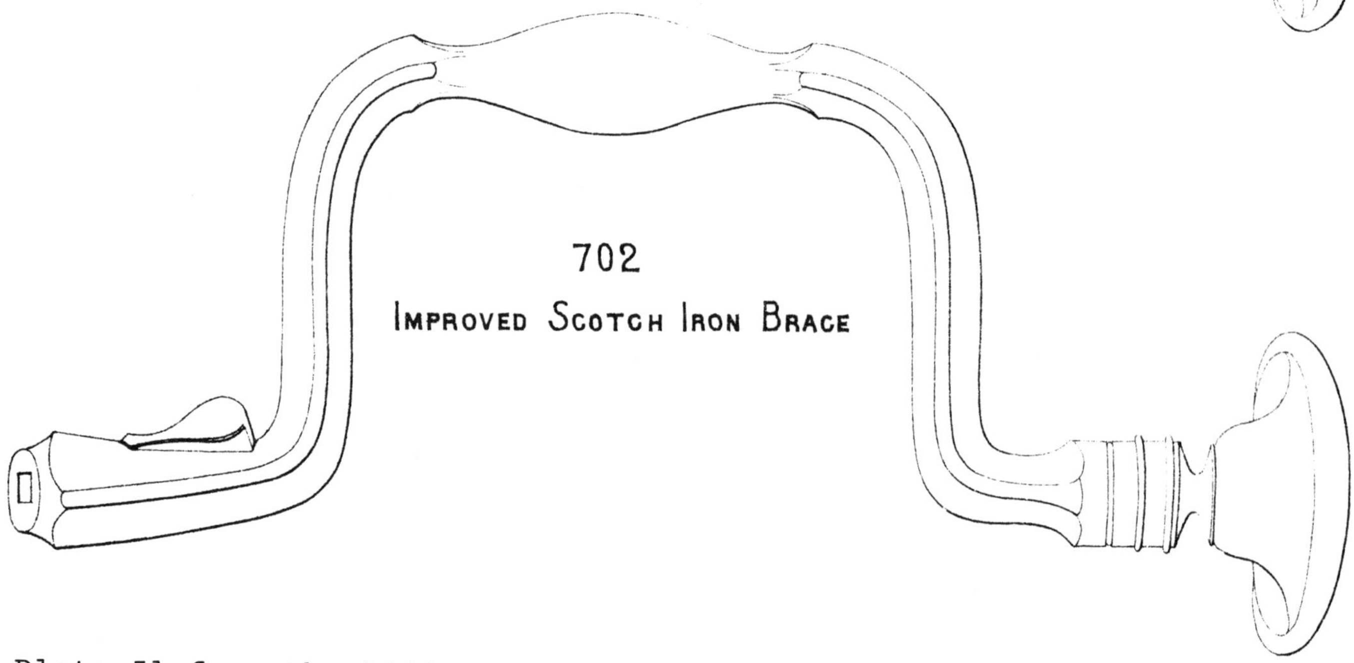

Plate 51 from the 1849 Marsh Brothers & Co Trade Catalogue

Archives, Local History Department, Central Library, Sheffield, England

Brass Emblem Plate in Head

Fig. 61 - Scotch Iron Brace Henry Brown Patentee

The tool stores and import sales' offices were all within the same district in New York City. Competition must have been keen. It must have been difficult to keep buying and and selling prices confidential. New York City was indeed a bustling activity for Sheffield tool sales.

It is apparent that considerable competition existed among various Sheffield joiner tool and brace makers during the period 1844-1851. This may have been a consequence of Brown & Flather advertising and sales at premium prices for their 'Anti-Friction Brace'. A listing of Registered Designs concerning braces appears in Table XXI and English Letters Patents 1845 - 1866 in Table XXII.

The Design Copyright Act of 1843 gave three years protection to the proprietors of non-ornamental designs on articles wholly or chiefly made from wood or metal. After 1850 such qualifying designs could be provisionally registered for a period of one year without effecting later full registration. Apparently a Registered Design was easier to obtain and less expensive than a Letters Patent. The latter at this date gave protection for a period of fourteen years. From 1844 through 1857 a total of fourteen Registered Designs concerning bit braces were issued. Table XXI lists the proprietors, numbers, dates of issue and titles of these designs. Most of these concerned methods of attaching the head to the stock or some feature of the pad holding the bit. From 1845 - 1866 eight Letters Patents were issued concerning braces. (Table XXII)

The first of these registered designs was granted to Thomas Moulson, tool manufacturer, September 27, 1844 (No.279, Class BT45/2) for an 'Improved Carpenter's Brace Head'. A brace having this design is shown in Figs. 62 & 63 and the drawing accompanying this design in Plate CLXXXI His only claim for a new feature was a brass ring, B, threaded to the neck which provided more support to the wooden cap. Around the outside of this ring was stamped: "T.Moulson, Proprietor, Registered Sept. 27, 1844, SANS RIVAL"[without rival].

After Henry Brown's 1844 design for a brace head, Fenton & Marsden were the next to be issued registration #702 (Class BT 45/4) on April 17, 1846. The drawing accompanying this is shown in Plate CLXXXII. Their claim was constructional with a split-conical washer, placed inside the brass neck, holding the neck securely to the spindle. Apparently this was not successful, as they were granted another somewhat similar, but more effective design No.986, March 4,1847, for a Brace Head. The details of this were described in an 1849 Sheffield Directory advertisement reproduced as Plate CLXXXII.This consisted of a two piece hardened steel collar, held securely by a ring, thus holding the brass neck to the spindle. The parts for this are illustrated in Fig. 65. . Apparently this was quite successful and was used for several years. Attention is called to the lever type chucks shown on the braces in Fig. 64. . One of these braces has the imprint of Fenton & Marsden, indicating this feature was used 1847-1849, which is the earliest known date for such a lever chuck.

Robert Marples received two registered brace pad designs: #845, Thumb Pad, October 15,1846 (Plate CLXXXIII),Figs. 67 & 68 and #1416, April 18,1848 (Plate CLXXXIV),Fig. 66. The latter was in the form of a slide catch. This was used extensively later with the Wm. Marples brass frame "Ultimatum' brace.

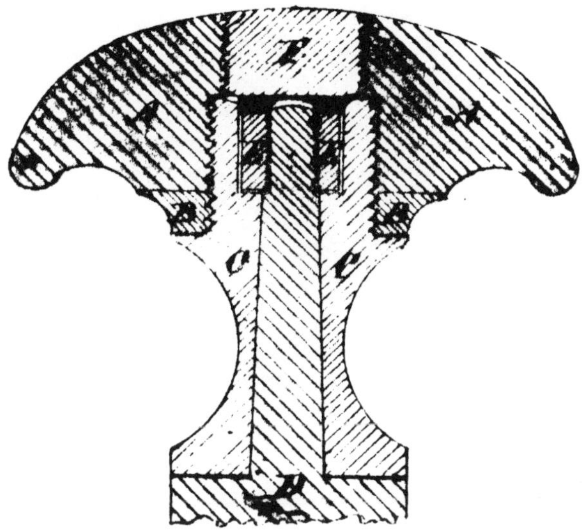

Plate CLXXXI
Drawing Accompanying
T.Moulson's Registered Design
#279, Sept. 27,1844

The above Drawing is a view of this improved Brace.

A is the wooden cap head.

B circular plate of brass with inside female thread

C Brass socket

D Spindle

E Nut

F Cap

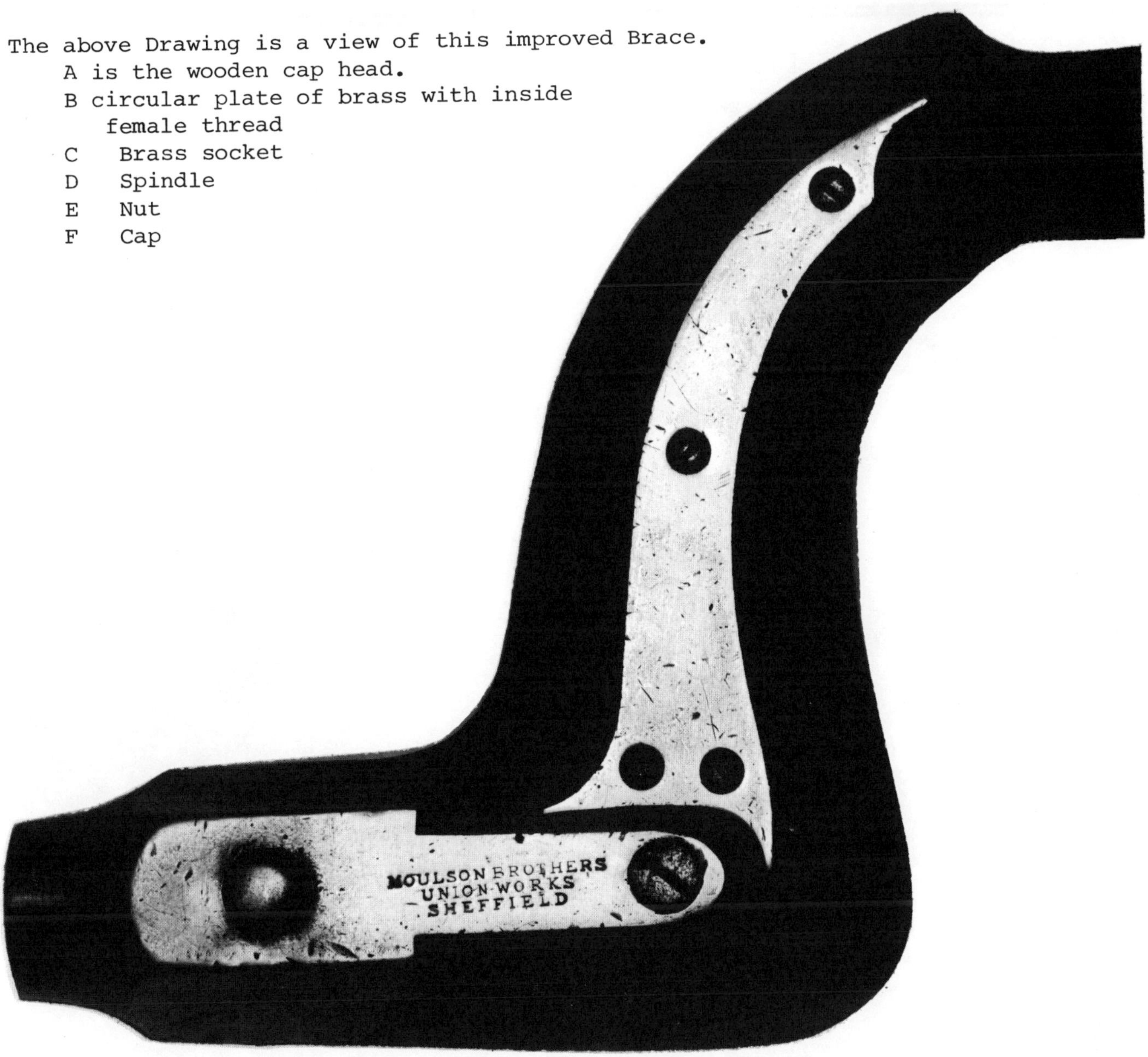

Fig. 62 - Thomas Moulson Registered Design No.279, Sept. 27,1844
Carpenter's Brace Head

Stamped Around Periphery of Plate between Head and Neck -

"*Registered - September 27,1844* - *SANS RIVAL* - Thomas Moulson, Proprietor

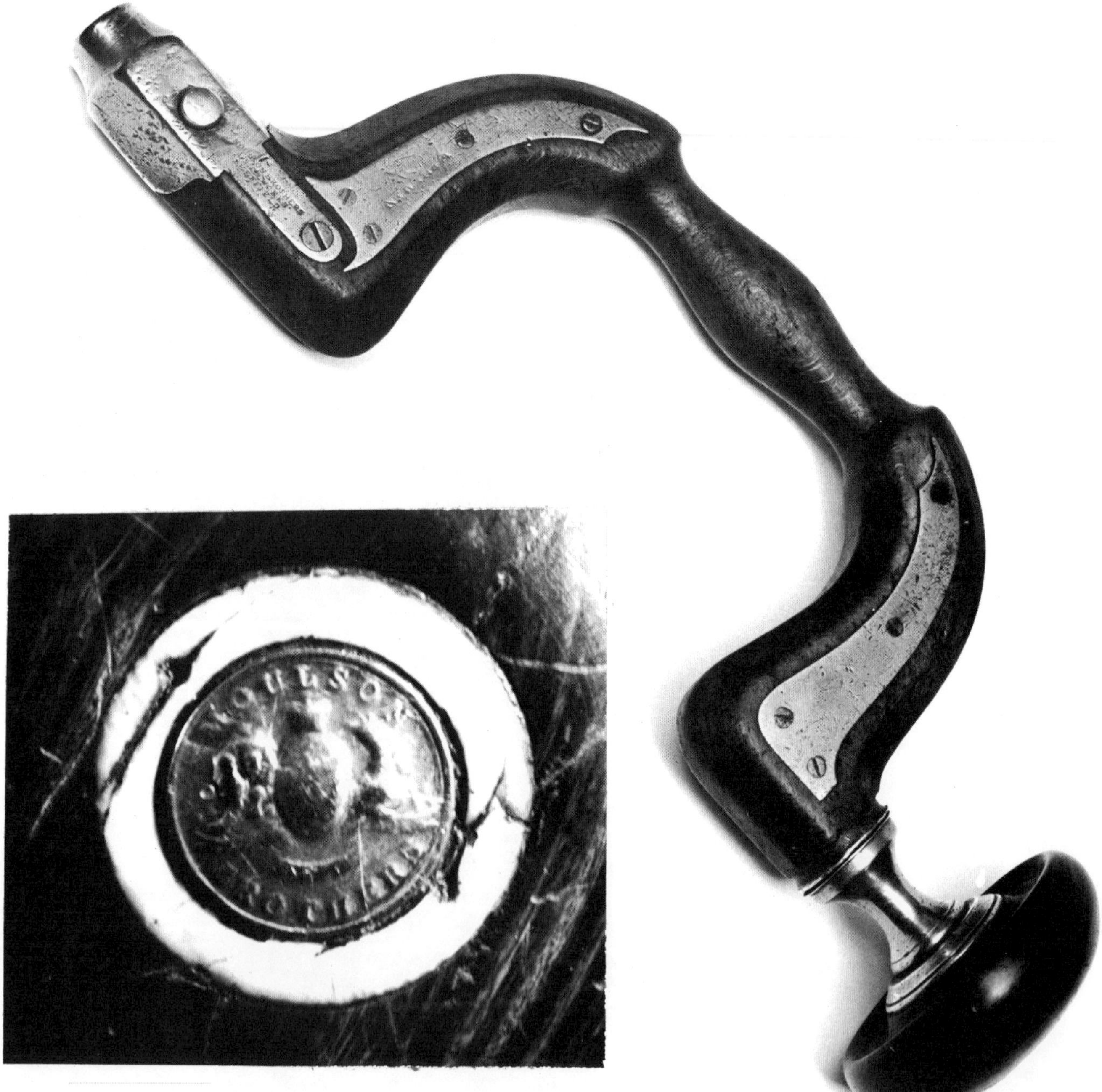

Emblem Plate Set In Brace Head

Fig. 63 - Thomas Moulson Registered Design No.279, Sept.1844
Carpenter's Brace Head

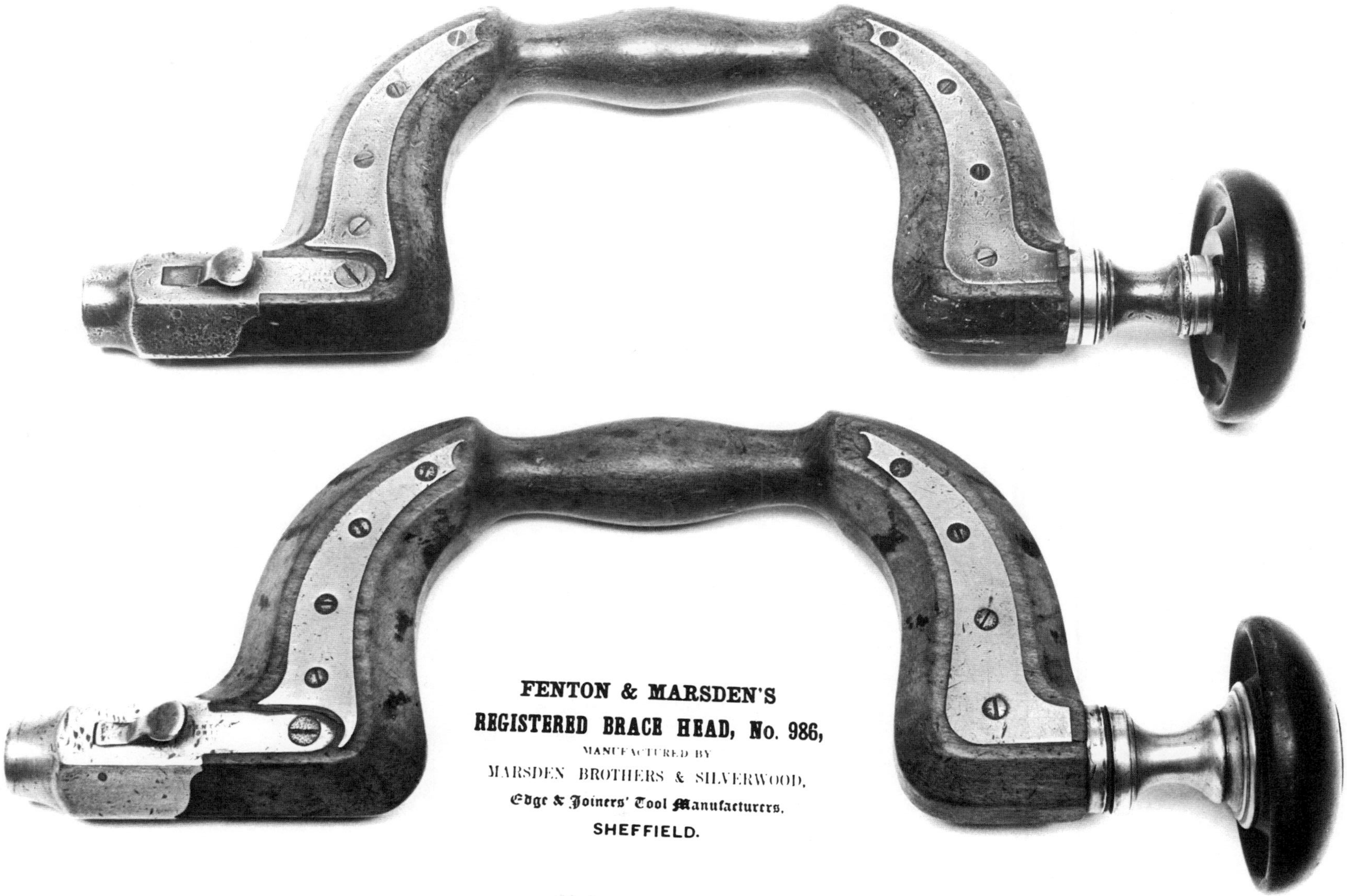

Fig. 64 - Plated Braces with Fenton & Marsden's Registered Design No. 986. Top - Marsden Brothers
Bottom - Fenton & Marsden

Table XXI - ENGLISH REGISTERED DESIGNS CONCERNING BIT BRACES

Order	Designer	Title	Date Registered	Registration Number
I	Thomas Moulson	Carpenter's Brace Head	Sept. 27, 1844	279
II	Henry Brown	Carpenter's Improved Brace Head	Nov. 2, 1844	306
III	Fenton & Marsden	Brace Head	April 17, 1846	702
IV	Robert Marples	Brace Pad	Oct. 15, 1846	845
V	Fenton & Marsden	Brace Head	March 4, 1847	986
VI	Bloomer & Philipps	Spring Brace Pad with Lever Thumb Bit	Oct. 10, 1847	1229
VII	Robert Marples	Brace Pad	April 18, 1848	1416
VIII	George Scofield	Improved Thumb Bit and Joint for Joiners' Brace Pad	April 20, 1848	1420
IX	George Horton	Joiner's Brace	Nov. 8, 1850	2528
X	Bloomer & Philipps	Spring Lever Brace Pad	June 11, 1851	244
XI	Joseph Cooper	Improved Joiner's Brace	Aug. 25, 1853	3505
XII	Henry Parker and Francis Thompson	Carpenter's Brace	May 4, 1855	100041
XIII	W. Adsetts	Double or Single Action Rotary Carpenter's Brace	May 19, 1856	3839
XIV	Robert Marples	Carpenter's Brace with Octagon Lever Pad	March 3, 1857	3954

[Note: All above resided in Sheffield,except Joseph Cooper from Birmingham]

Table XXII

ENGLISH LETTERS PATENTS CONCERNING BIT BRACES

Order	Patentee	Title	Date of Patent	Patent Number
I	John Bottom	Carpenter's Stocks or Braces	August 16, 1845	10,526
II	John Cartwright	Brace for Use of Carpenters and Others	December 16, 1848	12,377
III	James Chesterman	Improvement in Carpenter's Braces	November 13, 1849	12,843
IV	Charles Pilkington Thomas Pilkington and Abraham Pedigor	Joiner's Brace	December 20, 1852	1,113
V	Joseph Cooper	Improvement in Carpenter's Braces	September 4, 1855	504
VI	William Tucker	Bit Stocks or Auger Handles	March 3, 1859	570
VII	John Orvis	Braces and Tool Handles	October 23, 1865	2,735
VIII	Thomas A. Mathieson	Devices for Holding Boring Tools	March 12, 1866	744

* *

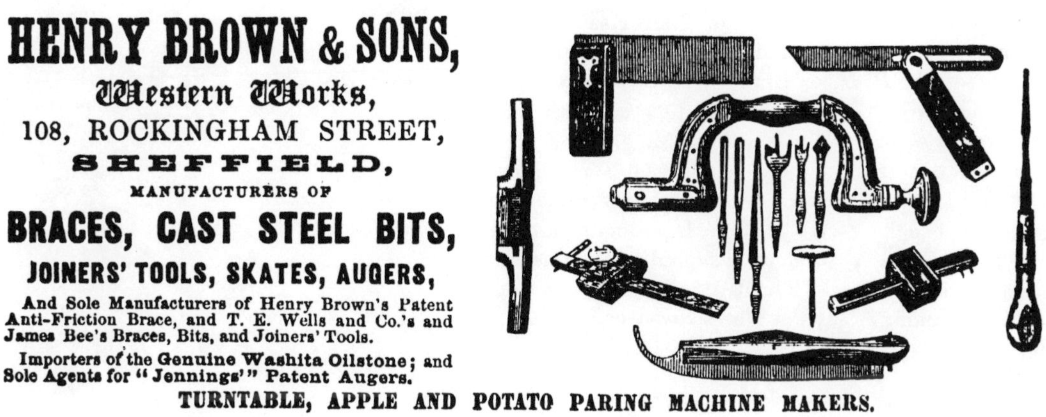

Henry Brown & Sons' 1854 Birmingham Directory Advertisement

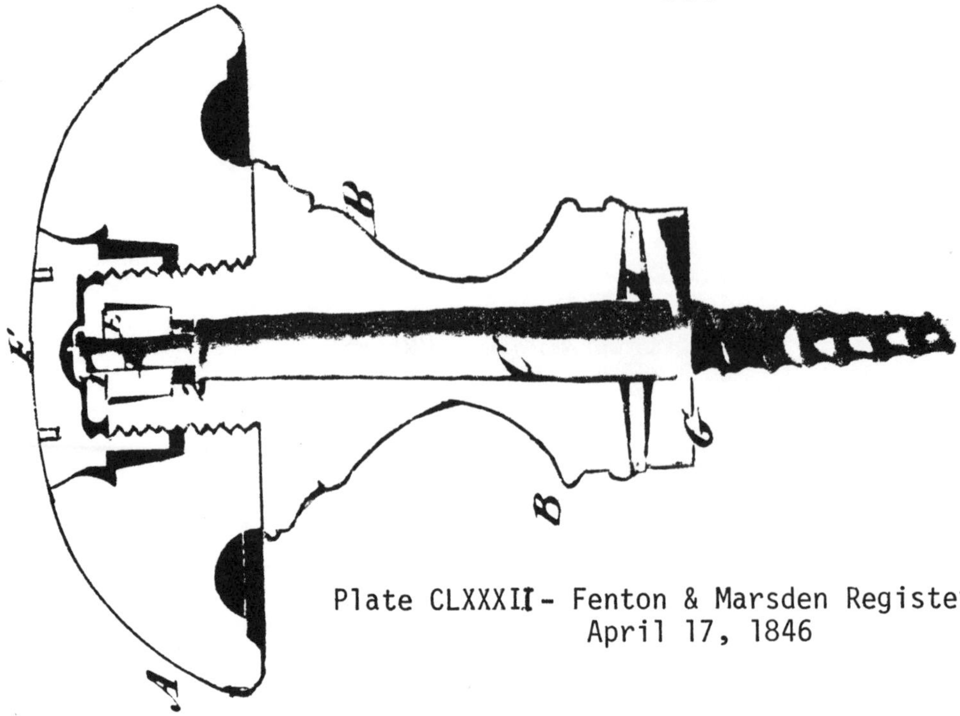

Plate CLXXXII - Fenton & Marsden Registered Design #702
April 17, 1846

FENTON & MARSDEN'S
REGISTERED BRACE HEAD, No. 986,

MANUFACTURED BY

MARSDEN BROTHERS & SILVERWOOD,

Edge & Joiners' Tool Manufacturers,

SHEFFIELD.

The Figure represents the Section of the Brace Head. *A* is a Spindle terminating at the top point *c*, and to be attached to the Stock of the Brace by a Screw, *b*; *D, D* is a Brass Neck, through which the Spindle *A* passes, and in which it turns freely; at the top of the Neck are placed two Screws, *d, d*. *E, E* is a Collar of hardened Steel, formed in two parts, as seen separately at *Fig.* 2, which is put through a recess in the neck *D*, and its central opening exactly fits the grove *a* of the Spindle. When this Steel Collar, *E*, is put in its place, it is held there by a Brass Collar, *e, e*, which is slipped over the neck *D*, and screwed down by the Cap *F*, which holds it to its place, the Cap *F* being prevented from working off by the Nut screwed on theNeck, *d*, having a reverse thread.

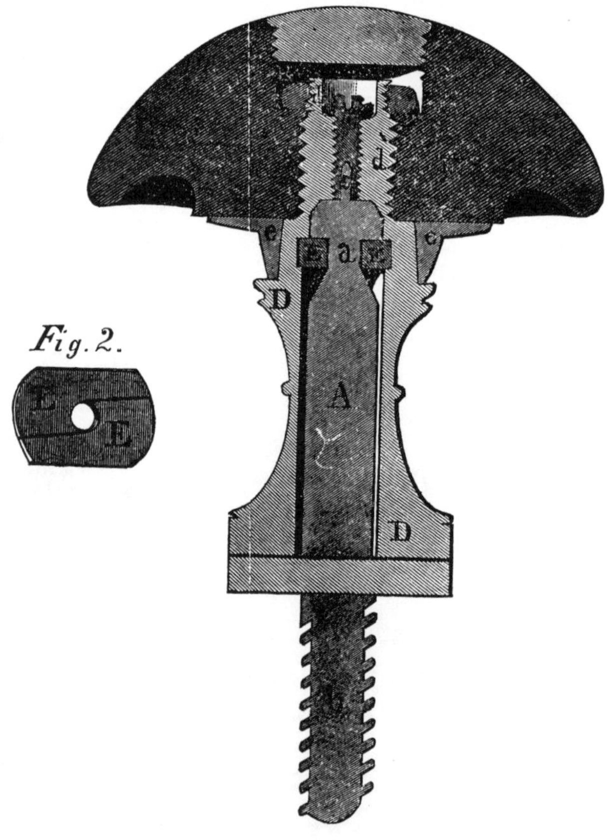

Fig. 2.

In offering the above Brace Head to the Public the Proprietors have great satisfaction to add that by such Improvements it will be found to SUPERSEDE ALL OTHERS hitherto manufactured. It possesses those advantages so long sought for, viz., SECURITY OF THE HEAD TO THE BRACE, EASE IN WORKING, WITHOUT FRICTION, and ultimate DURABILITY. This will be at once obvious to all Artizans; and claim that preference in their selections which by the immence saving of time and labour the above invention is deserving.

Plate CLXXXII - Fenton & Marsden Registered Design No. 986, March 4, 1847

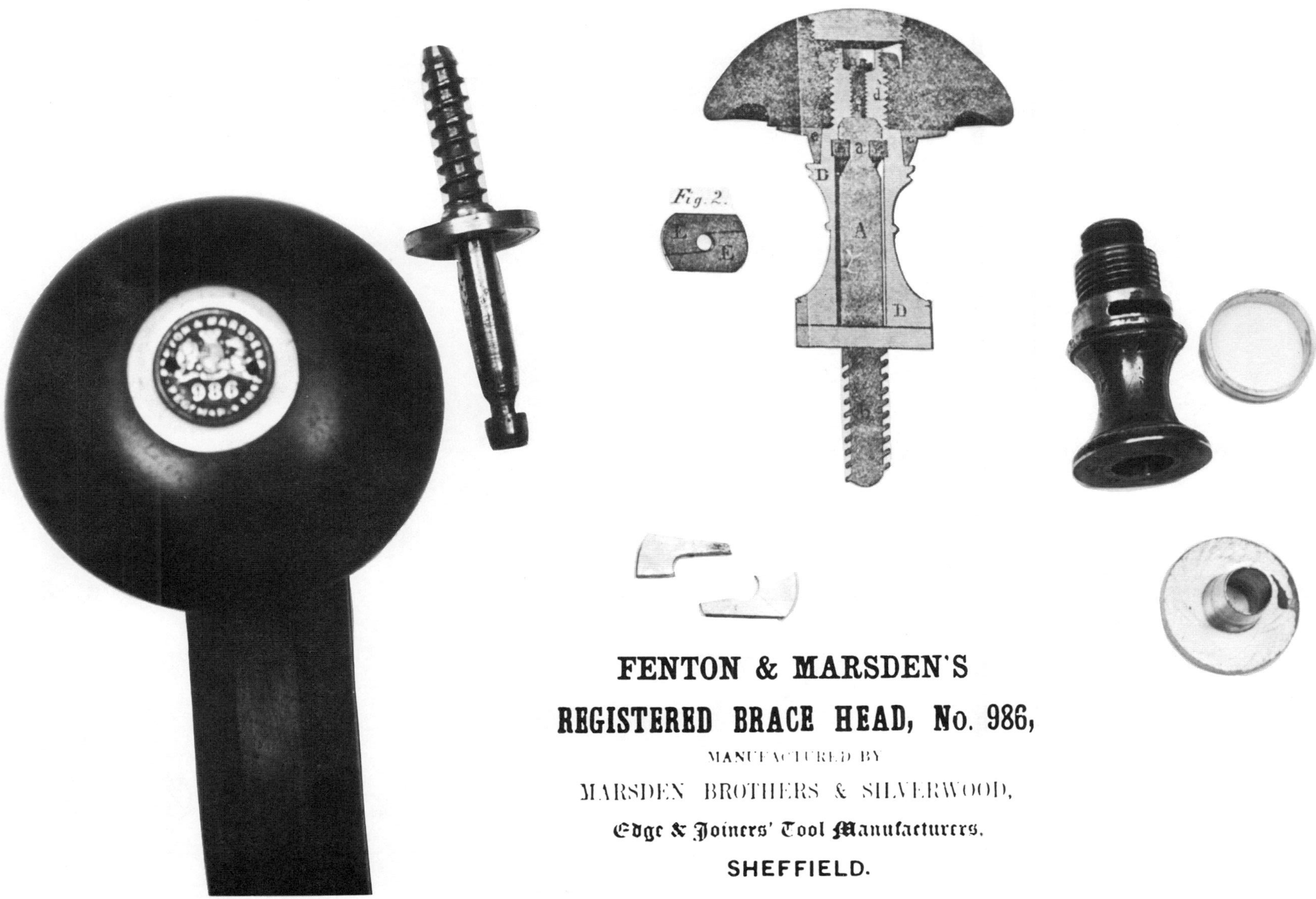

Fig. 65 - Parts of Brace Head for Fenton & Marsden Registered Design No. 986

Fig 1

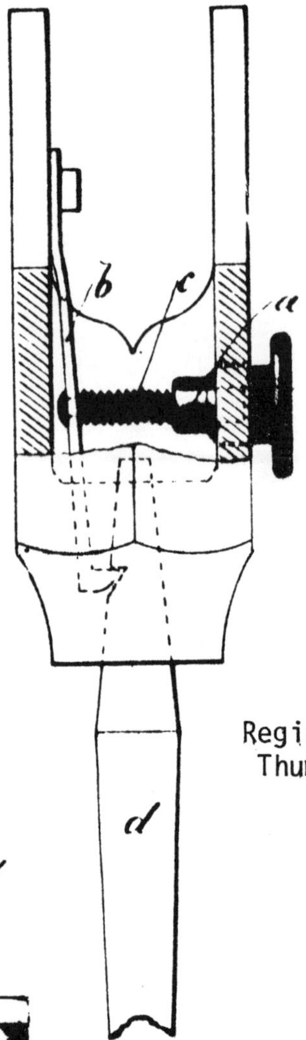

Fig 2

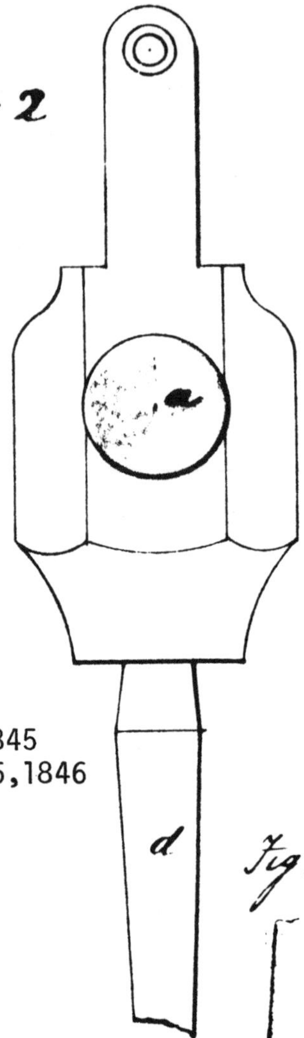

Plate CLXXXIII
Robert Marples'
Registered Design No. 845
Thumb Pad - October 15,1846

Fig 1

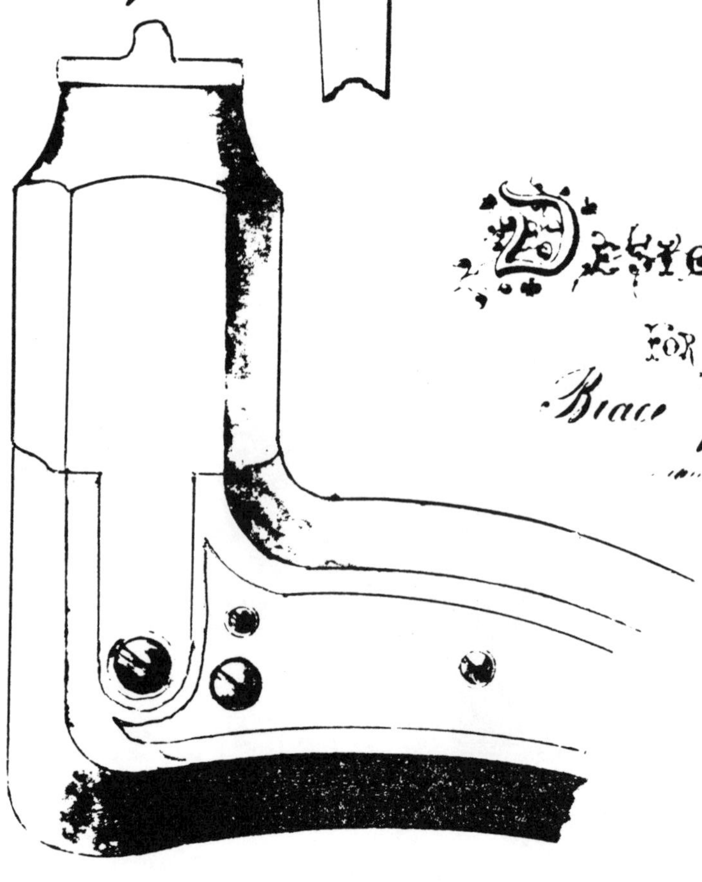

Design for Brace Pad

Fig 2

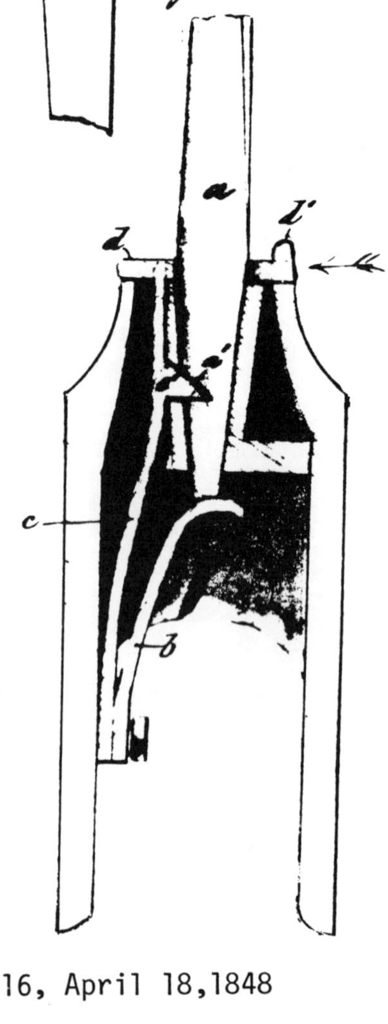

Plate CLXXXIV - Robert Marples' Registered Design No. 1416, April 18,1848

Fig. 66 - Robert Marples Registered Design No.1416, April 18,1848, Brace Pad

Fig. 67 Robert Marples Registered Design No.845, Oct. 15, 1846, Brace Pad

Fig. 68 - Robert Marples Registered Design No.845, Oct. 15,1846, Brace Pad

George Scofield, Sheldon Row, Willey Street, Wicker, Sheffield was issued a registered design #1420, April 20, 1848 (BT 45/8) for "Improved Thumb Bit and Joint for Joiner's Brace Pad to act on the Upright Pressure". The drawing accompanying this design is shown as Plate CLXXXV, along with Fig. 69 illustrating two views of this unique lever.[18] This is the earliest proto-type of the lever catch. It is suggested that Fenton & Marsden modified this design, placing the lever in a slot, pinned through the brace pad.

Bloomer & Philipps were issued two registrations for "Spring Brace Pad with Lever Thumb Bit": #1229. October 10,1847, Plate and No.244, June 11,1851, Plate CLXXXVI. The former is shown in Fig.70 This design noted "thereby rendering it much stronger than any former Invention". This is indeed open to question. The No.244 design claimed "thus in fitting, the stock is reduced very little. The arrangement is more simple and effective than any other in use and the stock is not weakened". Again very questionable. This design was most likely unsuccessful.

Indeed it must have been very difficult to supress information among these Sheffield firms concerning these developments. Much of the special brass castings for parts such as necks and pads were done by outside foundries which were readily accessible to all concerned. Most manufacturers there must have known what the others were doing and from whom they procured their parts. By studying the sources these parts could be readily obtained. Henry Brown, Fenton & Marsden (later Marsden Brothers), Robert Marples, Bloomer & Philipps and George Schofield all had competitive designs. This must have been very disconcerting to William Marples, another prominent joiner tool manufacturer. Apparently not being of inventive mind, but possessing a good business sense, William Marples may have decided to seek local inventive talent in his desire to obtain some unique and competitive brace design. He may have approached either James Chesterman, who had formerly worked with John Bottom of the Anti-Friction Brace Patent, or John Cartwright. In any event it was the latter who accomplished this task.

James Chesterman was first listed in the 1833 Sheffield Directory as patentee of the spring tape measure and inventor of the self-acting map and window blind roller. On January 11, 1842 he was granted Letters Patent #9214 for "Certain Improvements in Tapes for Measuring" jointly with John Bottom. In 1845 he was listed as working alone as "patentee and manufacturer". It was just previous to this time that Henry Brown had probably approached James Bottom to design a patent brace. By 1849 Chesterman had affiliated with a firm Cutts,Chesterman and Bedington, who manufactured his designs for roasting jacks and other designs. On November 13, 1849 Chesterman was granted Letters Patent No.12,843 "Carpenters' Braces and Other Tools for Drilling and Boring". The specification were presented in several pages with 25 drawings and eight claims. The reference to a carpenter's brace consisted of a bevel gear train with crank motion similar to the type shown in Fig. 71. He also showed a complicated bow brace with crank handle motion through a pair of straight gears. The rest of the claims were for metal drilling tools involving ratchet braces or more complicated devices. In 1852 the firm became Cutts, Chesterman & Co. and their Directory Advertisement noted among other products "boring machines, ratchet braces, etc." (Plate CLXXXVII) By 1859 the firm had dissolved, James Chesterman & Co. continuing at Bow Works, Nurserey Street.[19] Fig. 72 shows a ratchet brace.

Fig. 69a

George Schofield Registered Design No.1420 April 20, 1848. Improved Thumb Bit and Joint for Joiner's Brace - Pad to act on the Upright by Pressure. Fig. 1 - A,the spring; B,Upright acting on spring; C,joint Thumb Bit; D, Thumb Bit acting on the upright.

Fig.2 - Front of Pad with the Thumb Bit.

Plate CLXXXV

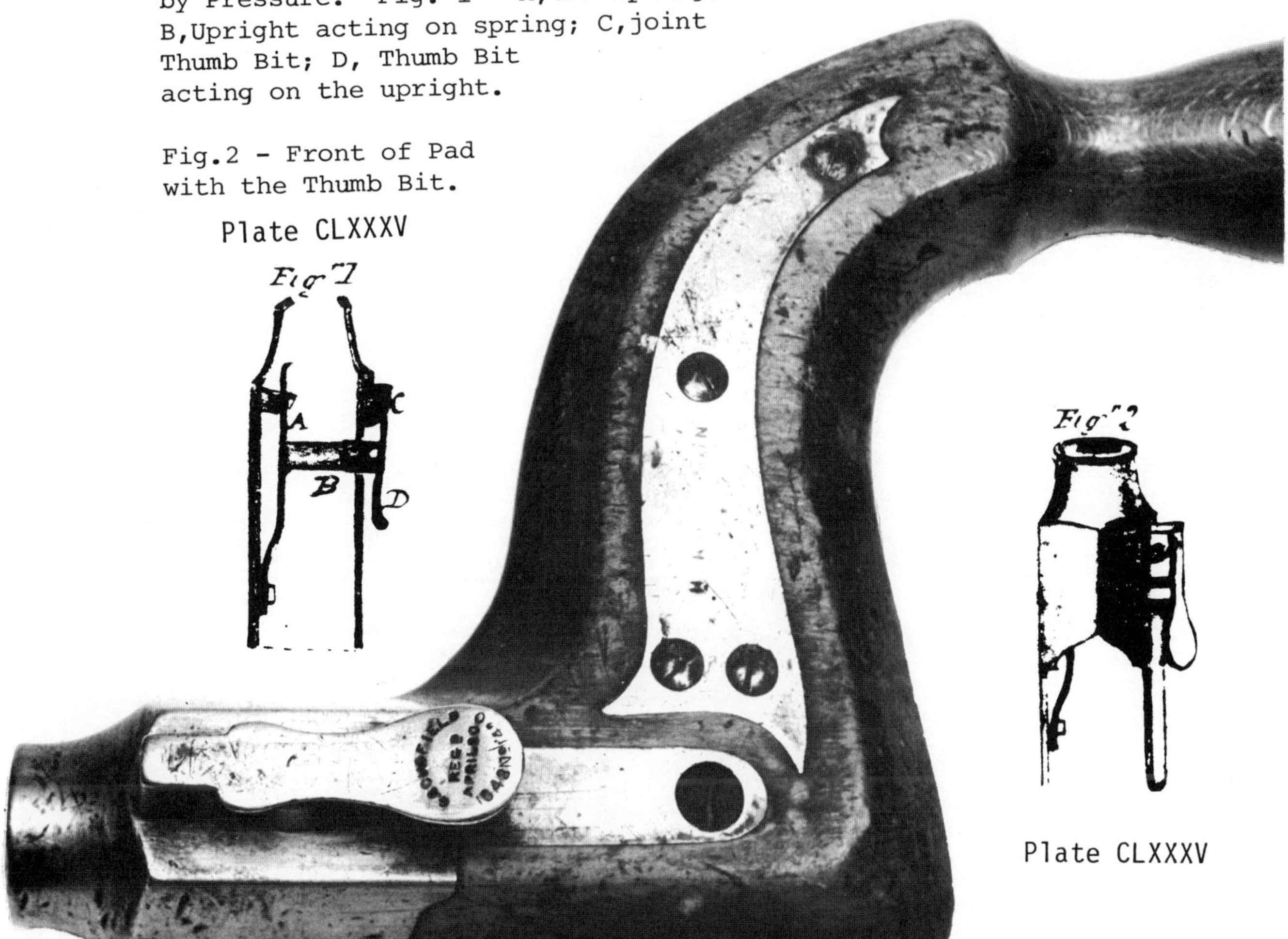

Plate CLXXXV

Fig. 69b - Registered Design by George Schofield, Shelton Row, Willey Street, Wicker, Sheffield, No. 1420, April 20, 1848

Fig: 2

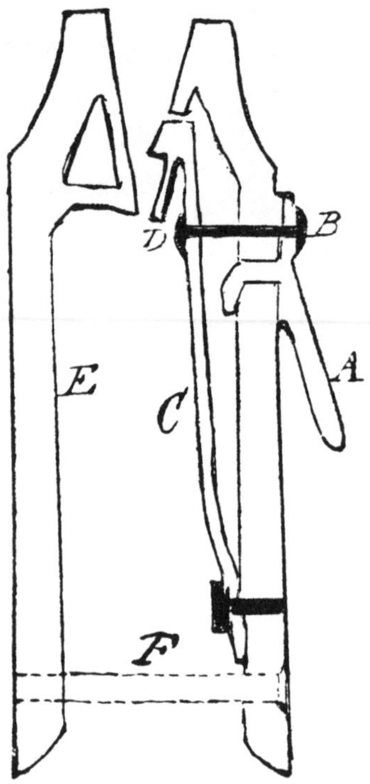

Fig: 3

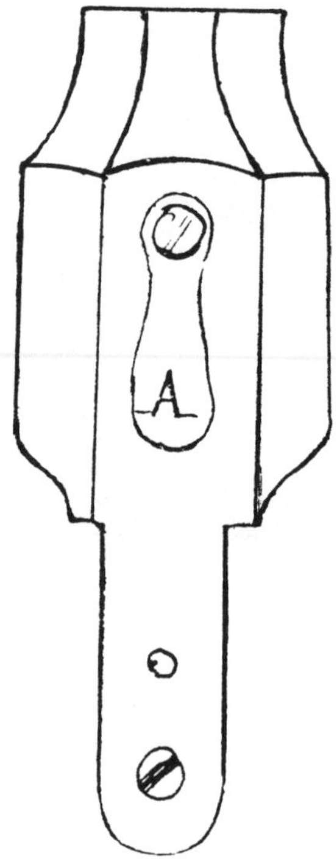

Drawing Accompanying Bloomer & Philipps Lever Thumb Brace Pad Registered Design #1229, October 10, 1847

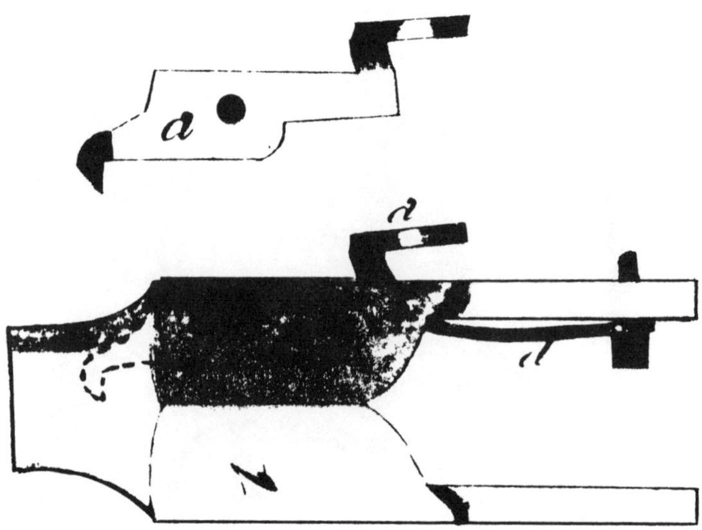

Drawing Accompanying Bloomer & Phillips Spring Lever Brace Pad Registered Design #244, June 11, 1851

Plate CLXXXVI - Bloomer & Philipps Brace Pad Registered Designs

Fig. 70 - Bloomer & Philipps Registered Design No.1229, October 10,1848
[Brace Pad and Emblem In Brace Head]

John Cartwright first appeared in the 1841 Sheffield Directory as a commercial traveller (salesman). In 1845-48 he was listed as a wood turner and hard wood dealer. On December 16, 1848 he applied for a patent concerning 'Brace for Use of Carpenters and Others', which was enrolled June 16, 1849 as Letters Patent No. 12,377. As this patent defined what subsequently became the William Marples ULTIMATUM BRACE the complete specifications to this are presented as Table XXIII and the accompanying illustrations as Plates CLXXXVIIIa,b & C. Discussion of this brace will follow later in this Chapter. Cartwright did not appear in the 1851 Directory, nor thereafter. It is assumed that he left Sheffield soon after acquiring this brace patent which he sold to William Marples, who thereby became Patentee.

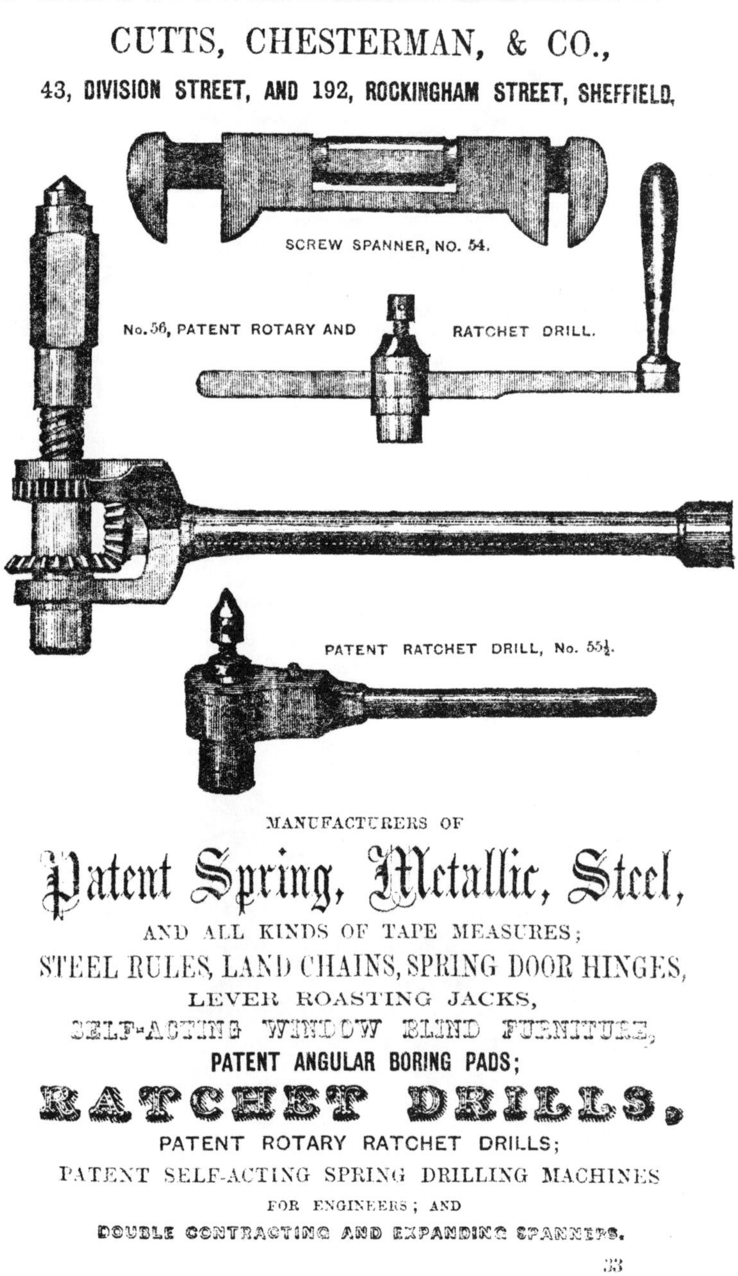

Plate CLXXXVII Cutts,Chesterman & Co.
1851 Advertisement,Sheffield Directory

Fig. 72 Ratchet Brace

Fig. 71 - Bevel Gear Crank Drill Braces - [top] Brass; [bottom] Iron

Patent No. 12,377 - Brace for Use of Carpenters and Others - John Cartwright,Sheffield

My invention consists of an improved Brace for the use of Carpenters and others in which certain parts are cast in metals and in manner as to allow of a moveable handle of horn or other suitable material being used and also having a more convenient mode of applying the revolving hand and a more convenient arrangement for releasing the bits from the spring catch in the pad; and in order that the Invention be most fully understood and readily carried into effect. I will describe the construction of the improved brace as made according to my invention.

Description of the Drawings

Figure 1 shews the side view of a brace made according to my invention. Figure 2 shews a section thereof. Figure 3 shews an end view looking from below upwards. Figure 4 shews the lower part or pad. The other Figures separate views of some of the details and in all Figure the same letters are used to indicate similar parts. a b c constitute the metal frame. Parts a and c of the brace are cast and are connected by the spindel b which carries the handle of horn or other suitable material. The casting a constitutes the upper part of the frame of the brace and the casting c constitutes the lower part of the frame of the brace, having at its lower part the pad holding the bits. The spindle b is square at its upper end and has cast upon it far upper part of the skeleton a the square part being properly prepared, as is well understood for retaining the square end secure in the casting. The other end of the spindle b also prefer to have square where the lower portion c of the frame is cast on it, but in casting on that end the square is to be prepared properly as is understood for allowing the part c of the brace to be removed and the spindle is prepared with a screw and a nut to hold all parts located when the brace is complete. The parts e e which constitute the filling of the frame may be of horn, gutta percha, wood or other suitable material. If of horn or gutta percha these parts may be made by pressing on the moulds as is well understood when moulding articles of these materials. f is the head of the brace. The parts are combined as shown in the drawing so that the head may freely rotate. The bits used in this brace are retained in the pad by a spring catch as it shown. The handle b' I prefer to be of horn though other material may be employed. I would state that the details of construction may be varied and I have by Figures 6, 7, 8, 9, & 10 shown a somewhat different construction though in substance the same principle of construction is retained the upper and lower parts a & b are castings connected by a stem or spindle for a handle, the chief difference being that the screw at the end of the spindle b has the part a cast thereon, the screw being properly prepared to allow its being unscrewed when applying the handle b'. When this arrangement in preferred the screw may be secured in the part a by a pin driven through the skeleton frame of this brace and I would state that it is not essential to the construction that the parts a and c for holding the bits in the pad may as heretofore or as shown in this Figure and in any convenient manner.

Having thus described the nature of my Invention and the manner of performing the same I would have it understood that I do not confine myself to the details as herein described so long as the peculiar character of my Invention be retained. But what I claim is the casting of the parts a and c and connecting them with a stem or spindle b for receiving a separate and moveable handle. I also claim the mode of construction and applying the heads as shown in Fig. 1, 2, & 11. I also claim the application of horn in the manufacture of hands for braces. And I also claim the mode of acting on the spring by the lateral movement of the nozzle as described. I also claim the making of the parts a & c of horn or gutta percha. And I also claim the making of braces with moveable handles of gutta percha or of horn on spindle b.

Enrolled the sixteenth day of June in the year of our Lord One thousand eight hundred and forty nine. Applied: December 16, 1848

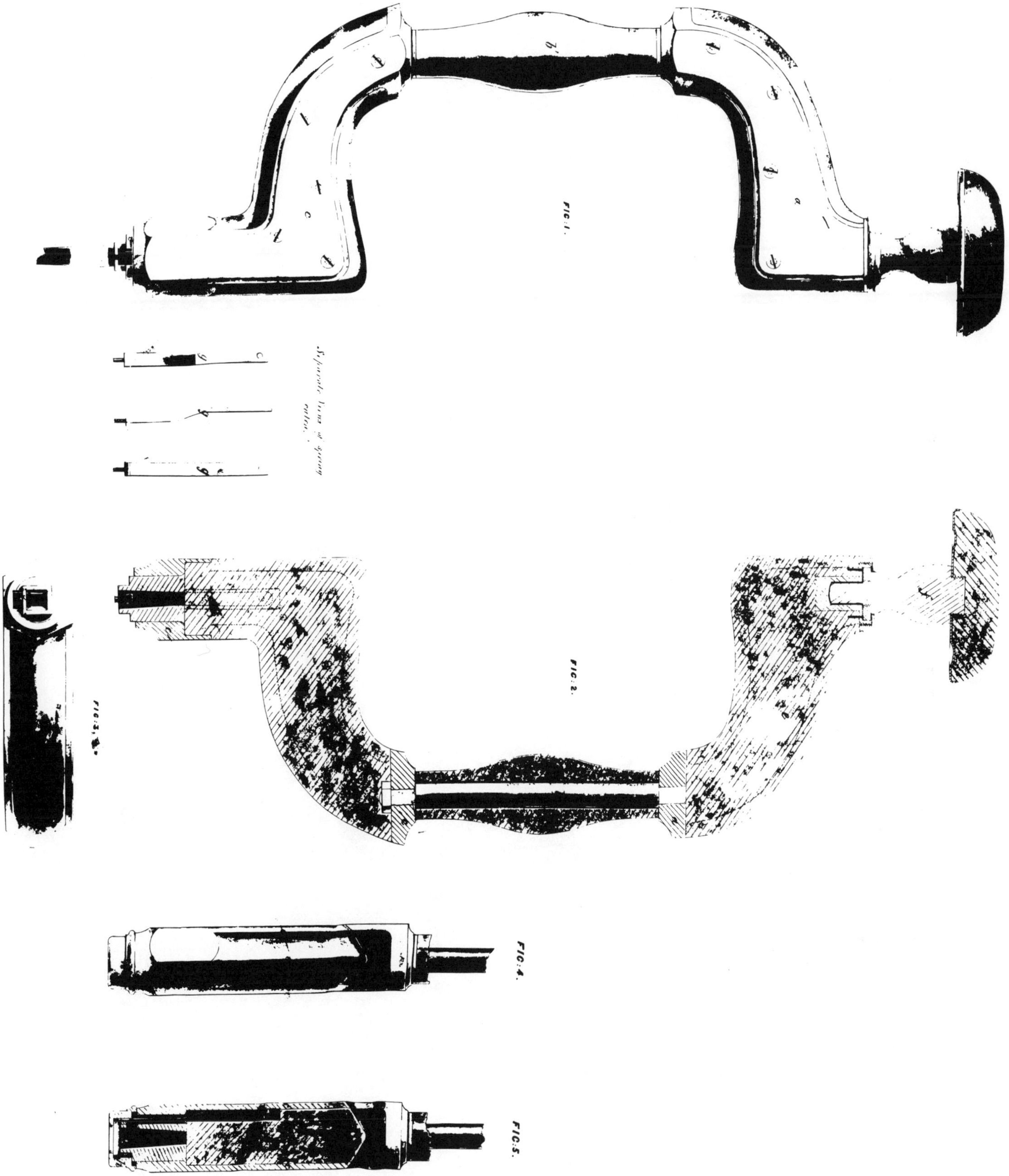

Plate CLXXXVIIIa - Drawing Accompanying Cartwright Patent 12,377

FIG: 8.

Section thro 2.2 Fig. 7

FIG: 9

FIG: 6.

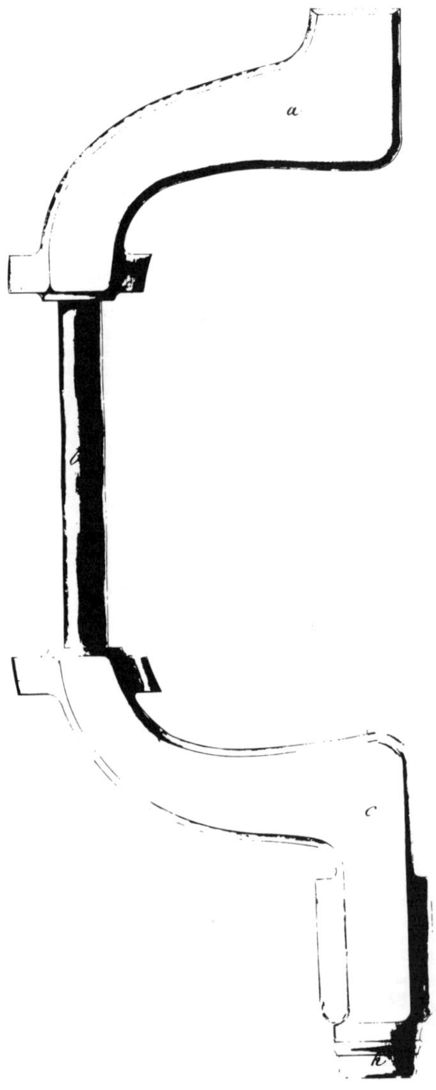

FIG: 7.

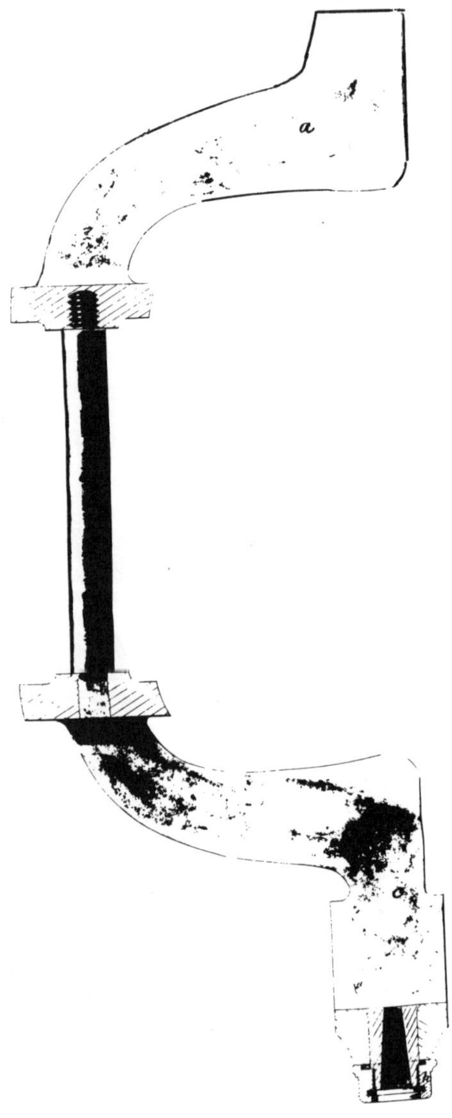

Section through 1,1. Fig 6.

FIG: 10.

Plate CLXXXVIIIb - Drawings Accompanying Cartwright Patent 12,377

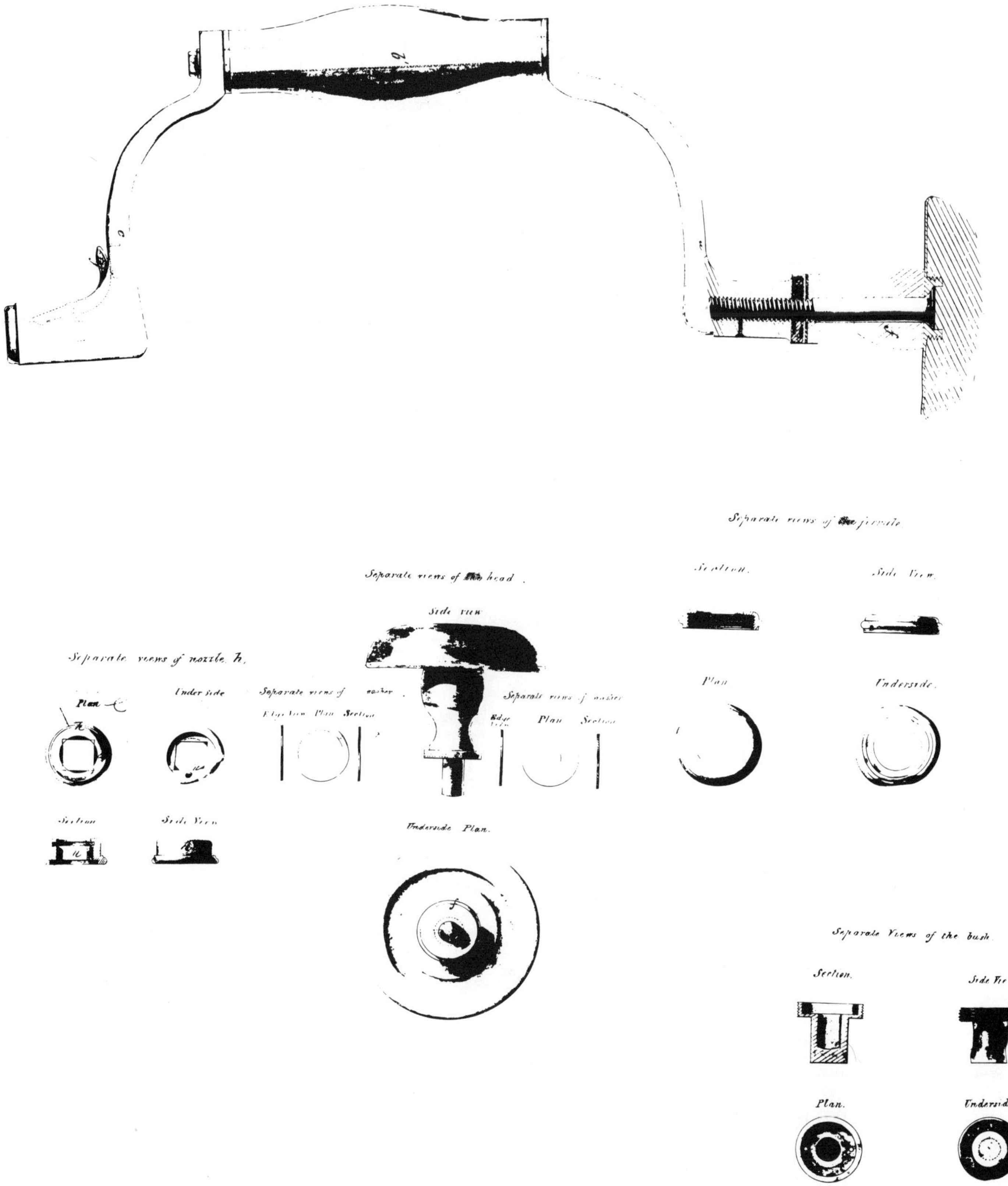

FIG: II.

Plate CLXXXVIIIc - Drawings Accompanying Cartwright Patent]2,377

At the Great Exhibition of the Works of Industry of all Nations held at the Crystal Palace in London, 1851, the following manufacturers exhibitied braces: -

Scotland	Sheffield
A. Mathieson, Glasgow	Henry Brown & Sons
A. Hannan, Glasgow	Bloomer & Philipps
Thos. A. Mathieson, Edinburgh	Morrison & Parker
C. & H. MacPherson, Edinburgh	William Marples
	Robert Marples
	Ward & Payne

The Catalogue of Exhibits recorded the following on display:

Henry Brown & Sons
1. Improved Plated Brace with Lignum Vitae Head
2. Iron Brace with Brass Head
3. Small Fancy Brace with Ivory Head
4. Patent Anti-Friction Brace with Ebony Stock and Ivory Head, Complete with Bits

Bloomer & Philipps
1. Kingswood Brace with Lever Thumb Bit
2. Ebony Brace with Lever Thumb Bit

Robert Marples
1. Improved Plated Brace in Boxwood
2. Registered Self-Acting Brace, Boxwood Inlaid with Pearl
3. Newly Invented Lever Brace, Beechwood, brass plated

William Marples
1. Kingswood Plated Brace with Straw Coloured Bits
2. Beechwood Unplated Brace
3. Brace with "improved pad" by which bits are introduced with much greater facility into a receptacle provided for holding them.

Henry Brown's usual advertising propganda accompanied his exhibit with the following note in regard to his Patent Brace: -

"The Improvements are that the Spindle works upon a Hard Steel Centre, instead of a Collar, which reduces the friction, and the Head is secured to the Neck by a nut screwed into the socket piece to prevent the head coming off, and produce the steady working of the Brace."

His firm was awarded an *"HONOURABLE MENTION"* for the quality of his braces as claimed in his 1851 Sheffield Directory Advertisement, noted at the left in Plate CLXXXIX.

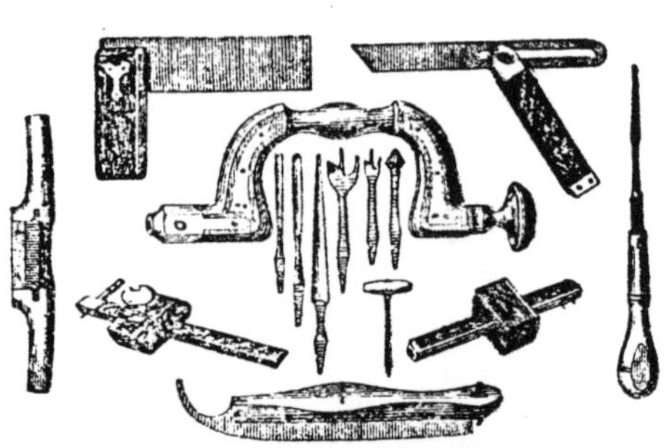

GREAT EXHIBITION

OF THE

WORKS OF INDUSTRY OF ALL NATIONS

1851.

President, H. R. H. Prince Albert.

Honourable Mention was made of the quality of **Henry Brown & Sons'** Patent and other BRACES, CAST STEEL BITS, JOINERS' TOOLS, SCREW BOXES, STOPS, SKATES, AUGERS, GIMBLETS,&c.; TOOL CHESTS for Amateurs, Emigrants and Mechanics, as well as all other Articles Manufactured by them.

—o—

THE ANTI-FRICTION BRACE

FOR WHICH

LETTERS PATENT HAVE BEEN GRANTED BY HER MAJESTY

IS STILL MANUFACTURED SOLELY BY

HENRY BROWN & SONS,

WESTERN WORKS,

SHEFFIELD,

Established, 1827.

Left: Plate CLXXXIX

HENRY BROWN & SONS 1851 SHEFFIELD DIRECTORY ADVERTISEMENT Noting Honourable Mention Award at 1851 London Great Exhibition.

It is surprising that four firms from Scotland exhibited their braces at the 1851 Crystal Palace, but there were none from Birmingham. Alexander Mathieson had been in the tool business at Glasgow since 1792 and offered a well rounded selection of high quality tools.[20] His son, Thomas Adam Mathieson, established a similar tool firm in Edinburgh about 1850. After his father's death, the two firms were combined as A.Mathieson & Sons, Ltd. A peculiarity of construction of these braces in Scotland was that the button was generally on the right side of the pad, as was the early practice at Birmingham, but not generally followed at Sheffield. Braces of this type made in Scotland also often had a circular ring cast around the front end of the pad. Both of these features are evident from Fig.73 illustrating three Mathieson braces. However, not all such braces made by Mathieson had this flange,as noted in Fig. 74 . A brace having imprint "Mathieson/Edinburgh" and with a very unusual chuck is shown in Fig. 75 . The pad is similar to the second Registered Design No.244 (1851) by Bloomer & Philipps. Perhaps Thomas Mathieson arrived at such a design after seeing the one shown by Bloomer & Philipps at the 1851 Exhibition. The brace at the top in Fig. 75 bears the owner's imprint of Richard Lord, which came from a collection of tools documented to be of early 19th Century origin.

The date at which William Marples first introduced the'Ultimatum' brass framed brace is not known. It is possible that the brace noted in the 1851 London Exhibition "with improved pad, by which the bits are introduced with much greater facility into the receptacle provided for holding them" might have been such. In all probability it must have required considerable time to tool up for production after acquiring the Cartwright Patent in 1849. The first dated known reference to this appears in the 1854 Sheffield Directory. His listings from 1852-1856 were as follows:

1852 - Marples,Wm. Joiners' Tools,skate,brace & bit mfr., 67 Spring Lane
1854 - Marples,Wm. manufacturer of joiners' tools, braces and bits, skates augers, gimlets & c. sole manufacturer & patentee of the patent metallic framed brace, Hibernia Works, 69 Broom Spring Lane
1856 - Marples, Wm. joiners' tools, brace & bits, &c.,Hibernia Works 27 Westfield Terrace

Such braces having emblem plates with the address of 67 Spring Lane are believed to have been the earliest manufactured and accordingly date from 1851 to 1855. After 1856 Wm. Marples' firm was at 27 Westfield Terrace. A rosewood brass framed brace of this early design is shown in Fig. 76 and the distinctive pad in Fig. 77.

This pad mechanism consisted of a catch attached to a hardened steel spring, the end of which fit into a hole in a slide washer, which was placed over the end of the pad and held securely by a threaded nut screwed on the end, held with a set screw. This design was only used for a short period and was soon followed by the expired registered design of Robert Marples, which was less expensive.

During the tenure of the Cartwright Patent it is believed that William Marples exclusively marketed this brass framed brace through 1863. Thereafter this was offered to the Sheffield trade. Marples must have initially made substantial investment to tool up for this item. Quite likely he used his influence among his assosiates in the trade to supress any new competitive designs from developing at Sheffield.

Fig. 73 - Three Mathieson Braces: Top - Glasgow; Middle - Edinburgh; Bottom - Glasgow

Fig. 74 - Brace with Brass Neck - A.Mathieson, Glasgow

Unusual Pad
MATHIESON/EDINBURGH

A.Mathieson/GLASGOW
With Owner Imprint: *Richard Lord*

Fig. 75 Two Mathieson Braces Made in Scotland

With the knowledge that both partners had sons that would soon be following in their footsteps in the business, Brown and Flather probably agreed to dissolve amicably about 1850. Henry Brown & Sons set up at 108 Rockingham Street, establishing Western Works. David Flather continued at Solly Works and changed the firm name to David Flather and Sons in 1855. Henry Brown's firm continued to manufacture the Patent Anti-Friction Brace, which both he and the Flather firm continued to advertise and sell through the 1860's. A brace of their manufacture is shown in Fig. 78. Note that the back side plate is stamped: "*H.BROWN & SONS/Sole Manufacturer of the/Patent ANTI-FRICTION BRACE/ John Bottom, Inventor*" and the front side plate: "*Henry Brown, PATENTEE*" and the button plate: *H.Brown & Sons* [with eagle design], and the tip of the brass pad: "*HENRY BROWN & SONS*". The brass plate in the head had: - "*HENRY BROWN, PATENTEE* [with the Royal Seal of the Lion and Unicorn].
A similar brace by D.Flather & Sons is shown in Fig. 79. The reference to *IMPROVED REGISTERED* imprinted on the brass pad must have been to Henry Brown's long expired Registered Design No.306 of 1844, but was still beneficial advertising. Note that the brass head plate stated: "*HENRY BROWN/ PATENTEE/ SHEFFIELD*" with an eagle design. This latter motif may have been designed with the American market appeal. Note that this is a different eagle design than the one used with the Scotch iron brace illustrated in Fig. 80. Both of these firms' advertisements in the 1851 Staffordshire Directory are shown in Plates CXC & CXCI.

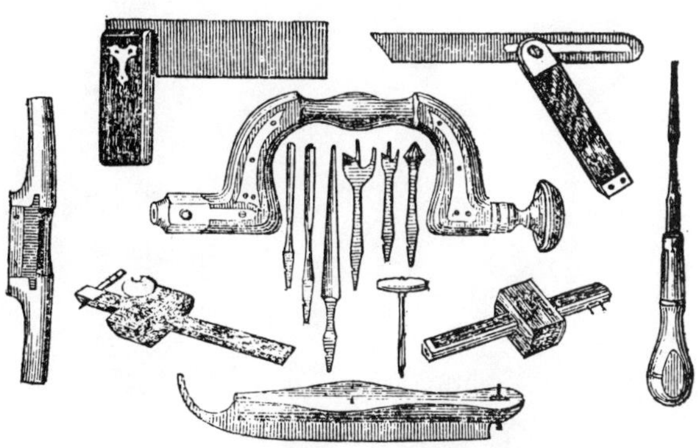

DAVID FLATHER,
SOLLY WORKS, SOLLY STREET, SHEFFIELD,
MANUFACTURER OF
BRACES, CAST STEEL BITS,
JOINERS' TOOLS,
SKATES, SCREW BOXES, PLOW STOPS,
TOOL CHESTS,
AUGERS, GIMBLETS, &c., &c.,
OF SUPERIOR QUALITY, AND BEST WORKMANSHIP;
AND MANUFACTURER OF
HENRY BROWN'S PATENT ANTI-FRICTION BRACE.
GRANTED BY HER MAJESTY'S ROYAL LETTERS PATENT.

Plate CXC - David Flather & Sons
1851 Staffordshire Directory Advertisement

HENRY BROWN & SONS,
WESTERN WORKS,
108, ROCKINGHAM ST., SHEFFIELD,
MANUFACTURERS OF
BRACES, CAST STEEL BITS,
JOINERS' TOOLS, SKATES, SCREW BOXES,
PLOW STOPS, TOOL CHESTS, AUGERS, GIMBLETS, &c.,
OF SUPERIOR QUALITY AND BEST WORKMANSHIP;
AND SOLE MANUFACTURERS OF
H. BROWN'S PATENT ANTI-FRICTION BRACE.
GRANTED BY HER MAJESTY'S ROYAL LETTERS PATENT.

THE ADVANTAGE OF THE PATENT ANTI-FRICTION BRACE ABOVE ALL OTHERS,
is its EASY WORKING, having NO FRICTION, as in other Braces. The Spindle and Socket Nut being of hardened Steel. The Brass Neck has a Nut screwed upon, or into, the Brass Socket Piece, thereby firmly securing the Ebony Cap of the Head, and preventing its coming off, causing a firm steady working of the Brace.

Plate CXCI - Henry Brown & Sons'
1851 Staffordshire Directory
Advertisement

Fig. 76 - William Marples Brass Framed Brace (top); Head and Emblem (bottom)

Fig. 77 - Pad Mechanism - Earliest Wm. Marples' Brass Framed Brace
[Collection of Henry Wing]

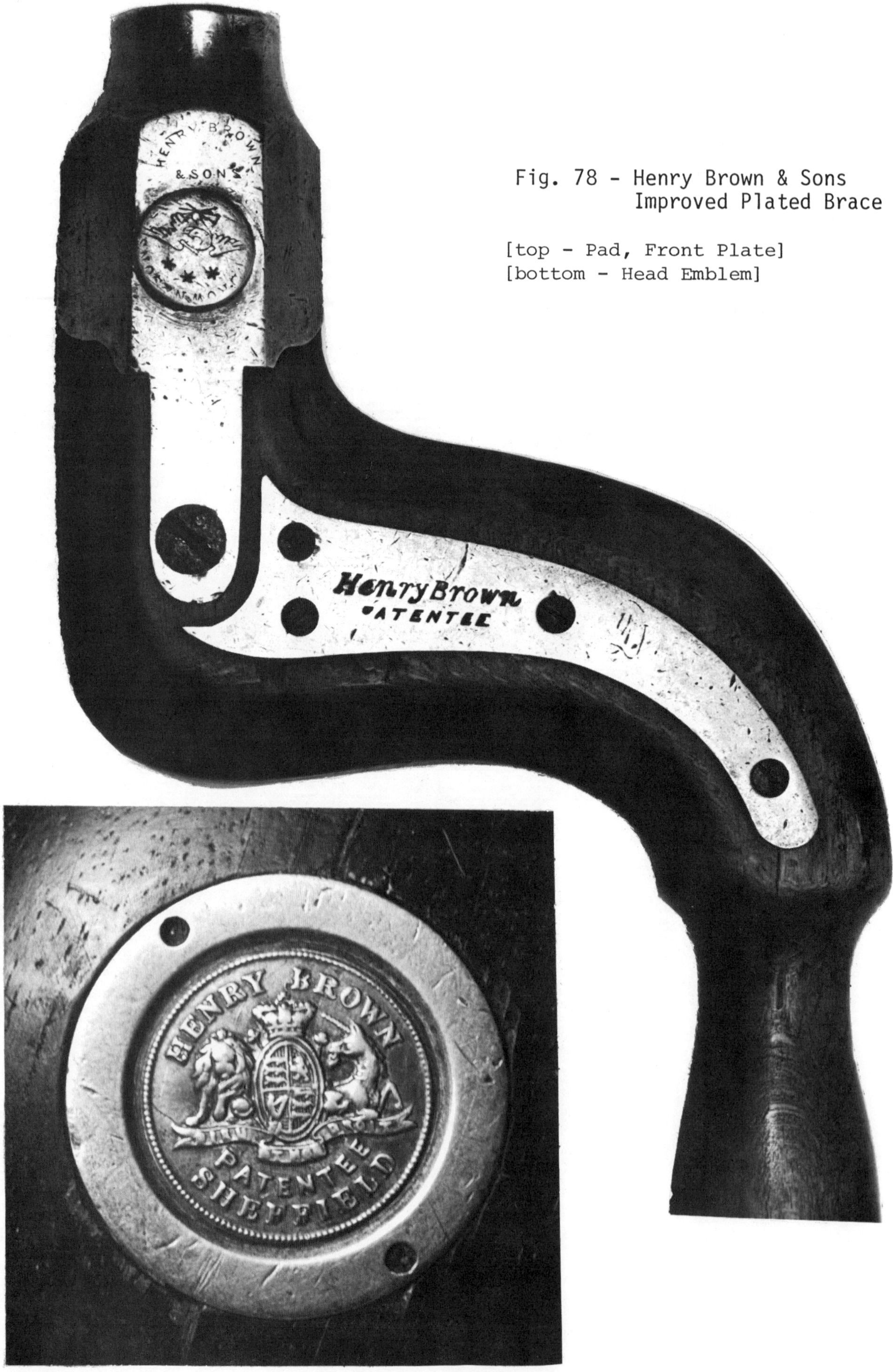

Fig. 78 - Henry Brown & Sons
Improved Plated Brace

[top - Pad, Front Plate]
[bottom - Head Emblem]

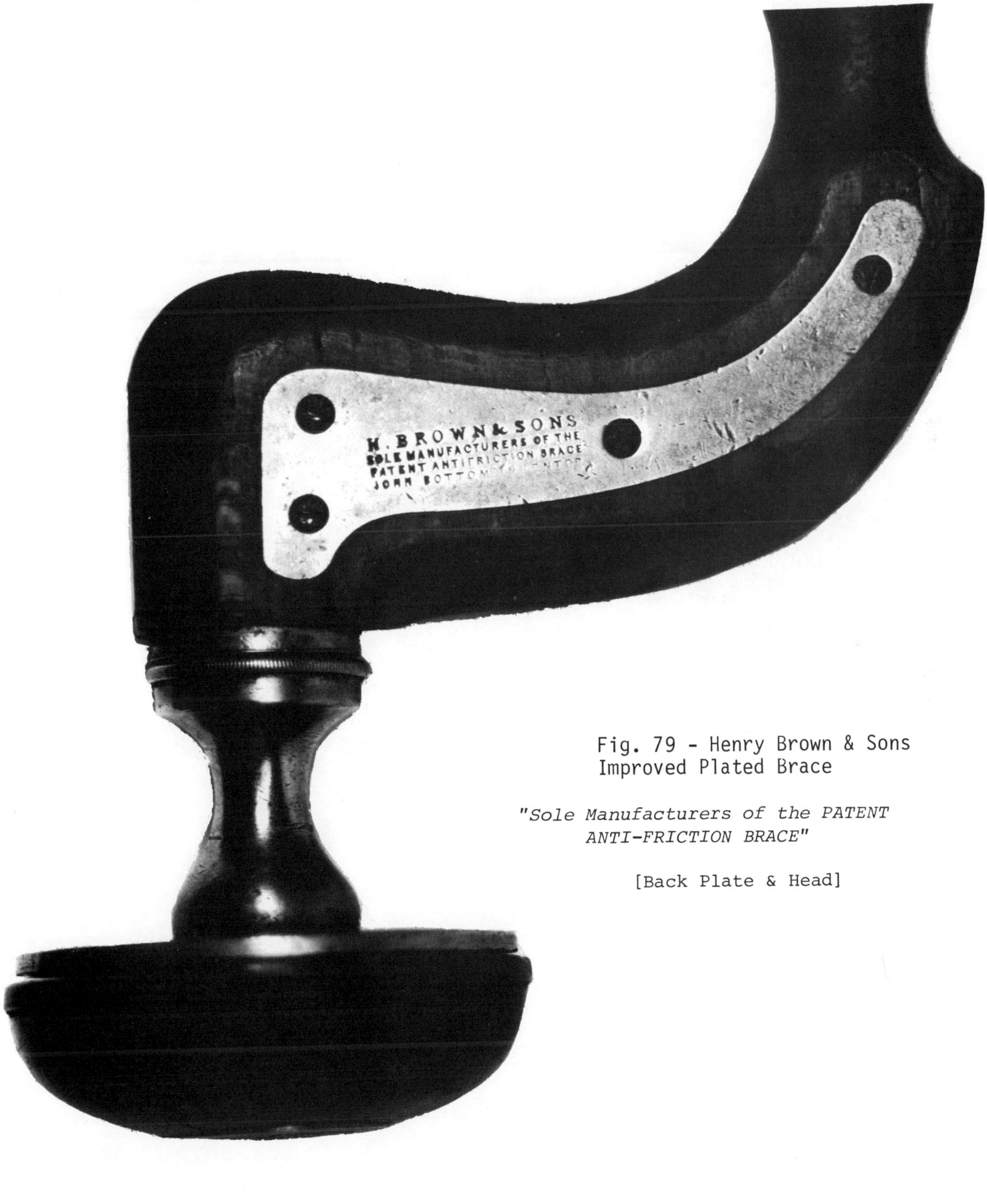

Fig. 79 - Henry Brown & Sons
Improved Plated Brace

"Sole Manufacturers of the PATENT ANTI-FRICTION BRACE"

[Back Plate & Head]

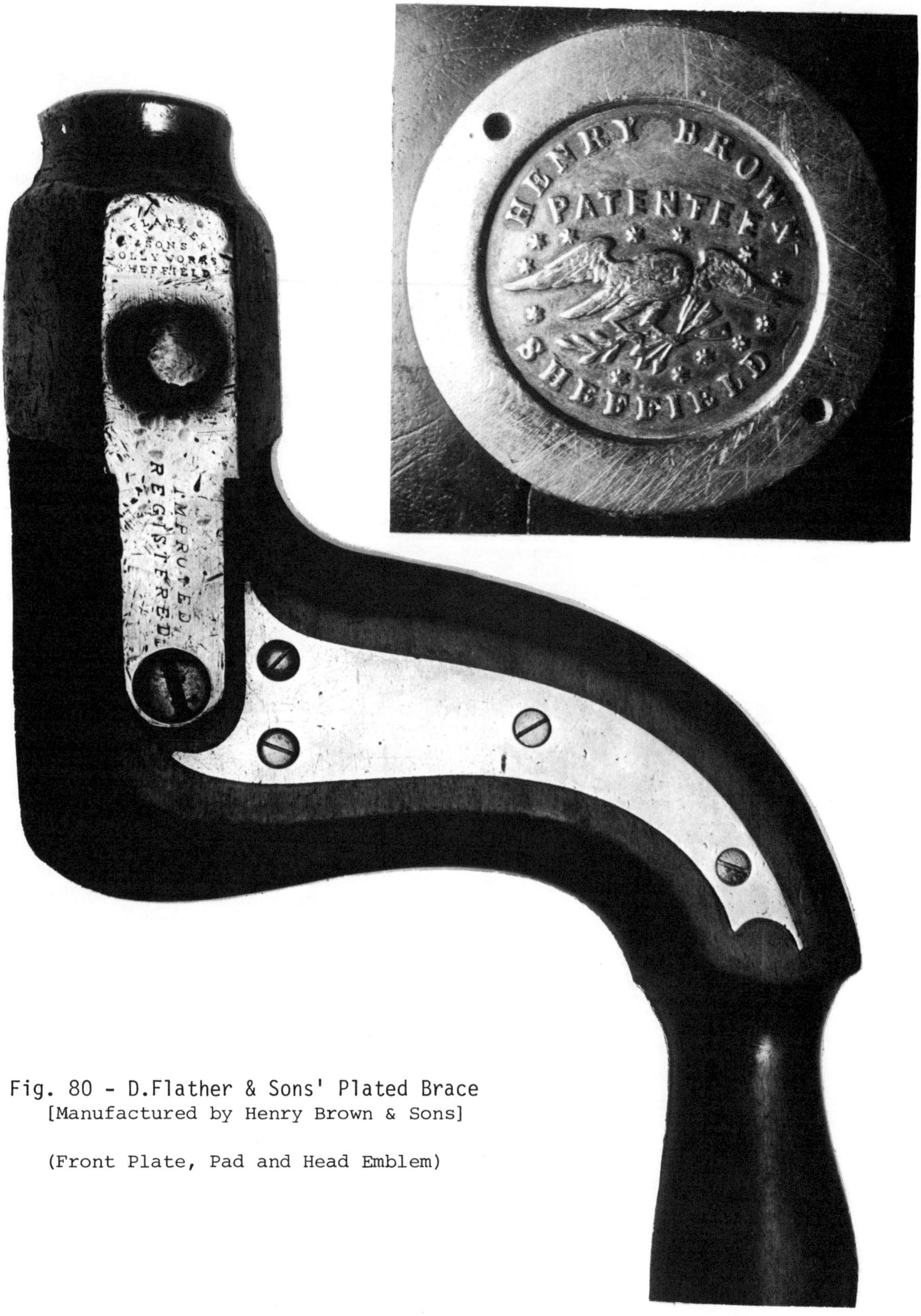

Fig. 80 - D.Flather & Sons' Plated Brace
[Manufactured by Henry Brown & Sons]

(Front Plate, Pad and Head Emblem)

Fig. 81 - Plated Beace by SCHOFIELD

The "improved plated" brace and the unplated "best" brace continued to be manufactured at Sheffield throughout the 19th Century. It is difficult to know which firms actually made these braces. Several concerns simply purchased these braces and applied their own imprints. A Check List of firms, both in Sheffield and Birmingham, appears as an Appendix to this Chapter. Figs. 81 - 87 illustrate several examples of these braces. Two types of brass heads were offered: hollow and solid. A hollow head, by an unknown maker, is shown in Fig. 88. This was cast integral with the neck, machined and fitted with a plug, screwed into the top. The solid head was made in two pieces, the flat head being screwed into the neck piece, as shown in Fig. 89. While this particular brace has an imprint "Wm. Greaves & Sons" , it is doubtful they made such braces, but probably purchased or traded for these to resell. Wm. Greaves were cutlery merchants and edge tool makers at the Sheaf Works, 1822-1849, succeeded by Eyre,Ward & Co. and T.Turton & Co.

Fig. 82 - Improved Plated Braces - (top) HAWKSWORTH; (bottom) SLATER
[Both probably made befoe 1841]

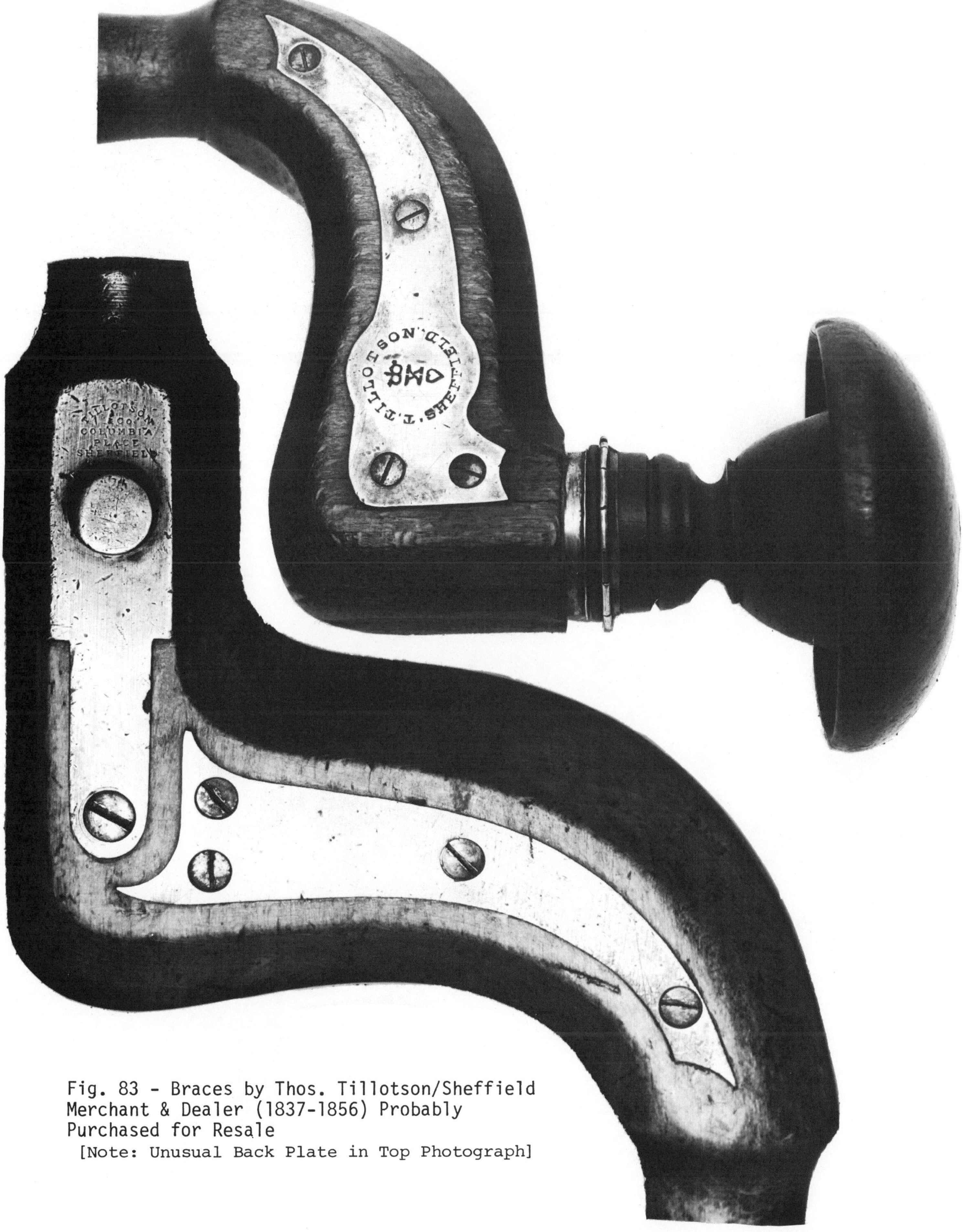

Fig. 83 - Braces by Thos. Tillotson/Sheffield Merchant & Dealer (1837-1856) Probably Purchased for Resale
[Note: Unusual Back Plate in Top Photograph]

Fig. 84 - Braces by Thos. Ibbotson
Top - Improved Plated Brace
Head Imprinted - *"Warranted not to turn off Thos. Ibbotson, MAKER"*
Bottom - Best Plain Brace - Imprinted on Wood Stock:
"Thos.Ibbotson/ Maker/Sheffield - WARRANTED GOOD"

Fig. 85 - Wm. Ash & Co.
(1822-1837)
Brass Plated Brace
This firm was well known for
high quality Plane Irons.

Fig. 86 Braces with Imprints
EYRE, WARD & Co. (top) successors to
Wm. GREAVES & Sons (bottom)

Fig. 87 Plated Brace by R.Groves & Sons
This firm was well known for quality saws. They probably traded saws for braces and applied their own imprint thereon.

Fig. 88 - Brass Hollow Head Plated Brace
[No Imprint]

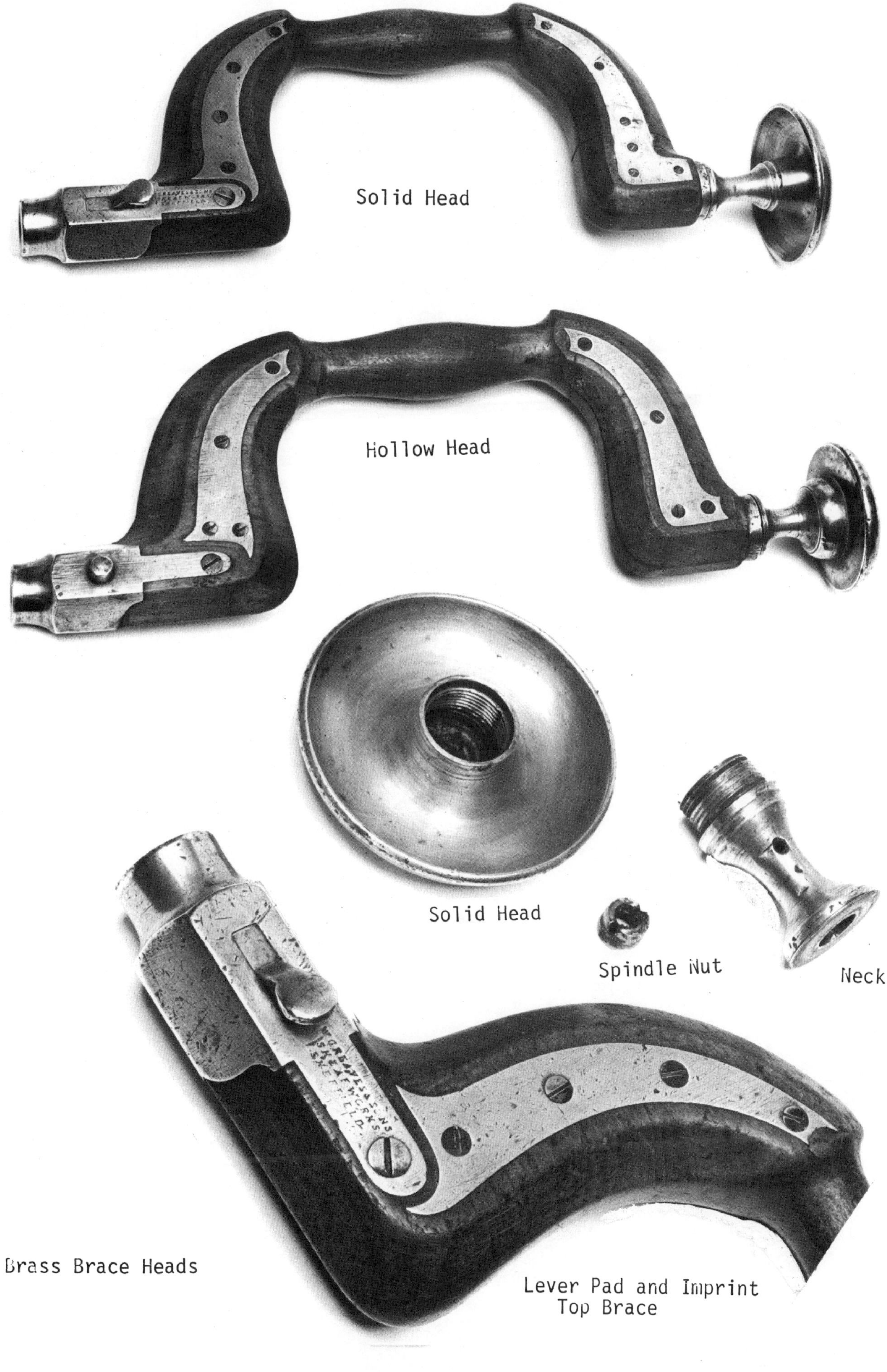

Fig. 89 Brass Brace Heads

George Horton of 18 Thomas Street, Sheffield, joiner tool maker, was granted Registered Design No. 2528, November 8, 1850 for a very unique and attractive brace (Plate CXCII , illustrated in Fig. 90 . The bows were cast in bronze and pinned to a spindle to which a rotating handle was attached.

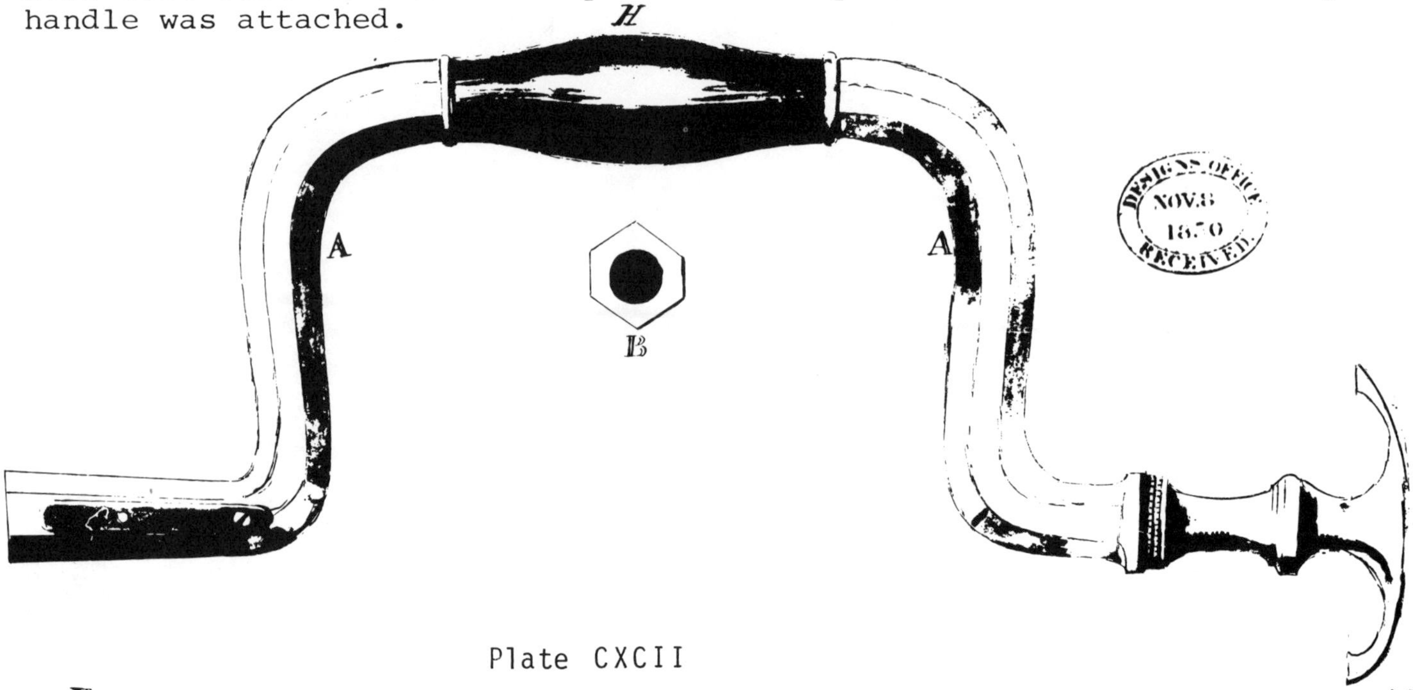

Plate CXCII

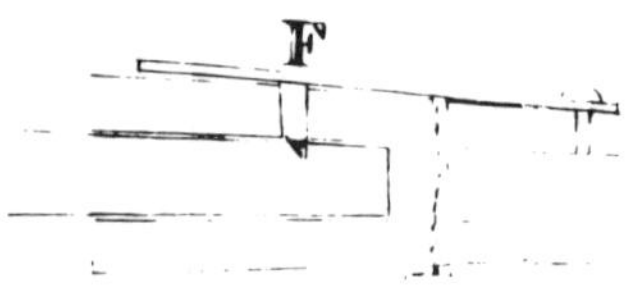

George Horton's
Registered Design No. 2528
November 8, 1850

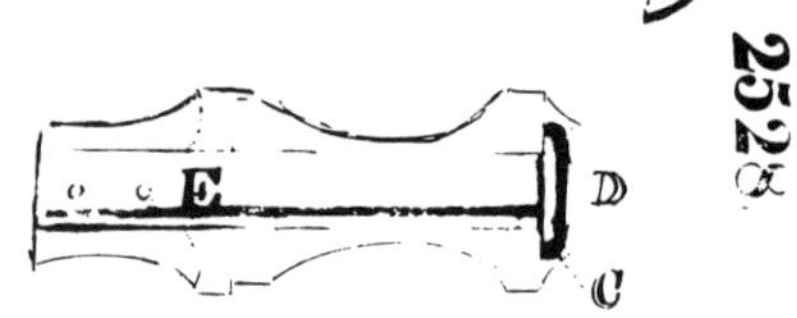

Specifications

1. The Article offered for Registration is a Joiner's Brace.
2. It is manufactured by George Horton, the proprietor, No.18 Thomas Street.
3. By the internal configuration of A.A.B. considerable lightness is obtained which is important to men carrying them a distance. By the bolt Head C and the Cup D firmness and durability are secured and by the Spring F being applied outside,the advantage is obtained, it can be easily replaced by men working a distance from home.
4. The Handle H is old, but the other parts are new in Shape.
5. It is for the Horton Bow and Handle I, the Bell Head and Cup D and the Spring F that are new in configuration and for which the proprietor wishes the Article to be Registered.

It is difficult to understand why this brace did not become an immediate success, but which would have seriously jepordized sales for Henry Brown and William Marples. It was attractive, functional, simple, durable and must have been inexpensive to manufacture. From the very few extant examples it is evident that these were not produced in large quantities, indicating that it was not made over a very long period. Horton was apparently a "little mester" of no great significance with little knowledge of marketing and probably with no financial backing. It is possible that both Henry Brown and William Marples prevented this from becoming a threat to their sales by using their influences among the trade. It certainly must have been less expensive to manufacture than the Ultimatum and no more costly than the Anti-Friction plated brace with brass head and neck. Something was 'rotten in Sheffield'. This was among the best braces made there, but it "died on the vine".

[Author's Collection]

Fig. 90 - George Horton's Registered Design Bronze Brace
[Collection of Jack Clouser]

"An Improved Joiner's Brace" was the subject of Letters Patent No.1113 applied for on December 20,1852 by Charles Pilkington, Thomas Pilkington and Abraham Pedigor. The Specification for this Patent, granted 30th April 1853 are shown in Table XXIV and the Drawings in Plates CXCIII Their claim that the brass side plates resulted in "appearance much improved" is indeed as questionable as there other claim "more useful and stronger" in comparison to the usual plated types. It certainly was less expensive to produce, since no routing was required on the sides to attach the plates. This "Patent Brass Mounted Brace" was first manufactured by the firm Pilkington, Pedigor and Storr at their Chester Works, Sheffield, 1854-1856. Their advertisement from the 1854 Sheffield Directory is shown in Plate CXCIV.

Plate CXCIV
1854 Sheffield Directory
Advertisement of
Pilkington, Pedigor & Storr
Joiner Tool Manuafctures

Note: *Sole Inventors and Manufacturers of the PATENT BRASS MOUNTED BRACE*

CHESTER WORKS.

PILKINGTON, PEDIGOR, & STORR,

MANUFACTURERS OF

Joiners' Tools, Braces, Bits, Skates, Augers, Gimblets,

&c.;

SOLE INVENTORS AND MANUFACTURERS OF THE

PATENT BRASS MOUNTED BRACE,

HEADFORD STREET, SHEFFIELD.

Two braces with their imprints are seen in Figs.91 and 92. One of these has a button thumb type release, similar to the Robert Marples Registration #845 of 1846, rather than the design indicated on the Patent Drawing. Another brace has been observed with only the imprint of Pilkington, Pedigor & Co., which indicates that George Storr must have left the firm, c1856. In 1858 the concern became C. & T. Pilkington, who continued through 1864 at Fitzwilliam Street. A brass mounted patent brace with their imprint as well as "Hargreaves, Smith & Co., New York" is shown in Fig. 93 . William Hargreaves is listed at 28 Eyre Lane in the 1854 and 1856 Sheffield Directories as "merchants and manufactures of table knives". In 1858 the listing became Hargreaves, Smith & Co. Wm. Hargreaves & Co. also appear in the 1854 and 1855 New York City Directories at 25 Cliff St. and "hardware and cutlery importers". In In the 1856 and 1857 NYC Directories this became Hargreaves, Smith & Co. "files, saws, tools, cutlery & Hardware" at the same address. J.Boake followed at this addres through 1863. A brace with imprint of C. & T. Pilkington and also "Made for J.Boake" is illustrated in Fig.94a&b,which on the basis of the above can be dated 1857. Since braces with these imprints are very rarely found in England, it is suggested that these manufactures made arrangements with W. Hargreaves and associates to market these in New York from whom they may have been distributed to other American dealers.[21] A conventional plated Sheffield brace with imprint of C. & T. Pilkington is shown in Fig. 95. This also has a thumb button of the Robert Marples' design.

Table XXIV

SPECIFICATIONS of CHARLES PILKINGTON, THOMAS PILKINGTON and ABRAHAM PEDIGOR
Letters Patent No. 1113 for JOINER'S BRACE dated 20th December 1852

Our improvements in the construction of a "brace" for the use of joiners, refer to the making, shaping and appyling those separate parts which are constructed of metal in such a manner that the part of the 'brace' which is made of wood is strengthened thereby; the "brace" being, by such arrangement of parts particularly referred to in the description, rendered more useful in its application as a tool than the ordinary "brace" in use, as also more easy of construction. The pad (a) is made larger than the pads of the "braces" at present in use, and is fitted and fixed on to the stock (b) differently. The ordinary pads are cast with ears or lugs, which are let into the wood and fastened with one screw only; in the "improved brace" the pad socket clasps the stock (b), the ears or sides (c,c), overlapping and fitting upon the wood, the front and back being alone let in flush; the pad (a) is secured firmly by two screws, and, instead of weakening, considerably strengthens the stock; it can be fitted and fixed with facility and economy of labour. The side plates (d,d) are laid on, not as present let into, the stock, and adpated to the bevel of the sides; thus the stock is strengthened in this part also, and in appearance much improved. The top is mounted with a brass clasp (c), corresponding in appearance with the sides and front of the pad (a), so as to secure uniformity of appearance at both ends of the stock. In form the cap (f) is different made convex instead of concave, and is thereby adapted to the form of the hand and to the relative position and office of the cap when in actual use. The brass or other material used for the neck (h) is continued higher up than in "braces" now made. In ordinary braces the cap (f) is made solid, but in the "improved brace", it is to be made hollow as shewn.

The advantages of our improvements are, that, by adoption of the various parts of the brace herein described, and the mode of fitting the same, a more useful and stronger brace is produced than the ordinary brace, and in the labour and cost of production a considerable saving is effected.

The improvements we claim as our invention are: the pad (a), and the introduction of the top clasp (e), and hollow cap (f); as also the modes of applying or fitting the same; together with the side plates (d,d), as herein described, and which may be applied separately, or otherwise, in the manufacture of joiner's braces.

* *

Comments on the above Patent -

1. The use of two screws for the pad may have actually weaked the stock rather than strengthen it, as claimed.
2. The 'laid on' plates were not as strong as the 'let in' conventional plated brace; the use of larger screws with the former may have weakened the stock
3. The 'laid on plates' did not make a "much improved appearance" as compared to the conventional 'let in' plate construction.
4. Undoutedly this patent brace was less expensive to produce as the "laid on' plates took less time to assemble than the 'let in' style. It is not known whether this Patent Brace sold at a lower price.
5. The claims of the above Patent were probably for sales' promotion.

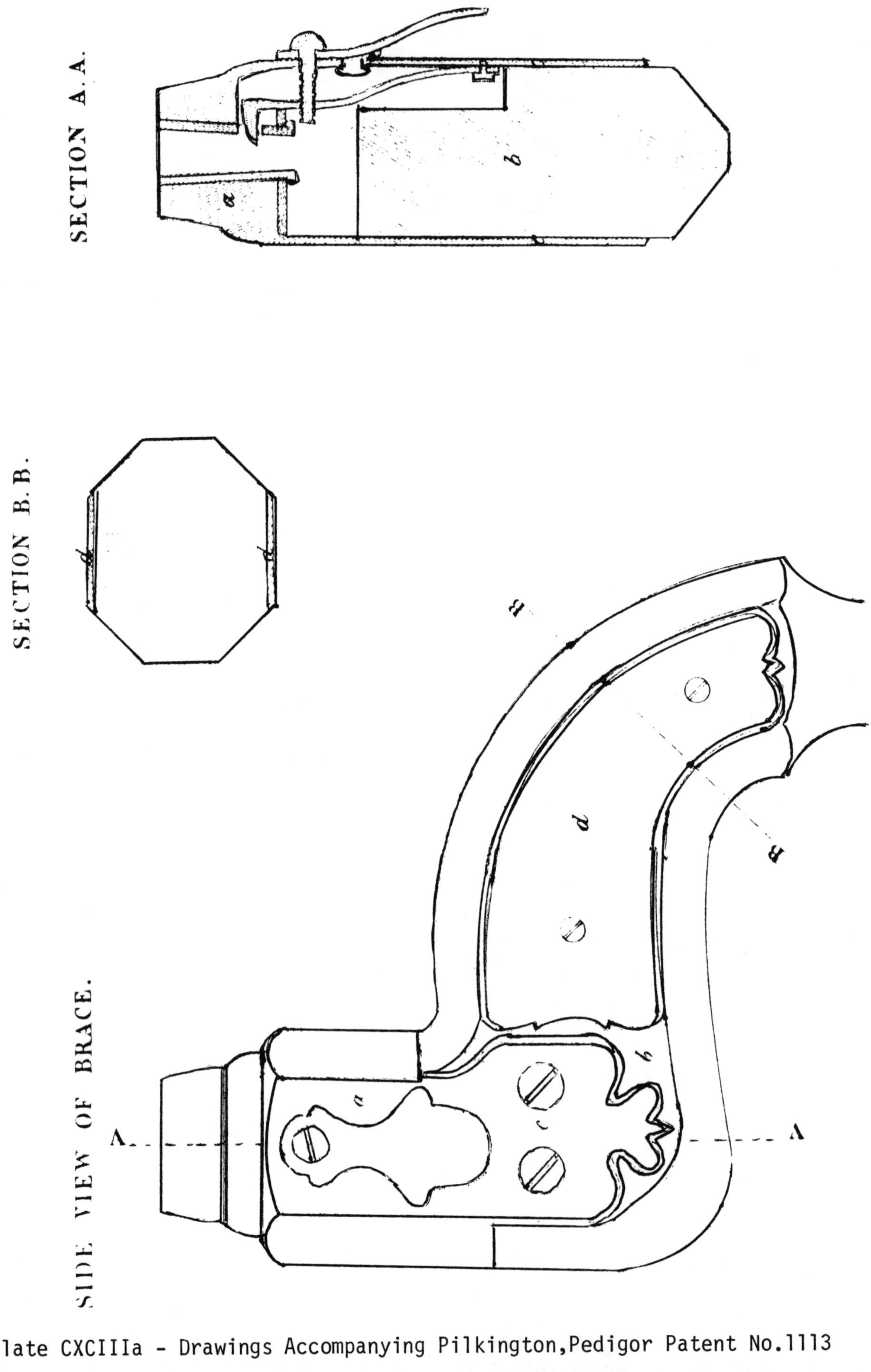

Plate CXCIIIa - Drawings Accompanying Pilkington,Pedigor Patent No.1113

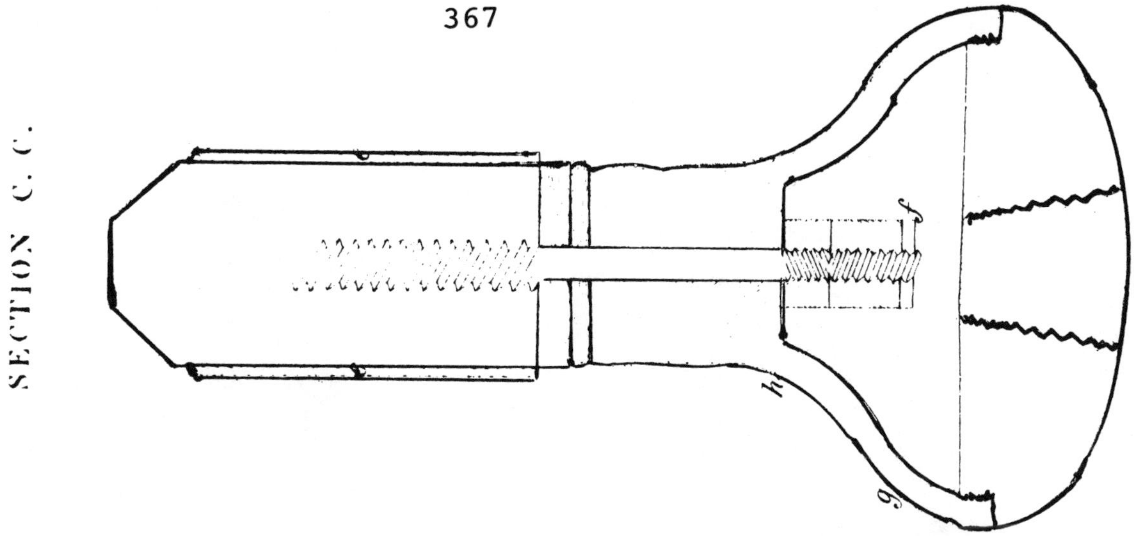

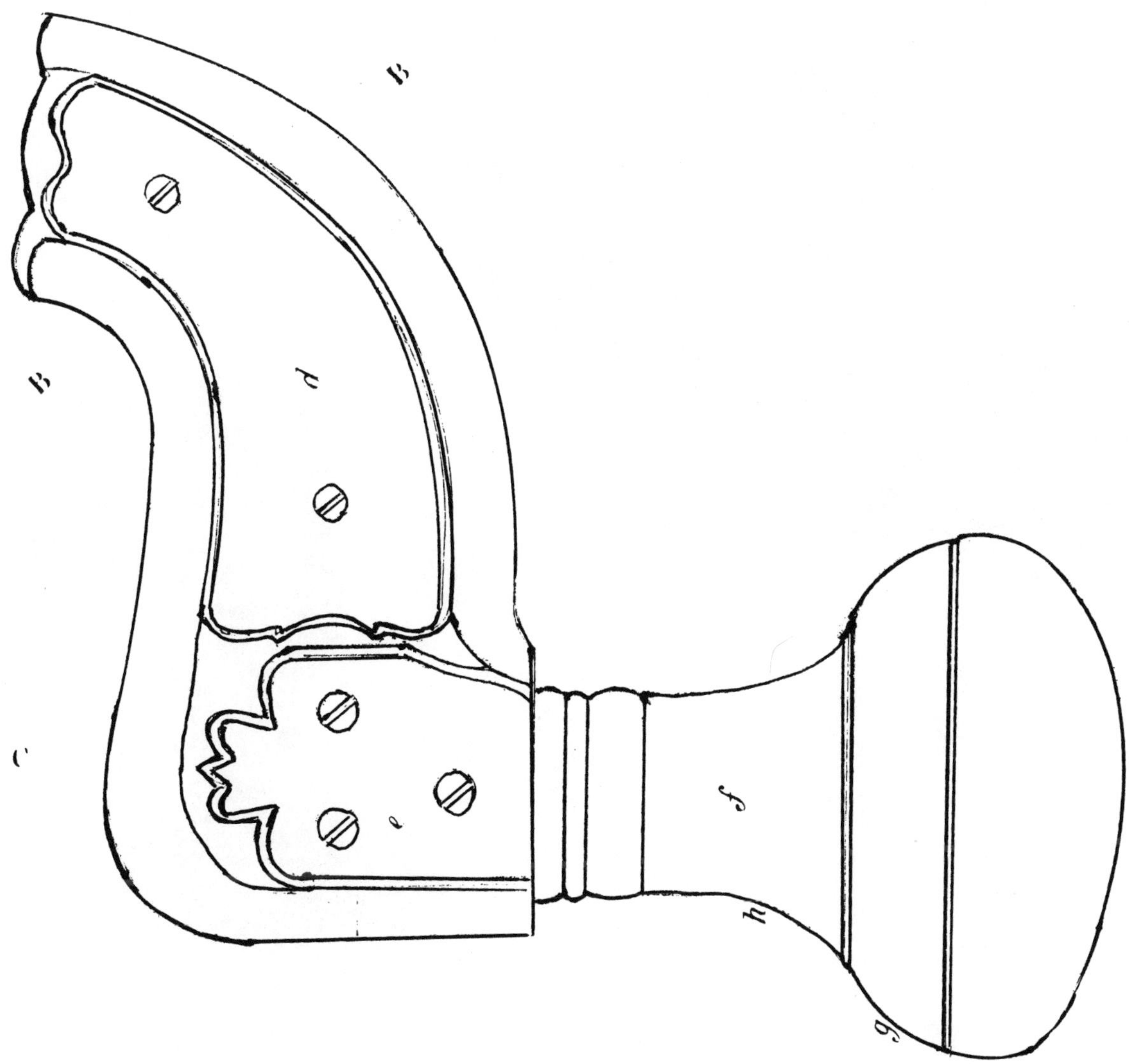

Plate CXCIIIb - Drawings Accompanying Pilkington, Pedigor Patent No. 1113
[These drawings were re-traced to improve legibility]

Fig. 91 - Pilkington,Pedigor and Storr Patent Brass Mounted Brace with Button Pad

Fig. 92 - Pilkington, Pedigor & Storr Patent Brass Mounted Brace with Lever Pad
[Anonymous Collection]

Fig. 94a - C. & T. Pilkington Patent Brass Mounted Brace made for J.Boake
[Collection of William Holden]

Fig. 94b - C. & T. Pilkington Patent Brass Mounted Brace made for J.Boake
[Collection of William Holden]

Fig. 93 - C. & T. Pilkington Brace with Hargreaves, Smith & Co.,New York Imprint
[Anonymous Collection]   (Note the Brass Head)

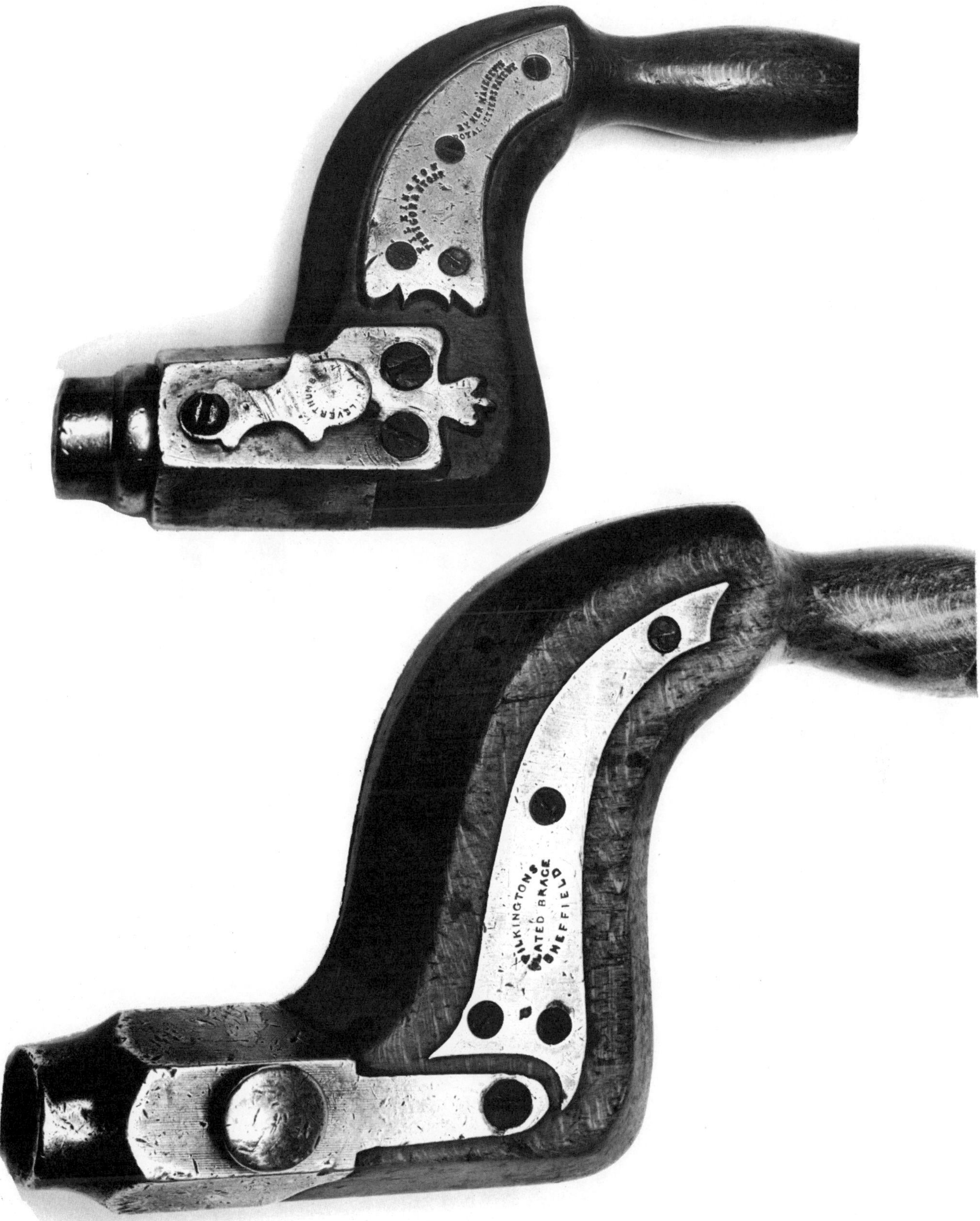

Fig. 95 - Pilkington, Pedigor & Storr Brace with Lever Pad (top)
Pilkington's [C.& T.] Plated Brace with Button Pad

Henry Parker of Rockingham St. and Francis Thompson of Westfield Terrace, Sheffield, trading under the style or name of Parker & Thompson (1854-1861) were granted Registered Design No. 100041, May 4, 1855, according to the drawing shown in Plate CXCV . Their only claims were for the ornamental designs of the pad, head and mouldings each side of the rotating handle. The last feature is the most unusual and this is among the very few known examples of an English wooden brace having a rotating handle.[22] This brace is illustrated in Fig. 96.

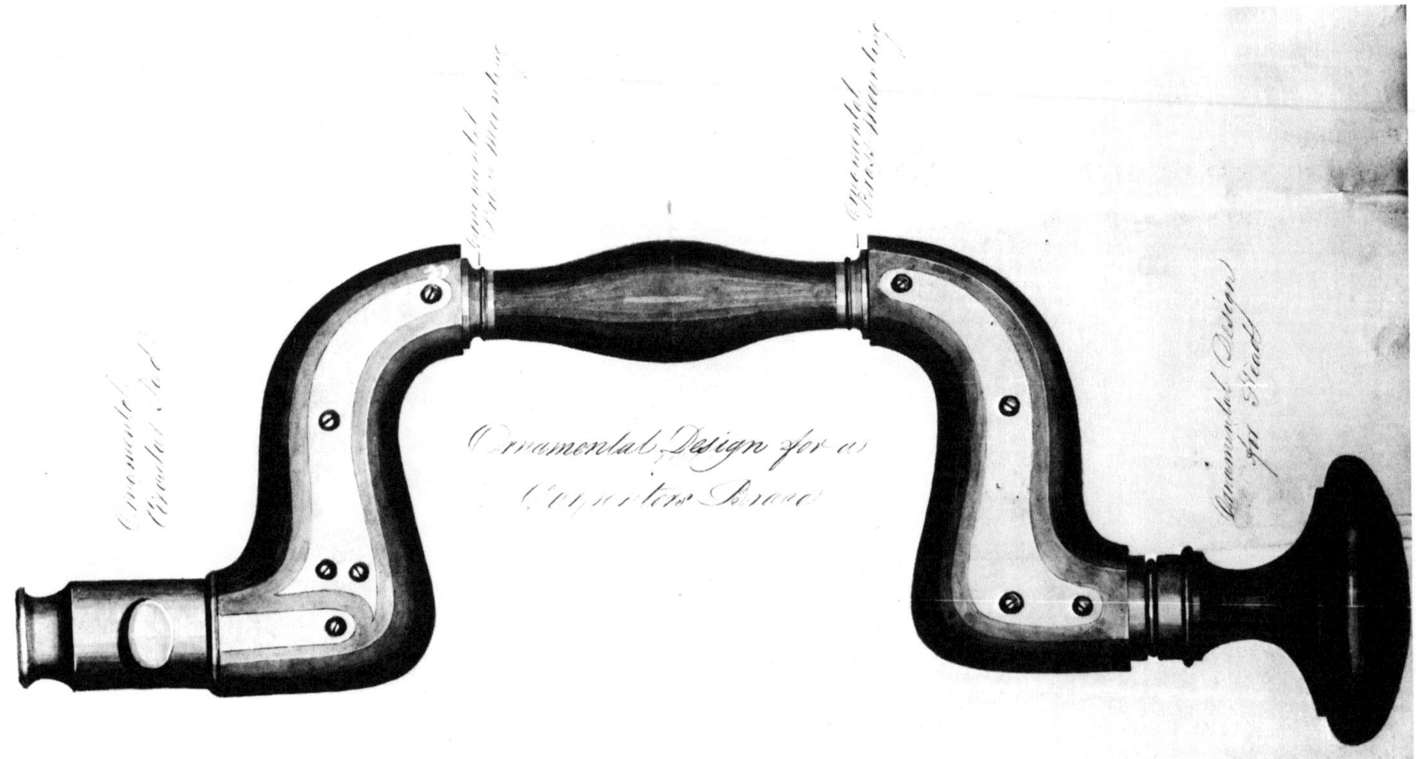

Plate CXCVI - Drawing Accompanying Parker & Thompson Registered Design No. 100041

The graceful appearance of the neck and head suggest Pilkington influence. The imprint on the most unsual brass pad, other than the Registration code, is that of"Wm. Hargreaves, Sheffield". It is possible that this could have been made by Thomas Pilkington , who continued alone until 1872 after dissolving partnership with his brother,Charles in 1864. In any event this brace appears to be more rare than any with Pilkington imprint, and the Wm. Hargreaves imprint suggests it was marketed in New York.

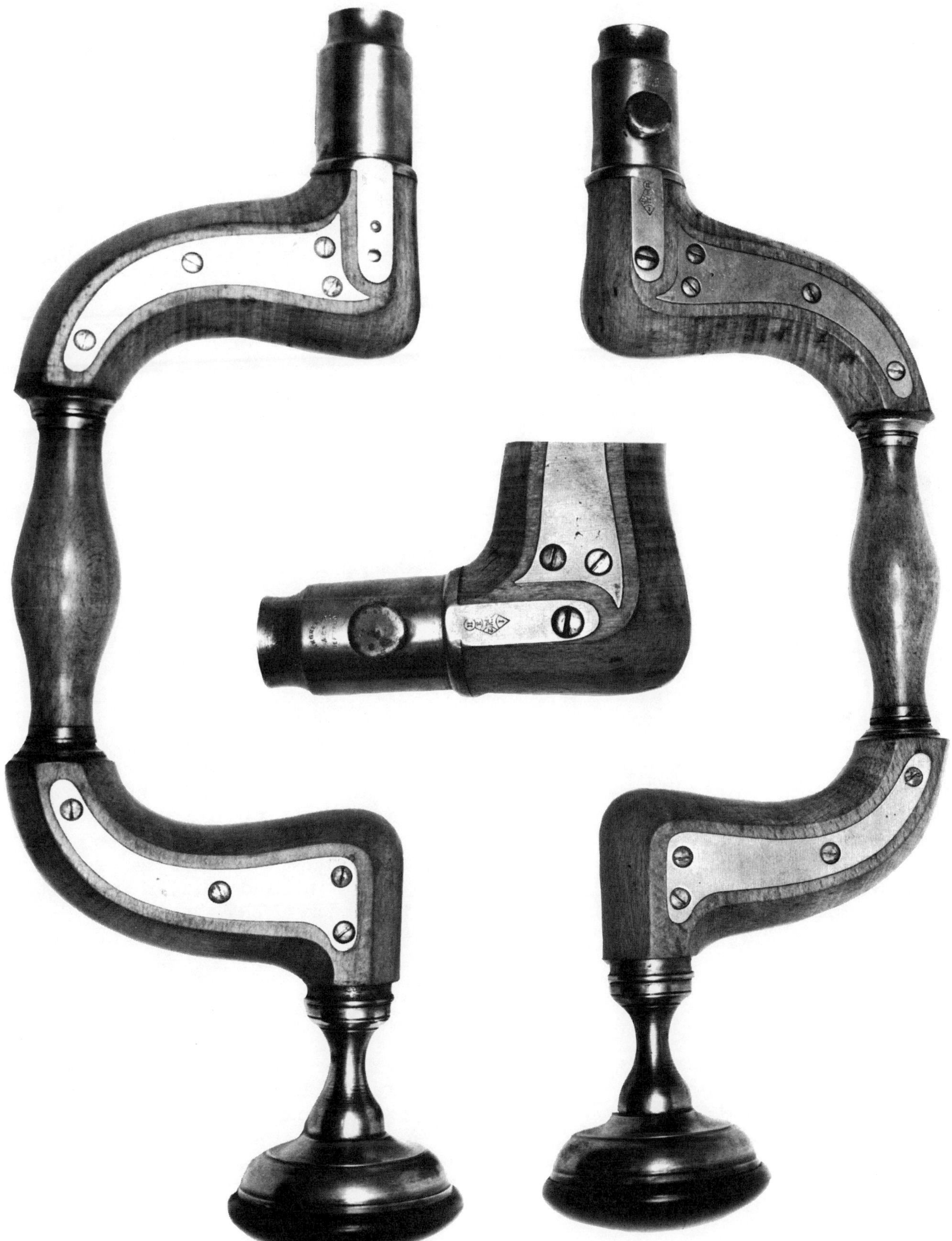

Fig. 96 - Parker & Thompson Registered Design No. 100041, May 4, 1855
Imprint of Hargreaves & Co./ Sheffield

Before continuing with further developments at Sheffield, some details concerning brace manufacturing at Birmingham will be presented. Joseph Cooper of Lionel Street, Birmingham was awarded Registered Design No. 3502, August 25, 1853, "Improvements for A Joiner's Brace". Among other features he claimed that the parabolic shape of the head spindle equalized the pressure and action on the head while in use; also a set screw in the stock held the axial screw securely. The details are shown in Plate CXCVII , which accompanied his Registration.

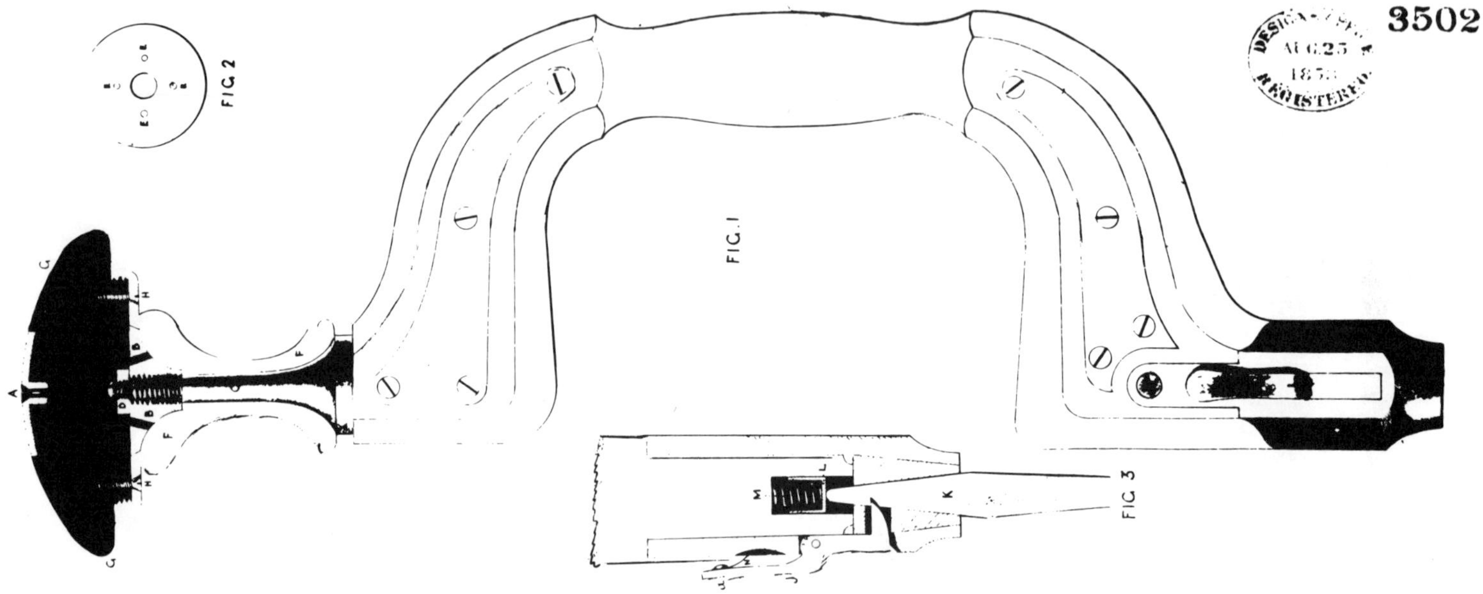

Plate CXCVII - Details of J.Cooper's Brace Registered Design No.3502

Later he incorporated some of these features with some new ideas, and was granted Provisional Protection in Patent No. 428, 27th February 1855 and regular 14 years protection in Letters Patent No. 504, March 7, 1855 for "Certain Improvements in Joiners' Braces" and also "In the Mode of Forming or Partially Forming the Various Bits used with this or Any Other Kind of Brace". The drawing accompanying this Patent, noting his claims for the brace , is shown in Plate CXCVIII. In addition to a design for a lever catch [This appears very similar to Bloomer & Philipps, Reg.Design #244] he claimed a unique acting lock button that pulled put, rather than the usual action of depressing to release the bit. This feature is shown in Fig. 97. Two braces with this design of catch are illustrated in Fig. 98. The emblem in the head of the brace at the left is shown in Fig. Fig. 99.

Advertisements of Joseph Cooper, Frederick Edmunds and John Newey, Birmingham bracemakers, appearing in various Birmingham Directories are presented in Plate CXCIX. A brace by Frederick Edmunds is illustrated in Fig. 101. Brace making continued at Birmingham throughout the 19th Century, but declined there after 1850 in comparison to Sheffield. A Check List of Birmingham Brace and Bit Manufacturers appears in the Appendix to this Chapter.

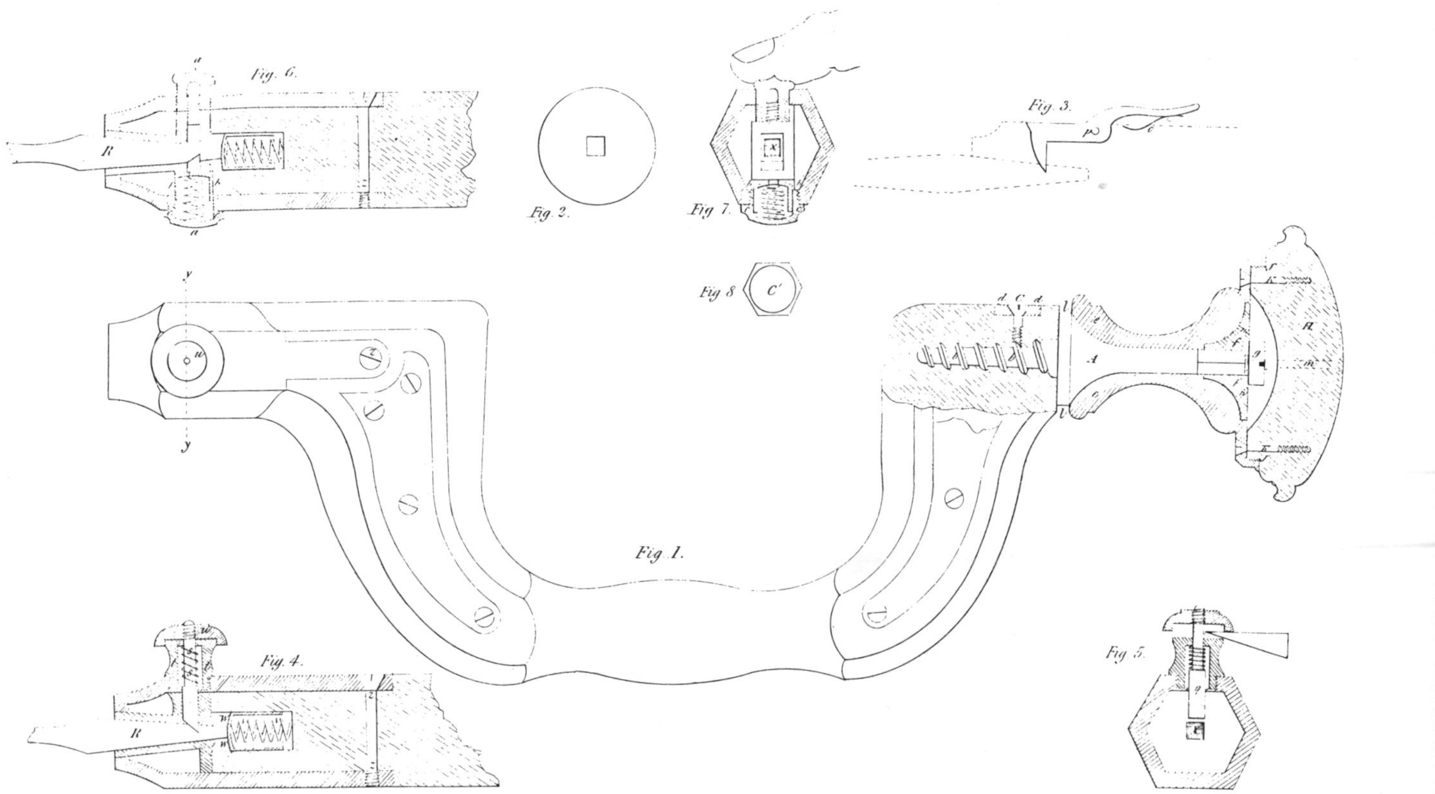

Plate CXCVIII - Drawing Accompanying Joseph Cooper's Brace Patent No. 504

Fig. 99 - Brass Emblem Cap and Ivory Ring Inset on Head of Joseph Cooper's Brace
(See Brace at left in Fig. 98)

Fig. 97 - Pull-out Latch - Joseph Cooper's Patent Brace [No.504, March 4,1855]

Fig. 98 - Two Joseph Cooper Braces. No. 504 Patent with Parabolic Brass Neck and Pull-out Latch (left); Wooden Head with Pull-out Latch on Pad (right) Note Trefoil Plates.

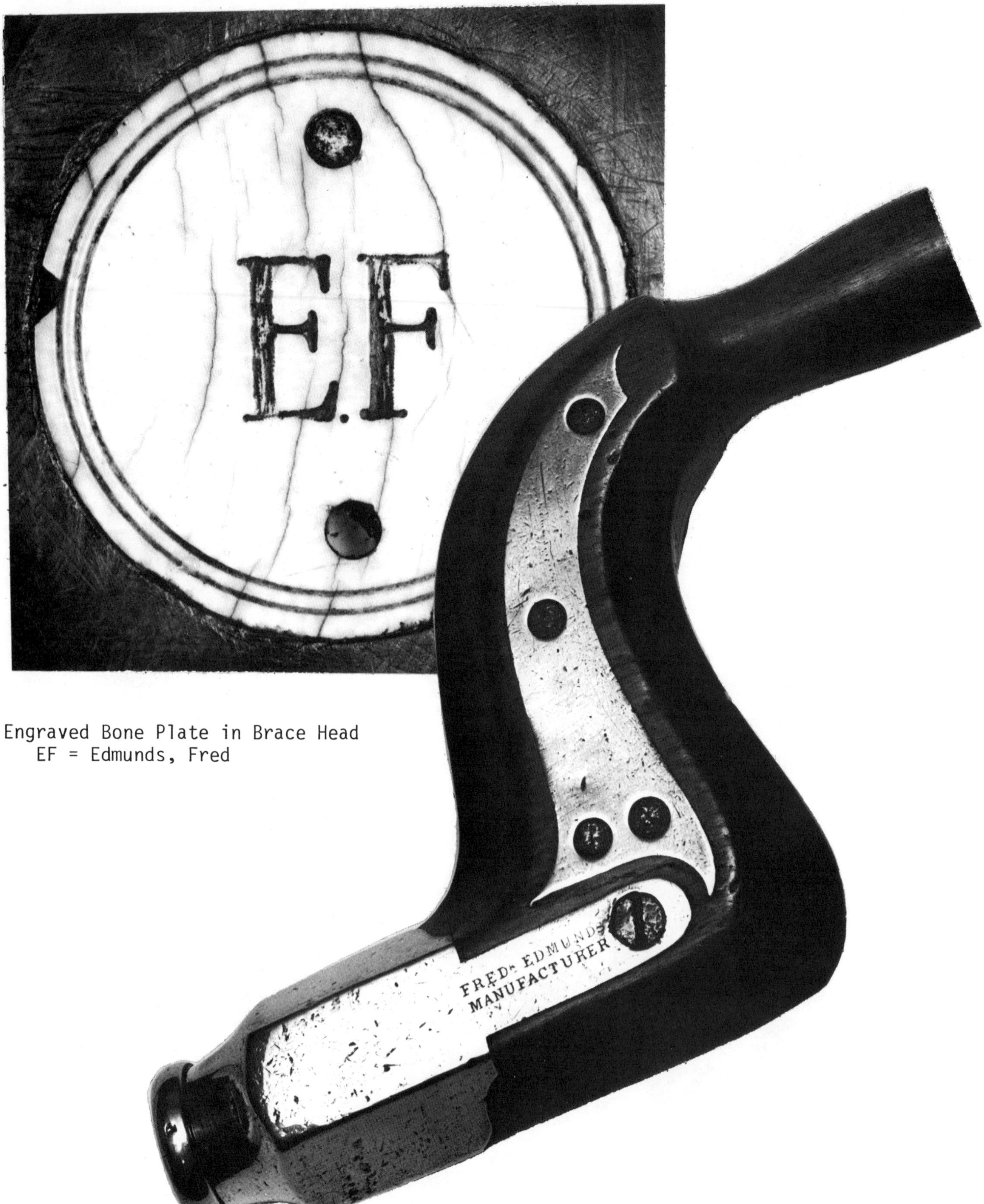

Engraved Bone Plate in Brace Head
EF = Edmunds, Fred

Fig. 100 - Plated Brace by Fred Edmunds, Birmingham
[Note Robert Marples #1416 Registered Design Brace Pad]

Fig. 101 - Braces by Palser, Birmingham

MIDLAND EDGE TOOL WORKS.

ESTABLISHED 1840.

Lionel Street, **JOSEPH COOPER & SON,** *BIRMINGHAM.*

PATENTEES AND SOLE MANUFACTURERS OF THE

Parabolic Head & Direct Action Spring Pad, for Carpenters' & Joiners' Braces.

ALSO MANUFACTURERS OF, IN EVERY VARIETY,

IRON AND WOOD BRACES	**LANCASHIRE TOOLS**	**TURNSCREWS**	**SQUARES**
BRACE BITS	**GIMBLETS**	**JOINERS' TOOLS**	**CHISELS**
AUGERS, for Ship & Railway Work	**BRADAWLS**	**SPOKESHAVES**	**PLANES**

All the above made according to Patterns adapted to every Market in the World.

Plate CXCIXa - Joseph Cooper & Son Advertisement 1880 Birmingham Directory

Plate CXCIXb - Joseph Cooper's Advertisement, 1858 Birmingham Directory

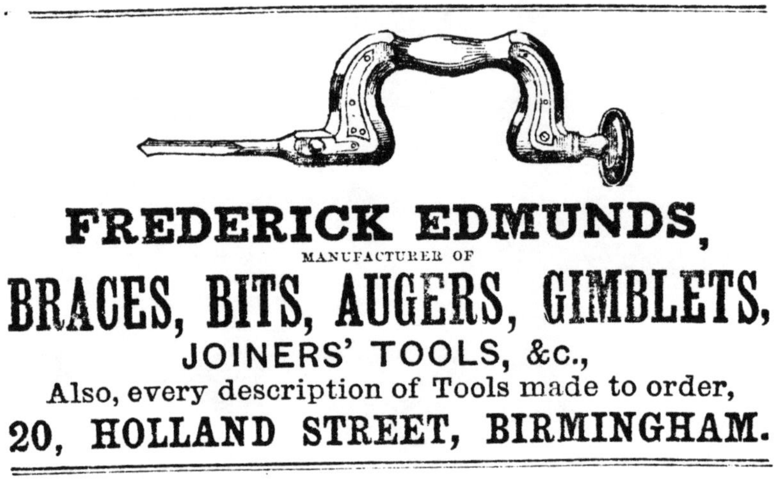

Plate CXCIXc - Frederick Edmunds' Advertisement, 1861 Birmingham Directory

Plate CXCIXd - John Newey Advertisement, 1863 Birmigham Directory

Returning to Sheffield, William Adsetts was first listed as an ivory, brass and wood turner at 3 Burgess Street in the 1852 Directory. Undoubtedly he was one of many who turned out conventional bit stocks to supply to the trade for fitting on heads and pads for completing into finished braces. On May 19, 1856 he was granted Registered Design No. 3839 "Double or Single Action Rotary Carpenter's Brace". This employed roller bearings within the head against a steel washer. The drawing and specifications for this are presented in Plate CC. His claim that this was *"the most practical invention ever yet introduced to Merchant, Manufacturers and all Persons using Braces"*; slightly exaggerated, but never the less it was a clever invention. In considering the scarcity of such extant braces, it does not seem to have been widely adopted. A brace of his design and manufacture is shown in Fig. 103. . The head has been removed and is shown with parts in Fig. 102 , also illustrating the type of turnscrew bit for remvoving the nuts from the spindle. In the 1859 Sheffield Directory he was listed at Pool Works, Burgess Street as "inventor and manufacturer of the improved double action rotary Carpenter's brace and wood turner of every description of handles". In 1864 he was listed at the same address, but simply as "ivory, wood and brass turner", with no reference to his invention.(See Plate CCI, 1852 Sheffield Directory Advertisement)

Robert Marples procurred his third Registered Design No. 3954 March 3, 1857 - An Improved Double Bound Carpenters' Brace with Octagon Lever Pad. The drawing and specification of this are presented in Plate CCII . He stated "The purpose and utility to be derived from the shape of the New Parts of this Design is, that the Brace or Pad is not so liable to be bruised or broken as the ordinary Brace; in addition to which, the Bits to be used with the same, may be applied and removed with more than ordinary facility". An example of this brace is shown in Fig. 107. The stock was beech. A thick plate (approximately 1/8") of cast brass gave continuous coverage to both side and the bottom of the stock. This may have been Robert Marples' competitive offering to the Wm. Marples' Ultimatum and the Pilkington "Brass Mounted Brace". Furnished with a conventional brass neck and Ebony head, it must have cost less than the former and no more than the latter. This was advertised in the 1861 and 1862 Sheffield Directories (See Plate CCIII) and also appeared in the 1867 Birmingham Directory. From the relatively few extant examples of this design, it would appear that relatively few were manufactured over this ten year period.[23]

A somewhat similar brace, having fully plated brass sides, but not around the bottoms of the bows, and with conventional button type pad, is shown in Fig. 106 . The imprint on one front plates states: "R.Groves & Sons/ BEEHIVE WORKS/ SHEFFIELD" and in script "*Impr'd PATENT*" No patent for such a brace is known and this wording was likely used for advertising propganda. In all probability R.Groves & Sons acquired this in trade for saws, their main manufacturing line. This firm certainly was not involved in joiner tool or brace manufacturing. While it was indeed an attractive brace, the plates were riveted through the sides and the numerous holes weakened the stock. Cracks developed at some of these locations during use.

A most unusual decorative brass plated brass, made by A.& W. Jinkinson (1854-1860) for T.Tillotson is shown in Fig.104. This was likely made for export trade to America.

Plate CC - - Drawing Accompanying William Adsetts Registered Design No. 3839
Double or Single Action Rotary Carpenter's Brace - May 19, 1856

Fig. 1 Represents a Lignum Vitae Brace Head in a finished state ready to be attached to the brace.
Fig. 2 Represents the Section of an improved Brass Neck Brace Head.
Fig. 3 Represents the Pin, on which the Head works; it is fastened at the top by two nuts and one being in reverse thread of the other renders it impossible for the Head to come off.
Fig. 4 Represents the surface on which the wheels are inserted.
Fig. 5 Represents the large nut with the two wheel inserted.
Fig. 6 Represents the screw or Stopper on which the Inventor's name and address are stamped.

The purposes of utility to which the shape or configuration of this Design for double or single action Rotary Brace Heads are that this can be applied with ease to both Lignum Vitae and Brass Neck Brace Heads for either wood or iron braces as shewn in the sectional parts above and the ease with which it can be worked renders it not only purely mechanical, *but the most practical invention yet ever introduced to Merchants, Manufacturers and all persons using Braces*. It will be seen on Reference to the Drawings that the evil so long complained of, viz. friction between the Brass Shoulders or Collars adjoining the stock is entirely done away with by the introduction of three steel wheels working on Steel Axels and running on a Steel radius, so that the greater the pressure in boring, the easier the head turns round, by introducing at the top, two wheels into the large nut, by which the Head is fastened on to the Stock, thereby giving it a back action, it is evident that either in boring or drawing the Bit, the workman has the advantage of the antifrictional movement and the head can be made to have either the double or single action to suit the purchaser. The great simplicity of the Design renders its utility more apparent than any elaborate statement which the Inventor might make and its advantages will be duly appreciated when brought into general use. To prevent fraud the Inventor's name with the registered number and date of registration will be struck on the screw stopper cap or head. The parts marked A are new in so far as regards shape or configuration, the rest being old and in use before.

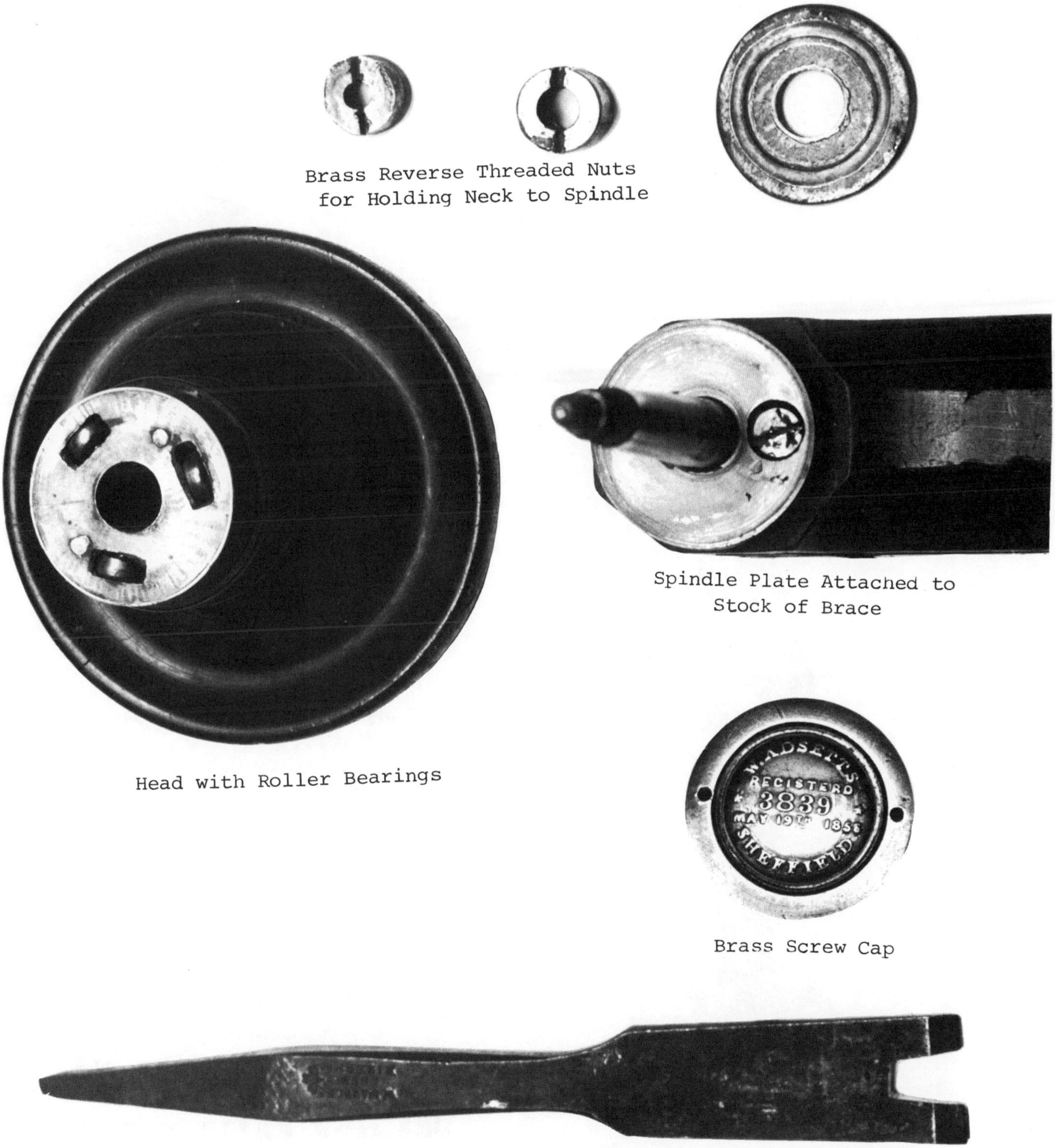

Fig. 102 - William Adsetts Brace Head Registered Design No. 3839, May 19,1856
Disassembled with Parts and Turn Screw for Turning Reverse Nuts.

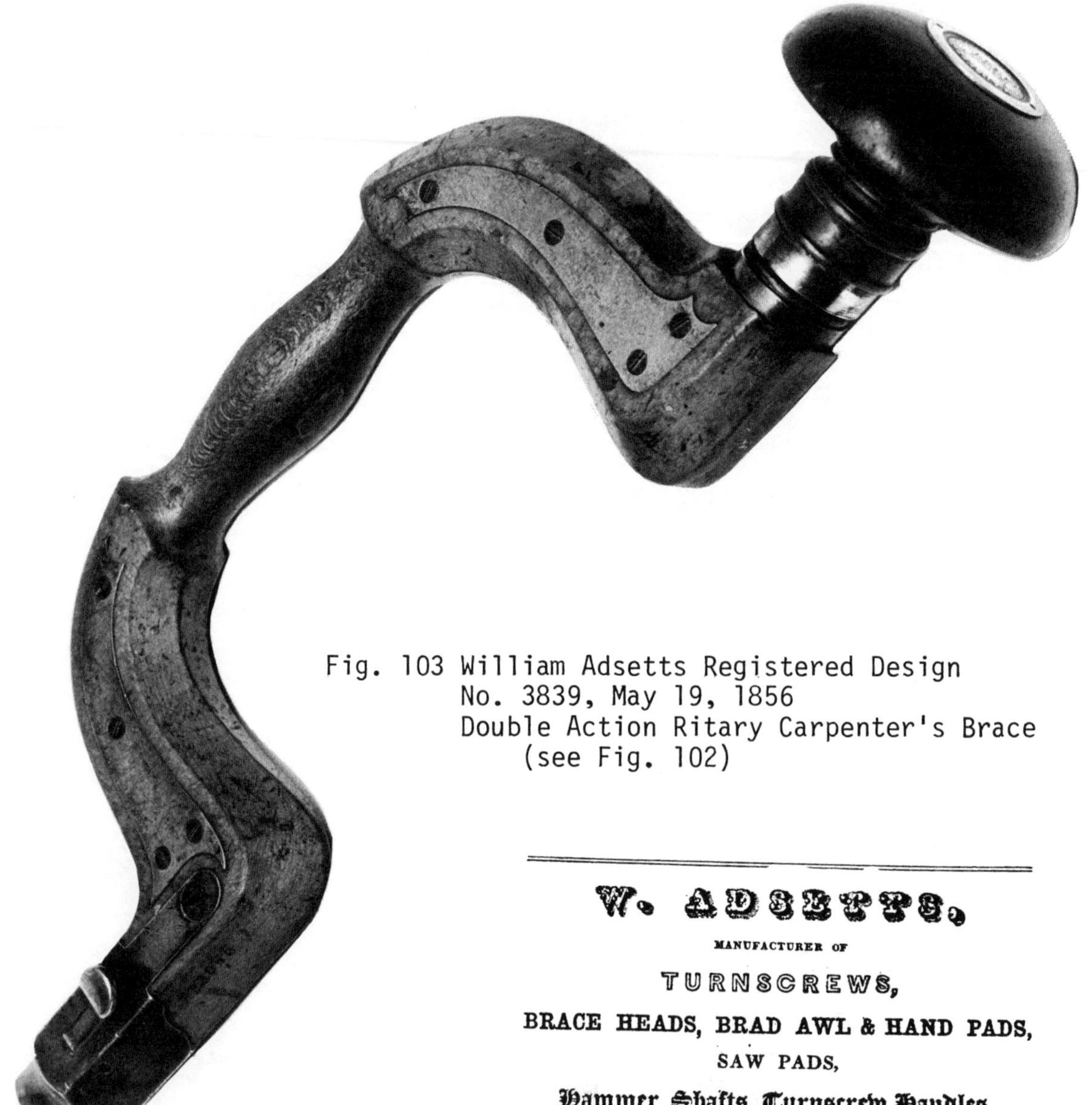

Fig. 103 William Adsetts Registered Design
No. 3839, May 19, 1856
Double Action Ritary Carpenter's Brace
(see Fig. 102)

W. ADSETTS,

MANUFACTURER OF

TURNSCREWS,

BRACE HEADS, BRAD AWL & HAND PADS,

SAW PADS,

Hammer Shafts, Turnscrew Handles,

BRAD-AWL HANDLES,

DRAWING KNIFE HANDLES,

And every Article Suitable for the Edge and Joiner Tool Trades;

POOL WORKS, TOP OF BURGESS STREET.

67

Plate CCI - William Adsetts' Advertsiement
1852 Sheffield Directory

Fig. - Decorative Brass-Plated Brace by A.&W. Jinkinson, Cadochan Works, Eldon Street, Sheffield - 1854 - 1860. Made for T.Tillotson & Co., Sheffield for New York Sale.

The cast brass plates are about 3/16" thick and all have slightly different designs. These are attached to the sides by four screws and were purely decorative and obviously made for the American market. Thomas Tillotson was a merchant at Sheffield, 1837-1860, dealing principally in table cutlery, but also sold joiner tools. According to New York City Directories he had a sales office at 90 Beekman Street, New York 1855-1860.

Fig. 104 - Decorative Brass Plate Brass by A. & W. Jinkinson for T.Tillotson
[Collection of Joseph Dziadul]

DESIGNS OFFICE MAR. 3 1857 REGISTERED.

3954

Improved Double-Bound CARPENTERS' BRACE WITH Octagon Compound-Lever Pads.

Robert Marples,

Hermitage Works,

SHEFFIELD,

Proprietor.

The purpose of utility to be derived from the shape of the New Parts of this Design is, that the Brace or Pad is not so liable to be bruised or broken as the ordinary Brace; in addition to which, the Bits to be used with the same, may be applied and removed with more than ordinary facility.

Description of Drawing.

Fig. 1, Presents a vertical side view of my Improved Brace, shewing that the angular parts, between the dotted lines A and B also C and D, are of an irregular octagonal shape, while that part of the Pad that extends from D to E is a regular Octagon.

The Plates F & G are continued around both sides, clipping the wood at h i, thus preventing the wood from splitting at this part. A top or edge view of these plates is exhibited in a detached view at Fig. 3. — Fig. 2, Represents a vertical Sectional view of the Pad, taken through the dotted line j & k, fig. 1; shewing the action of the Thumb-bit m on the Spring-catch n; p being a side view of the screw marked with the same letter in fig. 1

The parts of this Design for which protection is sought, in regard to the shape or configuration thereof, are the parts comprehended between the dotted lines A B and C & D which are New, the rest are old. —

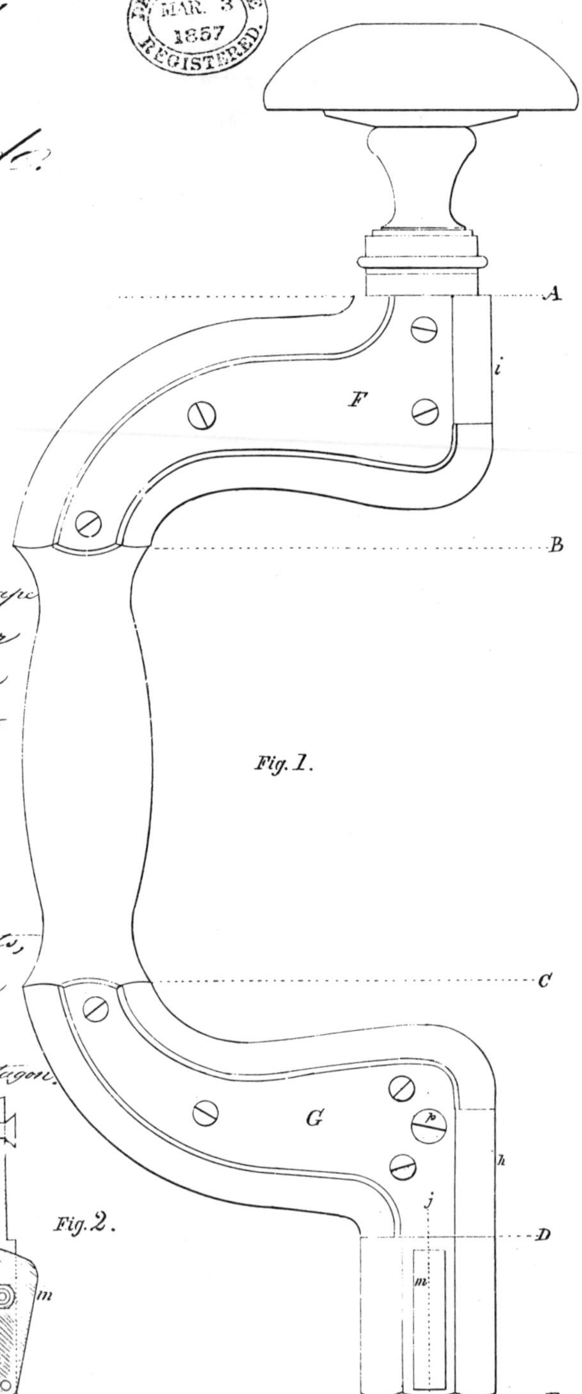

Fig. 1.

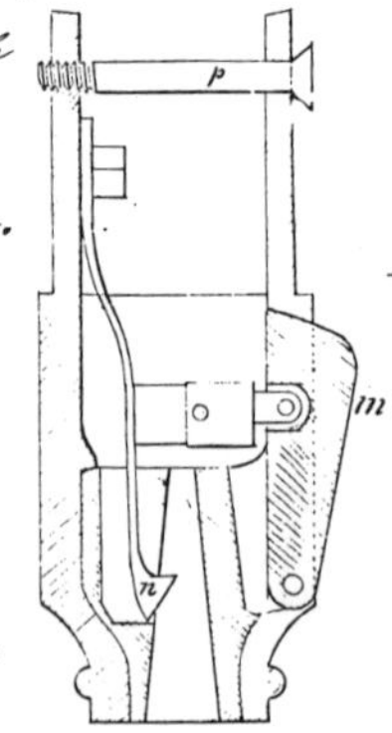

Fig. 2.

Fig. 3.

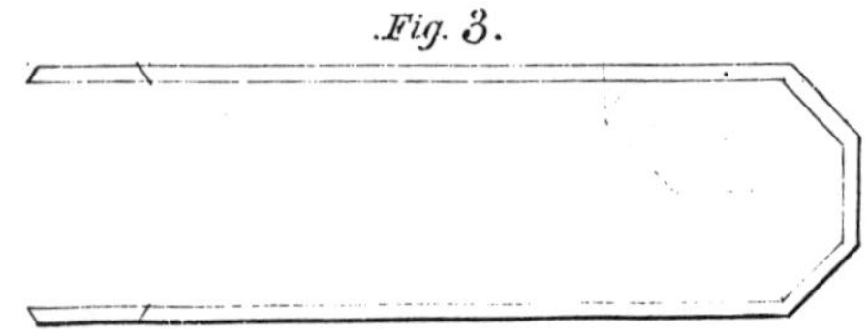

Plate CCII - Robert Marples' Registered Design #3954, March 3, 1857
"Improved Double Bound Carpenter's Brace with Octagon Lever Pad"

Brass Emblem Plate in Head

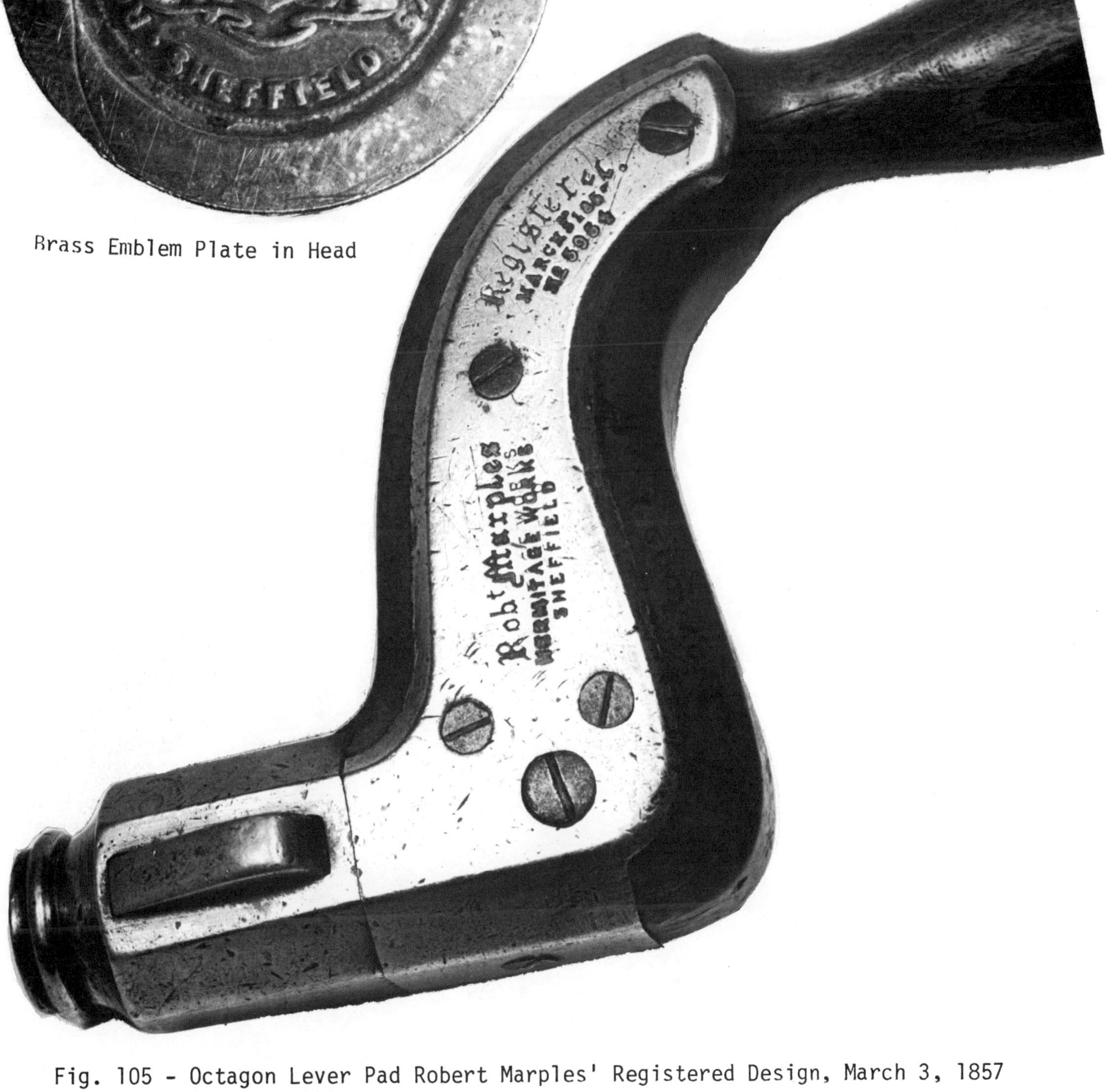

Fig. 105 - Octagon Lever Pad Robert Marples' Registered Design, March 3, 1857
(See Also Fig. 107)

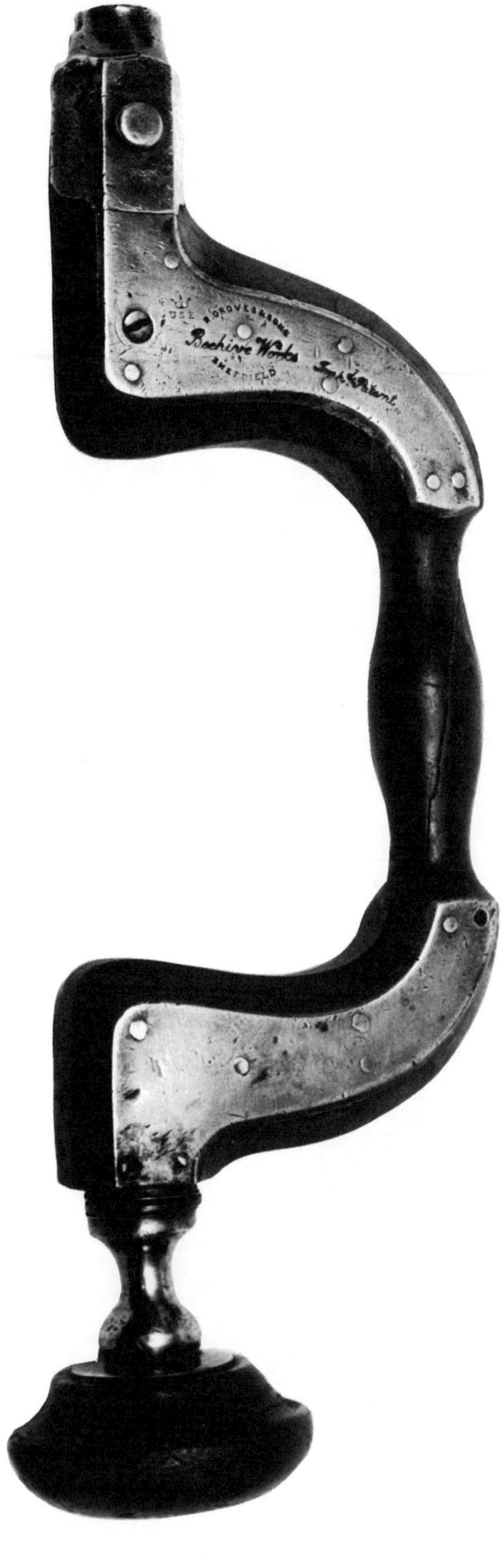

Fig. 106 - R.Groves & Sons
Full Plated Brass Brace

Fig. 107 Robert Marples'
Octagon Lever Pad Brace (See Fig.105)

ADVERTISEMENTS. 5

ESTABLISHED 1828.

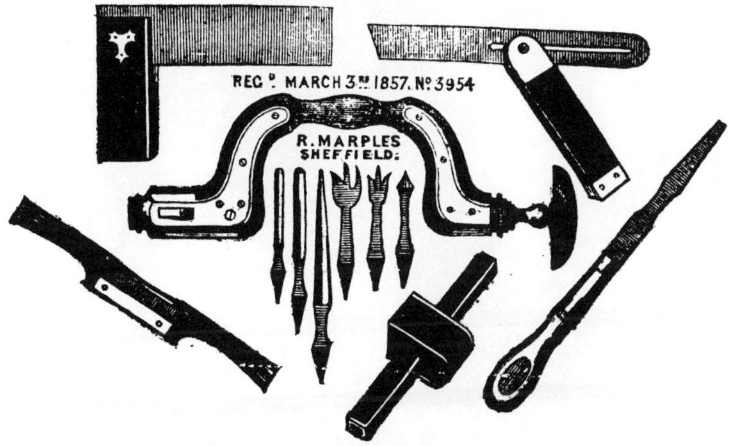

ROBERT MARPLES & SON,

MANUFACTURERS OF

JOINERS' TOOLS,

LIGHT AND HEAVY EDGE TOOLS,

COMPRISING

BRACES AND BITS,

CARPENTERS' SQUARES, BEVILS, GAUGES,

SPOKESHAVES, SPOKESHAVE IRONS,

TURNSCREWS, SPIRIT LEVELS, SCREW BOXES,

TOOL CHESTS,

JUVENILE TOOL CHESTS,

AUGERS, GIMLETS. SAW AND FRAMES, SAW PADS,

SAW SETS, HAND PADS, PRICKER PADS,

SCRAPERS, SKATES, HAMMERS,

CAST STEEL FIRMER CHISELS AND GOUGES, DRAWING KNIVES,

PLANE IRONS, SOCKET CHISELS & GOUGES, AXES, ADZES, &c.,

HERMITAGE WORKS,

HERMITAGE STREET,

SHEFFIELD.

Plate CCIII - Robert Marples & Son's Advertisment 1861 Sheffield Directory
[Note: Registered Design No. 3954, March 3, 1857 for Brace]

The introduction of the brass framed brace by Wm. Marples, which he called the "ULTIMATUM',[24] has been previously mentioned (p. 343) and illustrated (Fig. 76). With some later modifications this became a very important item in the joiner tool and brace manufacture at Sheffield, 1870 - 1890. The parts making up an Ultimatum brace are shown in Figs. 108a & b . The spring and brass locking cap were first assembled into the front bow casting. An iron spindle in the form of a bolt was driven through the circular hole at the top of the back bow.

Fig. 108a - Parts from a William Marples "Ultimatum" Brass Framed Brace

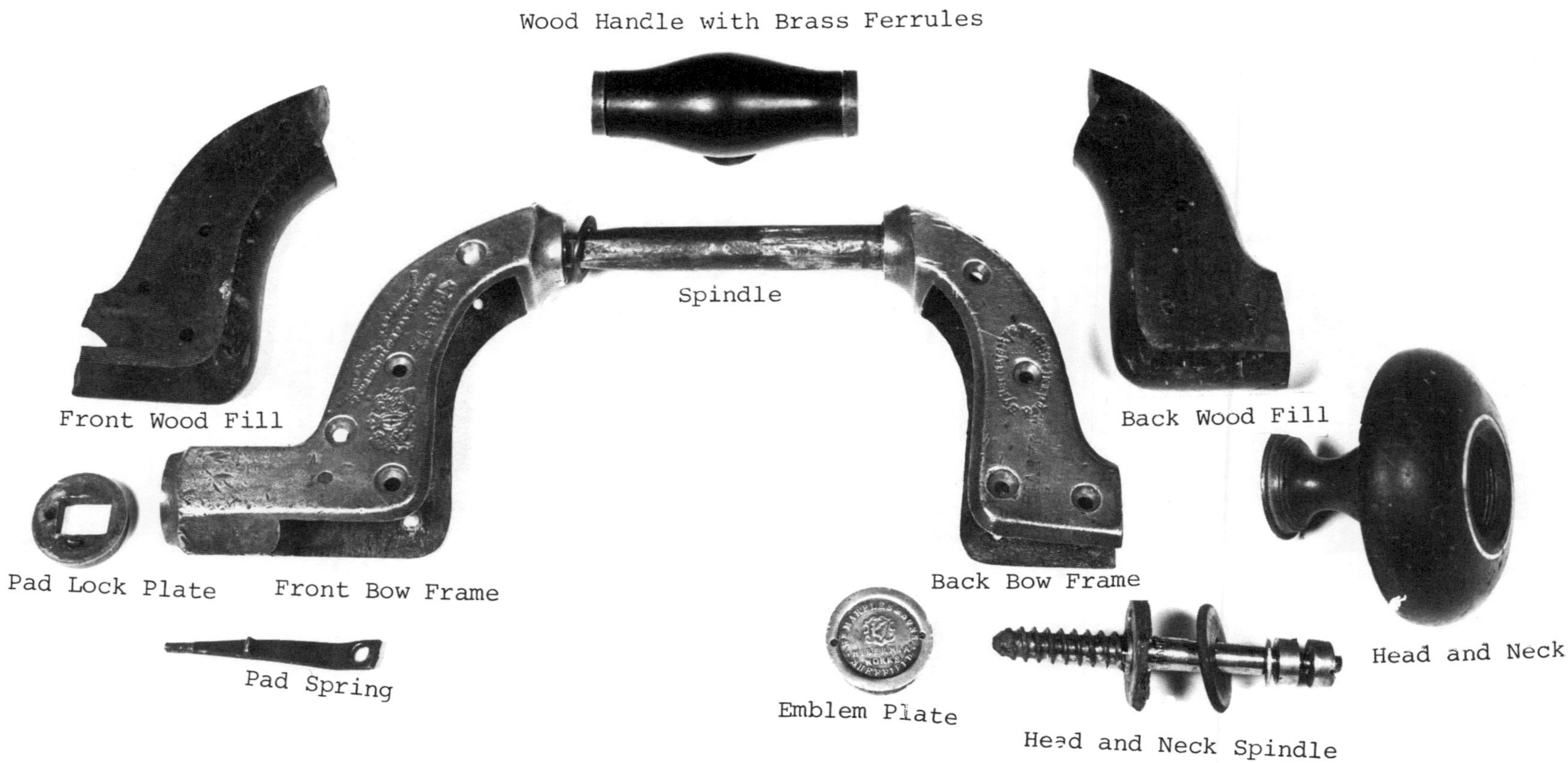

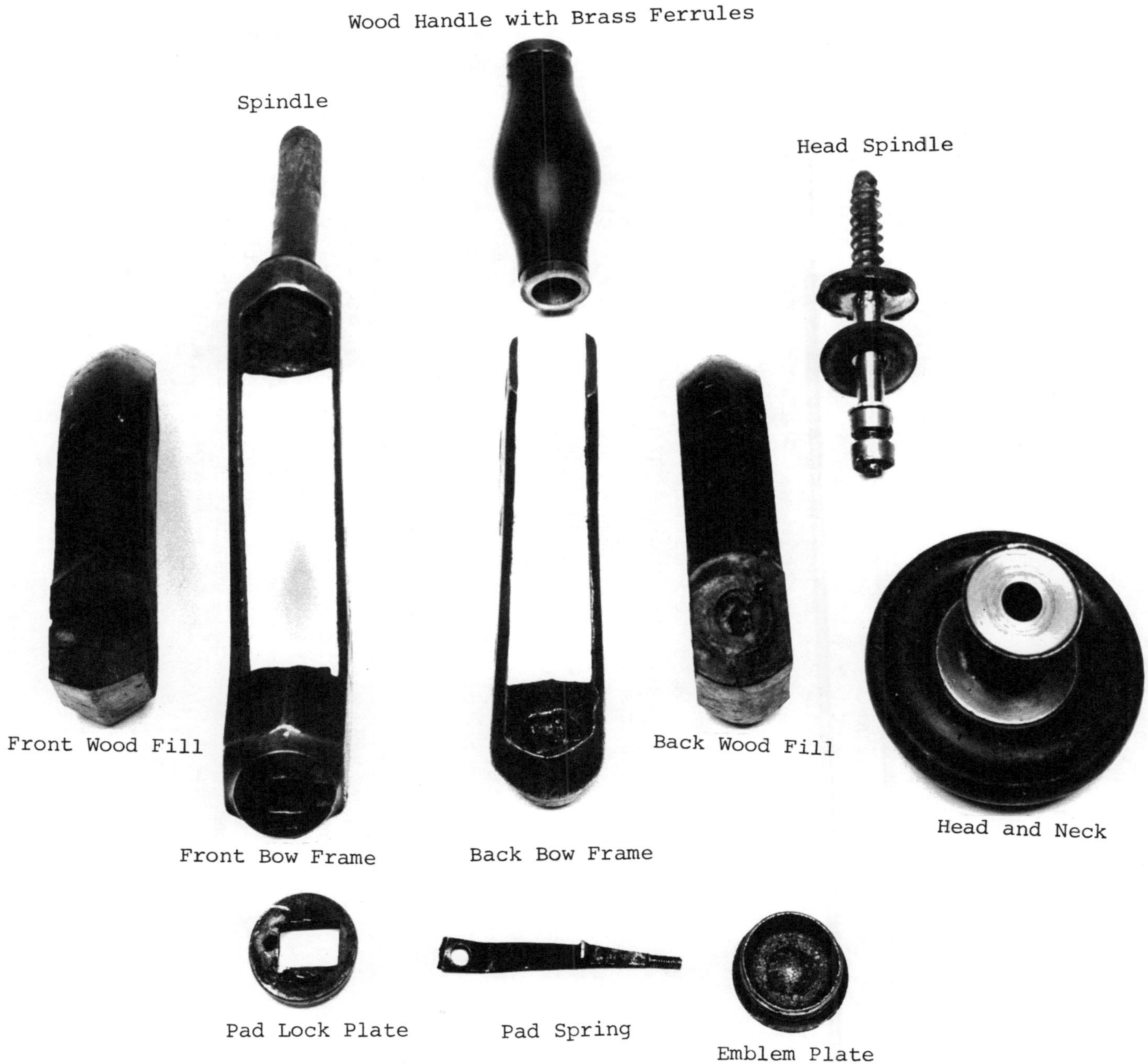

Fig. 108b - Parts from a William Marples "Ultimatum" Brass Framed Brace

The handle with brass ferrules and washers were slipped over the spindle. The front end was tapered to fit a square hole formed in the top of the front bow casting, driven into this section and peened on the end, thus connecting the two bows with the revolving handle. The wood fill was then fitted into the brass bows and held by screws. This must have required considerable expertise, especially fiting the front piece into the pad. The spindle for holding the head and neck was screwed into the back wood fill. The brass neck was held firm on this spindle by double reverse threaded nuts. The head was screwed on the neck and fitted with a brass cap.

The above method of joining the two bow sections by peening the bolt spindle was inferior to the procedure described in Cartwright's Patent. This specified fitting the front end and pinning it through the brass frame piece after drilling a small hole through this assembly. The back end of the spindle was threaded to fit a tapped hole in the back bow, or attached with a nut. The less expensive and inferior method followed by Marples was a weakness and often the joint loosened, resulting in wobble during use. This was later corrected in a brace made by J.Howarth & Sons in which both joints with the spindle and bow frames were pinned. This was called Soild Adjustment.

The Cartwright patent noted either horn or gutta percha as filling material between the plates, although stated that other material could be used. While it appears that the earliest were filled with rosewood, horn was also used. None have yet been found filled with gutta percha. Ebony soon became the most common material. Macassor ebony, often resembling rosewood, was frequently used. Beech was also used, possibly at such times that ebony was not available to complete a production run. These were offered at slightly less price, 20/ compared to 23/ for ebony, but now appear to be rarer than those available in ebony. Rosewood and box wood were available at the same price as ebony. Those made in box wood seem to be the rarest that survived.

The earliest known illustration of this brace appeared in the Wm. Marples & Sons Illustrated Catalogue, c1870, as item No. 340, "Wm. Marples & Sons Patent Ebony Brass-Framed Brace with Revolving Handle and Improved Pad". The 1870 Sheffield List had the same illustration, but this was referred to as the "Improved Ebony Brass-Framed Brace". The illustration of this brace appearing on the 1876 Sheffield List was made from the same cut, modified to show the thumb lever pad. All three of these are shown in Plate CCIV , and all were originally printed by Pawson & Brailsford. The prices of Braces and Braces with Bits from the 1870 Sheffield List are shown in Table XXV. There was a price increase in Bits as a consequence of the edge tool increase on July 1, 1872, but braces remained the same. The increases in prices of Bits on the 1876 Sheffield List are shown in Table XXV .

Apparently sometime after the Cartwright Patent expired, William Marples agreed to share this design among other Sheffield brace makers, possibly at first limiting this to associates belonging to the Cutlers' Company. William Marples continued to imprint "Sole Manufacturers of the Ultimatum Framed Brace". But the situation changed. The various stages in manufacture of the brass framed brass at Sheffield are classified in Table XXVI. The brass framed Trade Brace (III) was designed without imprint of maker so that a firm or tool dealer selling this could apply their own name. Thus the impression was created that such a brace was made by or specially for such merchants.(Fig 113)

Table XXVI Classification of Brass Framed Braces made at Sheffield

I. William Marples Ultimatum, c1850-1890-, Fig. 109
Design followed J.Cartwright Patent,1848-1862
Front Plate Insignia: William Marples Sole Manufacturer/of the Ultimatum Framed Brace/ SHEFFIELD Fig. 111
Seal of the Lion & Unicorm [two small circles below shield under crest line] If of small sweep and extra long, numbers appear below crest line]
Rear Plate inscription: "By Her Majesty's/Royal Letters Patent"
Head: Brass Insignia Plate and Inlaid Ivory Ring. Fig. 110

II. Robert Marples Ultimatum, c.1870-1904+ Fig. 109
Front Plate Insignia: Robert Marples/Hermitage Works/SHEFFIELD
Seal of the Lion & Unicorn [two tail lines below shield under crest line]
Rear Plate incription: ULTIMATUM Brass/Framed Brace
Brass insignia plate and ivory ring on Head. Fig.110

III. Trade Brace - made by William Marples & Sons or Robert Marples & Son c1870-1890, Fig. 113
Front Plate Insignia: Frequently no name of Merchant. Generally, but not always, has the Seal of Lion & Unicorn like Robt. Marples.
Rear Plate Insignia: METALLIC-FRAMED/PATENT BRACE
Head: Sometime EAGLE Design with"Warranted Superior"
Generally a plain brass disc with no inscription.

Peter Ross & Son,	Fig. 114	
Thomas Turner	Fig. 114	The Suffolk Brace
Smith & Scofield	Fig. 115	
Atkinson Brothers	Fig. 118	
J.T.Brufton & Co.	Fig. 115	
Alfred Ridge	Fig. 116	
John Wilson	Fig. 116	
R.S.W.Slack	Fig. 117	With All the Latest improvements

IV. Lever Type Pad - c.1876-1890+ [made by various firms or for the trade]
D.Flather & Sons [Improved Registered] Fig. 119
Marsden Brothers [Improved Brace] Fig. 120
I.Sorby Fig. 118
Colquhoun & Cadman Fig. 120

V. Miscellaneous Independently Made
James Howarth [Paragon Brace] Fig. 121
H. Pasley ["Ne Plus Ultra" Fig. 121

VI. Distinctive Modifications:
J.Howarth - Improved Solid Adjustment
Also for the Trade - E.Preston,London
Mawhood, Stephenson & Co.

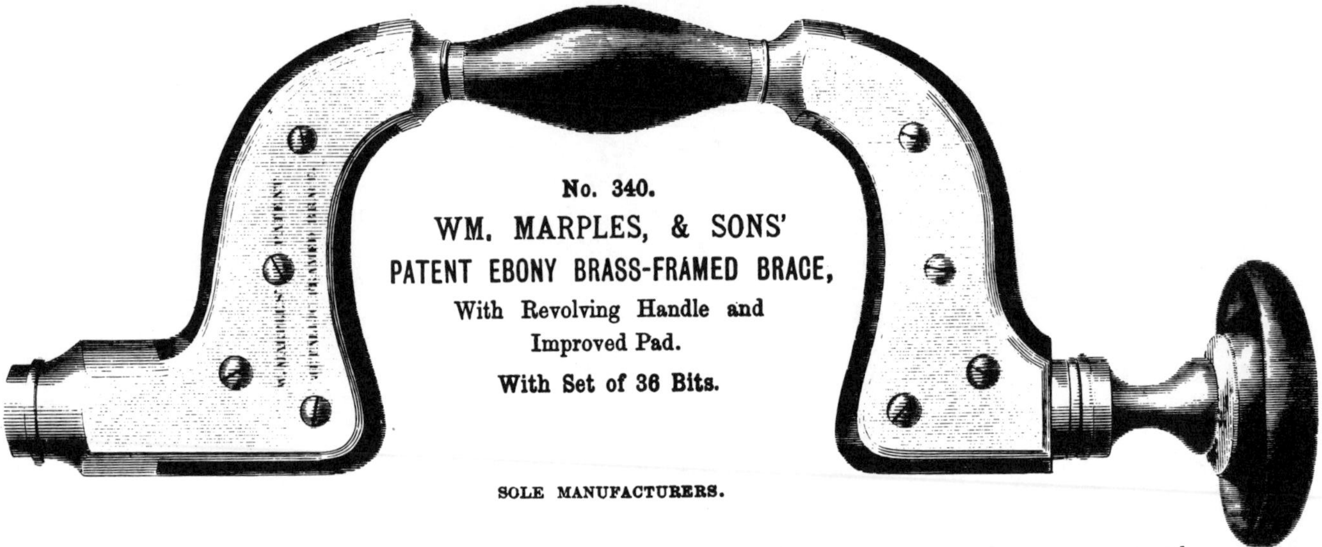

William Marples & Sons 1870 Catalogue [Page 9] #340 Wm. Marples Patent Ebony Brass-Framed Brace with Brass Neck. 23/ [no bits]; 20/ in Beech.

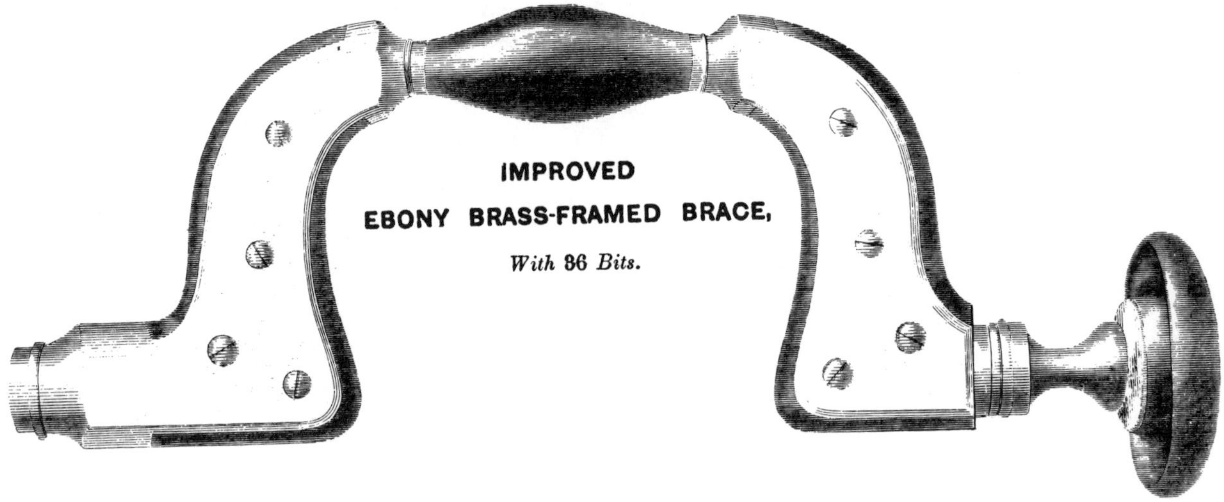

1870 Sheffield List [Page D2] #1764 Improved Ebony Brass-Framed Brace,#1503, 23/; [#1504, same in Boxwood & #1505, same in Rosewood, same prices, no beech listed]

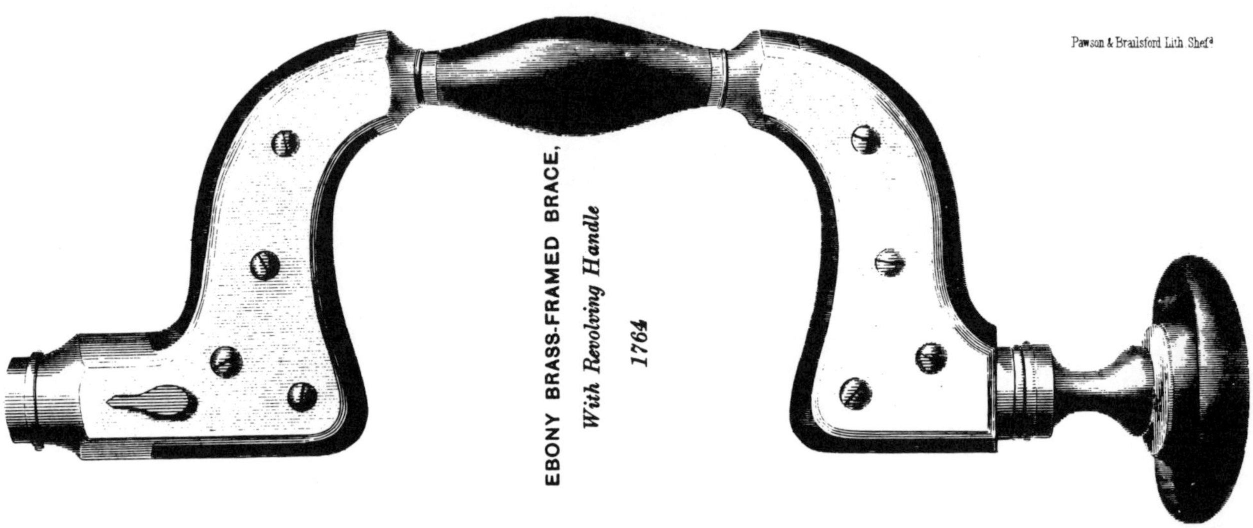

1876 Sheffield List [Page D1] #1764 Best Metallic-Framed Ebony Brace (Marples' Patent) with Brass Neck, 23/. [#1765,same in beech,20/] (Rosewood and Boxwood, not listed)

Plate CCIV - Pawson & Brailsford Illustrations of Brass-Framed Braces

WM MARPLES & SONS' BRACE & BITS.

By Her Majesty's Royal Letters Patent.

Table XXV

1870 & 1876 Brace Prices

Wm. Marples & Sons
1870 Catalogue [left]

1870 & 1876 Sheffield
List [below]

340 WM. MARPLES' PATENT METALLIC-FRAMED EBONY BRACE,
With Brass Neck Head, 23/

344 WM. MARPLES' PATENT METALLIC-FRAMED BEECH BRACE,
With Brass Neck Head, 20/

WM. MARPLES & SONS,

HIBERNIA WORKS, SHEFFIELD,

Are the SOLE MANUFACTURERS of the above Patent Brass-framed Brace, with Revolving Handle and Improved Pad.

1870 Sheffield List (Below)

BRACES.

No. of Article.			
1500	Best Plain Beech Braces	5/	each.
1501	Improved Plated Beech Braces	13/	„
1502	Improved Plated Beech Braces, with Brass Neck Head	15/	„
1503	Ebony Improved Framed Brace, with Improved Collar and Neck	23/	„
1504	Boxwood do. do. do. do.	23/	„
1505	Rosewood do. do. do. do.	23/	„
1506	Improved Scotch Iron Brace	14/	„
1507	Coopers' Braces	4/	„
1508	Best Coopers' Braces	5/	„
1509	Best Coopers' with Lignum Head	6/	„
1510	Iron Wagon Builders' Brace	10/	„

BRACES, with Bits complete.

No. of Article.	With	12	18	24	30	36	42	48	54	60	Bits.
1511	Best Plain Braces, with black Bits	13/8	16/	18/4	20/11	23/9	26/1	28/8	31/8	35/2	each.
1512	Do. do. with bright Bits	14/7	17/4	20/2	23/2	26/6	29/3	32/4	35/9	39/9	„
1513	Do. do. with straw coloured Bits	15/3	18/4	21/6	24/10	28/6	31/6	35/	38/9	43/1	„
1514	Improved Plated Braces, with black Bits	17/8	20/	22/4	24/11	27/9	30/1	32/8	35/8	39/2	„
1515	Do. do. with bright Bits	18/7	21/4	24/2	27/2	30/6	33/3	36/4	39/9	43/9	„
1516	Do. do. with straw coloured Bits	19/3	22/4	25/6	28/10	32/6	35/7	39/	42/9	47/1	„
1517	Improved Scotch Iron Braces, with black Bits	20/6	22/6	24/6	26/9	29/3	31/3	33/6			„
1518	Do. do. with bright Bits	21/5	23/10	26/4	29/	32/	34/5	37/2			„
1519	Do. do. with straw coloured Bits	22/1	24/10	27/8	30/8	34/	36/9	39/10			„
1520	Best Brace Heads, 24/ per doz.										
1521	Improved Brass Neck Brace Heads, 48/ per doz.										

No. of Article.	With	12	18	24	30	36	42	48	54	60	Bits
1764	Ebony Improved Brass Framed Brace, with Black Bits	27/8	30/	32/4	34/11	37/9	40/1	42/8	45/8	49/2	each
1765	Do. do. do. Bright do.	28/7	31/4	34/2	37/2	40/6	43/3	46/4	49/9	53/9	„
1766	Do. do. do. Straw-coloured do.	29/3	32/4	35/6	38/10	42/6	45/7	49/	52/9	57/1	„

1876 Sheffield List (below)

BRACES.

1764 BEST METALLIC-FRAMED EBONY BRACE, (MARPLES' PATENT) with Brass Neck Head, 23/ each.

No. of Article.		12	18	24	30	36	42	48	54	60 Bits
1764A	Do. with Cast Steel Black Bits	28/8	31/6	34/4	37/4	41/	43/6	47/	50/6	54/
„ B	Do. do. Bright Bits	29/3	32/4	35/6	38/9	42/6	45/7	49/4	53/2	56/10
„ C	Do. do. Straw Coloured Bits	30/	33/4	36/10	40/5	44/6	48/	52/	56/3	60/3

1765 METALLIC-FRAMED BEECH BRACE, (MARPLES' PATENT) with Brass Neck Head, 20/ each.

No. of Article.		12	18	24	30	36	42	48	54	60 Bits
1765A	Do. with Cast Steel Black Bits	25/8	28/6	31/4	34/4	38/	40/6	44/	47/6	51/
„ B	Do. do. Bright Bits	26/3	29/4	32/6	35/9	39/6	42/7	46/4	50/2	53/10
„ C	Do. do. Straw Coloured Bits	27/	30/4	33/10	37/5	41/6	45/	49/	53/3	57/3

Boxwood or Rosewood Braces, same price as Ebony.

If six Screw Auger Bits are included in any of the above Sets, the list price is 8/ extra.

Fig. 109 - "Ultimatum" Brass-Framed Braces: William Marples (top); Robert Marples (bottom)

Fig. 110 - Ebony Brace Heads "Ultimatums" of Fig. 109. William Marples (left); Robert Marples (right)

The various markings on the brass frame were applied by steel letter stamps using an embossing press. These were stamped on the cast frames before assembly after which the holes for the screws were drilled and countersunk. A peculiarity regarding the seal of the lion and unicorn,only present on William Marples' braces,is the presence of two small circles below the bottom border line. Also on William Marples' slightly longer and of smaller sweep,a number appeared stamped below the seal line. The significance of this is not known.

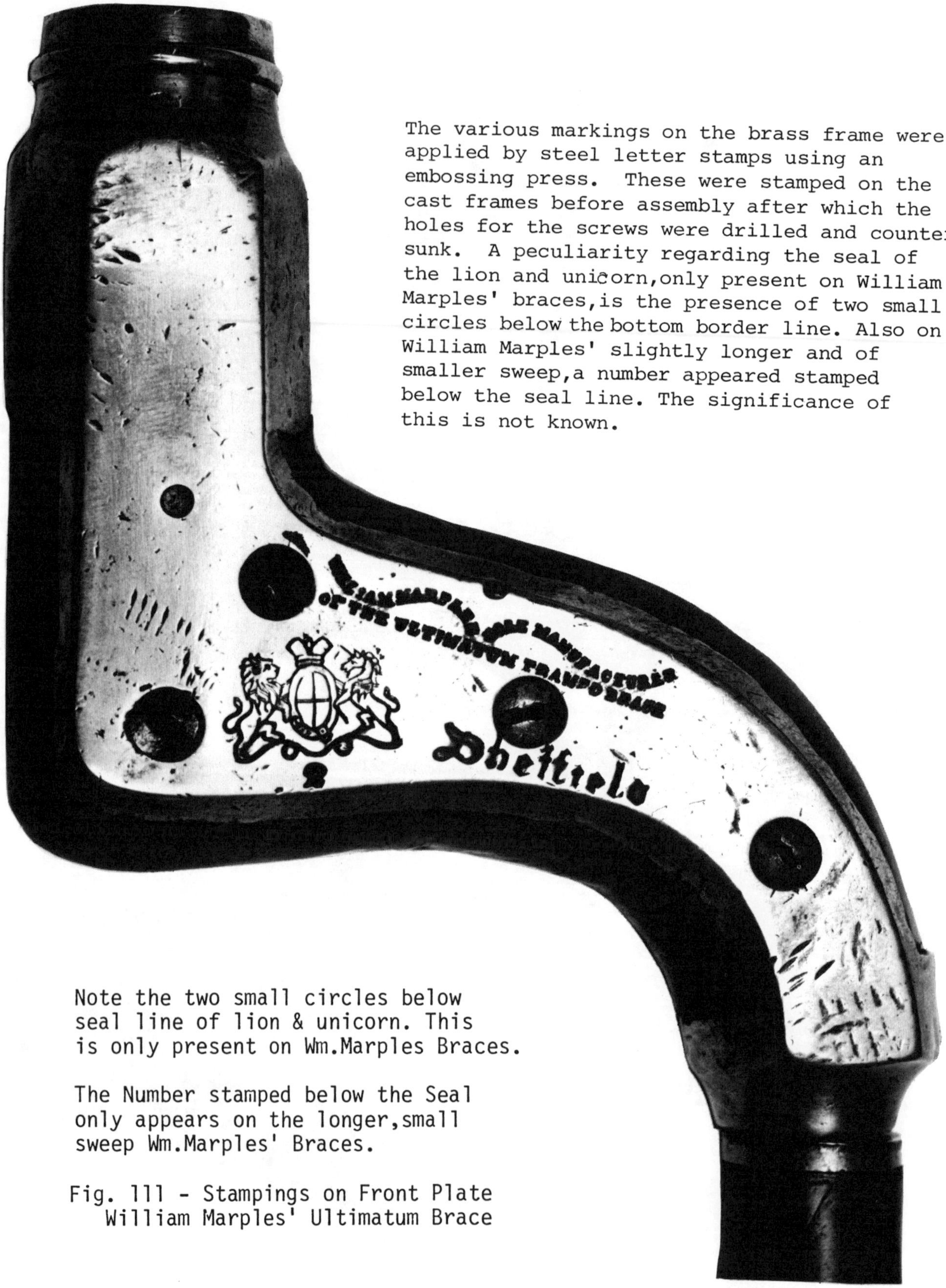

Note the two small circles below seal line of lion & unicorn. This is only present on Wm.Marples Braces.

The Number stamped below the Seal only appears on the longer,small sweep Wm.Marples' Braces.

Fig. 111 - Stampings on Front Plate William Marples' Ultimatum Brace

Fig. 112 - William Marples Beech Ultimatum
[Slightly longer and less sweep than usual]
(Note Numbers under the Seal)

Fig. 113 - Trade Brace
Ebony Brass-Framed

Fig. 114 - Trade Braces: Peter Ross & Sons (left); Thomas Turner (right)

Fig. 115 - Trade Braces: J.W.Brufton & Co (left); Smith & Schofield (right)
[Ebony] [Macassar Ebony]

Fig. 116 - Trade Braces: Alfred Ridge (left); John Wilson (right)

Fig. 117 - W.Kent [2nd Quality Wm. Marples & Sons] Ultimatum [left]
R.Slack Trade Brace [right]

Fig. 118 - I.Sorby (lever pad) [left]; Atkinson Brothers, Trade Brace [right]

Fig. 119 - B.Flather & Sons Brass-Framed Braces: Conventional Pad(left); Lever Pad (right)

Fig. 120 - Lever Pad Brass-Framed Braces: Marsden Brothers (left)
Colquhoun & Cadman, (right)

Fig. 121 - Brass-Framed Braces: Henry Pasley's "Own Manufacture" "Ne Plus Ultra"(left)
James Howarth & Sons "Paragon Brace" (right)

Wm. Marples & Sons

Ebony
Head Used with Beech Brace
[Fig. 112]

Fenton & Marsden
Registered Design #986

[Fig. 120]

Fig. 122 - Ebony Heads for Brass-Framed Braces

David Flather
Solly Works

[Fig.119]

Colquhoun & Cadman

[Fig. 120]

Fig. 123 - Brace Heads for Brass-Framed Braces

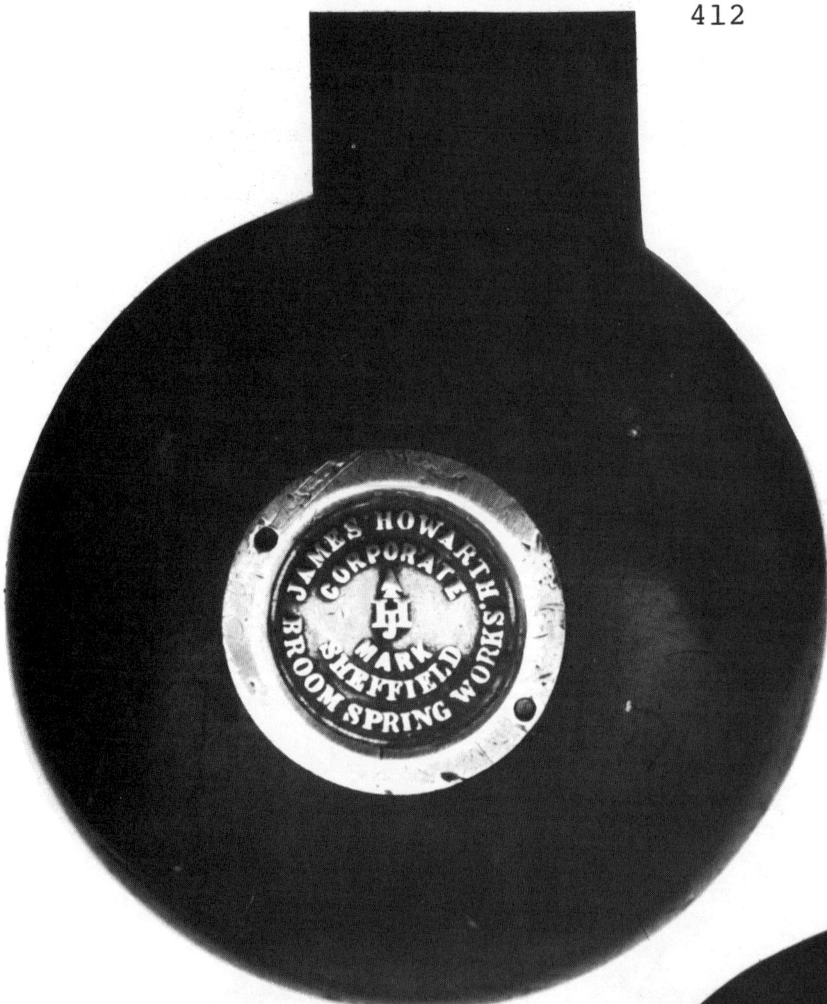

JAMES HOWARTH
Broomspring Works

Brace Head Fig. 121

WARRANTED SUPERIOR

Emblem often furnished
with Trade Braces.

Fig. 124 - Brace Heads for
Ebony Brass-Framed Braces

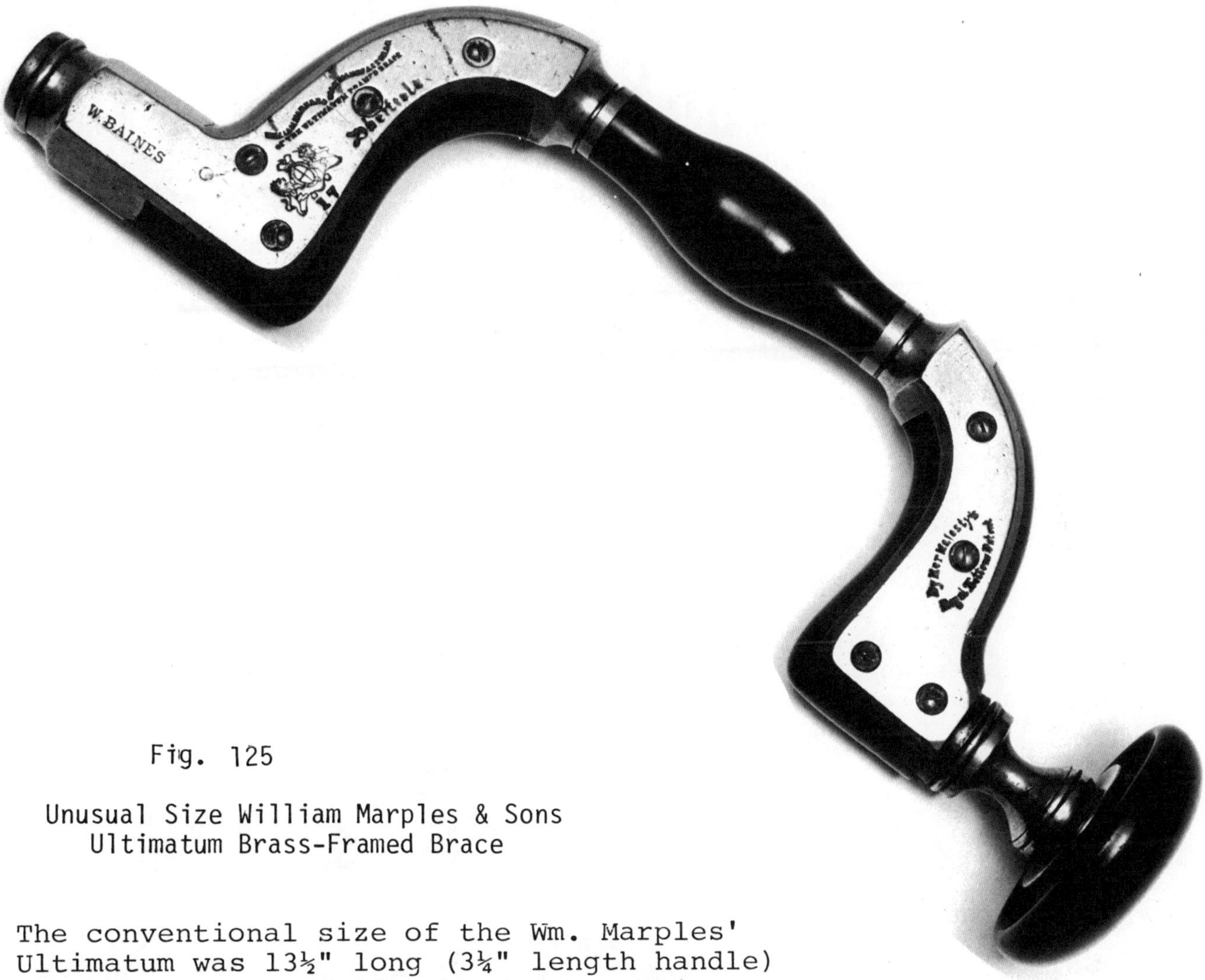

Fig. 125

Unusual Size William Marples & Sons
Ultimatum Brass-Framed Brace

The conventional size of the Wm. Marples' Ultimatum was 13½" long (3¼" length handle) and had a sweep of 8". The frame castings were 4½" high. The smaller sweep of 7" was 14½" long (3-3/4" length handle) and had frame castings 4" high. As previously mentioned, the smaller sweep was stamped with a series number below the "lion and unicorn" seal.

The above Wm. Marples' Ultimatum is a most unusual size. The front frame was for the larger size and the rear frame and handle were for the smaller size, thus producing a brace 14" in length. Since this was stamped with the number 17 below the seal on the front frame, it is suggested that this was assembled during a production run of making small sweep, longer braces. Possibly a larger front frame casting was substituted to complete production with a smaller rear frame casting.

ESTABLISHED 1828.

ROBERT MARPLES AND SON,

MANUFACTURERS OF

JOINERS' TOOLS.

GENTLEMEN'S & JUVENILE TOOL CHESTS FITTED UP COMPLETE.

LIGHT & HEAVY EDGE TOOLS, AUGERS, GIMLETS, PLANES, &c., &c.

ALSO MAKERS OF THE

ULTIMATUM METALLIC FRAMED BRACE, WITH REVOLVING HANDLE & IMPROVED PAD.

HERMITAGE WORKS,

HERMITAGE STREET, SHEFFIELD.

Plate CCV - Robert Marples & Son 1872 Birmingham Directory Advertisement

The earliest known advertisement for the *ULTIMATUM METALLIC FRAMED BRACE WITH REVOLVING HEAD & IMPROVED PAD* was not by William Marples, but by Robert Marples & Son. (Plate CCV). This first appeared in the 1872 Birmingham Directory, again in 1878 and finally in 1888 (Plate CCVI). In addition to making such braces for themselves, it is suggested that Robert Marples & Son may have also manufcatured the "Trade Brace' for other Sheffield joiner tool firms. An Ultimatum brace of their product is shown in Fig. 109. They were the only firm, other than William Marples & Sons to refer to such a brace as an ULTIMATUM. Incidentally there is no known immediate family relationship between William, Robert and Joseph Marples each operating independent joiner tool firms in Sheffield at the same date and all manufacturing braces.

ESTABLISHED 1828.

ROBERT MARPLES & SON,

TRADE MARK.

Granted by the Cutlers' Co., 1887.

MANUFACTURERS OF

JOINERS' TOOLS,

Light & Heavy Edge Tools, Augers, Skates, Gimlets, Planes, etc., etc.

TRADE MARK.

Granted by the Cutlers' Co., 1887.

GENTLEMEN'S AND JUVENILE TOOL CHESTS, TOOL RACKS AND TOOL CARDS FITTED UP COMPLETE, FOR HOME AND EXPORT.

ALSO MAKERS OF THE

ULTIMATUM METALLIC-FRAMED BRACE, With Revolving Handle & Improved Pad.

Hermitage Works, Hermitage Street,

SHEFFIELD.

TRADE MARKS—"**ROBERT MARPLES**," "**HENRY DIXON**," "**HENRY GREAVES**."

Plate CCVI - Robert Marples & Son 1888 Birmingham Directory Advertisement

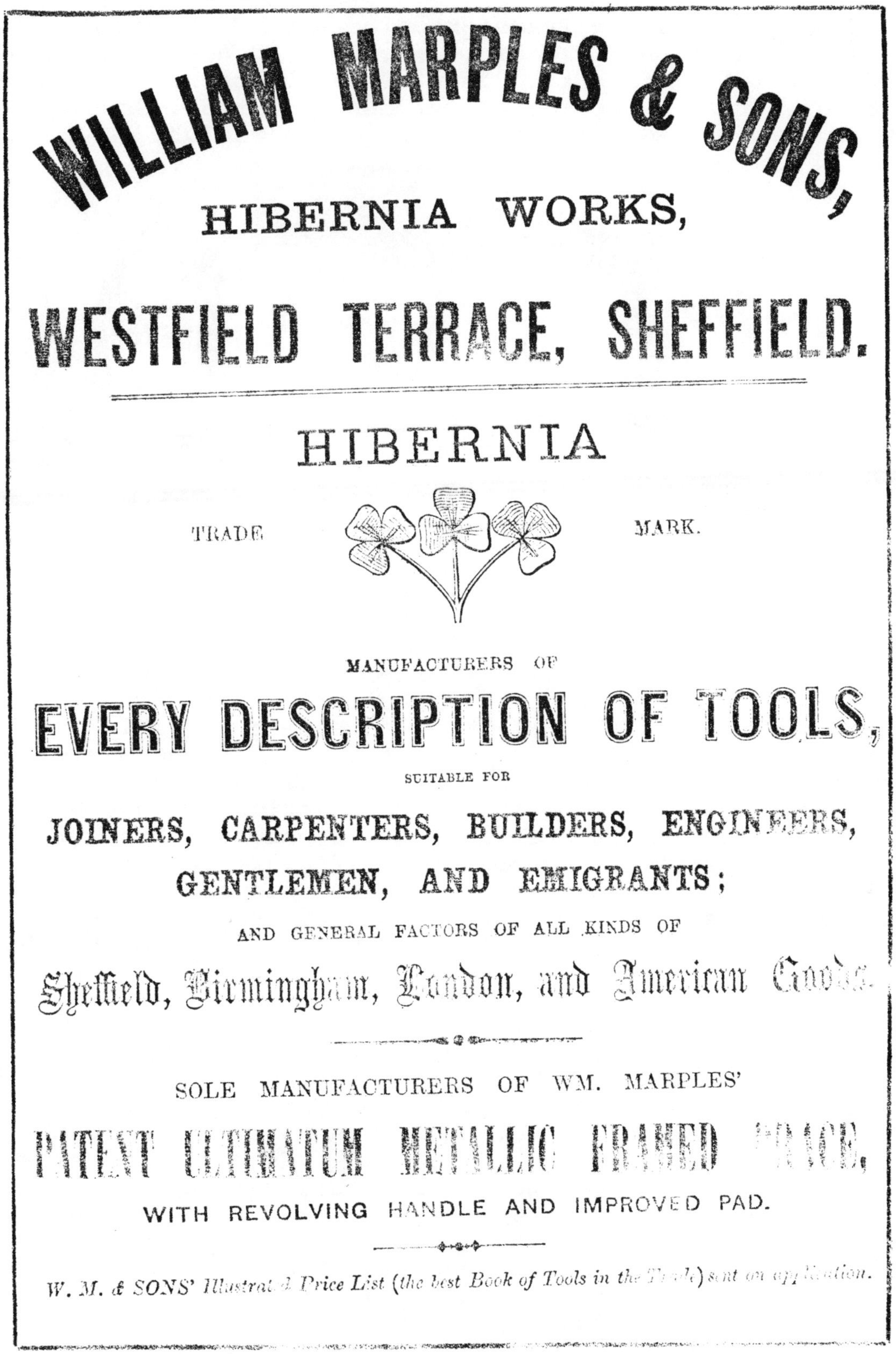

Plate CCVII - William Marples & Sons 1876 Sheffield Directory Advertisement

The first advertisement by Wm. Marples & Sons in a Sheffield Directory was in 1876. This was the only advertisment in which attention was noted that they were: *SOLE MANUFACTURERS OF WM. MARPLES' PATENT ULTIMATUM METALLIC FRAMED BRACE*. It was a true statement. The Patent had expired in 1862. They were the only firm that had manufactured this under such a patent, but in 1876 at least one other firm was also making such a brace.

The earliest known Trade Brace offered by Peter Ross & Sons, c1870-1871, Sheffield is shown in Fig. 114. Others are noted in Table XXVI. These were sold to firms throughout Great Britain. Several with imprints of firms in Scotland have been reported. Such a brace was offered and imprinted by B.Mills & Co., Philadelphia, USA (Fig. 126)[25]

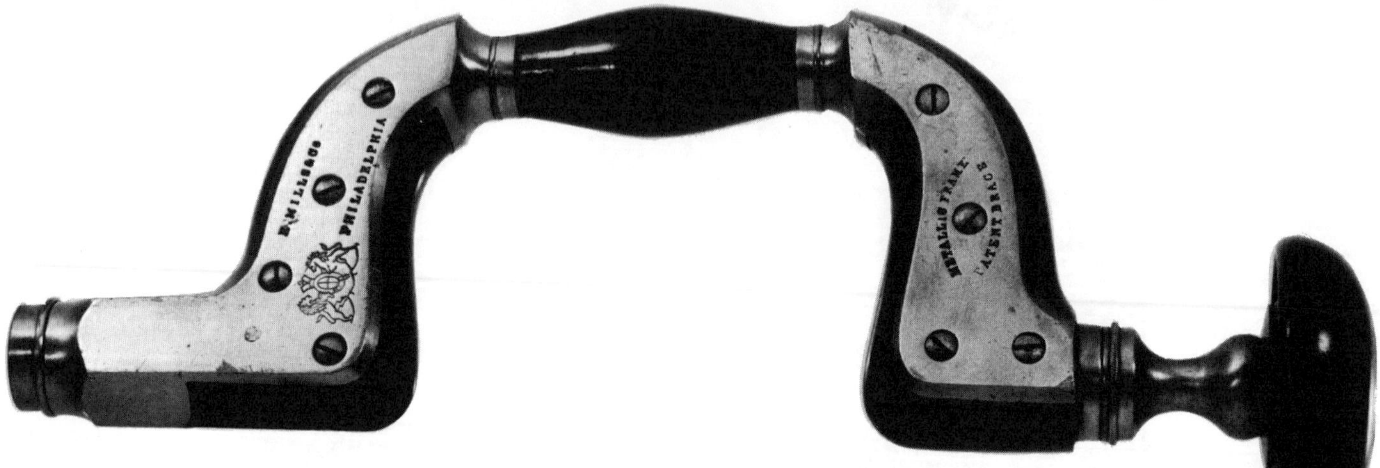

Fig. 126 - Brass Framed Trade Brace Imprinted B.Mills & Co, Philadelphia
[Courtesy of The National Museum of History and Technology, Washington, D.C.]
(Collection Item No. 236803.09)

By 1876 others appeared as makers (See Table XXVI). The thumb lever catch was shown on the 1876 Sheffield List. The price list for that List referred to this as 'Marples Patent'. (Table XXV) It is possible that the trade insisted on this qualifying statement, since Wm Marples & Sons still claimed to be "sole manufacturers of the Patent Brace". The Patent had expired 14 years earlier! Anyone with knowledge of brace construction at Sheffield could purchase brass frame castings and other parts from makers or any job foundry casting such parts and assemble their own brace. Undoutedly many did from 1870 - 1890. The same listings and prices were noted on the 1882 Sheffield List. While this Brass-Framed Brace was illustrated on the 1889 Sheffield List (same plate as in 1870), there were no prices listed, which indicated this product may have been withdrawn from general sales by this date.

Sometime after 1876 James Howarth & Sons developed an improved metallic brass framed brace which they marked "*IMPROVED SOLID ADJUSTMENT*". There were numerous improved features. The pad was made in the old style, and was easy to remove. Two lugs were cast in the pad body. The front frame was attached to the pad by screws into these lugs. This permitted removal of the pad assembly without taking out the front wood fill. The thumb lever was held in the pad by a single tapered pin easily removed. Thus either a defective thumb lever or the spring could be readily replaced, much easier than the old Marples' style. The joints of the frames with the handle spindle were properly pinned. The head spindle was attached to a thick brass plate which was held in the stock by screws, thus holding the head assembly securely to the stock. These features are apparent from examinations of Fig.127. An advertisement by James Howarth & Sons in the 1881 Sheffield Directory, illustrating this brace among other products, is shown in Plate

Similar braces were also marketed by William Preston & Co., London, 1863-1878.[26] Two such braces, one in boxwood, the other in ebony are shown in Fig. 128.

Fig. 127 - James Howarth & Sons Improved Solid Adjustment Brass Framed Brace

Fig. 128 - Improved Solid Adjustment (J.Howarth & Sons) with imprint Preston,London

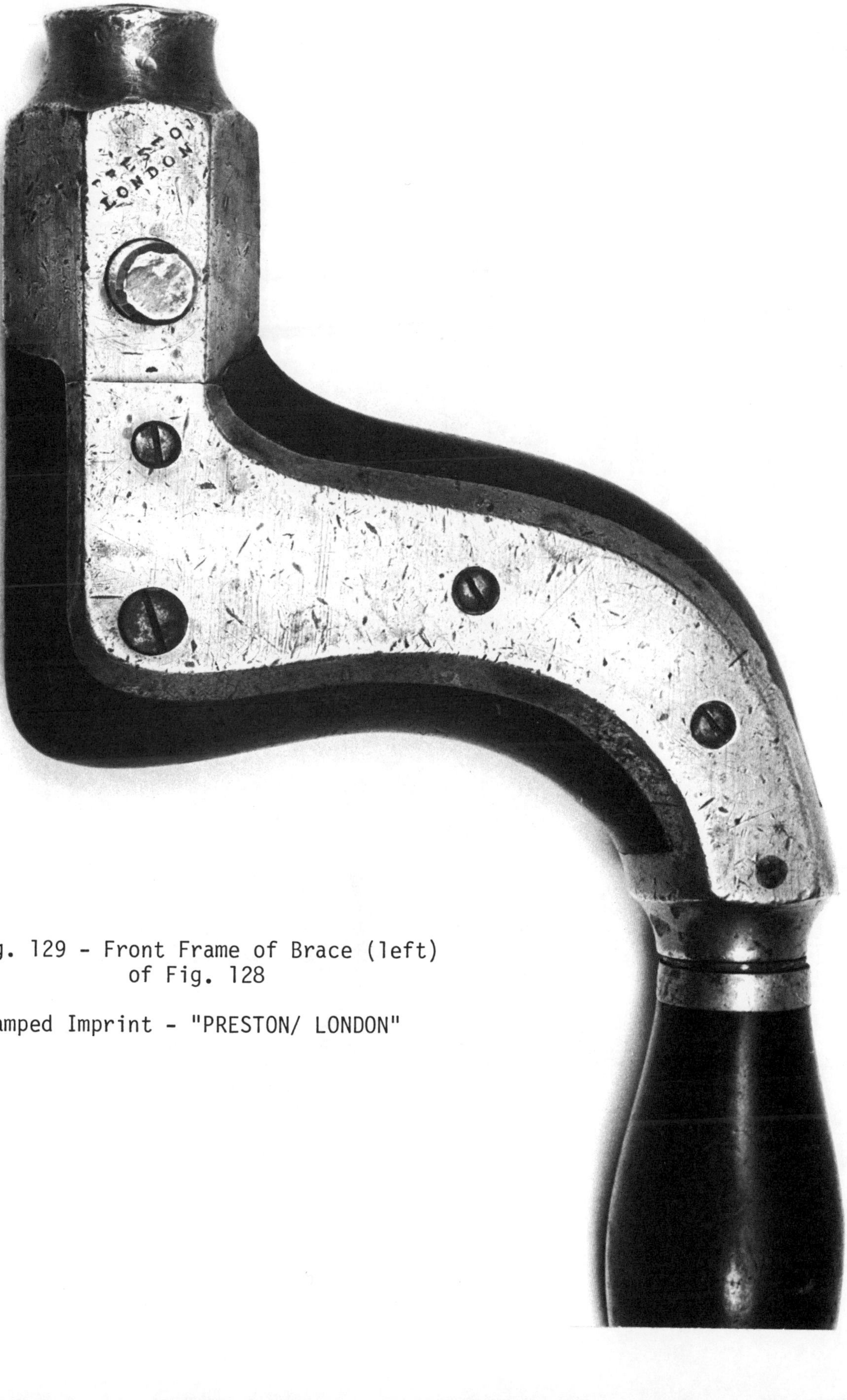

Fig. 129 - Front Frame of Brace (left)
of Fig. 128

Stamped Imprint - "PRESTON/ LONDON"

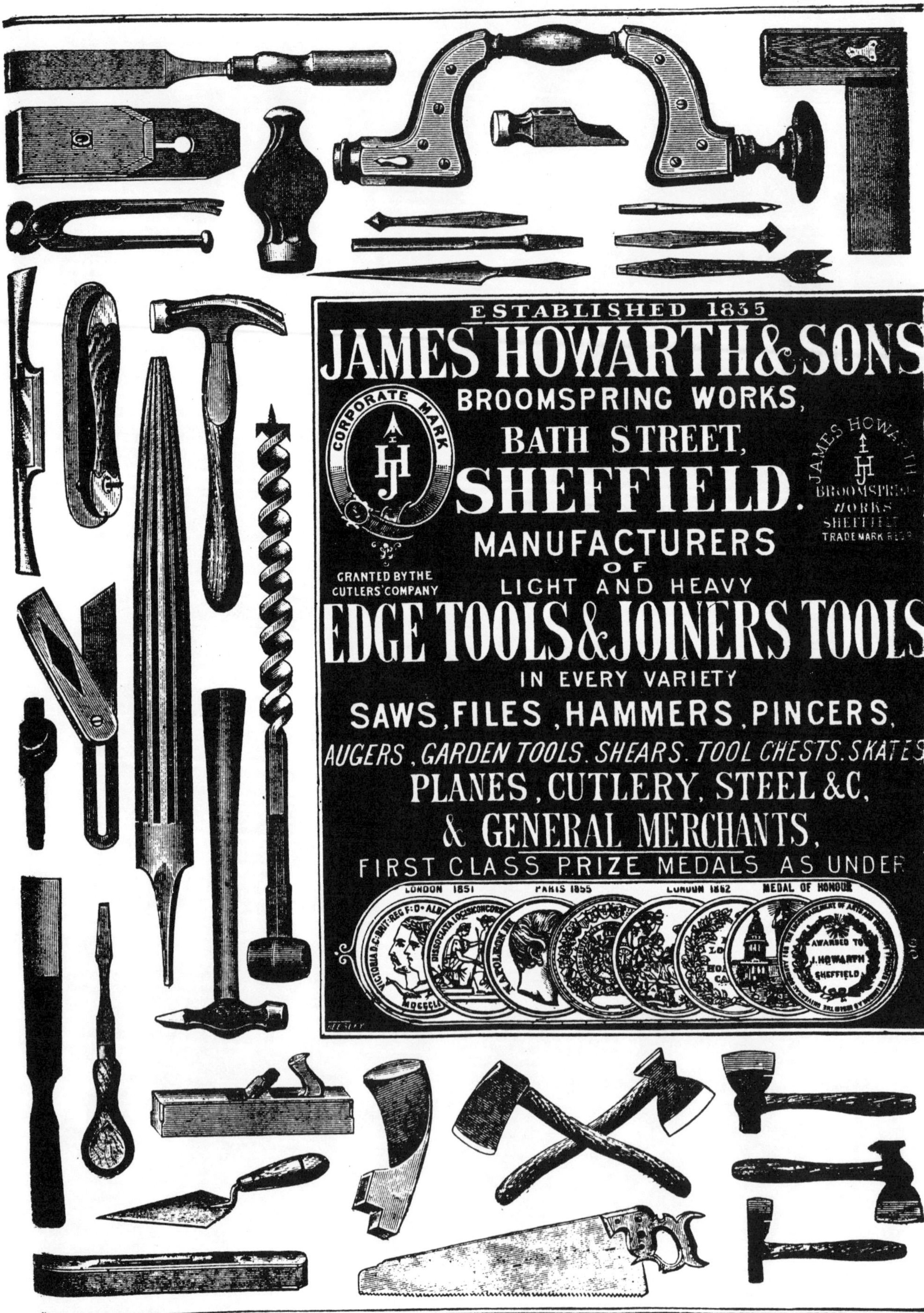

Plate CCVIII - James Howarth & Sons 1881 Sheffield Directory Advertsiment

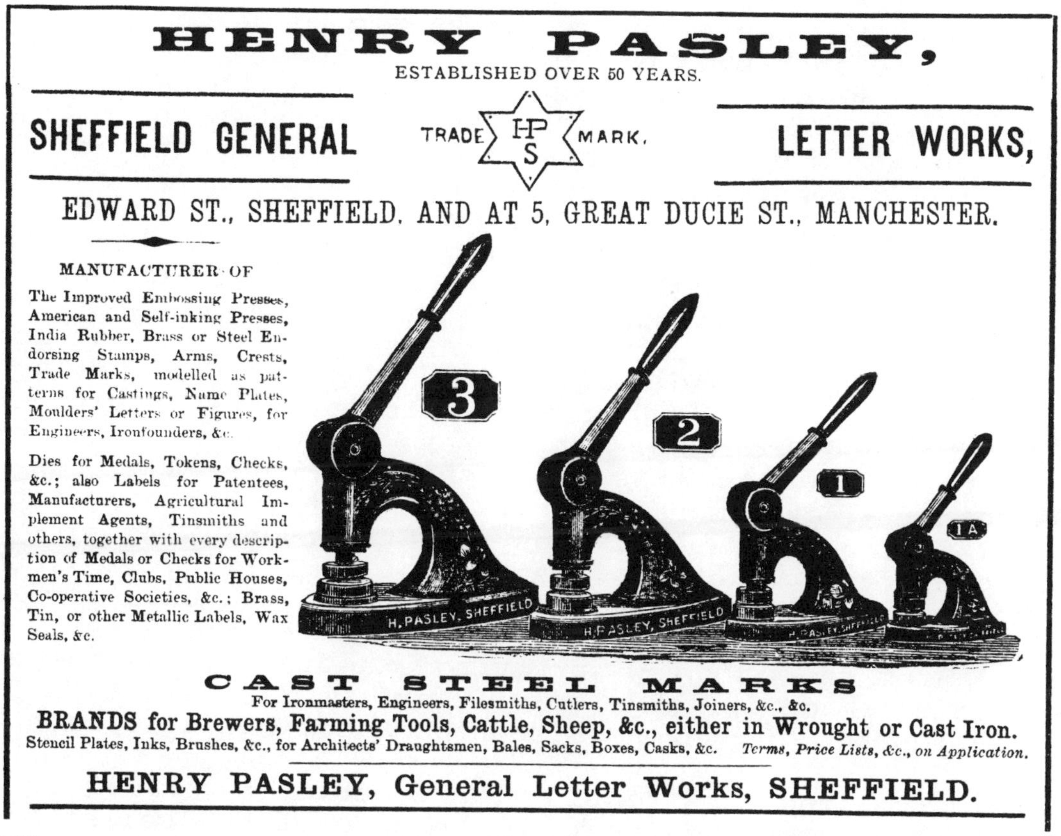

Plate CCIX - Henry Pasley's 1891 Advertisement in Sheffield Directory

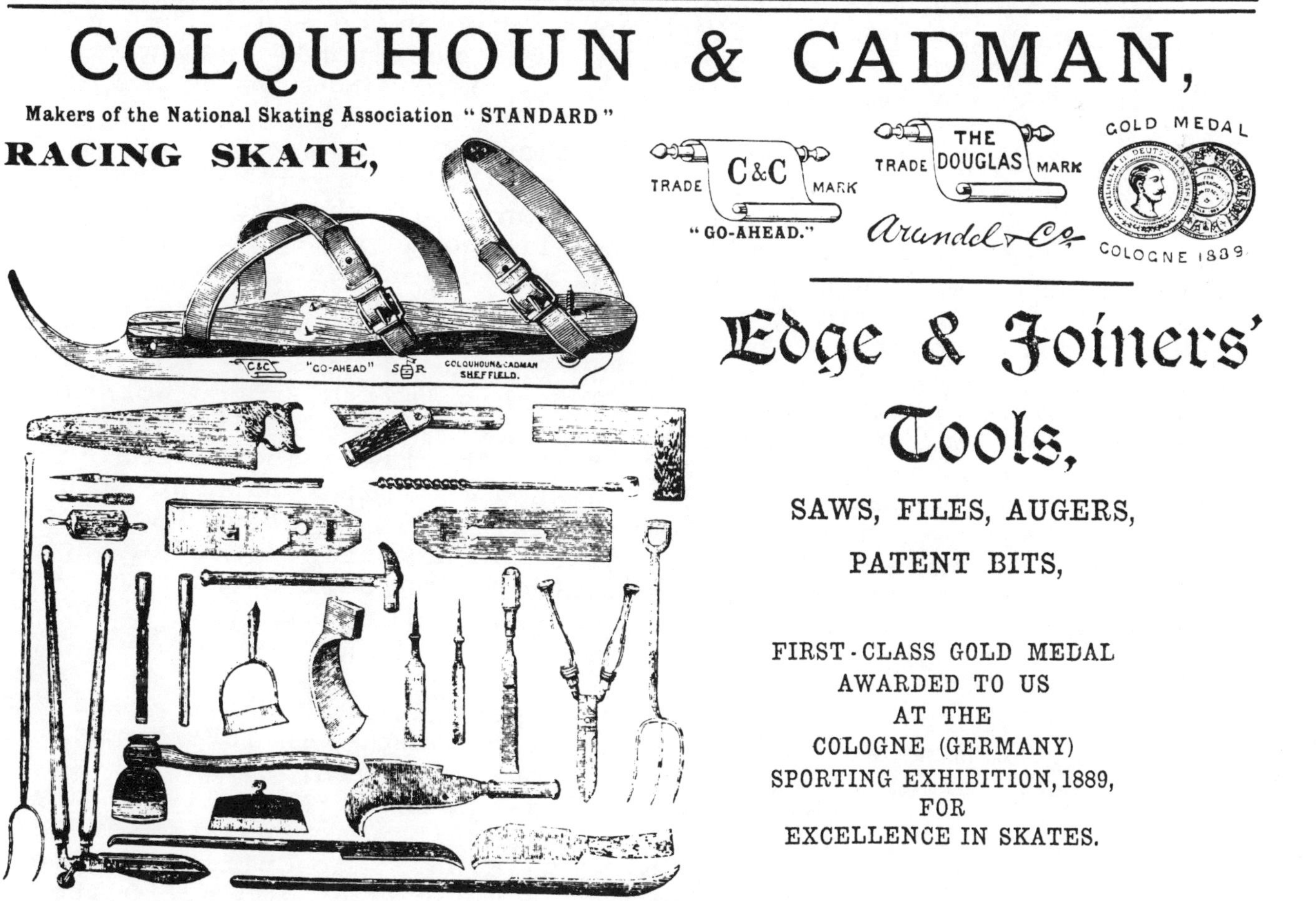

Plate CCX - 1891 Advertisement of Colquhoun & Cadman, *WHITE's HARDWARE TRADE MARKS*

A brass framed brace of entirely different construction is shown in Fig. 130. This has a central brass rib to which ebony plates have been attached to each side by cast brass screws. These were similar to those used for holding a saw handle. Cast in the head of these screws is the wording "*WARRANTED SUPERIOR*". Several of these have been found with the imprint "SIMS", and this name has been suggested as the possible maker.[27] Recently a brace of this type appeared in a London sales catalogue with the imprint *STEVENSON, MAWHOOD & Co*. This was a Sheffield firm listed at 2 Little Pond Street as merchants, manufacturers of steel, saws, tools &c., 1868-1879. This consisted of the partnership of John Parkinson Mawhood with Thomas Lowery Stevenson. Stevenson had previously been in partnership with William Sorby at the same address as Stevenson, Sorby & Co. After Mawhood, Stevenson & Co. dissolved in 1879, John Mawhood went into partnership with his brother, George, as Mawhood Brothers, a tool firm that lasted well into the 20th Century. Examination of this brace shows it to have a brass pad similar to those of Pilkington manufacture. It is therefore suggested that Thomas Pilkington made such braces for Stevenson, Mawhood & Co., and possibly other tool merchants. After C.& T. Pilkington dissolved in 1865, Thomas Pilkington worked alone in Sheffield until 1872. The date of c1870 is therefore suggested as when this brace may have been produced. From the very few surviving examples, it is judged that such braces were not made in large quantities. A Marples' type Trade Brace with imprint of Stevenson,Mawhood & Co. has also been observed.

Claims calling to attention the superiority of braces were often stamped on the plates, adjacent to the name of the seller or purported maker. This was sales propoganda, probably to compete with the Marples' *ULTIMATUM*. James Howarth & Sons applied *PARAGON* [conveying - Model of excellence]. (See Fig. 121) R.Slack on a Trade Brace stated: - "With all the latest improvements", of which there were none! Henry Pasley's "Own Manufacture" was stamped "*NE PLUS ULTRA*" [None any better] (Fig. 121)

Both Henry Pasley and his brother, Richard, were late entrants into the field of braces and joiner tools. Henry Pasley began work in Sheffield as a joiner in 1850. He soon added the occupation of brand maker to his trade. By 1856 he was listed as a mark figure and brand maker, adding embossing presses and stamps in 1860. In 1879 his listing in the Sheffield Directory was merchant and manufacturer of joiner tools, augers, gimlets, braces and bits, &c and copying and embossing presses, establishing the Sheffield General Tool and Letter Works on Edwards Street. After 1890 this became Brocco Works. It is not known whether he actually manufactured joiner tools. Quite likely he made marking stamps for many of the Sheffield tool firms. It is possible that he took in trade finished tools for payment of his stamps. He then may have applied his own stamp, thus marketing such tools under his own name. In all his numerous advertisements in Sheffield Directories, he only mentioned manufacturing cast steel marks, stamps, or embossing presses with no reference to making or selling any joiner tools. (See Plate CCIX)

Richard Francis Pasley first appeared in the 1876 Sheffield Directory as manager of the Spring Vale House. In 1879 he was listed as manufacturer of steel,table and pocket knives, scissors, saws, files, hammers, &c. at Anchor Works on Rockingham St. Actually he may have been a merchant of these products rather than a manufacturer. He advertised in 1892 as "The Sheffield Gilding Stamp Cutter" calling to attention his marks and stamps,but no mention of tools. Quite possibly he also acquired tools in trade for his stamps

Fig. 130 - Unique Brace-Frame Brace with Pilkington Type Pad
[Anonymous Collection]

Other late entries into the joiner tools and braces were the two firms John Wilson and Colquhoun & Cadman. As noted previously John Wilson had been a contract worker at William Marples & Sons. (pp.196-198). A trade brace with his imprint is seen in Fig. 116. Edward Colquhoun in 1879 was an edge tool maker employed at Marsden Brothers. The firm Colquhoun & Cadman was established in 1881, occupying Douglas Works on Arundel St. (See Advertsiment, Plate CCX) A brace with their imprint is shown in Fig. 120.

The production of the brass framed brace and its variants without doubt achieved the zenith in art of 19th Century Shefffield tools. Unfortunaely due to high manufacturing costs, it simply was not competitive with the more durable, lighter, more efficient and less expensive American iron brace. The latter were imported in large quantities after 1880, and the Sheffield brass framed brace diminished. It was relegated to the position of a presentation tool for honoring a retired craftsman. (See Fig. 131) The last known listing of the Ultimatum Brace was in the 1904 Robert Marples & Son Catalogue,offered at 23/. An American ratchet iron brace was priced a third of this.

Fig. 131 Engraving on Wm.Marples' & Sons Presentation Brace[Collection Kendall Bassett

The only prominent Sheffield joiner tool manufacturer not known to have offered a brass framed brace was Henry Brown & Sons. This is believed to have been due to professional rivalry between their Anti-friction Brace and the Marples' Ultimatum. About the time of the dissolution of Brown & Flather, c.1850, Henry Brown established his own sales' outlet in America. Thomas E. Wells, formerly of Wells & Bradshaw, Sheffield joiner tool manufacturers, went to Philadelphia and organized Brown and Wells. This also became known as T.E.Wells & Co. Braces with imprints of Brown & Wells and T.E.Wells & Co. are illustraded in Figs. 132 & 135. Attention is called that both of these have plates with rounded top extremities, rather than the pointed and curved design. While this may be co-incidental, the earliest form of this round designed plate is noted in a brace by T.E.Wells/Sheffield, c.1841. (See Fig. 134) It is possible that these braces were made in Sheffield and exported to Philadelphia. T.E.Wells & Co. sold out to an American firm of Booth & Mills in 1854, who continued manufacturing this type of brace with slight modifications at Philadelphia.[28] In 1857 Henry Brown & Sons opened a sales' office at New York City. This was managed by Henry Brown, jr. in 1858, while his brother, Samuel,remained in Sheffield with Henry Brown, sr.

The firm of Brown & Wells was listed in the Philadelphia (USA) Franklin Institute Fair Catalogue,1850, as "Manufacturers of Braces,Bits, Joiner Tools. &c" (See Below) They were reported to have exhibited tools at the Crystal Palace World's Fair in London in 1851. [*Great Exhibition Illustrated Catalogue*, London, Vol.II, 1852, p.1452[This firm is believed to have been Henry Brown in Sheffield and Thomas E. Wells at Philadelphia, which was also known as T.E.Wells & Co. This firm was sold out to Booth & Mills of Philadelphia in 1854, who continued to use the firm name of T.E.Wells & Co. along with their own through 1856. [See: "Premium Awarded by Philada. & New York Institutes BOOTH & MILLS, PHILADA." by Carl E. Bopp, *Chronicle of EAIA*,Vol.27, No.3,Sept.1974, pp.41-42]

A brace has been reported with imprint of Brown & Wells *" Sole Manufacturers of Henry Brown's Celebrated Antifriction Brace"* This substantiates the above suggested partnership. There is no record of the firm Brown & Wells in the Sheffield Directories.

BROWN & WELLS,

COR. EIGHTH AND WASHINGTON STS.

Manufacturers of

BRACES, BITS, JOINERS' TOOLS,

&c. &c.

All kinds of Bits made to order.

Advertisement in 1850 *Franklin Institute Catalgue* [Philadelphia,USA courtesy of Carl E. Bopp]

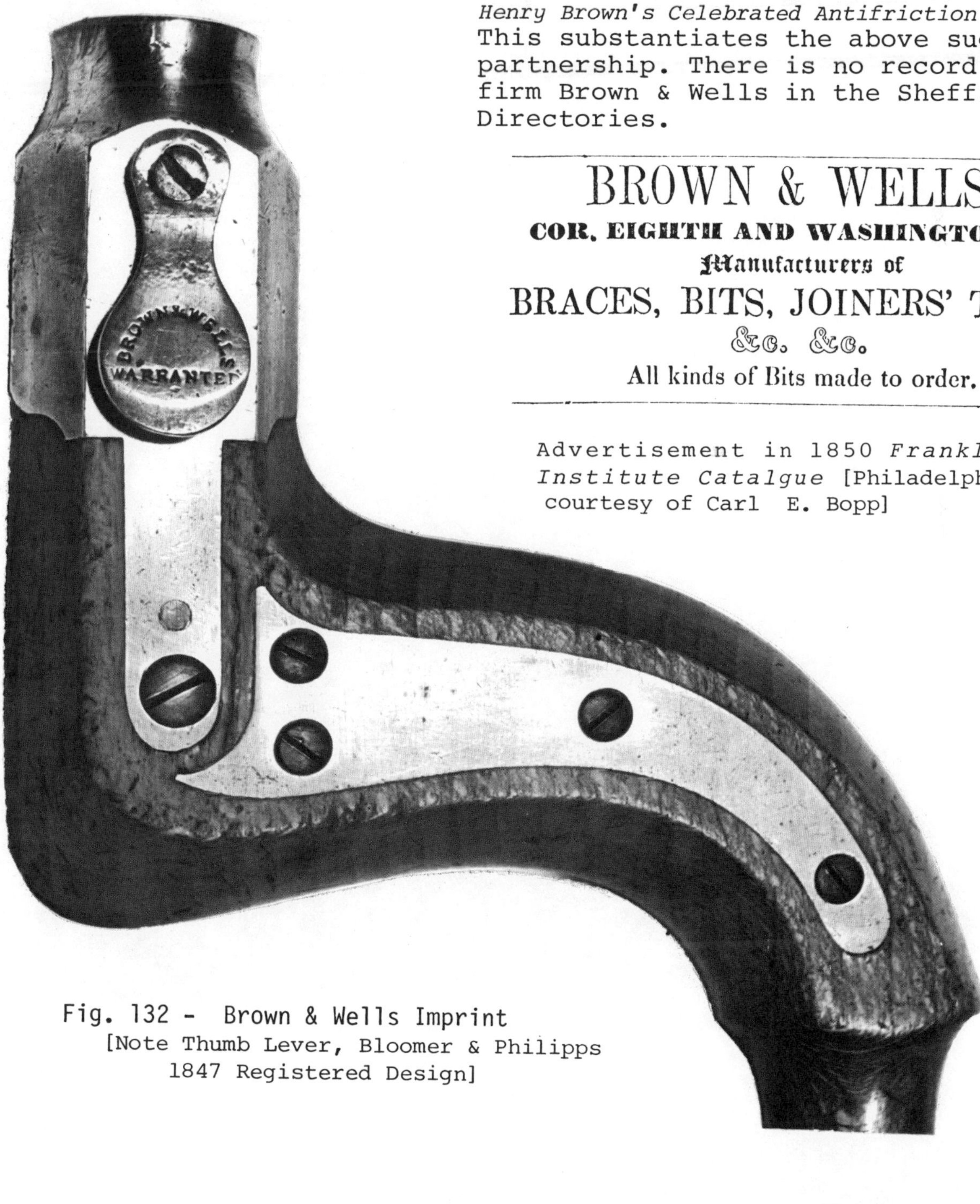

Fig. 132 - Brown & Wells Imprint
[Note Thumb Lever, Bloomer & Philipps 1847 Registered Design]

William Bradshaw's only listing in a Sheffield Directory was in 1841. He was noted as a joiner tool maker in the firm Wells & Bradshaw, which was the only listing of such a firm. A brace with imprint Wells & Bradshaw has not yet been reported.

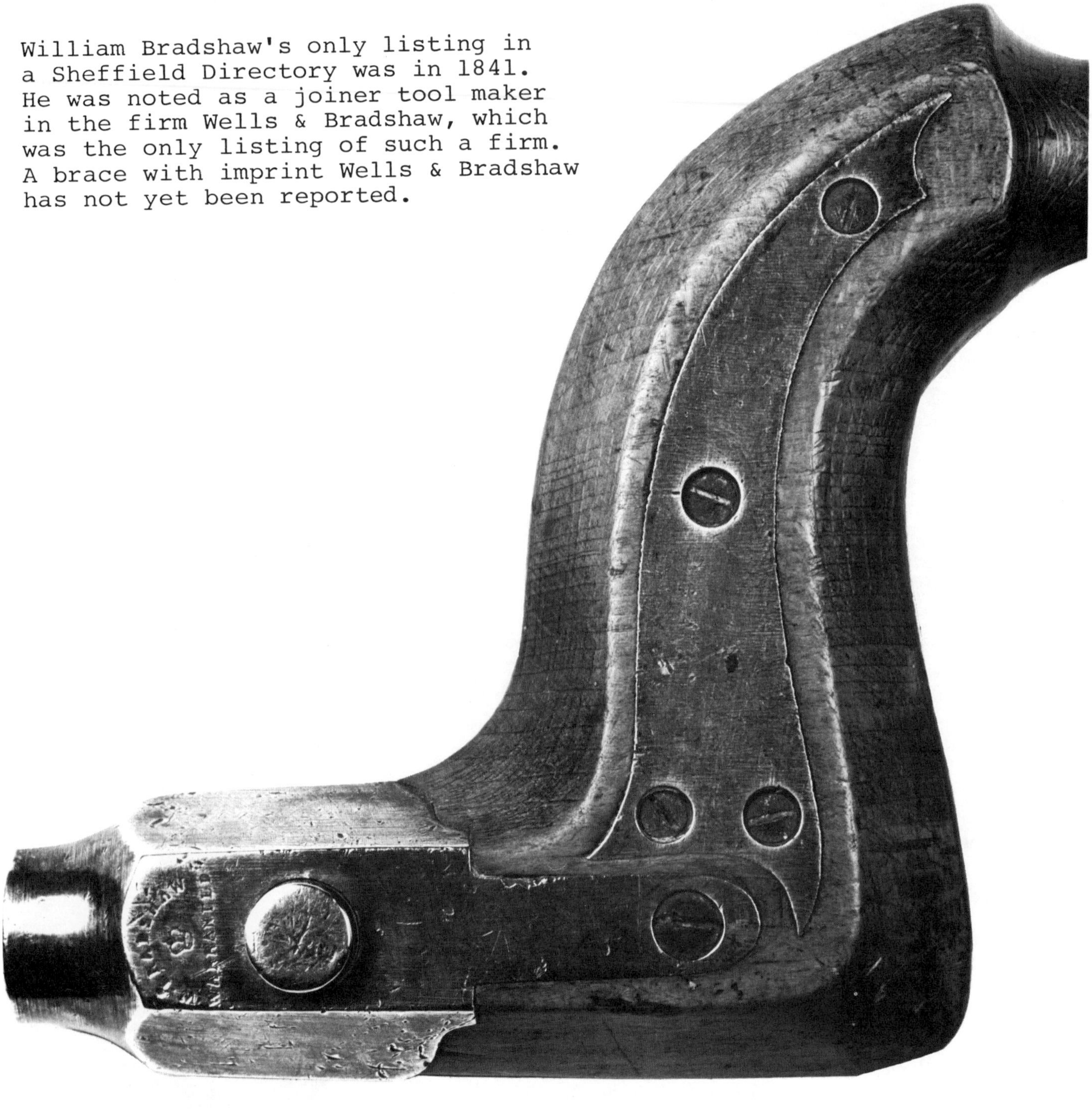

Fig. 133 - Plated Brace by William Bradshaw, Sheffield

Thomas Ebenezer Wells was only listed in the 1841 Sheffield Directory. He was noted as a partner in the firm of Wells & Bradshaw, joiner tool and brace makers.

He appeared in the Philadelphia (USA) Directory as a tool maker 1850 - 1853. He was a partner in Brown & Wells.

This is the earliest known Sheffield brace with the rounded top plate design, rather than the earlier curved and point sytle. This dates this feature to have appeared by 1841.

Fig. 134 - Plated Brace by T.E.WELLS/ Sheffield

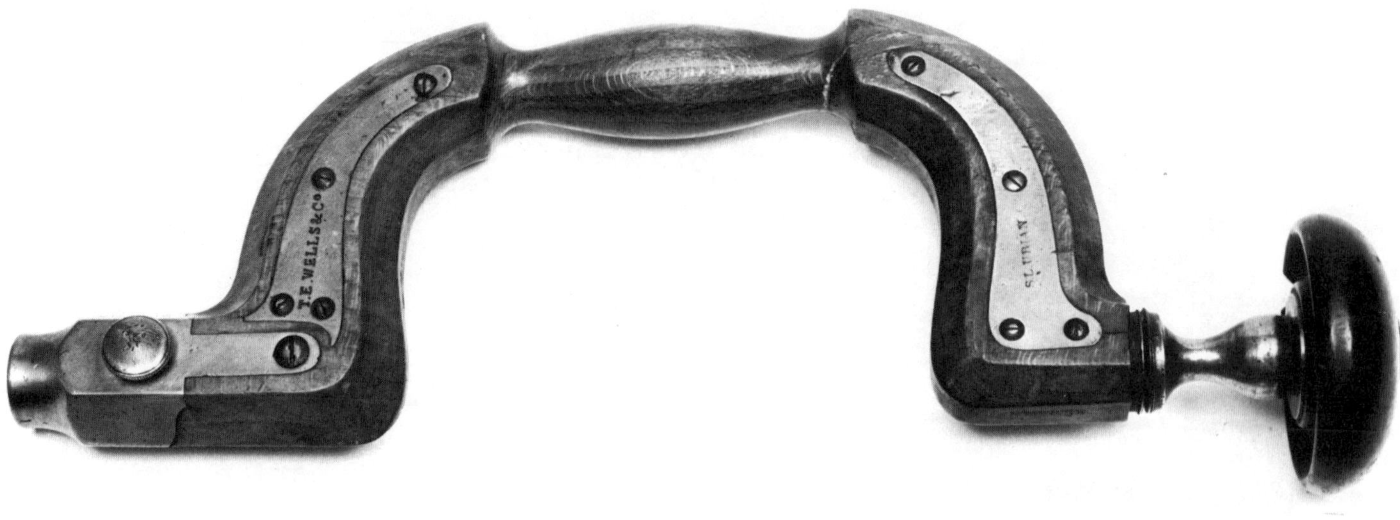

Fig. 135 - Plated Brace by T.E.Wells & Co., Philadelphia. Pennsylvania, USA

Herman Bowker was among those listed in the 1840-41 New York City Business Directory as a Wholesale Hardware and Cutlery Dealer. He continued to be listed in New York Directories through 1875. His listing in the 1849-50 NYC Directory stated: - "Importer of German Hardware and Cutlery, guns, pistols, fancy hardware, wrought nails, H.& R.Boker's best scythes constantly on hand". Apparently this firm also imported Sheffield braces. Two with imprint of H. & R. Bowker are illustrated in Figs. 136 & 137. The brace with the rotating handle is most unusual. Apparently H.Boker purchased some braces in Sheffield, sawed them in half and dowelled together two sections providing a spindle for supporting the rotating handle. In this instance it is apparent that the two sections are from different braces, as can be noted from the totally differnt grain of the beech in the two sections. Braces with the imprint H.Boker have also been noted. No listing for this person could be found in any 19th Century Sheffield Directory.

Another firm selling braces at New York City was Barton Brothers. This firm was listed in the 1849 Sheffield Directory with Edward Barton at 231 Glossup Road, Sheffield and his brother, William at New York. They were listed as Barton Brothers through 1855 at NYC, but 1856 - 1860 as William Barton & Co. It is quite likely that Edward procurred plated braces in Sheffield and sent them to William at NYC to resell. A brace with imprint Barton Brothers and *UTILE DULCI* [The useful with the agreeable] also IMPR'D. PATENT BRACE is illustrated in Fig. 138. The improvement and patent are not known, if any, but this may have been a brace made under the Bottom Patent by Henry Brown's firm.

A brace with imprint E.Windle/ *Premium Mobile* [First in easy use] / IMP'D PATENT is illustrated in Fig. 140 . There is nothing to indicate that this brace was made under any patent, except possibly the Bottom Patent controlled by Henry Brown. An unsual feature of this brace is that the push button on the pad is on the right side rather than the customary left. While no Windle could be found in a Sheffield Directory, such a name was located in the 1849-50 NYC Business Directory as an importer of Hardware & Cutlery. Another brace with this imprint has been observed having the 1847 Bloomer & Philipps registered design pad button. This had an eagle imprint with E.Windle & Co./Sheffield on the head cap. The eagle suggests made for American market.

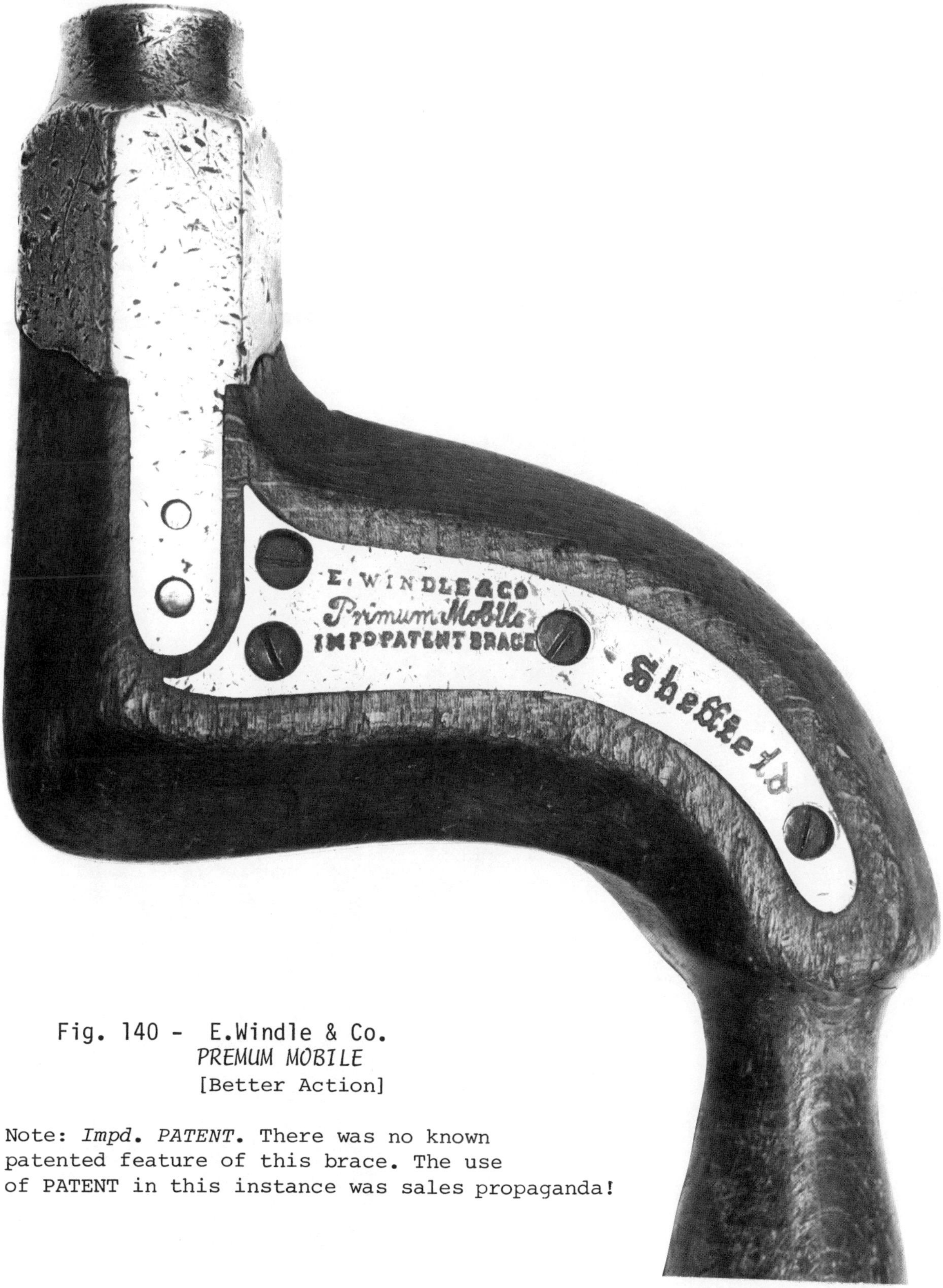

Fig. 140 - E.Windle & Co.
PREMUM MOBILE
[Better Action]

Note: *Impd. PATENT.* There was no known patented feature of this brace. The use of PATENT in this instance was sales propaganda!

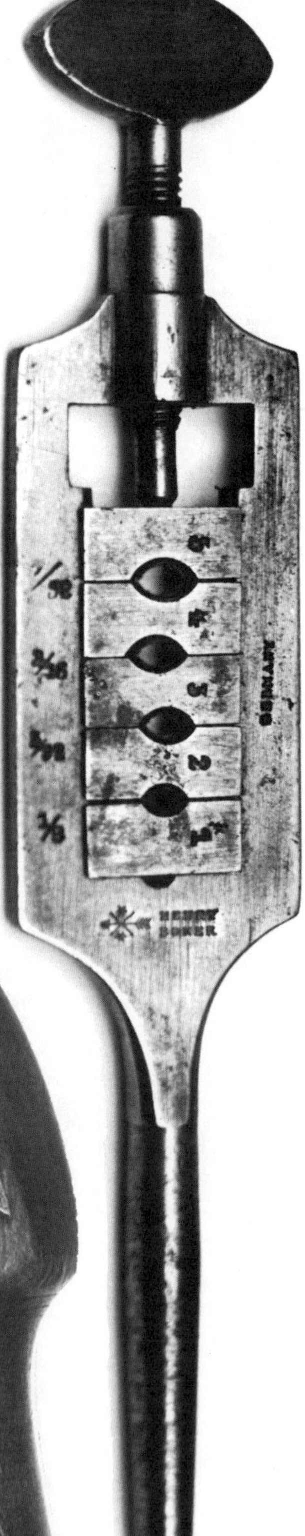

Threading Die

with imprint

HENRY BOKER
GERMANY

[*Collection of Henry Wing*]

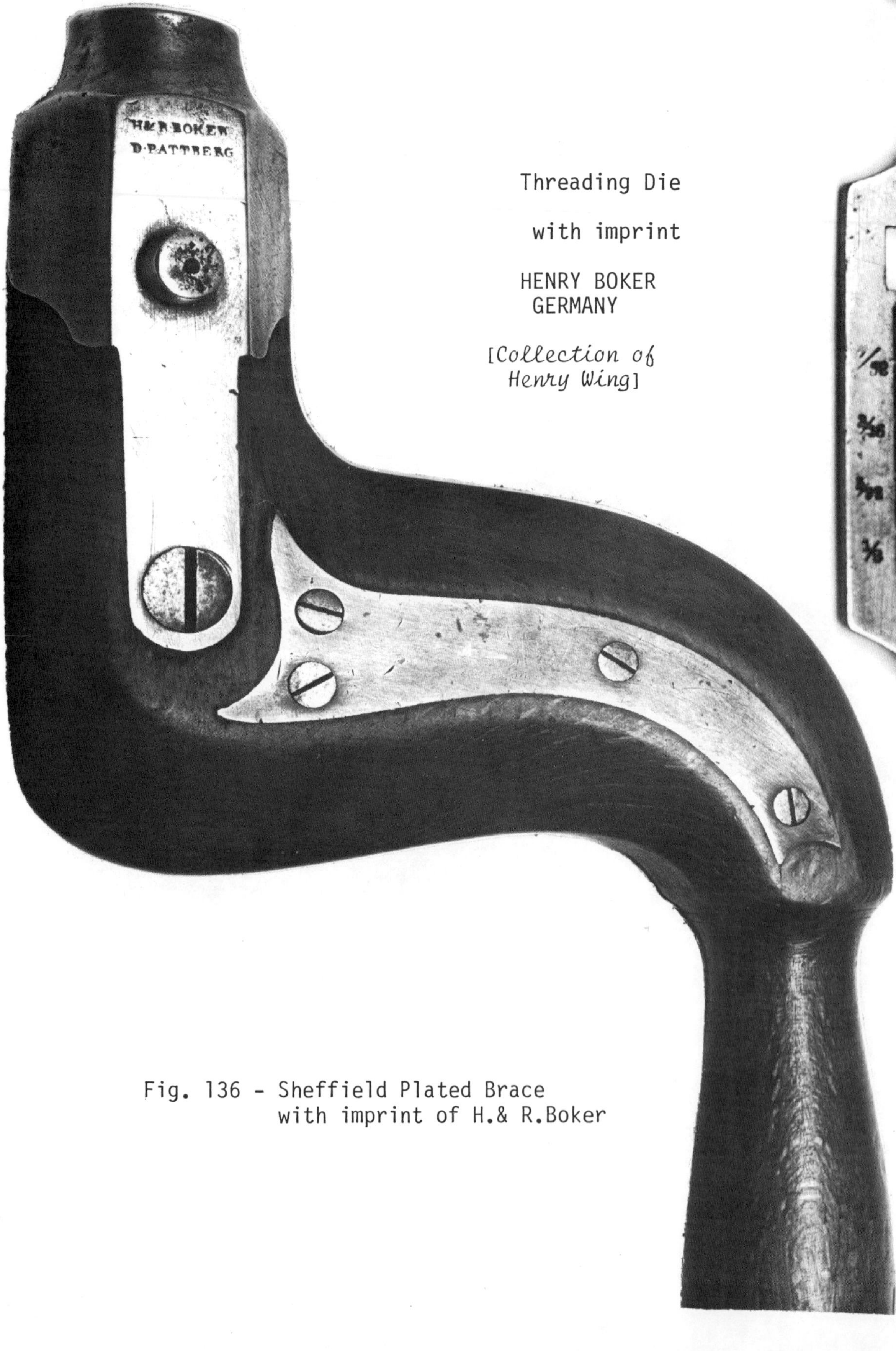

Fig. 136 - Sheffield Plated Brace
with imprint of H.& R.Boker

Fig. 137 - H.& R. Boker Imprint
Sheffield Brace with Rotating Handle

Fig. 138 - Barton Brothers Plated Brace

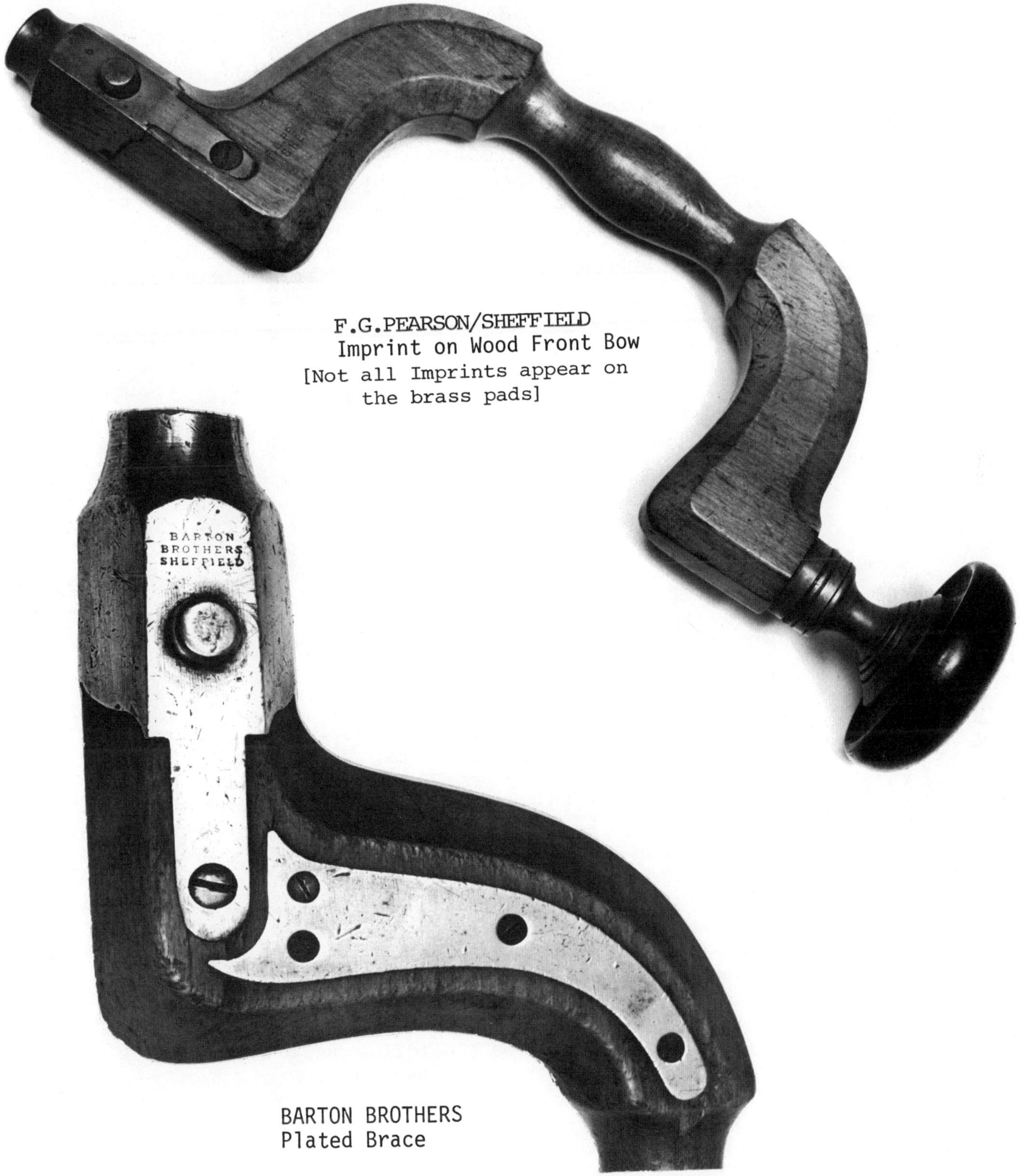

Fig. 139 - Barton Brothers' Brace (bottom); F.G.Pearson Brace (top)

In order to meet the trade competition of the brass framed brace, Henry Brown thought up a new advertising 'gimmick' for promoting his old Patent Anti-Friction Brace. His 1864 advertisement in the Sheffield Directory claimed their firm to be *inventors of the improved REGISTERED MECHANICS' BRACE*. (Plate CCXI) This was a new name for his old "Anti-Friction Patent Brace". At that date the Bottom Patent, under which this design had been protected, had expired five years previously. Nothing new had been added! A brace of this style and imprint of Henry Brown & Sons and *"The MECHANIC'S PATENT BRACE"* is illustrated in Figs. 141&142. Apparently he also convinced D.Flather & Sons that this was a marketable product and made such for their imprint and resale This had the same brass head plate with an eagle design noting that Henry Brown, Sheffield was the patentee of such design. D.Flather & Sons soon did one better on Henry Brown & Sons! They also marketed a fancy brace with trefoil plates with a side plate imprinted *"AMERICAN MECHANIC'S PATENT BRACE"*. (Fig.143) This had a head plate with the seal of the lion and unicorn and D.Flather & Sons imprint.

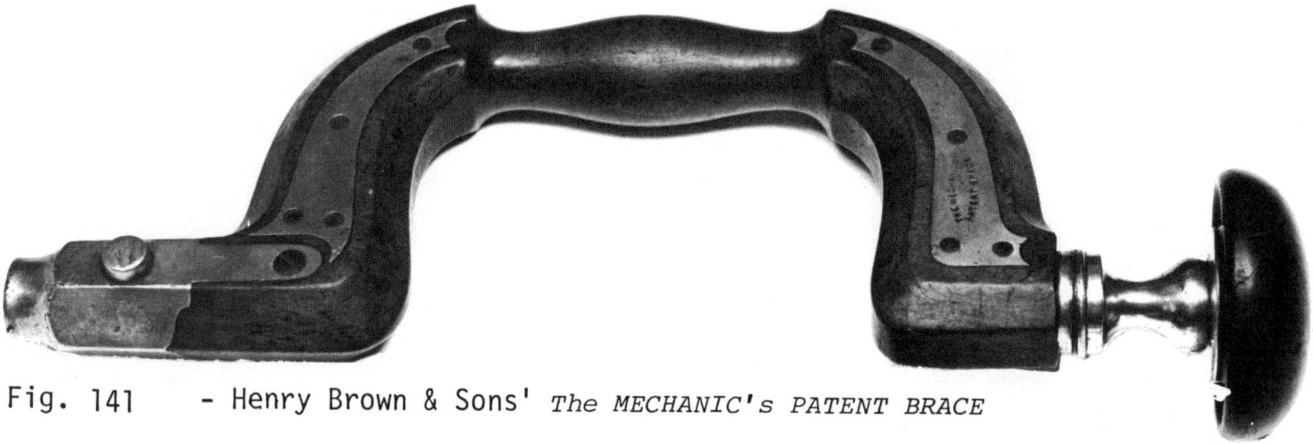

Fig. 141 - Henry Brown & Sons' *The MECHANIC's PATENT BRACE*

The 1864 advertisement of Henry Brown & Sons also called to attention that their firm offered tools manufactured by T.E.Wells & Co. and James Bee. The former concern was at Philadelphia, USA, 1850-1854 and the latter at Sheffield 1814-1828 and was the firm from whom Brown & Flather had purchased their business in 1828. Two plated braces with imprint of James Bee, probably made by Henry Brown & Sons during the period 1855-1870 are illustrated in Fig. 144 . Both have brass emblem plates denoting the imprint of Henry Brown & Sons. It is suggested that firm desired another trade imprint to offer in the American market, and undertook the imprint of James Bee, to which they were entitled, having purchased this right several years previously.

David Flather & Sons noted in their advertisements: (Plate CCXII)

> *"We are the Sole Proprietors of the following Marks: -*
> *DAVID FLATHER & SONS, D.FLATHER & SONS, F.Walter & Co.*
> *H.HAWKE, and P.HAWKE & Co."*

The significance of the F.Walter and Hawke marks are not known, but these may have been brace firms acquired by D.Flather & Sons. Braces with the imprints of H.Hawke and F.Walter are shown in Figs. 146 & 147 These were obviously made by D.Flather & Sons, as the imprint Solly Works appeared in the head emblems. In spite of an apparent excellent relationship with Henry Brown & Sons, D.Flather & Sons could not be detered from offering the brass framed brace. Two different styles with their imprint have been noted. (Fig. 119)

HENRY BROWN & SONS,

WESTERN WORKS,

108, ROCKINGHAM STREET, SHEFFIELD.

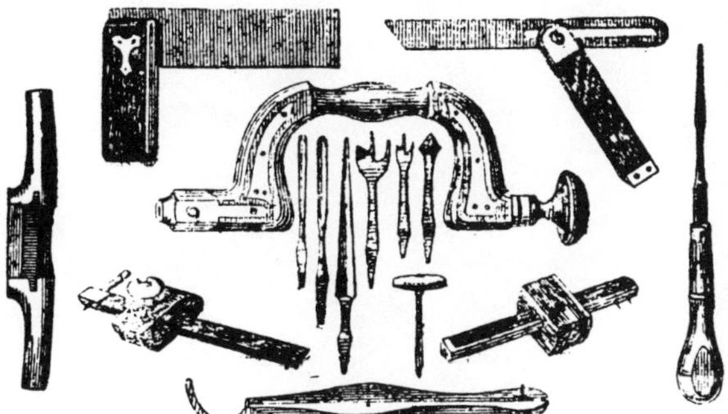

MANUFACTURERS OF

EDGE & JOINERS' TOOLS,

SKATES, BRACES,

CAST STEEL BITS,

Tool Chests

AND SCREW BOXES,

Augers, Gimblets,

BOBBIN

AND OTHER BORING BITS

Made to Pattern;

AND SOLE MANUFACTURERS OF

HY. BROWN'S PATENT

ANTI-FRICTION BRACE,

AND

T. E. WELLS & Co.'s AND

JAMES BEE'S

BRACES, BITS, AND

JOINERS' TOOLS.

INVENTORS OF THE IMPROVED

REGISTERED MECHANICS' BRACES,

IMPORTERS OF AMERICAN AUGER BITS,

Hardware, and Arkansas and Washita Oilstones.

Sole Agents for Jennings' American Augers.

Plate CCXI - 1864 Sheffield Directory Advertisement of Henry Brown & Sons

ESTABLISHED 1817.

DAVID FLATHER & SONS,

MANUFACTURERS OF

JOINERS' TOOLS,

BRACES, BITS, EDGE TOOLS, FILES, SAWS, SKATES, AUGERS,
GIMLETS, TOOL CHESTS, GARDEN SHEARS, &c., &c.

Also, David Flather's Improved Registered Plated Braces,

AND GENERAL MERCHANTS.

AMERICAN PATTERN GRIP AND RATCHET BRACES. EVERY DESCRIPTION OF SCREW AND MACHINE BITS.

We are the Sole Proprietors of the following Marks:—"DAVID FLATHER & SONS," "D. FLATHER & SONS," "F. WALTER & CO.," "FLATHER BROTHERS," "H. HAWKE," and "P. HAWKE & CO."

INDENTS PROMPTLY EXECUTED.

Solly Works, Solly Street, SHEFFIELD.

Plate CCXII - 1893 Advertisement of D. Flather & Sons - Sheffield Directory

Fig. 142 - Henry Brown & Sons "The MECHANICS' PATENT BRACE"

Fig. 143 - D.Flather & Sons "THE AMERICAN MECHANICS' PATENT BRACE"

Fig. 144 - Brace by Henry Brown & Sons. Imprint: "JAMES BEE/Sheffield"

Fig. 145 - Brace by Robert Marples & Son. Imprint" "DIXON/Hermitage Works/Sheffield"

Fig. 146 - Brace by D.Flather & Sons. Imprint: "H.HAWKE/Solly Works/Sheffield"

Fig. 147 - Brace by D.Flather & Sons.
Imprint: "F.Walter/SOLLY WORKS/Sheffield"

The policy of using several different trade marks seemed to proliferate. Possibly it was considered to be beneficial as to prestigious advertising. Robert Marples called to attentionin their advertisement in the 1878 Birmingham Directory use of three different trade marks: - *ROBERT MARPLES, HENRY DIXON, HENRY GREAVES*. These were previously shown in an 1888 advertisement. (Plate CCVI). In their 1904 Trade Catalogue different qualities were assigned to these marks: Robert Marples, Special Quality; H.D.Marples, 1st Quality; Henry Dixon, 2nd Quality; Robert Henry, 3rd Quality; and Henry Greaves, Common Quality. (See Plate CCXIII).Henry Dixon Marples was the son of Robert, but the identity of Henry Greaves is not known. A brace with the imprint *DIXON/ Hermitage Works/ Sheffield*, is shown in Fig. 145.

William Marples & Sons noted seven trade marks in their 1928 Centenary Catalogue: W.Marples & Sons, 1st Quality; John Moseley & Son, planes; W.Kent, 2nd Quality; J.Dixon, Third Quality; E.Henry, Third Quality; and J.Robertson, Third Quality. (See Plate CCXIV) The significance of these are not known. J.Mosley was the plane business acquired in 1890. William Kent Marples and Edwin Henry Marples were sons of William Marples the founder of the firm.

The 1864 advertisement of Henry Brown & Sons called to attention their firm was "Sole Agents for Jennings' American Augers". The introduction of this double twist auger bit with a spur on each cutting lip revolutionized the art of wood boring. This was a patent issued to Russel Jennings of Deep River, Connecticut. (US Patent #12,318, Jan. 30, 1855) Initially such bits were about 8 inches long and made in increments of 1/16" from 1/4" to 1" and in 1/8" increments from 1" to 1½" diameters. These permitted deeper, cleaner and larger holes to be drilled than possible with English centre bits, but created demands for stronger braces with improved holding devices. Numerous patents incorporating such features soon followed in the United States. A forged iron brace with a split clamp chuck, tightened by a wing screw, patented by Spofford became very popular in the early 1860's. This was made by Fray Co. at Bridgeport, Connecticut. However this was soon surpassed by a design patented by Barber and made by the Millers Falls Manufacturing Co., near Greenfield, Massachusetts. Releasing,changing and securing the bits was readily accomplished by hand turning a threaded tubular sleeve, fitted over the jaws. A further improvement was made early in the 1870's incorporating a ratchet device which permitted intermittant action. The superiority of the American brace had been achieved. By 1880 large numbers were being imported to Sheffield tool dealers.

Another auger bit that influenced the necessity for iron braces was developed by Ransom Cook of Saratoga Springs, N.Y. (US Patent #8162, June 17, 1851) He stated in his specification: - "The nature of my improvement consists of giving the lips or cutting edge of boring implements a curved or gouge shape at their extremities". This was extensively manufactured in the form of brace bits by James Swan Co., successor to Douglas Manufacturing Co. in 1856 at Derby, Connecticut. A pattern very similar to this was developed in England, known as Gedge's Pattern, which was manufactured at William Gilpin, Snr. & Co., Wedge's Mills, Cannock, Staffordshire, England, c1868.

Gedge' and Jenning's Patterns and other auger bits are illustrated in Plate CCXV from Ward & Payne Catalogue, c1911. American Pattern Brace Bits are seen in Plate CCXVI from an 1883 Wm Marples' Catalogue.

THE FOLLOWING ARE OUR TRADE AND CORPORATE MARKS.

Special Quality.

ROBERT MARPLES
Hermitage Works,
SHEFFIELD.

AND

REGISTERED TRADE MARK
HERMIT

Used on Goods of Special Qualities only, together with Corporate Mark.

1st Quality.

H. D. MARPLES,
Hermitage Works,
SHEFFIELD.

AND

REGISTERED TRADE MARK
HERMIT

Used on Goods of First Quality, together with Corporate Mark.

2nd Quality.

HENRY DIXON,
Hermitage Works,
SHEFFIELD.

AND

REGISTERED TRADE MARK
HERMIT

Used on Goods of Second Quality only, together with Corporate Mark.

3rd Quality.

ROBERT HENRY,
Hermitage Works,
SHEFFIFLD.

Common Quality.

HENRY GREAVES,
Hermitage Works,
SHEFFIELD.

"HERMIT, SHEFFIELD." *ESTABLISHED 1828.*

ROBERT MARPLES & SON,

65, Hermitage Street, SHEFFIELD, England.

Plate CCXIII - Robert Marples & Son Trade and Corporate Marks
[1904 Robert Marples & Son Trade Catalogue]

All Tools Manufactured by William Marples & Sons, Ltd., are produced under the best conditions and under the direct supervision of the Directors, and are branded with the Company's name and Trade Mark.

REGISTERED TRADE MARKS.

1st. QUALITY.	2nd. QUALITY.	3rd. QUALITY.
W. Marples & Sons.	W. Kent.	E. Henry & Co.
John Moseley & Son.	J. Dixon.	J. Robertson.
"Hibernia"		

Established in Sheffield 1828.

All Tools in this Catalogue are of British Manufacture.

Plate CCXIV - William Marples & Sons Registered Trade Marks
[1928 Centenary Trade Catalogue of William Marples & Sons]

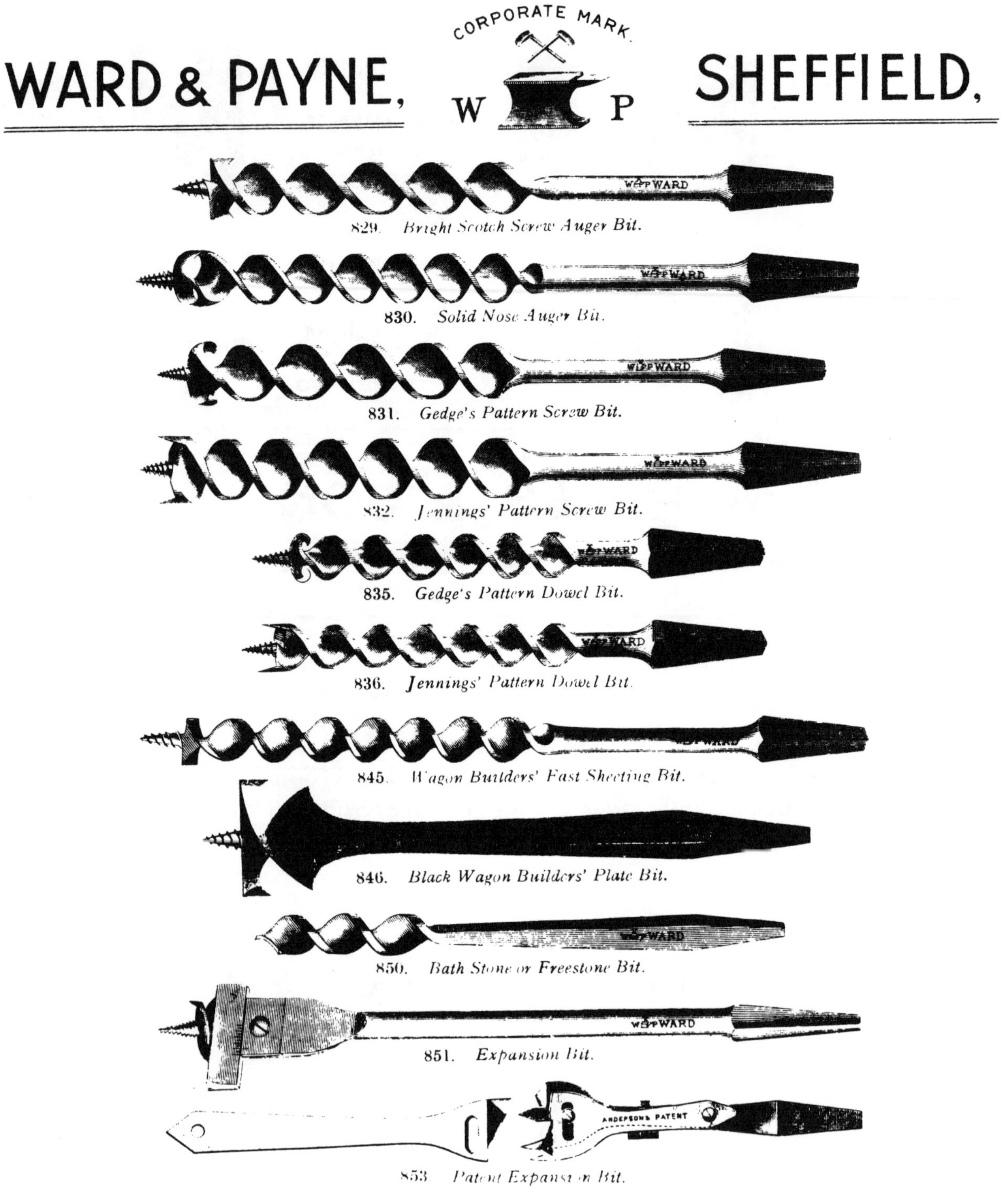
WARD & PAYNE, CORPORATE MARK W P SHEFFIELD,

829. *Bright Scotch Screw Auger Bit.*

830. *Solid Nose Auger Bit.*

831. *Gedge's Pattern Screw Bit.*

832. *Jennings' Pattern Screw Bit.*

835. *Gedge's Pattern Dowel Bit.*

836. *Jennings' Pattern Dowel Bit.*

845. *Wagon Builders' Fast Sheeting Bit.*

846. *Black Wagon Builders' Plate Bit.*

850. *Bath Stone or Freestone Bit.*

851. *Expansion Bit.*

853. *Patent Expansion Bit.*

Plate CCXV - Various Auger Bits, Ward & Payne Trade Catalogue, c1911

PATENT BORING BITS.

DOUGLAS SCREW AUGER BITS.

DOUGLAS BRIGHT SPUR SCREW AUGER BITS—

3/16	1/4	5/16	3/8	7/16	1/2	9/16	5/8	11/16	3/4	13/16	7/8	15/16	1	1 1/8	1 1/4	1 3/8	1 1/2 in.
18/-	16/-	16/-	17/-	17/-	18/-	21/-	24/-	26/-	29/-	31/-	34/-	38/-	42/-	52/-	58/-	63/-	73/- doz.

DOUGLAS BRIGHT SPUR DOWEL BITS—

3/16	1/4	5/16	3/8	7/16	1/2	9/16	5/8	3/4 in.
18/-	16/-	16/-	17/-	17/-	18/-	21/-	24/-	29/- doz.

JENNINGS' CELEBRATED SCREW AUGER BITS.

SECTION OF CUTTER.

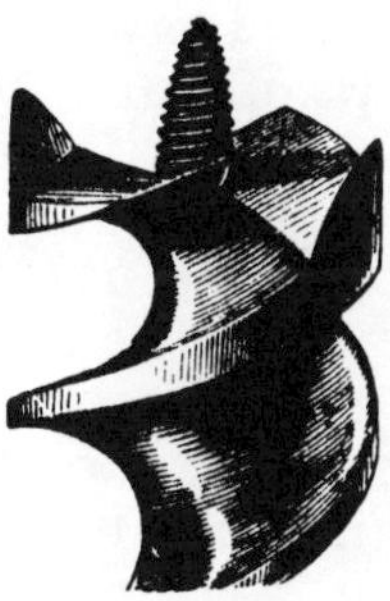

JENNINGS' PATENT BIT.

JENNINGS' CELEBRATED DOUBLE-SPUR SCREW AUGER BITS—

1/4	5/16	3/8	7/16	1/2	9/16	5/8	11/16	3/4	13/16	7/8	15/16	1	1 1/8	1 1/4	1 3/8	1 1/2 inch.
20/-	23/-	25/-	29/-	31/-	34/-	36/-	40/-	44/-	48/-	52/-	56/-	60/-	70/-	80/-	90/-	100/- per dozen.

JENNINGS' DOWEL OR SHORT AUGER BITS ......

1/4	5/16	3/8	7/16	1/2	9/16	5/8 inch.
20/-	23/-	25/-	29/-	31/-	34/-	36/- per dozen.

JENNINGS' CARRIAGE OR LONG AUGER BITS, 18-in. long ....

3/8	7/16	1/2 inch.
42/-	48/-	52/- per dozen.

N. B.—We are Sole Agents for Great Britain for Jennings' Bits. There is no Auger Bit equal to it, being very fast cutting and boring a clean hole without splintering the wood. Many imitations have been introduced from time to time, but the best ones have been complete failures. We recommend dealers to only sell the genuine article.

CLARK'S PATENT EXPANSIVE BITS.

CLARK'S PATENT EXPANSIVE BITS, small size, boring 1/2 to 1 1/2-in. holes ... ... 11/- each.
Do. do. do. do. large do. do. 7/8 to 3-in. do. ... ... 15/- do.

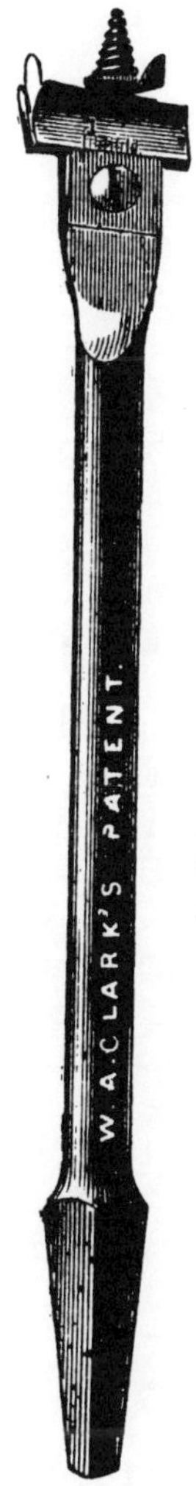

CUTTERS FOR ABOVE.

No. 1—Small size, for small Bits ... ... 2/- each. | No. 3—Small size, for large Bits ... ... 3/4 each.
„ 2—Large size, for small Bits ... ... 2/6 „ | „ 4—Large size, for large Bits ... ... 3/10 „

The sale of these expansive Bits increases annually with us. They are easily adjusted, and bore fast and clean.

Plate CCXVI - American Patent Boring Bits [p.42] *W.. Marples & Sons Catalogue of American Tools, Hardware and Novelties,1883* [Collection of Donald Wing]

The need for an iron brace in England had long been realized. While blacksmith made iron braces (See Fig. 148) had been available in the 18th Century, the first commercially manufactured iron brace made on a large scale was the "Scotch Iron Brace". (See Item #702, Marsh Brothers 1849 Catalogue, p. 318) As noted previously, these were available at Sheffield in 1830 by Hanny & Kinder. These sold at the same price as the Improved Plated Brace, 13/, and with brass necks at 16/. The forged bit socket was furnished with a thumb lever catch. (See Fig. 149) A larger iron brace, generally with a 14" sweep, known as "Waggon Iron Brace" appeared on the 1870 and 1876 Sheffield List. (See Fig. 150) These were furnished with a wing screw for locking the bit in the socket. While these iron braces were certainly more rugged than the wooden, brass plated or brass framed Sheffield braces, the pad, being of relatively soft wrought iron, was subject to deformation with difficulty in maintaining tight fits with interchangeable bits.

William Tucker of Blackstone, Massachusetts was granted English Letters Patent No. 570, March 3, 1859 for a method of fixing a bit in an iron stock. The drawing and specification of this Patent are shown in Plate CCXVII. Tucker was granted U.S. Patent No. 23,326, March 22, 1859 for this same invention. An American brace of this design is illustrated in Fig. 152. It is not known whether any braces of this type were made in Great Britain.

John Orvis of West Roxbury, Massachusetts was granted provisional protection at the Patent Office in London for his invention "Improvements in the Construction of Braces" (No.2735, October 23, 1865). His claim was for placing a ring on the arm over a hinged conically aperture joint for reception of the bit. No application of this mechanism is known to have been made in Great Britain. Orvis was not issued a U.S. Patent for this. However Obed Peck of Windsor, Vermont had been granted U.S. Patent No. 40,275, October 13, 1863 for a similar idea.

Thomas Adam Mathieson, tool manufacturer of Glasgow and Edinburgh and son of the renown Alex.Mathieson, was granted Letters Patent No. 744, dated 12th March 1866, for an invention "Improvements in Holding Device for Boring Tools". This was acknowledged to be "a communication from abroad by Cyrus Avery, residing at Tunkhannock [Pennsylvania] United States of America". The patent claim was for a "loose spring key", made of steel, "fitted to slide in a slot" where the tang of the bit passes when the tool is in the socket. The drawing accompanying the patent is shown in Plate CCXVIII. This device on a Scotch iron brace is illustrated in Fig. 153.. The imprints "*Patent SPRING LOCK*" with the Crown Seal of Victoria can be noted on the enlargement of the head bow, and "*A.MATHIESON, Glasgow & Edinburgh*" with Saracen mark on the tail bow.

The Civil War between the States (1861-1865) had interrupted export sales of Sheffield tools to America. The demand had been seriously curtailed. After the War United States rapidly became independent of foreign supplies of tools. Domestic manufacture supplied the major tools. Examples were: The Bailey Iron Plane by Stanley Rule & Level Co.; Collins' axes, Disston's saws; Buck Brothers' chisels; Spofford braces by Fray and Jennings' augers. Many of these products were competitive with English tools and were imported to Sheffield tool dealers in substanial quantities.

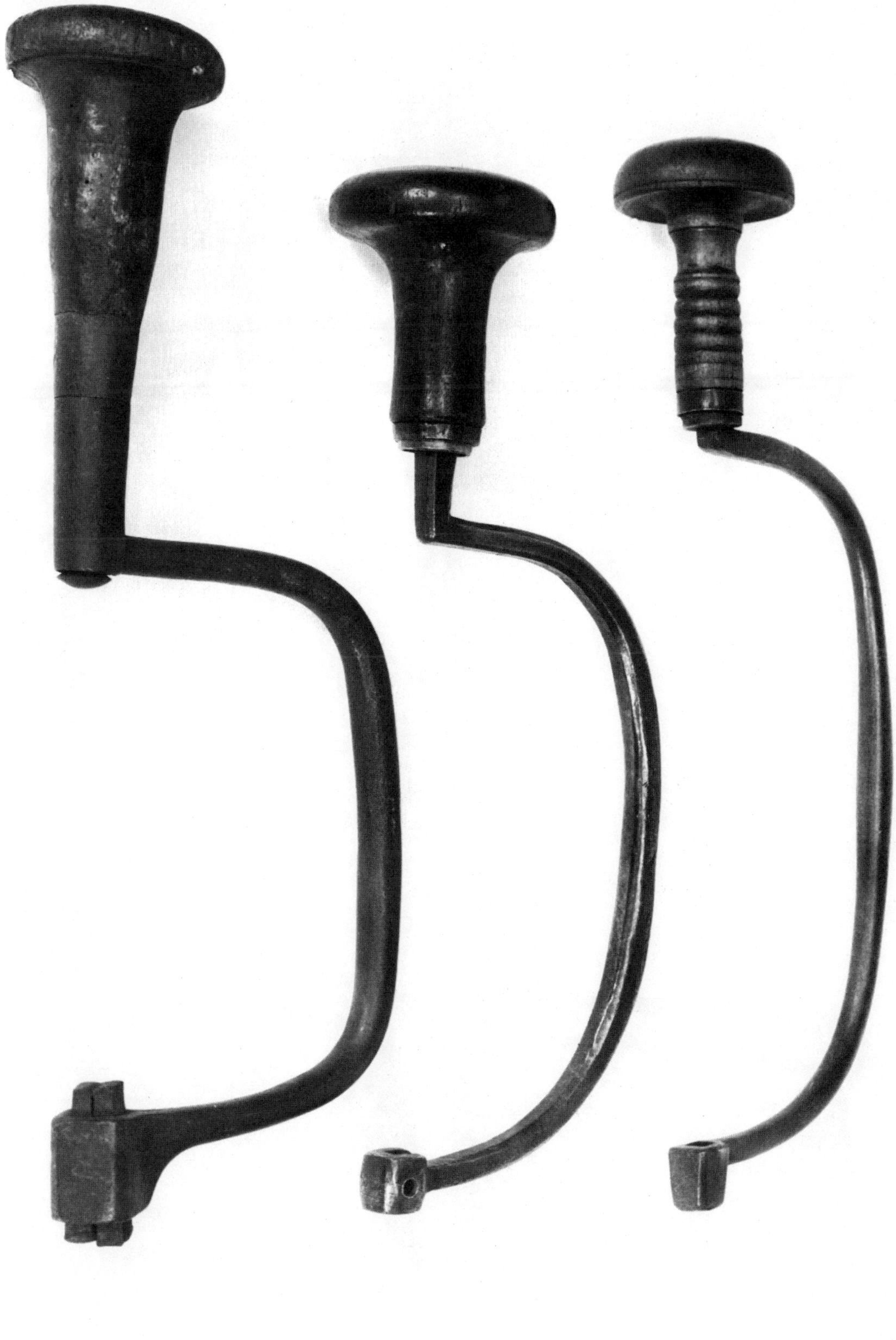

Fig. 148 - Blacksmith Forged Iron Braces

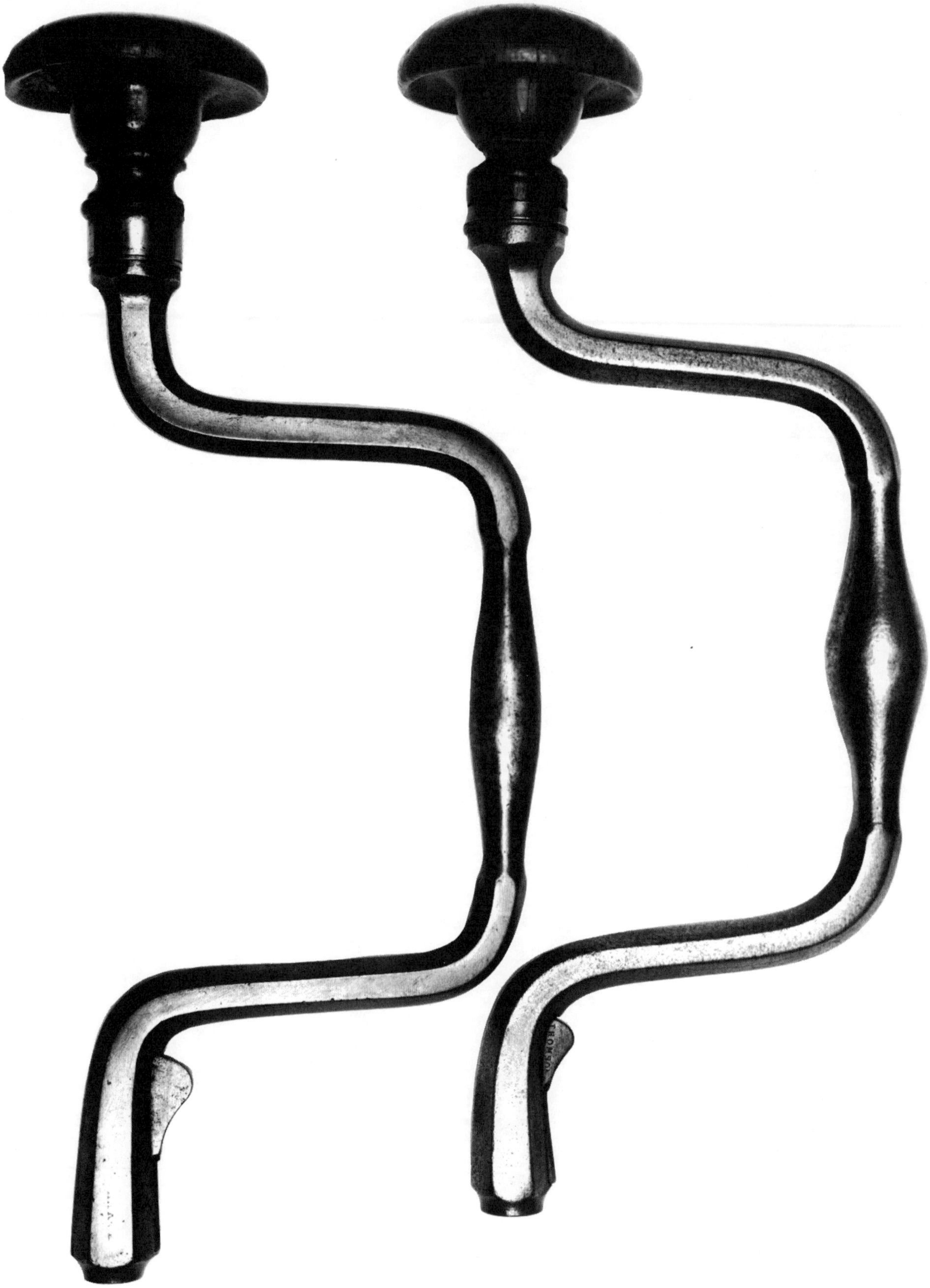

Fig. 149 - "SCOTCH" Iron Braces

Fig. 150 - Forged Iron "WAGGON" Braces

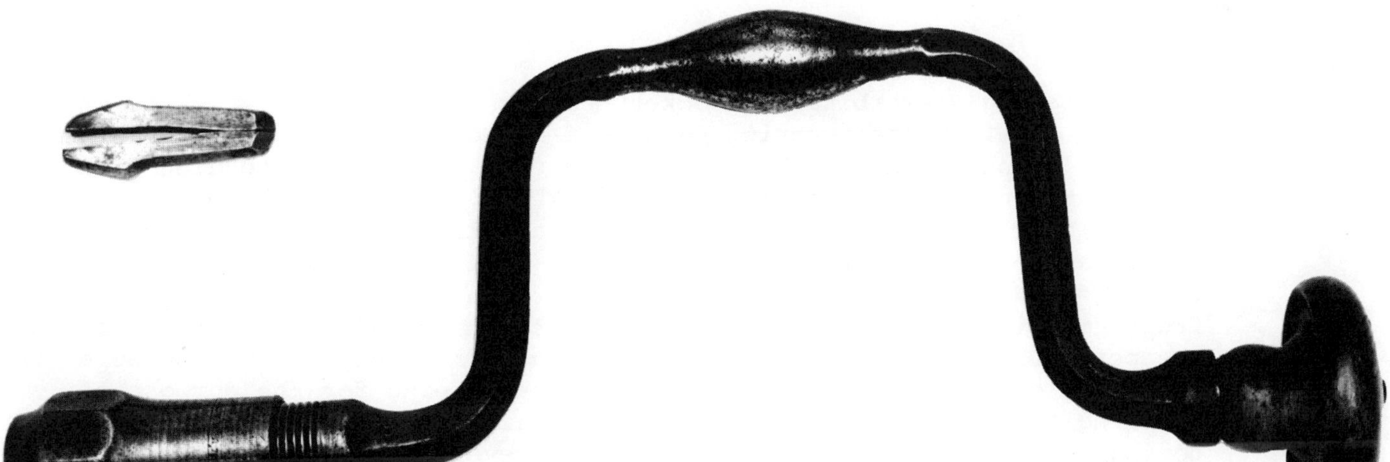

Fig. 151 - "Scotch" Iron Brace Furnished with American Type Jaw Chuck

Plate CCXVII

William Tucker's Letters Patent N°. 570

Bit Stocks or Auger Handles March 3,1859

"The nature of my invention consists in a peculiar application and arrangement of a screw and segmental head or its eqivalent with reference to the bit stock socket, and for the purpose of enabling a person to fix or hold a boring or other tool therein.

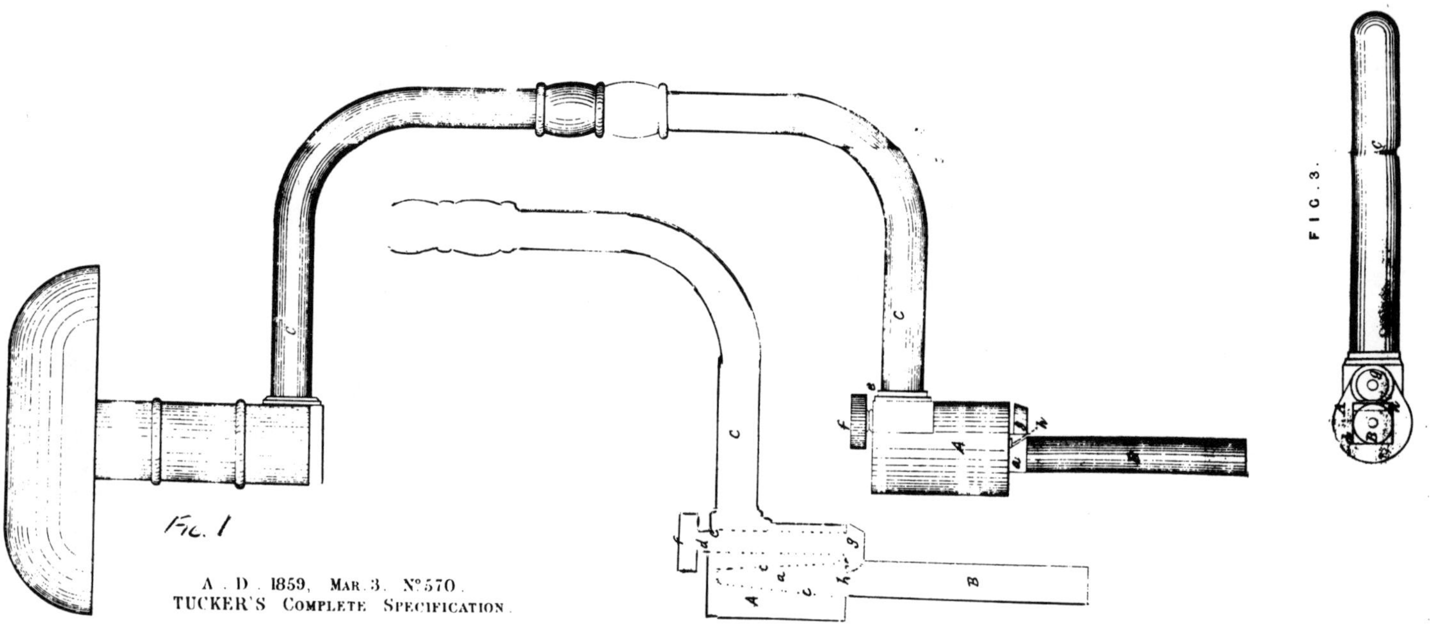

In the Drawing A exhibits the socket head or part of a bit stock C, which not only receives the tapering part a of a tool or bit B, but projects from the bell crank of the stock, as shown in the Drawings. Extending down through the head A and alongside the socket c, I arrange a female screw d, into which a male screw e, provided with a milled head f, is to be screwed, and so as to project from each end of the said head A, as shown in the Drawings. On the lower end of the male screw a button or segmental conical frustrum g is affixed, so as to rotate with the screw. The curved part of the periphery of the button projects further from the axis of the screw than the chord or straight portion of it, the same being for the purpose of enabling the curved part to enter an angular notch h made in the bit or boring tool while the male screw is being revolved. By turning the screw the button will enter the said notch and be drawn upwards, so as to act against the upper part of the notch and draw the tool or bit close up into the socket c, and there hold it until the screw is turned backward, so as to carry the button out of the notch. This simple Invention admits of a bit or boring tool being fastened in the socket of its stock or handle with greater facility and security"

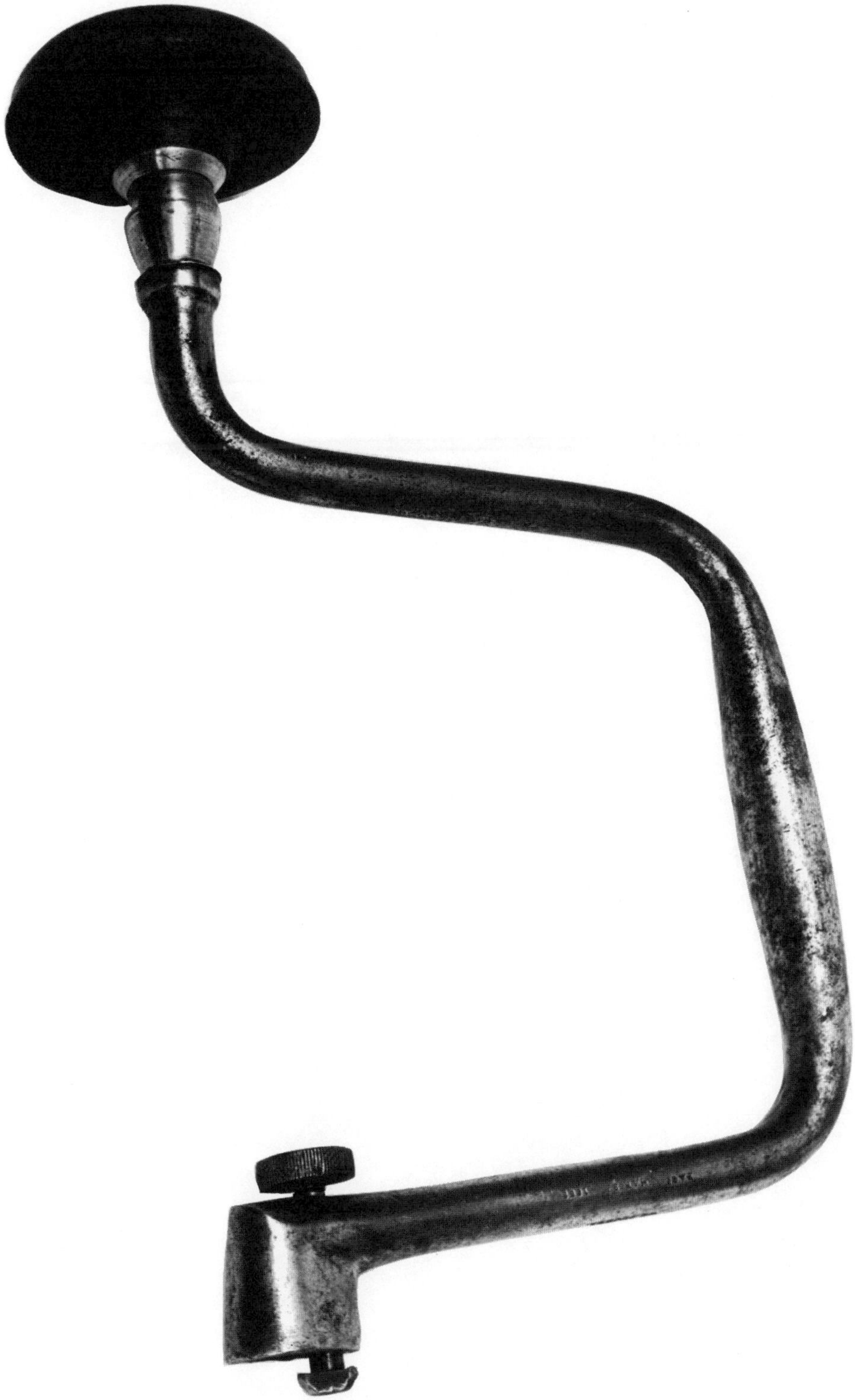

Fig. 152 - William Tucker's Patent Brace - American Manufactured

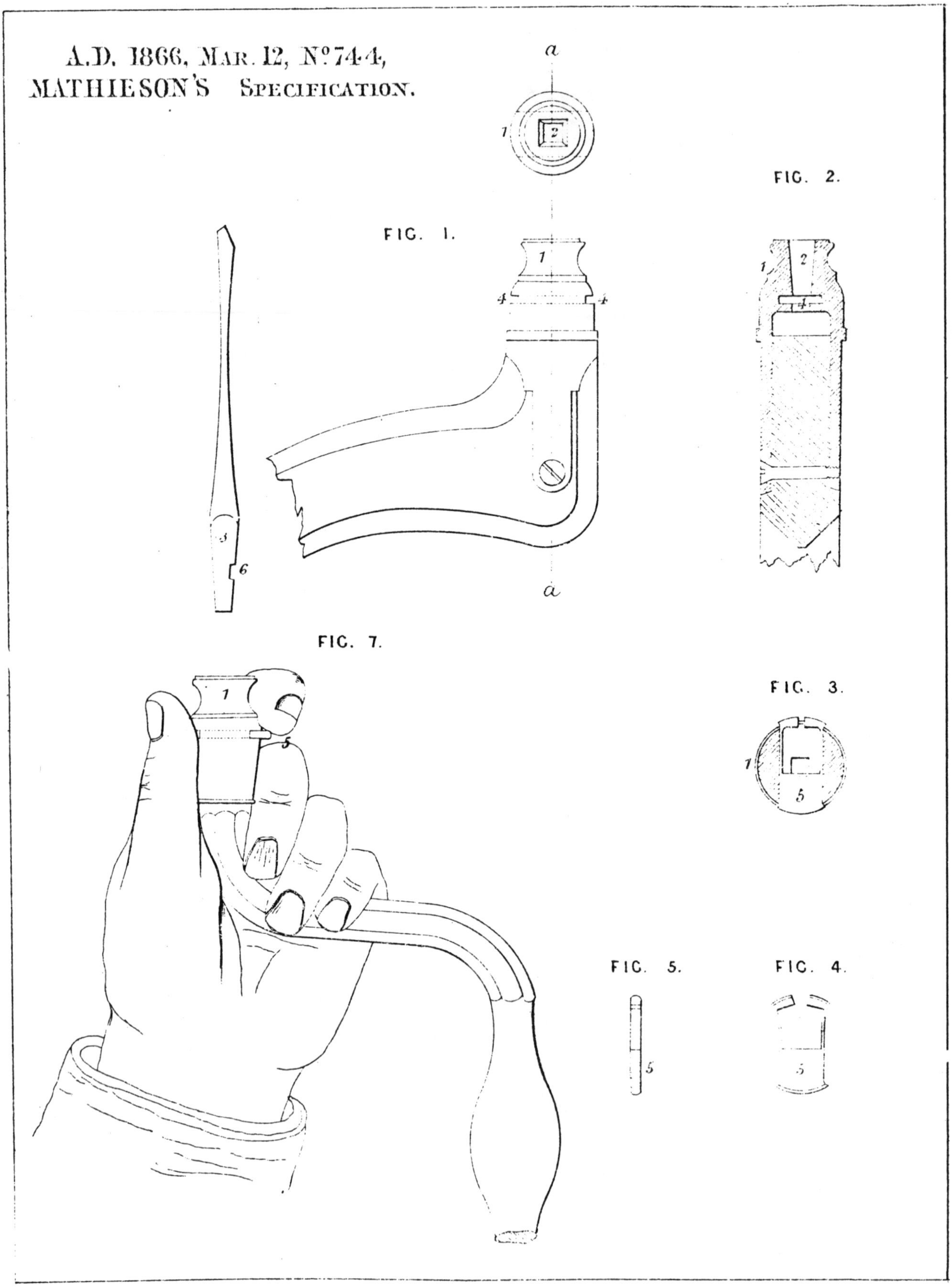

Plate CCXVIII - Thomas Mathieson's Brace Patent Drawing

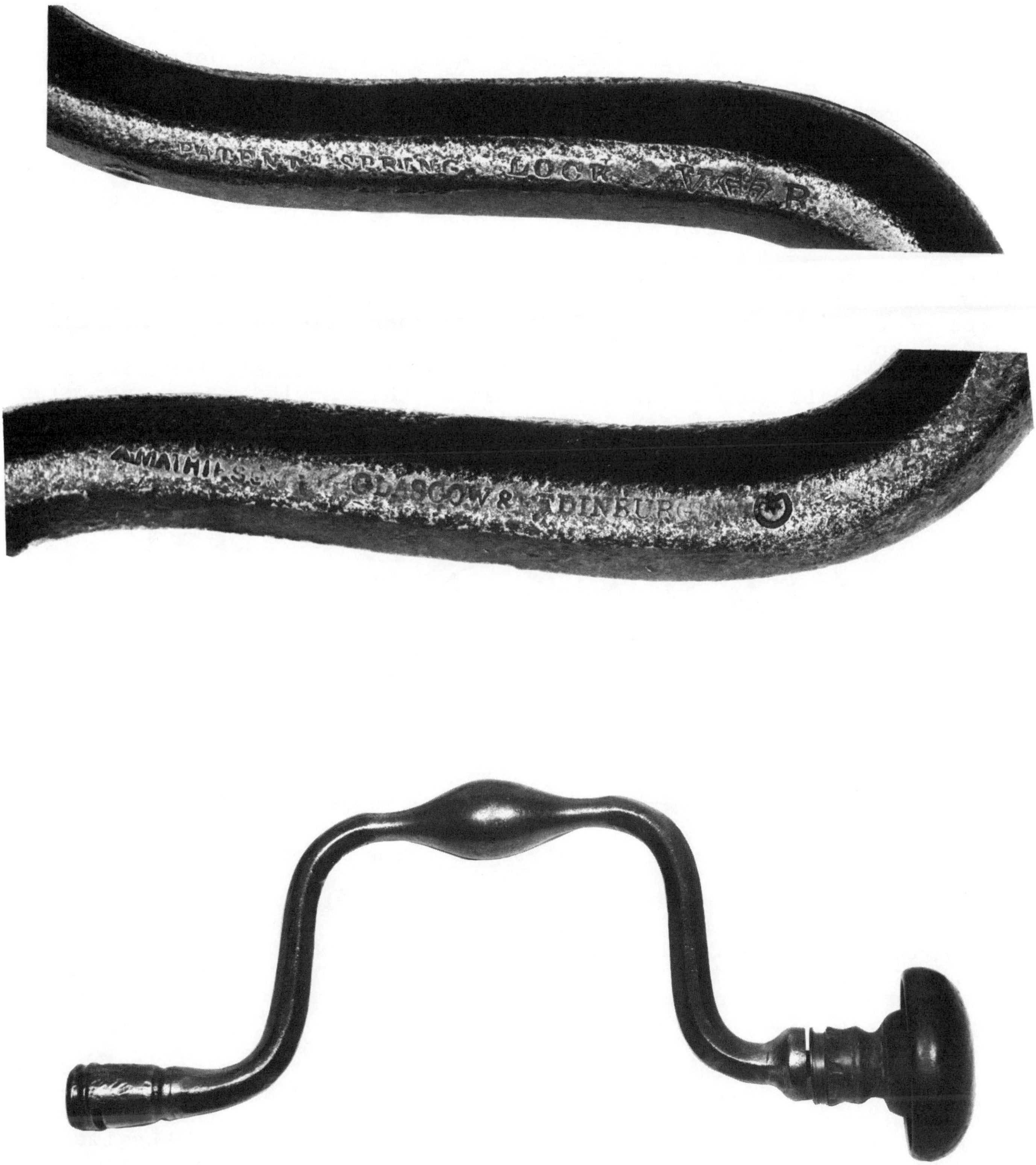

Fig. 153 - Thomas Mathieson's Patent Spring Lock Brace

William Marples & Sons soon realized the merits of American tools. Their 1881 advertisment in the Sheffield Directory called to attention a large line of foreign tools; also noted offices in New York City and Deep River, Connecticut. (See Plate CVI, p.146) On January 1, 1883 this firm issued *No.2 Price List of AMERICAN TOOLS, HARDWARE & NOVELTIES,* 144 pages. Two pages from this Catalogue are shown: Plate CCXVI, Patent Boring Bits, and Plate CCXIX, Iron Braces. A note regarding Jennings' Auger Bits stated: - "We are sole agents for Great Britain for Jennings' Bits. There is no Auger Bit equal to it, being very fast cutting and boring a clean hole without splintering the wood. Many imitations have been introduced from time to time, but the best ones have been complete failures. WE RECOMMEND DEALERS ONLY SELL THE GENUINE ARTICLE."
Apparently at this date they had taken away this agency from their old rivals, Henry Brown & Sons.

The same 1883 Marples' Catalogue offered a series of "The Cheapest Set of Brace" with 6, 12, 18 or 24 bits. (Plate CCXX). This type of brace was manufactured in large quantities in Germany. This became known as "Common Ball" or"Thumb Screw" brace. Apparently these were also made in Sheffield, noted on the 1876 Illustrated List as items No. 3150 & 3151 (Plate CCXXI). Three of this style braces are illustrated in Fig. 154. Early in the 20th Century this "cheap" brace became known as "Six Penny Brace" due to its low selling price.[30]

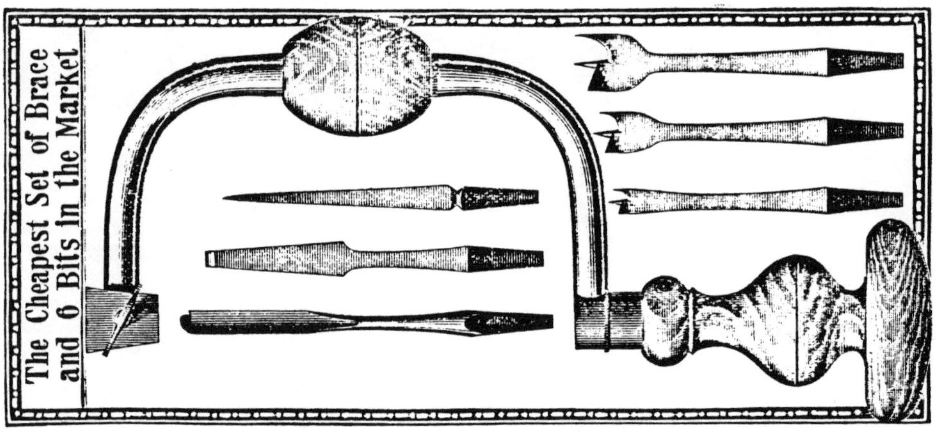

Plate CCXX - Cheapest Brace, *1883 Wm. Marples & Sons Catlogue of American Tools*

Two additional American braces, Nickel Plated (probably made by Millers Fals Manufacturing Co.,Massachusetts), illustrated in this *1883 Catalogue,* are shown in Plate CCXXII. The accompanying note: *"NB - WE WOULD CALL SPECIAL ATTENTION TO THESE BRACES. - THEY ARE SUPERIOR IN QUALITY AND FINISH TO ANY IN THE MARKET"*, conclusively proves that William Marples & Sons were convinced of the American superiority in braces at this date.

Not only had the production of English styles of bit braces peaked by 1880, but rapidly declined as the 19th Century drew to a close. The American jaw chuck could not be equaled. Adaption of this was soon to follow in England. Instead of a round nurled pattern the English style adopted was octagonal. (See Fig. 151). These new patterns, along with the old styles of braces were illustrated in Jinkinson & Sons' Catalogue c1890. (Plate CCXXIV) Prices of these are shown in Table XXVII. The English styles were outmoded and overpriced. The end was near. However, these were still available at the end of the Century, as noted in Rd. Melhuish & Sons 1899 Trade Catalogue, *Tools for the 20th Century*. (PlateCCXCV).

Fig. 154 - "Cheap" Iron Braces

IRON BRACES.

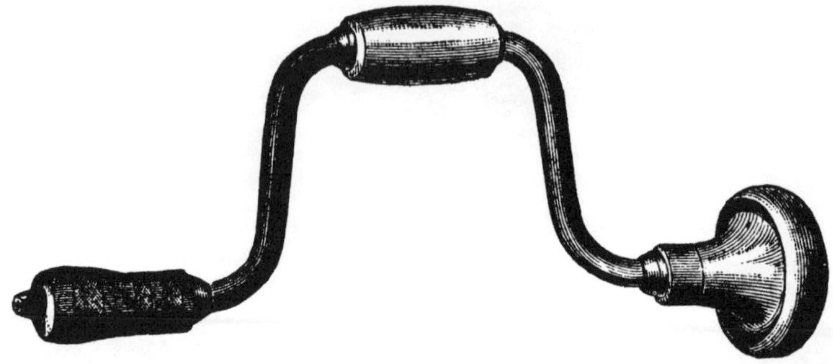

No. C—AMERICAN IMPROVED IRON BRACES, 8-in. Sweep, with Whitewood Head and Handle ... 4/- each.

„ **C**— Do. do. do. do. 10-in. do. do. do. ... 4/6 „

No. 129—COMMON RATCHET IMPROVED IRON BRACES, 10-in. Sweep, with Whitewood Head and Handle . 7/- each.

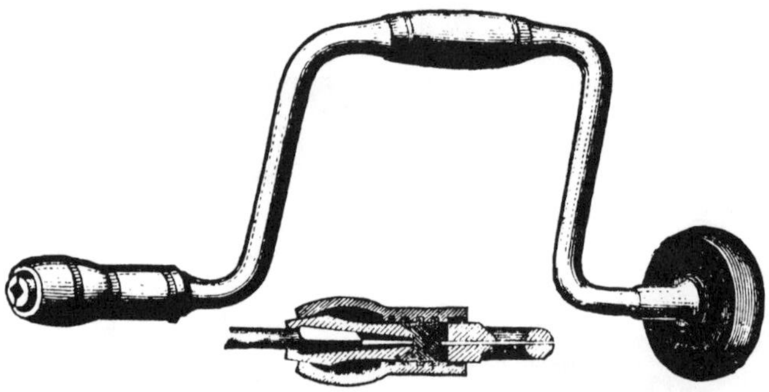

No. 100.

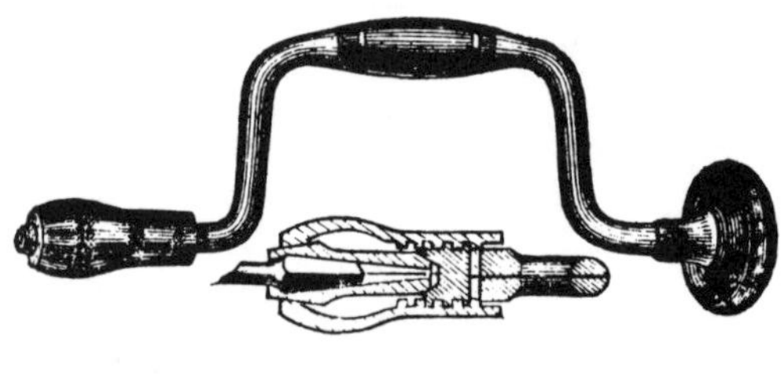

No. 80.

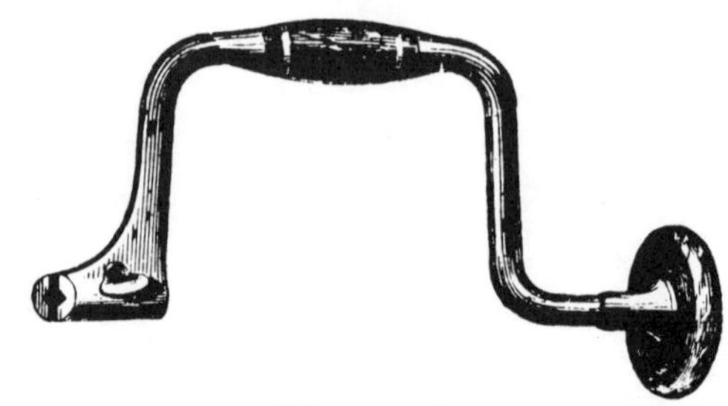

No. 108.

No. 80—SPOFFORD'S PATENT BRACE, 8-in. Sweep, with Hardwood Head and Handle, 9/- each.

„ **100**— Do. do. do. 10-in. do. do. do. do. 10/- „

„ **108**— Do. do. do. 8-in. do. do. do. do. 9/- „

„ **110**— Do. do. do. 10-in. do. do. do. do. 10/- „

Plate CCXIX - American Iron Braces [p.43] *1883 Wm. MARPLES & SONS CATALOGUE OF AMERICAN TOOLS, HARDWARE and NOVELTIES* [Collection of Donald Wing]

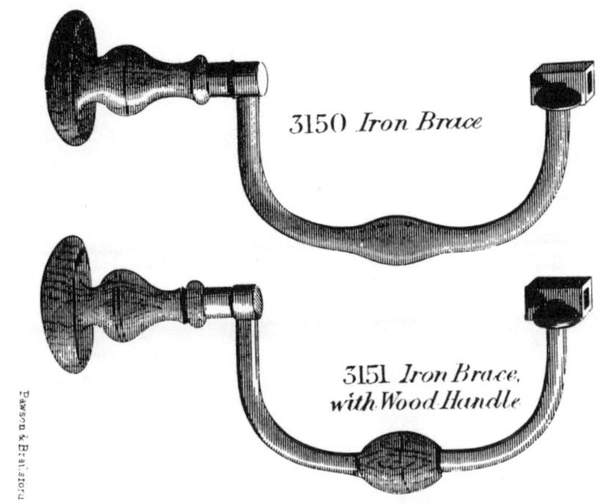

Plate CCXXI - "Cheap" Iron Braces from 1876 Illustrated Sheffield List

IRON BRACES. 45

EXTRA QUALITY IMPROVED IRON BRACES & DRILLS.

NICKEL PLATED.

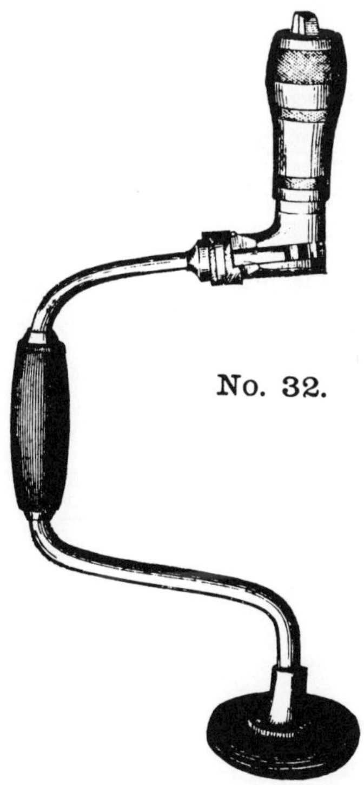

No. 32.

N.B.—We would call special attention to these Braces.—They are superior in quality and finish to any in the Market.

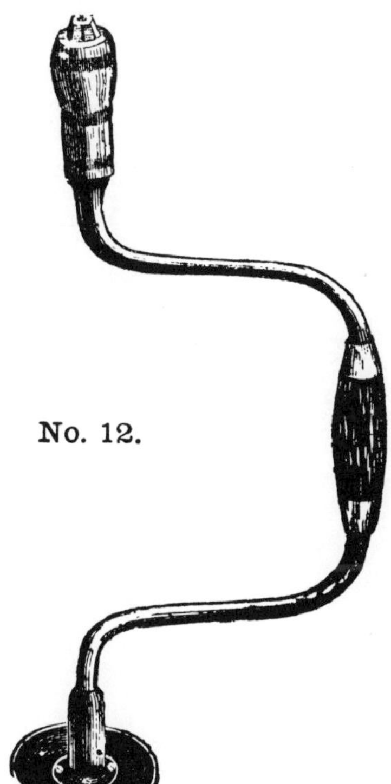

No. 12.

Plate CCXXII - Nickel Plated Iron Plated Braces from 1883 Wm. Marples & Sons Catalogue

Best Braces, with Bits Complete.

919 BEST PLAIN BRACES, LIGNUM HEAD, without Bits, 9/ each.

No.	With	12	18	24	30	36	42	48	54	60	Bits.
920	Cast Steel Black Bits, Sharpened, and Bright Insides	14/8	17/6	20/4	23/4	27/	29/6	33/	36/6	40/	each.
921	C. S. Bright Bits	15/3	18/4	21/6	24/9	28/9	31/7	35/4	39/2	42/10	„
922	Do. Straw Coloured Bits ...	15/11	19/4	22/10	26/5	30/9	33/11	38/	42/2	46/2	„

If with Improved Ebony Head and Brass Neck, 2/ each extra to Lignum Head.

If with all Ebony Head, 1/ each extra to Lignnm Head.

923 IMPROVED PLATED BRACES, LIGNUM HEAD, without Bits, 13/ each.

No.	With	12	18	24	30	36	42	48	54	60	Bits.
924	C. S. Black Bits, Sharpened, and Bright Insides... ...	18/8	21/6	24/4	27/4	31/	33/6	37/	40/6	44/	each.
925	Do. Bright Bits	19/3	22/4	25/6	28/9	32/9	35/7	39/4	43/2	46/10	„
926	Do. Straw Coloured Bits ...	19/11	23/4	26/10	30/5	34/9	37/11	42/	46/2	50/2	„

923A IMPROVED PLATED BRACES, with Improved Brass Necks, without Bits, 15/ each.

No.	With	12	18	24	30	36	42	48	54	60	Bits.
924A	C. S. Black Bits, Sharpened, and Bright Insides... ...	20/8	23/6	26/4	29/4	33/	35/6	39/	42/6	46/	each.
925A	Do. Bright Bits...	21/3	24/4	27/6	30/9	34/9	37/7	41/4	45/2	48/10	„
926A	Do. Straw Coloured Bits ...	21/11	25/4	28/10	32/5	36/9	39/11	44/	48/2	52/2	„

927 LIGNUM HEAD IMPROVED SCOTCH IRON BRACES, without Bits, 14/ each.

No.	With	12	18	24	30	36	42	48	54	60	Bits.
928	C. S. Black Bits, Sharpened, and Bright Insides... ...	18/10	21/9	24/9	27/9	31/9	34/	37/9	41/6	45/	each.
929	Do. Bright Bits	19/5	22/7	25/11	29/2	33/6	36/1	40/1	44/2	47/10	„
930	Do. Straw Coloured	20/1	23/7	27/3	30/10	35/6	38/5	42/9	47/2	51/2	„

927A IMPROVED SCOTCH IRON BRACES, with Improved Brass Necks, without Bits, 16/ each.

No.	With	12	18	24	30	36	42	48	54	60	Bits.
928A	C. S. Black Bits, Sharpened, and Bright Insides... ...	20/10	23/9	26/9	29/9	33/9	36/	39/9	43/6	47/	each.
929A	Do. Bright Bits	21/5	24/7	27/11	31/2	35/6	38/1	42/1	46/2	49/10	„
930A	Do. Straw Coloured Bits ...	22/1	25/7	29/3	32/10	37/6	40/5	44/9	49/2	53/2	„

Boxwood or Rosewood Braces, 12/ each extra to Plated Beech Braces.
Ebony Ditto 16/ do. do. do.

930B PATENT METALLIC-FRAMED EBONY BRACE, with Brass Neck Head, 23/ each.

No.	With	12	18	24	30	36	42	48	54	60	Bits.
930C	C.S. Black Bits	28/8	31/6	34/4	37/4	41/	43/6	47/	50/6	54/	each.
930D	C.S. Bright Bits	29/3	32/4	35/6	38/9	42/6	45/7	49/4	53/2	56/10	„
930E	C.S. Straw Coloured Bits ...	30/	33/4	36/10	40/5	44/6	48/	52/	56/3	60/3	„

930F PATENT METALLIC-FRAMED BEECH BRACE, with Brass Neck Head. 20/ each.

No.	With	12	18	24	30	36	42	48	54	60	Bits.
930 G	C.S. Black Bits...	25/8	28/6	31/4	34/4	38/	40/6	44/	47/6	51/	each.
930 H	C.S. Bright Bits	26/3	29/4	32/6	35/9	39/6	42/7	46/4	50/2	53/10	„
930 I	C.S. Straw Coloured Bits ...	27/	30/4	33/10	37/5	41/6	45/	49/	53/3	57/3	„

Boxwood and Rosewood Braces, same price as Ebony.

Brass Collars for Sash Bits, 1/6 per pair extra.

If six Screw Auger Bits are included in any of the above Sets, the list price is 8/ extra.

0930 Malleable Iron Braces, Common Lignum Head ... ... 6/6 each
0931 Ditto do. Strong do. ... ... 7/ „
0932 Ditto do. Brass Neck and Black Heads ... ... 7/6 „

No.		5½	6	6½	7	8	inch sweep.
0933	Wagon Builder's Iron Brace	9/	9/	9/6	10/	11/	each.

0934 Common Ball Braces, 15/ per doz.; if with 18 Bright Bits on Cards, 66/ per doz.
0935 Ditto Iron do. 28/ „
0936 Ditto Ball do. with 18 Bright Bits on Cards, 90/ per doz.

No.		7	8	9	10	inch sweep.
0937	American Pattern Braces, Common No. 1 ...	3/				each.
0938	Ditto ditto Better Quality 2 ...		4/		4/	„
0939	Ditto ditto Best „ 3 ...		9/		9/	„
0940	Ditto ditto Best „ 4 N.P.		10/		10/	„
0941	Ditto Common Ratchet 5 ...		8/		8/	„
0942	Ditto Best „ 6 ...		11/6		11/6	„
0943	Ditto Best „ 7 N.P.		12/6		12/6	„

Table XXVII - Price of Braces, Page 73 *AMOS JINKINSON & SONS CATALOGUE,* c.1890
[Collection of Sheffield Central Library]

See Plate CCXXIV

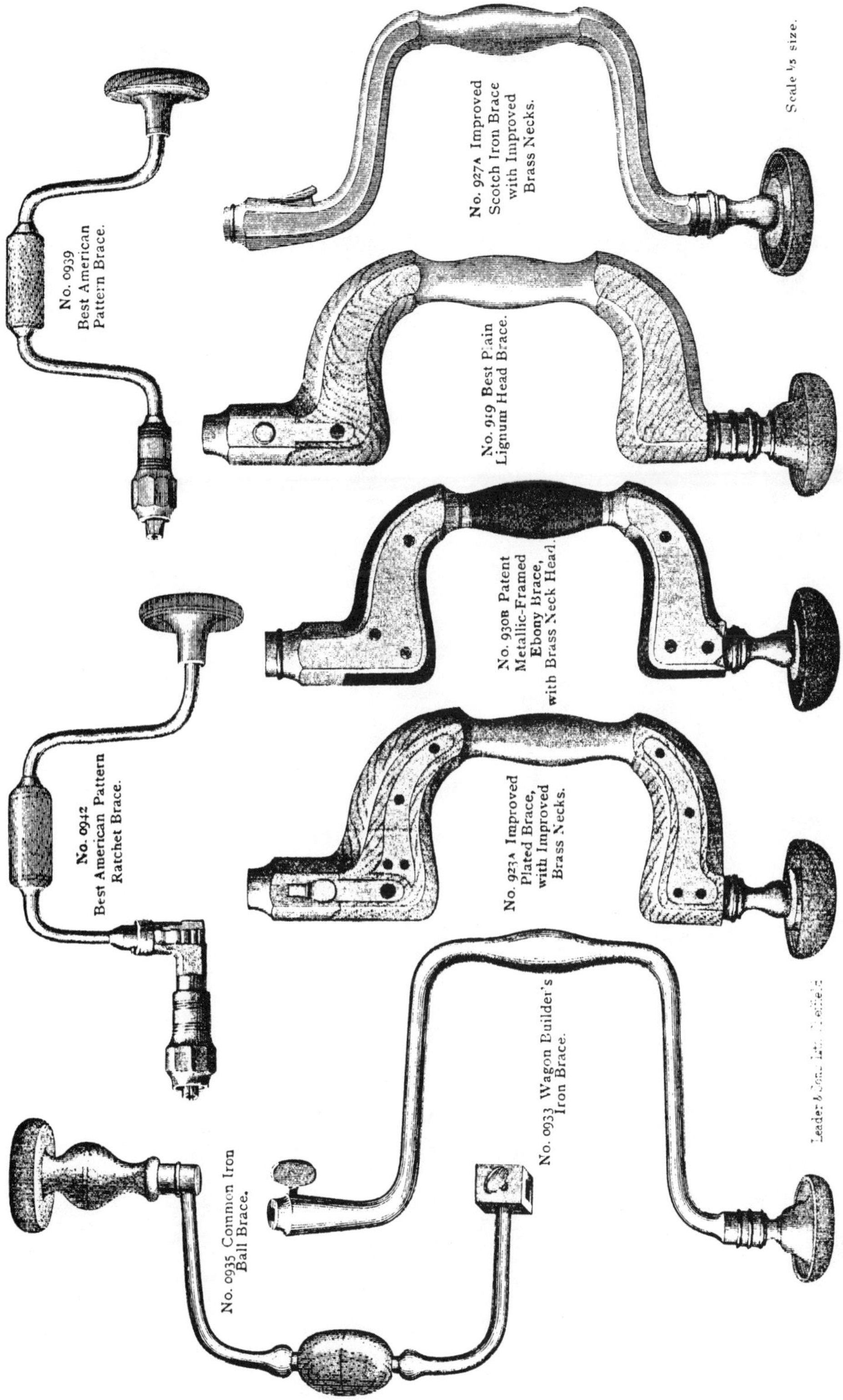

Plate CCXXIV - Braces from Page 72 *AMOS JINKINSON & SONS CATALOGUE*, c1890
[Collection of Sheffield Central Library]

The above illustration was the same used in the *D.FLATHER & SONS TRADE CATALOGUE* c.1900, which was reproduced as page 100, Fig.159, R.A.Salaman, *Dictionary of Tools used in the woodworking and allied trades,c1700-1970* [London, 1975]

Patent Metallic Framed Ebony Brace.

No. 346.

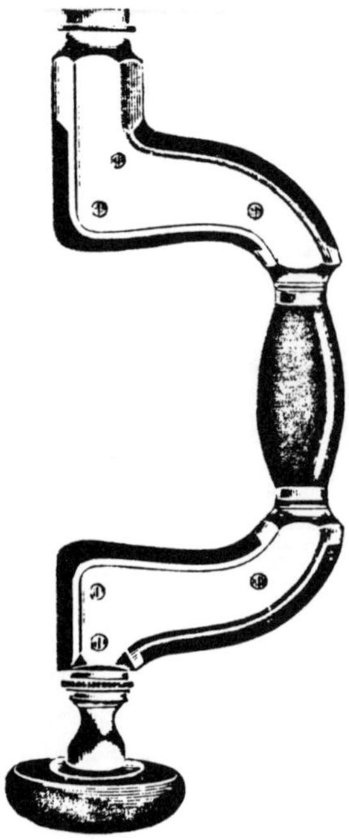

With Brass Neck Head, and 36 Straw coloured Boring Bits.

Well-assorted Shapes and Sizes.

Price ... ... **30/-** complete.

Waggon Builders' Iron Brace.

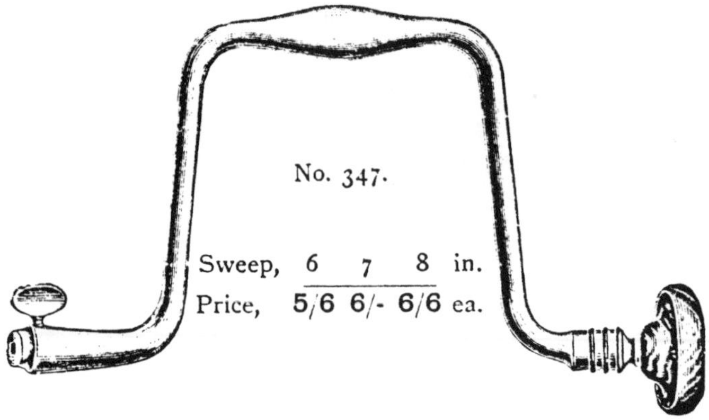

No. 347.

Sweep,	6	7	8 in.
Price,	**5/6**	**6/-**	**6/6** ea.

Improved Scotch Iron Braces.

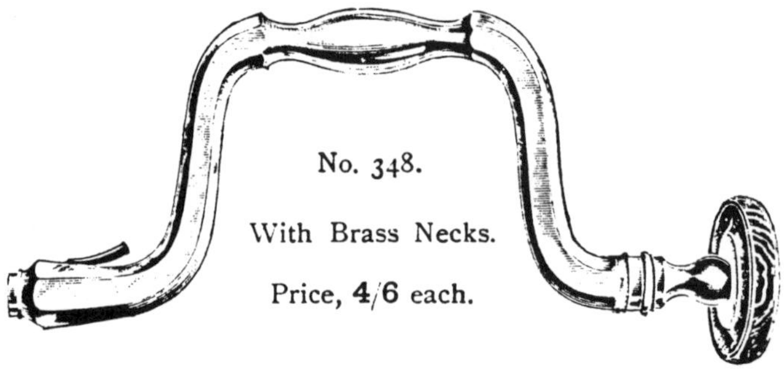

No. 348.

With Brass Necks.

Price, **4/6** each.

Best Plain Beech Brace.

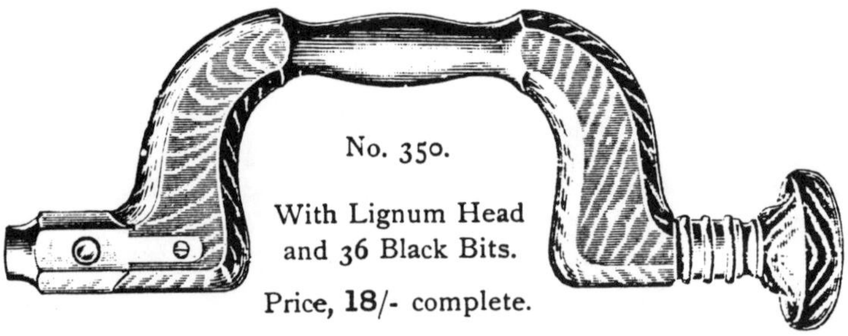

No. 350.

With Lignum Head and 36 Black Bits.

Price, **18/-** complete.

Best Plated Beech Brace.

No. 349.

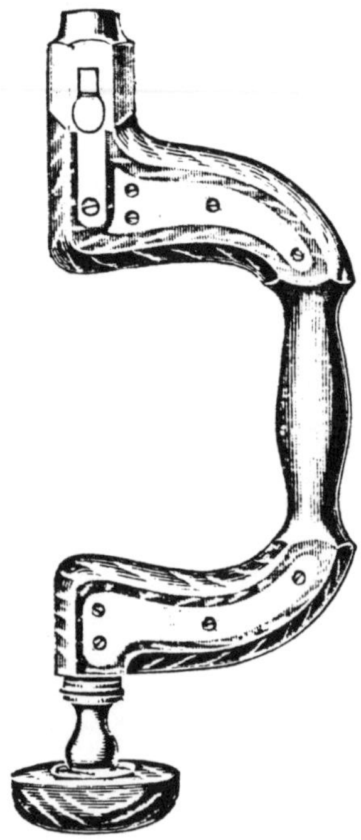

With Brass Neck, Head, and 36 Bright Bits.

Well-assorted Shapes and Sizes.

Price ... **24/-** complete.

Sets of Assorted Brace Bits.

Set of ...	24	36	Bits.
Best Quality Straw-coloured Boring Bits, Well-selected Shapes and Sizes ...	—	**14/-**	per set.
„ „ Bright Bits ...	**8/6**	**12/6**	„
„ „ Black Bits ...	**7/6**	**11/-**	„

A large assortment of Melhuish's Hand and Breast Drills are shown in our No. 10, Metal Workers' Section of Catalogue.

Plate CCXXV - English Braces from Page 418 *TOOLS for the 20th CENTURY*
Rd. Melhuish & Sons, 1899 Trade Catalogue, London [Author's Collection]

These English wooden braces of the 19th Century were indeed items of great beauty. It is virtually impossible to trace the identity of all firms who actually made such braces. A check list of Birmingham and Sheffield firms follows in the Appendix. This is by no means inclusive, nor does this in all cases indicate that a particular listed firm was a manufacturer. A few advertisements of Sheffield firms of probablle brace makers are shown in Plates CCXXVI-CCXXVIII. In conclusion a few more examples of wooden brace are shown in Fig. 155 through Fig. 162 . These significant examples did not fit into the subject matter in the chronology generally followed in the text.

Fig. 155 - Brass Plated Brace by J.T.Brufton/Sheffield
[made during last quarter of 19th Century]

WILLIAM H. FURNISS,

MANUFACTURER OF ALL KINDS OF

BRACES, BRACE BITS, ENGINE BITS, BOBBIN BITS, AND

TABLE AND BUTCHERS' STEELS,

171½, EYRE STREET, SHEFFIELD.

N.B.—REPAIRS, &c., ON PAGE 146 IS AN ERROR.

1876

WM. BURKINSHAW & SON,

BURGESS WORKS, UNION LANE, SHEFFIELD,

MANUFACTURERS OF

Joiners' Tools, Braces & Bits, Gimlets, Tool Chests, Skates,

GARDEN SHEARS, HAMMERS, PINCERS, &C., &C.

SHIP CAULKING MALLETS, IN LIGNUM VITÆ AND BOXWOOD.

General WOOD TURNERS, in Hard and Soft Woods.

1876

NATHANIEL HOLDEN,

MANUFACTURER OF

Gimlets, Braces & Bits, Augers,

BRADAWLS, &c.

MITCHELL STREET WORKS, ST. PHILIP'S ROAD,

SHEFFIELD.

1879

Plate CCXXVI - Advertisements from Sheffield Directories

TRADE MARKS.

CHARLES PARKIN,

MANUFACTURER OF

Braces and Bits, Joiners' Tools, Edge Tools.

AUGERS, SKATES, TOOL CHESTS, &c.

32, ROCKINGHAM STREET, SHEFFIELD.

1876

WILLIAM RIDGE,

MANUFACTURER OF

GIMLETS, BRACES, BITS,

AUGERS, SPOKESHAVES, SQUARES,

GAUGES, TOOL CHESTS, &c.

SHEFFIELD LANE, near SHEFFIELD,

AND CARVER LANE, SHEFFIELD.

1868

ESTABLISHED 1790.

ALFRED RIDGE & SONS,

MANUFACTURERS OF

JOINERS'. & EDGE TOOLS,

BRACES, BITS, GIMLETS, AUGERS,

GEDGE'S AND JENNINGS' BITS, &c.

EXCHANGE TOOL WORKS, ROCKINGHAM ST.,

SHEFFIELD.

1891

Plate CCXXVII - Advertisements from Sheffield Directories

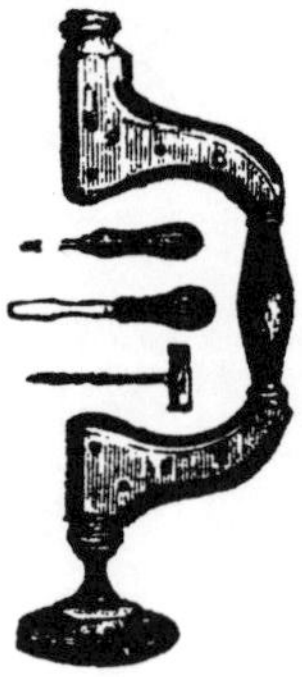

WILLIAM COTTAM,

SHOREHAM TOOL WORKS,

Shoreham Street, SHEFFIELD,

MANUFACTURER OF

Brace Bits, Gimlets, Turnscrews, Bradawls,

PATENT HAND PADS,

AND

GENERAL WOOD TURNER.

1883

ESTABLISHED 1857.

JOHN BAXBY,

MANUFACTURER OF EVERY DESCRIPTION OF

BORING TOOLS, AUGERS, BRACES & BITS.

ALL KINDS OF

MACHINE BITS,

Gedge's & Jenning's Augers and Auger Bits,

GIMLETS, TURN-SCREWS, JOINERS' TOOLS, &c.

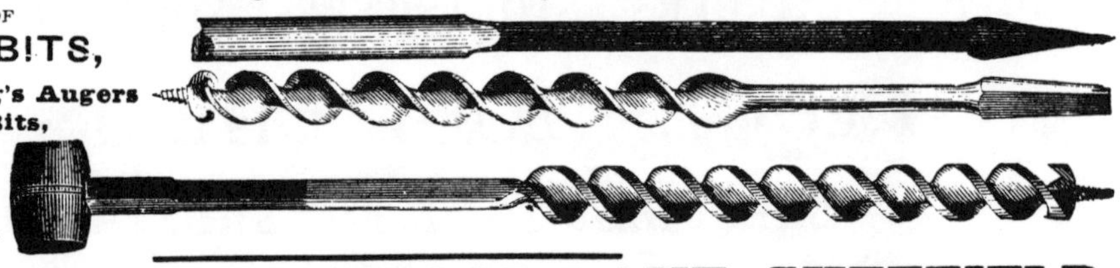

DILIGENCE WORKS, 114, BROAD LANE, SHEFFIELD.

1881

AMOS JINKINSON,

MANUFACTURER OF ALL KINDS OF

Scotch Gedges, Jenning's Ship Carpenters' & Machine

AUGERS AND BITS;

GIMLETS, BRACES & BITS, JOINERS' TOOLS, EDGE TOOLS, &c.

CARDIGAN AUGER WORKS,

Back of 88, Division street, & Westfield terrace, Sheffield.

1879

Plate CCXXVIII - Advertisements from Sheffield Directories

Note: - THIS HEAD WARRANTED
NOT TO WORK OFF

Fig. 156 - Joseph Marples "1st TRIAL" Sheffield Plated Brace

Note: PATENT LEVER stamped on Lever. [No known Patent in this instance covering this feature. Former expired reigistered design. Wording advertising propaganda

Fig. 157 - J.Askhams' Plated Braces/ Sheffield, c1856-1862

This is another example of an unlisted brace manufcaturer who probably traded for braces. He was a merchant and spring knife manufacturer and may have sold braces in New York with his knives.

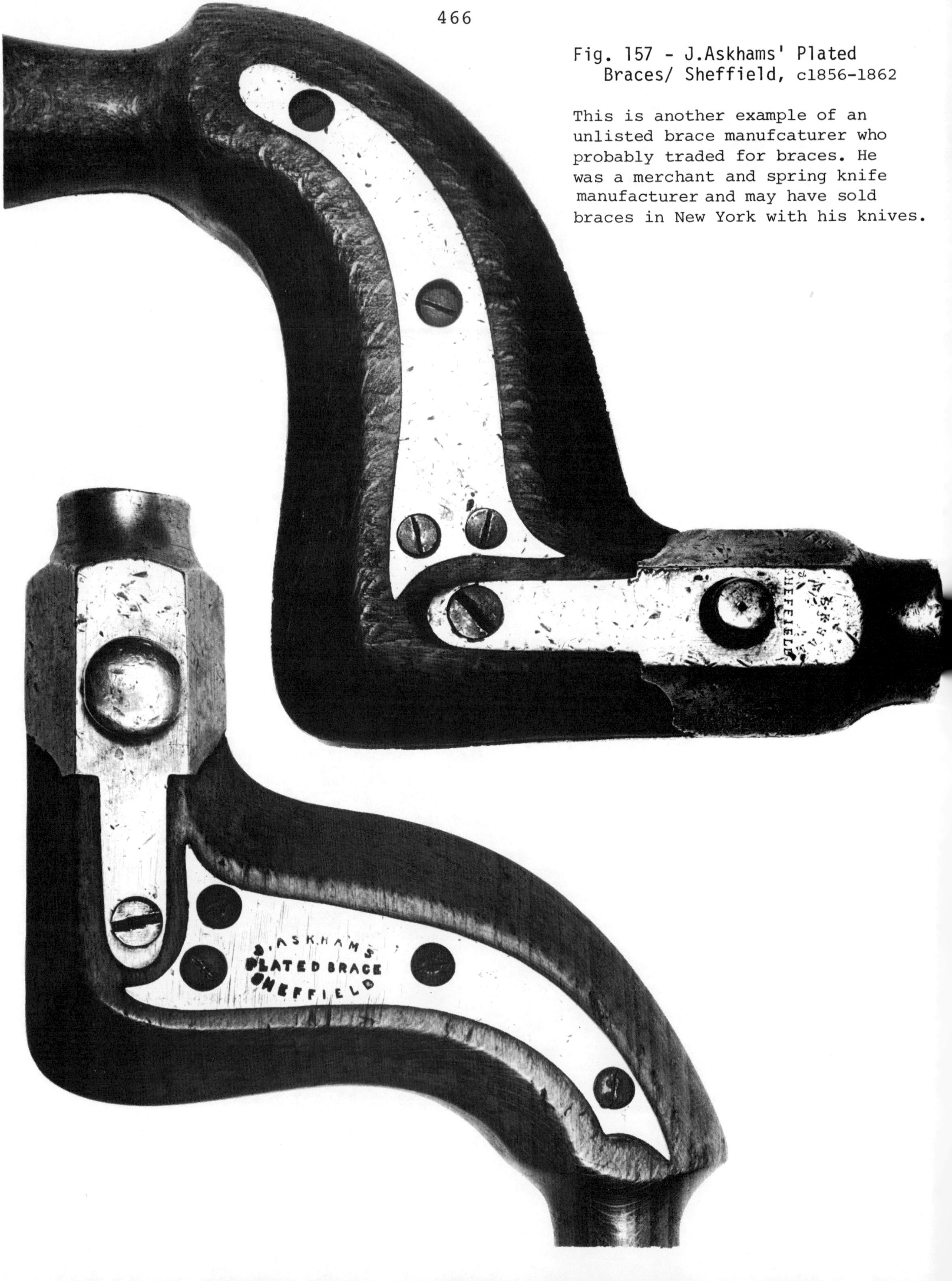

Fig. 158 - THREE-QUARTERS SIZE SHEFFIELD BRACES

Bottom: 3/4 Size Rosewood Brace, Maker Unknown, 11-1/2" long x 6" Sweep.

Middle: 3/4 Size Beech by Henry Brown & Sons. 11-1/2" long x 6" Sweep.

Top: Conventional Size Plated Brace, Wm. Marples & Sons, 14-1/2" long x 7" Sweep.

Fig. 159 - James Howarth, c1860
Plated Brace with Lever Pad

Fig. 160 - William Jackson, Sheaf Island Works
Sheffield

Plated Brace with Button Pad [No known PATENT, but former Registered Design]

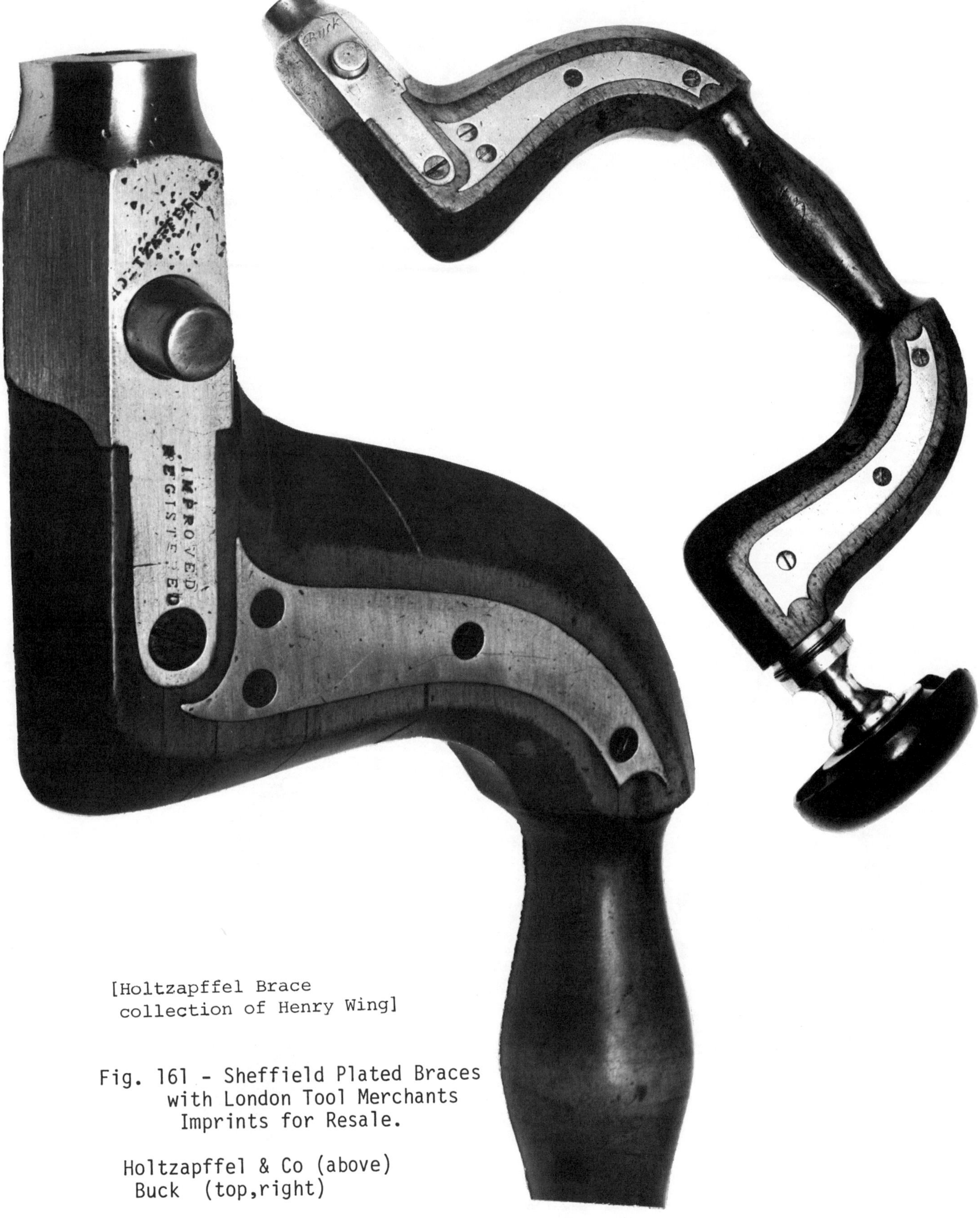

[Holtzapffel Brace
collection of Henry Wing]

Fig. 161 - Sheffield Plated Braces
with London Tool Merchants
Imprints for Resale.

Holtzapffel & Co (above)
Buck (top,right)

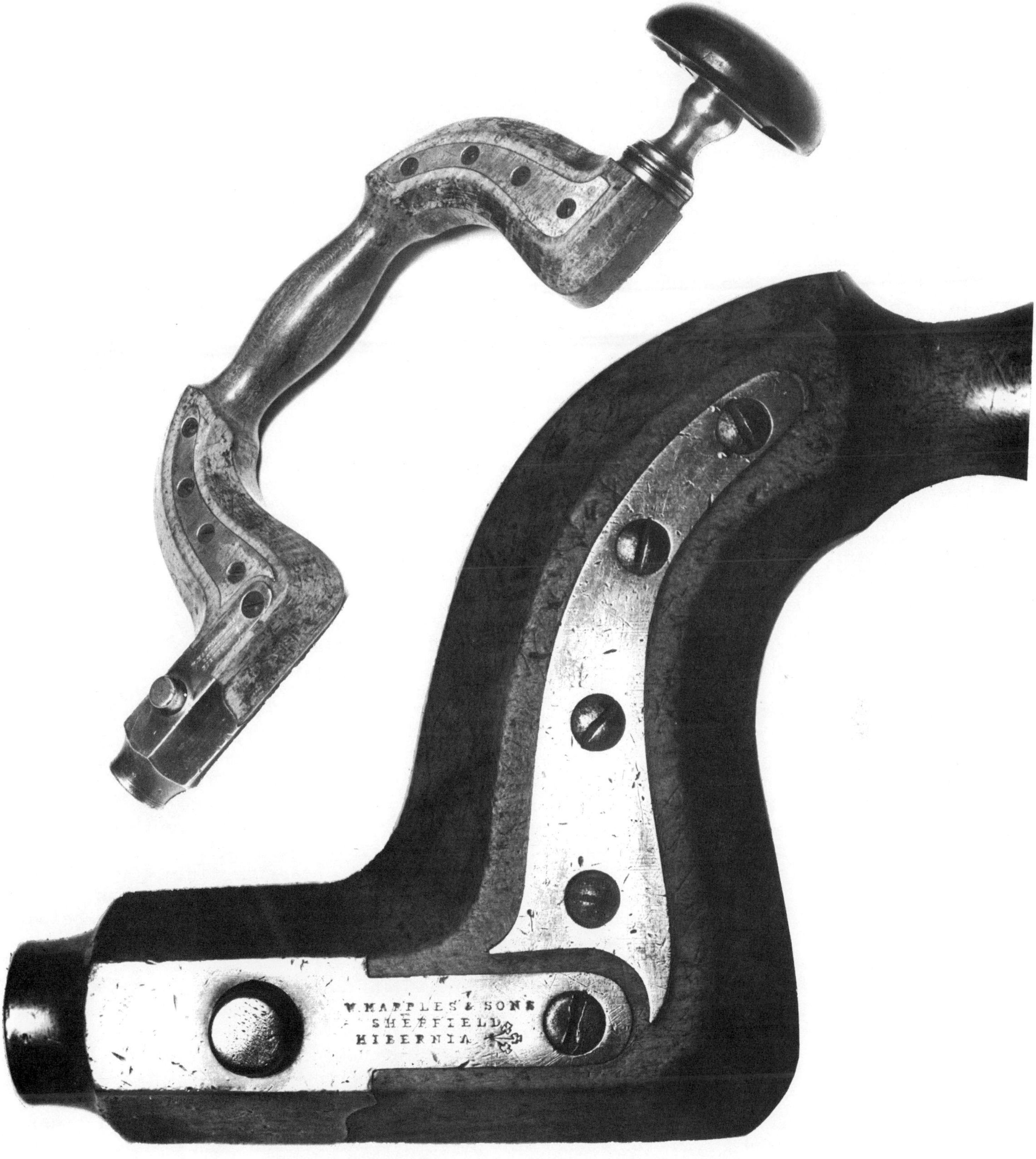

Fig. 162 - WILLIAM MARPLES & SONS, Hibernia Works, Plated Brace

1. A brief history of the development of boring tools has been written by Frank Preston, *Man Maketh Holes,* published by Stanley Tools Ltd., Sheffield. A very good presentation appeared in R.A.Salaman, *Dictionary of Tools*, pp.92-100.
2. W.L.Goodman, *The History of Woodworking Tools,* [London,1964], p.175.
3. Joseph Moxon, *Mechanick Exercises,* [Praeger Reprint,New York,1970], p.94
4. Plate 142 *R.Timmins & Sons Pattern Book [Tools for Trades & Crafts],* p.192
5. *ibid,* Both are illustrated Plate 142, p.193
6. *Wyke Catalogue of Tools for Watch & Clock Makers,*[Charlottsville,Va,1978],p.99.
7. W.L.Goodman, *British Planemakers from 1700,* [2nd edition,1978], p.147
8. *ibid,* p.139
9. See *Tools for Trades & Crafts,* pp.196-197.
10. *ibid,* p.190 [Original in Dept. of Prints,Victoria & Albert Museum, London].
11. *ibid,* Plate 29, p.191.
12. *The Complete Edition of Oxford English Dictionary,* [1971], p.2674.
13. See *Tools for the Trades & Crafts,* p.43.
14. *ibid,* Fig.3, p.9.
15. *ibid,* pp. 10-13.
16. ibid, p.192. Suggested to have been 1840.
17. S.Pollard, *Three Centuries of Sheffield Steel,* [Sheffield,1954],p10 & p.17.
18. Surviving braces with this feature are rare. In all probability the awkward projection of the lever mechanism resulted in breakage.
19. The firm became quality producers of rules and tapes and merged to form Rabone-Chesterman in Birmingham in 1964.
20. See *Alex.Mathieson & Sons Illustrated Price List of Woodworking Tools.* [2nd edition, 1979, Ken Roberts Publishing Co., Fitzwilliam, NH]
21. At the date of this writing, 25 braces with these imprints are known to the author and only two of these were found in England.
22. This is more rare than the Pilkington types.
23. During my early research this brace was thought to be a proto-type of the Ultimatum, but since it was later,this was actually a competitive brace.
24. The name *ULTIMATUM* was obviously developed by William Marples to convey "the last word" or "unsurpassable", but was not copyrighted.
25. This brace was erroneously described as being made by E.Mills & Co.,Phila.,USA by Peter Welsh,*Woodworking Tools 1600-1900* [Contributions from the Museum of History & Technology, Washington,DC,Paper #51, p.217, 1966]. The seal of the Lion & Unicorn clearly identifies this as of British manufacture.
26. W.L.Goodman, *Op.cit,British Planemakers from 1700,* p.144.
27. Arnold & Walker, *Catalogue No.2* [London,1975], p.9.
28. Carl E. Bopp,"Premium Awarded by Philada. & New York Institutes,Booth & Mills Philada", *Chronicle of EAIA,* Vol.27, No.3, Sept.1974, pp.4]-42.
29. William Gilpin, auger maker, Wedges Mills, Cannock, near Lichfield, Staffordshire was granted Patent No.3448, June 11, 1816 for a method of forging augers. It was my pleasure to visit this plant in 1979 and witness auger bits still being manufactured there.
30. R.A.Salaman, *op.cit, Dictionary of Tools, etc.,* p.95.
31. In the 1849 Sheffield Directory there were 56 entries for "horn pressers" Typical were: "William Ashall, hornpresser & cutler,17 Sims croft; Samuel Andron & Son, hornpresser and Britannia metal ware manufacturer, 12 Holly Court".
32. In 1830 there were 18 Brass Founders listed in the Sheffield Directory. By 1876 this number had increased to 50. There was no secrecy of what was being produced. If one foundry was not making a particular item, it was not difficult to determine where such was being made and order the same. Among the 31 brass founders listed in the 1857 Sheffield Directory was John Brown at 9 Church Road (late E.Cockayne] "brass fender manufacturer, manufacturer of brasses for millwrights, engineers, machinists, JOINER TOOLS, and all other brass work; whitesmith; locksmith and bell hanger & gas fitter, &c."
33. Awarded a prize. *Illustrated Exhibitor* [London,1851],p.xxxii, Class XXI,Edge Tools

List of 19th Century Sheffield Brace and Bit Manufacturers

[Compiled from Sheffield Directories]

ABBEY, Charles	-1828-1830-	CREW, John	-1881-1888-
ALLEN, John	-1837-	CUTTRAM, William	-1879-1889-
ALLEN & GREENWOOD	-1834-	DALTON, P. & Co.	-1849-1852-
ASH, William	-1822-1837-	DEAKIN, George	-1852-
ATKINSON BROTHERS	-1875-1884-	EDLEY. William	-1834-1852-
ATKINSON, Charles	-1859-1865-	ELLIS, G. & Son	-1879-1884-
BARLOW, John	-1876-1884-	ENGLAND, John	-1876-
BARLOW & HOBSON	-1860-1872-	FELLOWS, William	-1837-1865-
BARTON BROTHERS	-1849-1852-	FENTON & MARSDEN	-1841-1848-
BAXBY, John	-1860-1872-	FLATHER, David	-1852-1854-
BAXBY, William	-1876-1884-	FLATHER, D. & Sons	-1856-1900-
BEE, James	-1814-1828-	FLETCHER, John	-1833-1839-
BIRDS, George	-1854-1861-	FOSTER, William	-1876-1884-
BIRDS & RIDGE	-1889-	FOX BROTHERS	-1891-1900-
BIRTLES, William	-1889-1896-	FURNIS, Charles	-1849-
BLOOMER, George & Co.	-1852-	FURNISS, William H.	-1879-1884-
BLOOMER, John	-1830-1852-	GAMBLE, Arthur	-1891-
BLOOMER & PHILIPPS	-1849-1852-	GLOSSOP, C.	1884-1889-
BRADSHAW, William	-1841-	GREEN, William	-1834-1876-
BROWN, Henry & Sons	-1852-1900-	GREENWOOD, Frederick	-1837-
BROWN & FLATHER	-1828-1849-	GUION, Henry	-1879-1889-
BRUFTON, John T.	-1876-1900-	HANNAY & KINDER	-1830-
BURKINGSHAW, William	-1876-1891-	HARRIS, Frederick	-1876-1884-
CHAPMAN, James	-1868-1900-	HARRISON, James	-1859-1854-
CHUMBLEY, EDLEY & HAWKSWORTH	-1830-	HARTLEY, Henry	-1859-1868-
COOPER, Joshia	-1888-1900-	HASLAM, George	-1834-1854-
COOPER & MOULSON	-1879-1884-	HAYWOOD, CLARKE & Co.	-1849-
COTTAM, William	-1879-1889-	HAWKSWORTH, Henry	-1834-1841-

HELLIWELL, Isaac	-1834-1841	PARKER & THOMPSON	-1854-1861-
HOBSON, Joseph	-1830-1854-	PASLEY, Henry	-1889-1900-
HOLDEN, Nathaniel	-1876-1884-	PASS, Horatio	-1837-
HOLMES, Henry	-1871-1900-	PEARSON & GRIFFITH	-1841-
HORTON, David	-1879-1884-	PILKINGTON, Thomas	-1865-1871-
HORTON, George	-1849-1864-	PILKINGTON, Chas.& Thos.	-1857-1864-
HORTON BROTHERS	-1881-1884-	PILKINGTON,PEDIGOR & STORR	-1854-1856-
HOWARTH, James & Sons	-1865-1900-	PLATTS, Joseph	-1822-
HURST, John & Sons	-1849-1871-	RIDGE, Alfred	-1854-1884-
IBBOTSON, Thomas & Co.	-1841-1900-	RIDGE, Alfred & Son	-1888-1893-
JACKSON, William	-1879-1889-	RIDGE, George	-1854-1891-
JINKINSON, Amos	-1863-1900-	RIDGE, John	-1854-1865-
JINKINSON, Amos & William	-1854-1860-	RIDGE, Septimus	-1854-
MANSFIELD, Richard	-1856-1863-	RIDGE, Thomas	-1876-1879-
MARPLES, Joseph & Son	-1849-1883-	RIDGE, William	-1856-1893-
MARPLES, Robert	-1841-1860-	RIDGE BROTHERS	-1879-1891-
MARPLES, Robert & Son	-1861-1900-	RIDGE & PEARSON	-1860-
MARPLES, William	-1833-1860-	RIDGEWAY, William	-1891-1896-
MARPLES, William & Sons	-1861-1900-	RIDGEWAY, William & Sons	-1898-1972-
MARPLES & MANSFIELD	-1854-	RIXHAM, C.H.	-1860-1884-
MARSDEN BROTHERS & Co.	-1854-1896-	ROCK, Thomas	-1852-1879-
MARSDEN BROS.& SILVERWOOD	-1849-1852-	ROCK & Son	-1852-
MAW & STALEY	-1823-1828-	ROSS, Peter	-1856-
MILLINGTON, John	-1876-1890-	ROSS, Peter & Sons	-1860-1871-
MORRISON, John	-1876-1890-	ROUND, Elizabeth	-1837-1839-
MORRISON & PARKER	-1845-1852-	ROUND, George	-1845-1852-
MOULSON BROTHERS	-1830-1860-	ROUND, William	-1822-1841-
PARKIN, Charles	-1861-1879-	SANSOM, George & Co.	-1879-
PARKIN, Chas. & Son	-1881-1896-	SAUNDERSON, William	-1825-1828-
PARKIN, John & Co.	-1834-1837-	SAVILLE, Charles	-1833-1852

SCHOFIELD, George	-1822-1857-
SCHOFILED, George & John	-1830-1841-
SCHOFIELD, G. & Son	-1839-
SCHOFILED, Isaac	-1865-
SENIOR, Samule	-1828-1830-
SHEPHERD, George	-1830-
SILVERWOOD & MARSH	-1857-
SISSONS, John	-1830-
SLACK, Robert	-1865-1879-
SLATER, Joseph	-1822-1841-
SLATER, Thos. & Jos.	-1841-
SMITH, Samuel & Sons	-1888-1895-
SMITH & SCHOFILED	-1881-1884-
SORBY, John & Sons	-1822-1890-
SORBY, Robert & Son	-1863-1900-
STALEY, Thos. & Wm.	-1830-1837
STONES, Frederick	-1837-1841-
STOREY, george	-1900-
SWINCOW & MANUEL	-1845-1852-
TAYLOR & GREENWOD	-1841-
THOMPSON, Frank	-1859-
THOMPSON, Henry	-1876-1884-
WALES, Thomas & Sons	-1893-1900-
WELLS, Thos. E.	-1841
WELLS & BRADSHAW	-1841-
WARD & PAYNE	-1861-1900-
WHEATCROFT, John	-1825-1828-
WHEATCROFT, Jno. & Sons	-1830-
WILSON,J.,MARSDEN BROTHERS	-1898-1900-
WISLON, Samuel	-1849-1856-
WORSELY, Daniel & Co.	-1879-
WRIGHT, John & Sons	-1828-1830-
WRIGHT BROTHERS	-1841-
WRIGHT, James	-1837-

* *

Imprints appearing on Sheffield Braces, but whose names are not listed as Brace Manufacturers in the Sheffield Directories.

R.Adfers, no data
Adsetts, William, wood turner, brace head manufacturer, [see text]; 1854-1859
Ashton & Jackson, no data
Askhams, J., merchant and knife manufacturer, 1873
Boker, H., German importer with New York sales offices, 1840-1875
Bower, Chester Works, no data
Colquhoun & Cadman, joiner tool makers, 1881-1900-
Dixon, [trade mark of R.Marples & Sons]
Eyre,Ward & Co., table knife manufacturers, successors to W.Greaves & Sons,c1850
Groves. Richard & Sons, saw makers
Greaves, Wm. & Sons, Sheaf Works, edge tool manufacturers, 1822-1849
Hargreaves, Wm., merchant with New York Offices, Hargreaves & Smith [see text]
Hawke, H. [trade mark of D.Flather & Sons]
Kent, Wm. [trade mark, 2nd Quality of Wm. Marples & Sons]
Kirk, Peter, auger and bit manufacturer, 1862-1876
Pasley, Richard D. - letter and stamp manufacturer [see text]
Stephenson,Mawhood & Co., merchabts, 1868-1872 [see text]
T.Tilotson & Co., merchants and dealers
T.Turner & Co., merchants and table knife manufacturers, Suffolk Works
S. Walker, not data [plated brass with imprint "Pro Bono Publico" Impd. Patent
Walter, F, trade mark of D.Flather & Sons
Windle, E., no data [imprint "Primum Mobile], see text

In all probability numerous other names will be reported after publication of this work.

List of 19th Century Birmingham Brace & Bit Makers
[Compiled from Birmingham Directories]

ALLEN, William	-1835-	HOPKINS, T. & C.	-1846-1863-
ATKINS & SONS	-1846-1877-	HORTON, Thomas	-1816-1835-
BETTS, W.	-1849-	JOHNSON, Edward	-1831-
BIRKIN, Thomas	-1835-	JONES, Thomas F.	-1846-
BLACKMORE, John	-1828-1831-	LANE, GIBBS & LANE	-1866-
BLACKMORE, Thomas & John	-1829-	LEADBEATER, Samuel	-1835-1863-
BLACKMORE & SONS	-1816-1820-	LEADBEATER & AINSWORTH	-1872-1880-
BLACKBURN, Thomas	-1854-1858-	MOORE, Isaac	-1828-1853-
BRIDGEWATER, Timothy	-1816-1820-	MOSS, William	-1835-
BRITTEN, James	-1828-	NEELD, Thomas	-1816-1820-
BRITTAIN, William	-1829-1880-	NEWEY, John	-1849-1864-
BROWN, Henry	-1864-1871-	NICKLIN, Edw.	-1828-1849-
BURTON, Charles	-1846-	PALMER, John & Son	-1879-1880-
BUSHNELL, William	-1852-1858-	PALSER, Abel,Thos.& Jas.	-1854-1856-
CHAMBERLAIN, Thomas	-1846-	PALSER Bros.	-1858-
CLAMP, William	-1846-	PALSER, A.	-1863-1880-
CLARK, M.	-1849-1864-	PALSER, James	-1852-1853-
COATES, Joseph	-1846-	PALSER, Edward	-1828-1831-
COHEN, Adolph	-1842-	PALSER, William	-1835-1853-
COOPER, George	-1835-1853-	PALSER, William, jr.	-1846-1865-
COOPER, G. & Sons	-1849-1867-	PARKER, J.	-1849-
COOPER, Joseph	-1846-1880-	PERKS, John	-1846-
COOPER, J & Son	-1880-1886-	PERKS, Thomas	-1846-
COOPER, J. & Sons	-1849-1860-	PILL, W.	-1860-1863-
COOPER, T.	-1866-	PITT, W.	-1854-1877-
COOPER, Thomas	-1828-1831	POUNTNEY, C.	-1861-1866-
CROWSHAW, J.	-1860-1861-	RENCER, George	-1852-1853-
DREW, John	-1829-	REYNOLD, John	-1852-1856-
EDMUNDS, Frederick	-1855-1880-	ROSS, Thomas	-1831-1870-
EDMUNDS, Henry	-1846-1853	ROSS, Thomas & Sons	-1876-1877-
EDMUNDS, William	-1849-	SMITH, J.	-1846-
EMERY, William	-1846-1863-	STANDLEY, J.	-1846-
EVANS, J.	-1876-1877-	STEPHENTON, James	-1828-1831-
EVANS, W. & J.	-1863-	STEPHENTON, Joseph	-1835-1858-
FALLOWS, William	-1846-1853-	STEVELEY, J.	-1849-1856-
FLEWITT, Thomas	-1846-	TAYLOR, S. & D.	-1818-1820-
FREETH, B. & Co.	-1818-1820-	THOMAS, Richard	-1846-1864-
GLASCOTT, Joseph	-1816-1820-	TIMBRELL, Charles	-1842-
GLASSCOT, E.	-1831-	TERRY, J.	-1846-
GRIFFIN, William	-1828-	TIMMINS, R. & Sons	-1835-1880-
GRIFFIS, William	-1831-1853-	WALKER, George	-1846-
HADEN, S. & Sons	-1864-	WHITE, William	-1816-1820-
HANDY, Charles	-1828-1846-	WHITEHOUSE, Thomas	-1849-1858-
HANDY & WOOLLEY	-1852-1861-	WHOOLEY, W.	-1855-1880-
HEAP, J.	-1877-	WYNN, William & Cornelius	-1816-1880-
HOPKINS, H. & C.	-1861-1862-		

Summary and Conclusions

At the beginning of the 19th Century supremacy of edge and joiner tool manufacturing shifted from Birmingham to Sheffield. The principal factors were the growth of the crucible steel melting and a fully developed cutlery trade at Sheffield. The quality of tool steel products there was probably the best in the world during the 19th Century.

Considerable trade in finished goods between manufacturers, merchants and suppliers obscured the actual makers of many tools. Imprints of a firm name do not necessarily indicate that tools were actually made by such a firm or person. Price fixing prevailed among tool manufacturers at Sheffield throughout the 19th Century. Discounts to the trade depended upon volume of business.

A registered design was not a patent, although often claimed as such long after protection had expired. In such instances it appears to be a prestigeous advertising claim. The word "patent" was freely used with no legislation against it until late in the 19th Century.

The so called "Scotch" beech brace with brass pad was introduced at Birmingham, probably late in the 18th Century. At Sheffield this became known as "Best Plain Brace". The plated brace was introduced at Sheffield, c1829, and remained in style throughout the remainder of the 19th Century. The brass framed brace (Ultimatum) was introduced by William Marples, c1854. This was more of a presentation piece than a practical joiner's tool.

Artistic designs in combination with exotic colored woods with brass trim led to attractive products. These designs had profound influences on tool lines manufactured in United States. The "Golden Age" of export of tools to United States was 1840-1860. After the Civil War development in metal planes, braces and saws in America resulted in serious competition and a decline in sales of Sheffield tools.

While there was considerable innovative talent in Great Britain offering improvements, the manufacturers in general were not receptive to promoting new designs. It may have been a marketing problem.

The quality of workmanship and this form of art of these English tools, 1850-18880, will probably never be equaled on a commercial scale.

The growth of the Sheffield tool industry is summarized in Table XXVIII.

Table XXVIII - Firms Listed in Sheffield Directories, 1821 - 1900

	1821	1828	1837	1849	1871	1900
Brace & Bit Manufacturers	5	11	13	20	10	11
Brass Foundries	8	18	18	31	52	38
Edge Tool Manufacturers	40	44	53	63	77	85
Joiner Tool Manufacturers	10	20	32	41	56	60
Rule Makers	1	2	2	2	5	-
Saw Makers	43	60	78	97	104	126
Skate Manufacturers	7	10	10	11	19	-

INDEX

I - Names and Firms

II. SUBJECT MATTER

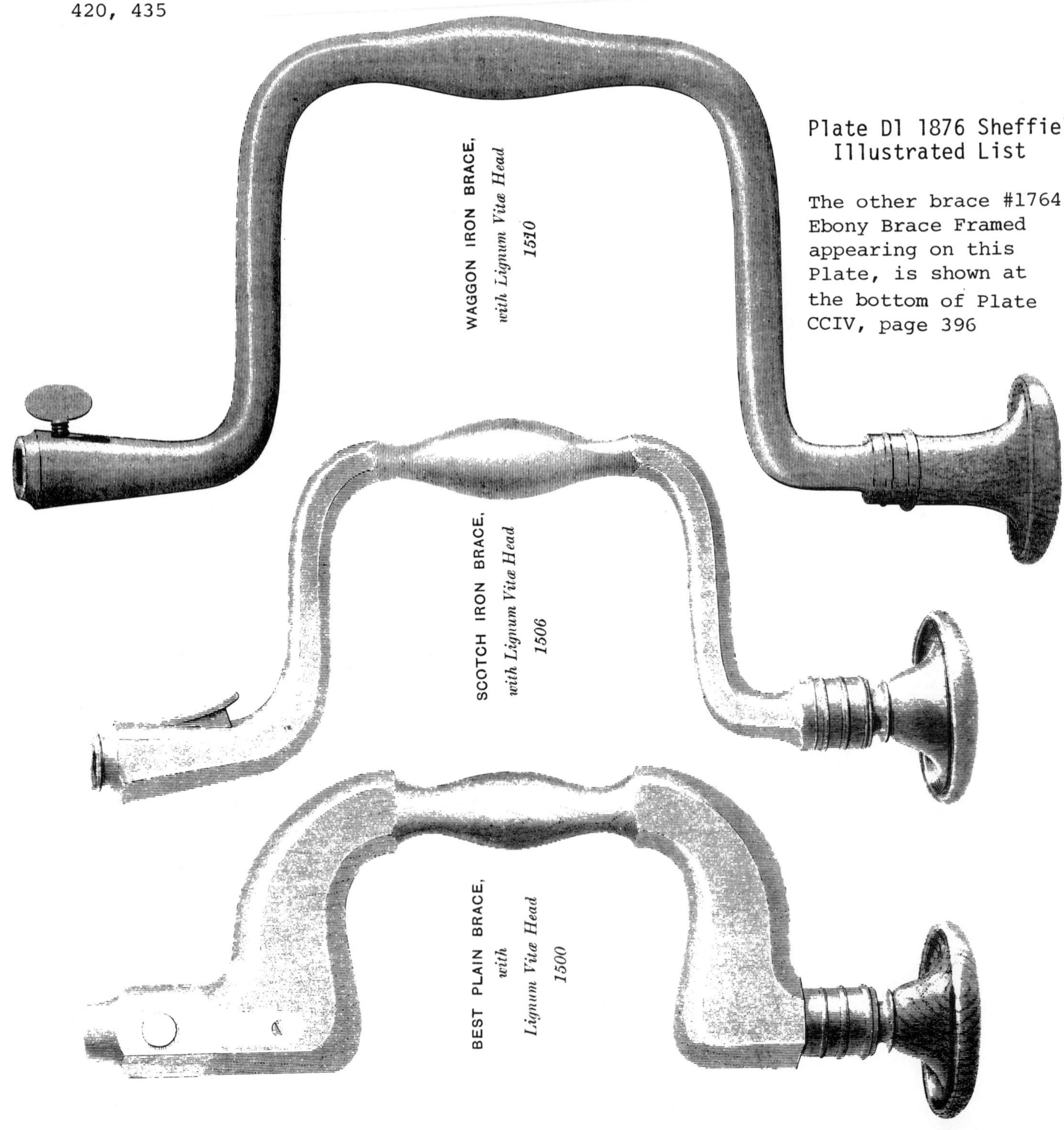

Plate D1 1876 Sheffield Illustrated List

The other brace #1764 Ebony Brace Framed appearing on this Plate, is shown at the bottom of Plate CCIV, page 396